FLUID FLOW

Pumps, Pipes and Channels

FLUID FLOW

Pumps, Pipes and Channels

By
Nicholas P. Cheremisinoff

ANN ARBOR SCIENCE
THE BUTTERWORTH GROUP

Second Printing, 1982

230 Collingwood, P.O. Box 1425, Ann Arbor, Michigan 48106

Library of Congress Card Catalog Number 81-68034
ISBN 0-250-40432-X

Library of Congress Cataloging in Publication Data

Cheremisinoff, Nicholas P.
Fluid flow.

Includes bibliographical references and index.
1. Pumping machinery–Fluid dynamics. 2. Pipe–Fluid dynamics. 3. Channels (Hydraulic engineering) I. Title.

TJ901.C47	620.1′064	81-68034
ISBN 0-250-40432-X		AACR2

Manufactured in the United States of America

Butterworths, Ltd., Borough Green, Sevenoaks, Kent TN15 8PH, England

PREFACE

Fluid flow is a fundamental area in all manufacturing, process design, and, particularly, chemical, petroleum and allied industries. Design of fluid flow systems is the basic building block of chemical manufacturing. If a pump is the heart of such systems, the pipes, channels and conduits that carry fluids are the arteries. The methods described in this book are important to 10,000 chemical manufacturing and processing firms, employing more than 1.1 million people, with sales of more than $100 billion. The chemical industry alone represents 7.5% of the U.S. gross national product (GNP), and enjoys an annual growth rate nearly twice that of the overall GNP.

The Romans employed lead pipes to collect their water, and the ancient Chinese conveyed natural gas through bamboo tubes. However, centuries passed before the science of fluid mechanics was born. Gradually, fundamental laws governing the mechanics of flow phenomena began to be understood and applied to engineering problems. Sanitary conditions created a demand for the water distribution systems that have become ubiquitous throughout the developed nations. Discoveries of uses for oil and natural gas gave impetus to the design and construction of millions of kilometers of additional pipelines. In addition, millions of kilometers of additional piping networks and channels are employed in scores of manufacturing plants and refineries throughout the world, and in pollution control, utilities, etc.

Our knowledge of fluid mechanics has become exceedingly great, with principles that not only affect fluid transport, but also numerous process designs. It is not possible to design rationally for heat and/or mass transfer operations without understanding the fluid mechanics of the process. To this end, there have been scores of investigations, papers and textbooks devoted to various aspects of flow phenomena. Unfortunately, few books have been devoted entirely to industrial applications, and even fewer provide design procedures and recommendations for even the most common flow situations that engineers often encounter.

This volume originated as a guide for process engineers. Basic calculation methods, design practices and recommendations, in addition to definitions and theorems, are included. This book is not intended to be a definitive work. Indeed, many important flow situations, such as slurry flows and two-phase flows, are not treated in this volume. Instead, the book is meant to be an introductory and practical reference text to standard process flow problems. There are three types of readers who may find this volume useful: (1) engineering students in fluid mechanics; (2) engineers who desire a refresher course in principles and standard calculations; and (3) process engineers in first job positions.

The book is divided into two sections. The first is devoted entirely to reviewing basic fluid mechanics principles. Although illustrative examples are included in the discussions, no problems are given at the end of this section. If used as a textbook, the instructor must provide his or her own problems to supplement Section 1. The second part of the book applies these principles to problem solving, and a collection of problems and selected solutions by subject are presented in Chapter 2.12. There are a number of chapters in the second section that are specific in nature and provide valuable information most relevant to the process engineer.

Finally, there are several contributed chapters and appendixes prepared by experts. I wish to acknowledge each of these contributors. Without the efforts of these individuals and organizations, this volume simply would not have been possible.

Nicholas P. Cheremisinoff

Nicholas P. Cheremisinoff is employed by Exxon Research and Engineering Company in Florham Park, New Jersey. He received his BS, MS and PhD degrees in chemical engineering from Clarkson College of Technology and was an instructor there in 1976–1977.

Dr. Cheremisinoff has contributed to the industrial press and is the author of several engineering reference books. Among his research interests are the study of flow phenomena of multiphase systems and the development of biomass based energy resources. He is a member of a number of professional societies including Tau Beta Pi, Sigma Xi and AIChE.

ACKNOWLEDGMENTS

The following individuals and organizations are respectfully acknowledged for their contributions to this volume:

The technical staff of Peerless Pump, Eastern Operations, Indianapolis, Indiana (Peerless Pump, an Indian Head Co., Montebello, CA), for preparation of Chapter 2.5, "System Analysis for Pumping Equipment."

Dr. Gerhard Vetter, of LEWA-Herbert Ott GmbH & Co., Stuttgart, Germany (a division of American LEWA, Inc., Natick, MA), for preparation of Chapter 2.6, "Oscillating Displacement Pumps."

Mr. Ray Battilana, of Durametallic Corp., Kalamazoo, MI, for preparation of Chapter 2.7, "Mechanical Seals for Pumps."

The technical staff of Durametallic Corp., Kalamazoo, MI, for preparation of Appendix A, "General Conversion Factors and Physical Properties Data."

The American Petroleum Institute (Washington, DC), for permission to use Appendix B, "Tables of Recommended SI Units and Conversions."

Dr. Ramesh Gupta, of Exxon Research and Engineering Co., Florham Park, NJ, for his assistance in preparing Chapter 1.7, "Introduction to Dimensional Analysis."

My heartfelt thanks are also extended to the many companies and individuals too numerous to list here who contributed photographs and technical information.

Finally, a special thanks to Paul N. Cheremisinoff, of New Jersey Institute of Technology, for his suggestions and encouragement.

CONTENTS

Section 1
Principles of Fluid Flow

Section 2
Calculations and Design Basics for Incompressible Flows

SECTION 1

PRINCIPLES OF FLUID FLOW

SCOPE AND SECTION OVERVIEW

To the chemical engineer, fluid mechanics is the backbone of his education. However, process engineering draws from the ranks of many engineering disciplines, which often do not stress the importance of this subject in a single course or series of courses. This book attempts to summarize major fluid mechanics principles and apply them to basic process flow situations encountered in plant operations. More specifically, the overall thrust of this volume is to provide guidelines useful in selecting and sizing pumps and pipelines, and in handling channel flow problems. This includes criteria for selecting and using proper flow instrumentation and troubleshooting pumping operations. Emphasis, then, is placed on incompressible flows, although as noted in the Foreword, not all types of incompressible flow situations are addressed. Admittedly, the book is limited as a general text and, if used so, is recommended only as an introduction for undergraduate chemical engineers, or for nonchemical engineers taking a course in fluid mechanics. The second section does, however, provide some challenging problems that may stimulate graduate student thinking.

The material presented was compiled from three sources: (1) the author's lectures given to undergraduate and graduate classes at Clarkson College of Technology and New Jersey Institute of Technology; (2) a collection of design notes compiled over the last six years working in industry; and (3) contributed chapters and notes provided by several organizations and companies.

This first section presents an overview of important principles in fluid mechanics. There are seven chapters covering Systems of Units, Fluid Statics and Pressure Measurement, Properties of Fluids, the Equations of Motion and Energy, Laminar Flows, Turbulent FLows and an introduction to Dimensional Analysis. The importance and coverage in each of these chapters is noted in some brief introductory remarks to each chapter.

One point that should be noted up front is the use of consistent units in calculations. Among practising engineers there is a serious problem in adjusting to the SI system of units. The transition to a universal standard of units is inevitable; however, it may be some time before "the old school" (of which I include myself) of English unit users completely surrenders. Consequently, engineers must be familiar with a dual system. The first chapter explains the SI system of units in detail and Appendix B provides conversions from English to SI units. This Appendix was prepared by the American Petroleum Institute, which should be commended for its efforts to achieve industry acceptance of SI. In an attempt to please all readers, a dual system is used throughout the book and solutions to selected problems are given in both English and SI.

A final note is that the two sections are independent. That is, the more experienced reader can go directly to Section II. Chapter 2.1 essentially summarizes major points brought out in the first section and relates them to basic design procedures. It is recommended that Chapter 1.1 ("Notes on Systems of Units") be read first before proceeding to Chapter 2.1. Each section is provided with its own table of contents and lists of figures and tables to enable users to gain quick access to data/information needed for specific problem solving.

CONTENTS FOR SECTION 1

LIST OF FIGURES FOR SECTION 1

LIST OF TABLES FOR SECTION 1

CHAPTER 1.1

INTRODUCTION AND NOTES ON SYSTEMS OF UNITS

In the discussions to follow, we will use an analogy of constructing a building. The construction will have four phases: the first phase is to clear the land and to prepare the initial groundwork; the second is to lay a foundation; the third, to erect a support structure; and the last to construct the building's shell and facing. The groundwork will consist of experimental observation of flow phenomena, which for our purposes already has been done by an outside contractor. The foundation of the building will be the ability to record and communicate these observations in an orderly, consistent manner. The support structure will be the principles and governing equations of fluid mechanics developed from observing phenomena. Finally, the finished shell will be resultant process design practices stemming from fluid mechanics principles.

We start then with defining systems of units. By the use of known and accepted units of measurement, a mechanism for reporting observations of physical phenomena and developing methodology for harnessing the benefits of these phenomena is established. Definitions or units of measurement are arbitrary in origin. Often, different investigators working in a particular area elect to use their own standards. The scientific and engineering communities have attempted to formulate and institute a policy of an International System of Units (SI) through the International Organization of Standardization (ISO).

The transition to SI has advanced considerably over the past decade. The Metric Conversion Act of 1975 (PL 94-168) has been enacted, declaring the coordination and planning of increasing use of the metric system (SI) in the United States to the government policy. A memorandum by the Assistant Secretary of Commerce for Science and Technology in the *Federal Register* of October 26, 1977 (pp. 56513 and 56514, vol. 42, no. 206) interprets and modifies SI for the United States. In addition, the

Act also provides for the establishment of the U.S. Metric Board to coordinate voluntary conversion.

Despite this increased activity the complete changeover to SI has not materialized fast enough. Consequently, in this transition phase, practicing engineers, scientists and students must be familiar with several systems of units: English-British Units, cgs and SI. It is not our intention to take the bull by the horns and present the preferred SI system to problem solving as we feel that the material contained herein would not be as widely used as it should by practicing engineers. Rather, we shall emphasize application of both major systems, English and metric, to practical problem solving in the hopes that the student will become fluent in both.

THE INTERNATIONAL SYSTEM OF UNITS (SI)

SI is the abbreviation for the International System of Units (Le Système International d'Unités). The prelude to SI was the cgs (centimeter-gram-second) system of metric units. By contrast, SI is based on the meter, kilogram and second as the fundamental units.

The SI system utilizes three classes of units. The first is referred to as *base units*. By convention, these are dimensionally independent. The second are *supplementary units*, which are used to measure plane and solid angles. *Derived units* are those formed by algebraic combinations of base units, supplementary units and other derived units. There are specific names and symbols assigned to the units in each class.

The real advantage of SI is the elimination of conversion factors within the system. That is, all derived combinations are in terms of unity. For example, the derived unit of power is the *watt*, which in base units is defined as 1 joule of work completed in 1 second of time.

Base Units

There are seven base units, each considered to be dimensionally independent with specific definitions. These are:

- meter (m)–for length
- kilogram (kg)–for mass
- second (s)–for time
- ampere (A)–for electric current
- kelvin (K)–for temperature
- mole (mol)–for the amount of a substance
- candela (cd)–for luminous intensity

Table 1 gives the definitions of each of these units.

Table 1. SI Base Units and Their Definitions (courtesy of the American Petroleum Institute - API Pub. 2564)

Quantity	Name	Symbol	Definition
Length	Meter (or metre)	m	The meter is the length equal to 1 650 763.73 wavelengths in vacuum of the radiation corresponding to the transition between the levels $2p_{10}$ and $5d_5$ of the krypton-86 atom. (Eleventh CGPM, 1960, Resolution 6)
Mass	Kilogram	kg	The kilogram is the unit of mass (not force); it is equal to the mass of the international prototype of the kilogram. (Third CGPM, 1901, Resolution 3) This international prototype, made of platinum-iridium, is kept at the International Bureau of Weights and Measures. A copy of the international prototype is maintained by the national standards agency of each major country. The kilogram is the only base unit defined by an artifact and is the only base unit having a prefix.
Time	Second	s	The second is the duration of 9 192 631 770 periods of the radiation corresponding to the transition between the two hyperfine levels of the ground state of the cesium-133 atom. (Thirteenth CGPM, 1967, Resolution 1)
Electric Current	Ampere	A	The ampere is that constant current which, if maintained in two straight parallel conductors of infinite length, of negligible circular cross section, and placed 1 meter apart in vacuum, would produce between these conductors a force equal to 2×10^{-7} newton per meter of length. (CIPM, 1946, Resolution 2 approved by the Ninth CGPM, 1948)
Temperature	Kelvin	K	The kelvin, unit of thermodynamic temperature, is the fraction 1/273.16 of the thermodynamic temperature of the triple point of water. (Thirteenth CGPM, 1967, Resolution 4) The unit kelvin and its symbol K are used to express an interval or difference of temperature. (Thirteenth CGPM, 1967, Resolution 3) In addition to the thermodynamic temperature, Celsius temperature (formerly called Centigrade) is widely used. The degree Celsius (°C), a derived unit, is the unit for expressing Celsius temperatures and temperature intervals. Celsius temperature, *t*, is related to thermodynamic temperature, *T*, by the following equation: $t = T - T_0$ Where: T_0 = 273.15 by definition. The temperature interval 1°C equals 1 K exactly.

Table 1, continued

Quantity	Name	Symbol	Definition
Amount of Substance	Mole	mol	The mole is the amount of substance of a system that contains as many elementary entities as there are atoms in 0.012 kilogram of carbon-12. When the mole is used, the elementary entities must be specified and may be atoms, molecules, ions, electrons, other particles or specified groups of such particles. (Fourteenth CGPM, 1971, Resolution 3)
Luminous Intensity	Candela	cd	The candela is the luminous intensity, in a given direction, of a source that emits monochromatic radiation of frequency 540×10^{12} hertz and that has a radian intensity in that direction of 1/683 watt per steradian. (Sixteenth CGPM, 1979)

Supplementary Units

At present, there are only two supplementary units in the SI system. Both of these are purely geometric:

- radian (rad)–for the unit of plane angle
- steradian (sr)–for the unit of solid angle

The radian is the plane angle between two radii of a circle that cuts off, on the circumference, an arc equal in length to the radius (see Figure 1A). The steradian is the solid angle that, having its apex in the center of a sphere, cuts out an area of the surface of the sphere that equals that of a square with sides of length equal to the radius of the sphere (see Figure 1B).

Derived Units

These are expressed algebraically in terms of base units having mathematical symbols for multiplication and division. A number of derived units have been given special names and assigned symbols. Examples of derived units and those given special names are given in Table 2.

Also, there are several other units that are widely used but not part of SI. These include the minute, hour, day and year as units of time; degree, minute and second of arc (in addition to the radian); the metric ton (1000 kg); the liter (1 cubic decimeter); the nautical mile; and the knot. These may be used along with SI units.

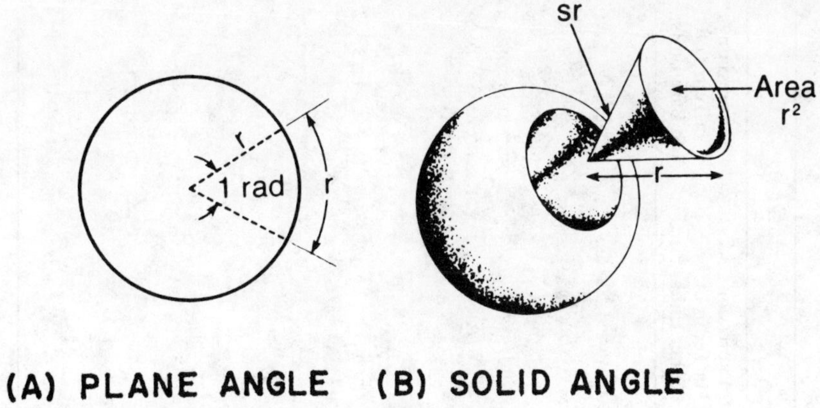

Figure 1. Illustrates the definitions of radian and steradian.

DECIMAL MULTIPLES AND USE OF LETTER SYMBOLS

Table 3 lists prefixes for decimal multiples and submultiples of SI units. The symbol of a prefix is considered to be combined with the unit to which it is directly attached, thus forming with it a new unit symbol that can be raised to a positive or negative power and can be combined with other unit symbols to form the symbols for compound units.

Distinction between upper case and lower case symbols is important as shown by the following examples:

- M = mega = 10^6
- m = milli = 10^{-3} (where m is a prefix)
- N = newton
- n = nano = 10^{-9}

When a compound unit includes a unit symbol that is also a symbol for a prefix, care must be taken to avoid confusion. As an example, the unit newton·meter should be written as N·m to avoid confusion with mN (millinewton).

Symbols can be used together with modifying subscripts and/or superscripts. Superscripts are often used to designate a place in space or time, or a constant or reference point. Superscripts can be used to designate a dimensionless form, a reference or equilibrium value, or mathematical identification such as an average value, derivative, tensor index, etc.

Table 4 gives a list of commonly used symbols and their definitions by category. Table 5 lists the Greek alphabet for proper pronunciation of Greek symbols noted in Table 4.

Table 2. Examples of SI-Derived Units

Quantity	SI Unit			
	Name	Symbol	Expression in Terms of Other Units	Expression in Terms of SI Base Units
Frequency	Hertz	Hz		s^{-1}
Force	Newton	N		$m \cdot kg \cdot s^{-2}$
Pressure	Pascal	Pa	N/m^2	$m^{-1} \cdot kg \cdot s^{-2}$
Energy, work, quantity of heat	Joule	J	$N \cdot m$	$m^2 \cdot kg \cdot s^{-2}$
Power, radiant flux	Watt	W	J/s	$m^2 \cdot kg \cdot s^{-3}$
Electric potential, potential differences, electromotive force	Volt	V	W/A	$m^2 \cdot kg \cdot s^{-3} \cdot A^{-1}$
Electric resistance	Ohm	Ω	V/A	$m^2 \cdot kg \cdot s^{-3} \cdot A^{-2}$
Conductance	Siemens	S	A/V	$m^{-2} \cdot kg^{-1} \cdot s^3 \cdot A^2$
Area	Square meter	m^2		
Volume	Cubic meter	m^3		
Speed, velocity	Meter per second	m/s		
Acceleration	Meter per second squared	m/s^2		
Density	Kilogram per cubic meter	kg/m^3		
Concentration (of amount of substance)	Mole per cubic meter	mol/m^3		
Specific volume	Cubic meter per kilogram	m^3/kg		
Luminance	Candela per square meter	cd/m^2		

Table 3. List of SI Prefixes for Forming Decimal Multiples and Submultiples

Prefix	Symbol	Factor by Which Unit is Multiplied
tera	T	10^{12}
giga	G	10^{9}
mega	M	10^{6}
kilo	k	10^{3}
hecto	h	10^{2}
deca	da	10
deci	d	10^{-1}
centi	c	10^{-2}
milli	m	10^{-3}
micro	μ	10^{-6}
nano	n	10^{-9}
pico	p	10^{-12}
femto	f	10^{-15}
atto	a	10^{-18}

Table 4. Commonly Used Symbols and Definitions

	Symbol	Unit or Definition
General Symbols		
Acceleration	a	m/s^2
of gravity	g	m/s^2
Base of natural logarithms	e	
Coefficient	C	
Difference, finite	Δ	
Differential operator	d	
partial	δ	
Efficiency	η	
Energy, dimension of	E	J, N · m
Enthalpy	H	J
Entropy	S	J/K
Force	F	N
Function	ϕ, ψ, χ	
Gas constant, universal	R	To distinguish, use R_o
Gibbs free energy	G, F	G = H − TS, J
Heat	Q	J
Helmholtz free energy	A	A = U − TS, J
Internal energy	U	J
Mass, dimension of	m	kg
Mechanical equivalent of heat	J	Unity, dimensionless
Moment of inertia	I	$(m)^4$
Newton law of motion, conversion factor in	g_c	Unity, dimensionless
Number		
In general	N	
Of moles	n	
Pressure	p	Pa, bar
Quantity, in general	Q	

Table 4, continued

	Symbol	Unit or Definition
General Symbols		
Ratio, in general	R	
Resistance	R	
Shear stress	τ	Pa
Temperature		
Dimension of	θ	
Absolute	T	K (Kelvin)
In general	T, t	°C
Temperature difference, logarithmic mean	$\overline{\theta}$	°C
Time		
Dimension of	T	s
In general	t, τ	s, hr
Work	W	J
Geometrical Symbols		
Linear dimension		
Breadth	b	m
Diameter	D	m
Distance along path	s, x	m
Height above datum plane	Z	m
Height equivalent	H	m
Hydraulic radius	r_H, R_H	m, m^2/m
Lateral distance from datum plane	Y	m
Length, distance or dimension of	L	m
Longitudinal distance from datum place	X	m
Mean free path	λ	m
Radius	r, R	m
Thickness		
In general	B	m
Of file	B_f	m
Wavelength	λ	m
Area		
In general	A	m^2
Cross section	S	m^2
Fraction free cross section	σ	
Projected	A_p	m^2
Surface		
Per unit mass	A_w, s	m^2/kg
Per unit volume	A_s, a	m^2/m^3
Volume		
In general	V	m^3
Fraction voids	ϵ	
Humid volume	ν_H	m^3/kg dry air
Angle	α, θ, ϕ	
In x, y plane	α	
In y, z plane	ϕ	
In z, x plane	θ	
Solid angle	ω	
Other		
Particle-shape factor	ϕ_s	

Table 4, continued

	Symbol	Unit or Definition
Intensive Properties Symbols		
Absorptivity for radiation	α	
Activity	a	
Activity coefficient, molal basis	γ	
Coefficient of expansion		
Linear	α	$m/(m \cdot K)$
Volumetric	β	$m^3/(m^3 \cdot K)$
Compressibility factor	z	$z = pV/RT$
Density	ρ	kg/m^3
Diffusivity		
Molecular, volumetric	D_V, δ	$m^3/(s \cdot m), m^2/s$
Thermal	α	$\alpha = k/C_p m^2/s$
Emissivity ratio for radiation	e	
Enthalpy, per mole	H	J/kmol
Entropy, per mole	S	$J/(kmol \cdot K)$
Fugacity	f	Pa, bar
Gibbs, free energy, per mole	G, F	J/kmol
Helmholtz free energy, per mole	A	J/kmol
Humid heat	c_s	$J/(kg\ dry\ air \cdot K)$
Internal energy, per mole	U	J/kmol
Latent heat, phase change	λ	J/kg
Molecular weight	M	kg
Reflectivity for radiation	ρ	
Specific heat	c	$J/(kg \cdot K)$
At constant pressure	c_p	$J/(kg \cdot K)$
At constant volume	c_v	$J/(kg \cdot K)$
Specific heats, ratio of	γ	
Surface tension	σ	N/m
Thermal conductivity	k	$(J \cdot m)/(s \cdot m^2 \cdot K)$
Transmissivity of radiation	τ	
Vapor pressure	p*	Pa, bar
Viscosity		
Absolute or coefficient of	μ	$Pa \cdot s$
Kinematic	ν	m^2/s
Volume, per mole	V	$m^3/kmol$
Symbols for Concentrations		
Absorption factor	A	$A = L/K^*V$
Concentration, mass or moles per unit volume	c	kg/m^3, $kmol/m^3$
Fraction		
Cumulative beyond a given size	ϕ	
By volume	x_v	
By weight	x_μ	
Humidity	H, Y_H	kg/kg dry air
At saturation	H_s, Y^*	kg/kg dry air
At wet-bulb temperature	H_w, Y_w	kg/kg dry air
At adiabatic saturation temperature	H_a, Y_a	kg/kg dry air
Mass concentration of particles	c_p	kg/m^3
Moisture content		
Total water to bone-dry stock	X_ϕ^*	kg/kg dry stock

Table 4, continued

	Symbol	Unit or Definition
Symbols for Concentrations		
Equilibrium water to bone-dry stock	X^*	kg/kg dry stock
Free water to bone-dry stock	X	kg/kg dry stock
Mole or mass fraction		
In heavy or extract phase	x	
In light or raffinate phase	y	
Mole or mass ratio		
In heavy or extract phase	X	
In light or raffinate phase	Y	
Number concentration of particles	n_p	number/m^3
Phase equilibrium ratio	K^*	$K^* = y^*/x$
Relative distribution of two components		
Between two phases in equilibrium	α	$\alpha = K_i^*/K_j^*$
Between successive stages	β	$\beta = (y_i/\nu_i)_x/(x_i x_i)_{n+1}$
Relative humidity	H_R, R_H	
Slope of equilibrium curve	m	$m = dy^*/dx$
Stripping factor	S	$S = K^*V/L$
Rate Symbols		
Quantity per unit time, in general	q	
Angular velocity	ω	
Feedrate	F	kg/s, kmol/s
Frequency	f, N_f	
Friction velocity	u^*	$u^* = (\tau_w \rho)^{1/2}$, m/s
Heat transfer rate	q	J/s
Heavy or extract phase rate	L	kg/s, kmol/s
Heavy or extract product rate	B	kg/s, kmol/s
Light or raffinate phase rate	V	kg/s, kmol/s
Light or raffinate product rate	D	kg/s, kmol/s
Mass rate of flow	w	kg/s, kg/hr
Molal rate of transfer	N	kmol/s
Power	P	W
Velocity, in general	n	m/s
Revolutions per unit time	u	m/s
Longitudinal (x) component of	u	m/s
Lateral (y) component of	v	m/s
Normal (z) component of	w	m/s
Volumetric rate of flow	q	m^3/s, m^3/hr
Quantity per unit time, unit area		
Emissive power, total	W	W/m^2
Mass velocity, average	G	$G = w/S$, $kg/(s \cdot m^2)$
Vapor or light phase	G, $\overline{G}$	$kg/(s \cdot m^2)$
Liquid or heavy phase	L, $\overline{L}$	$kg/(s \cdot m^2)$
Radiation, intensity of	I	W/m^2
Velocity		
Nominal, basis total cross section of packed vessel	v_s	m/s
Volumetric average	V, $\overline{V}$	$m^3/(s \cdot m^2)$, m/s
Quantity per unit time, unit volume		
Quantity reacted per unit time, reactor volume	N_R	$kmol/(s \cdot m^2)$

Table 4, continued

	Symbol	Unit or Definition
Rate Symbols		
Space velocity, volumetric	Λ	$m^3/(s \cdot m^3)$
Quantity per unit time, unit area, unit driving force, in general	k	
Eddy diffusivity	δ_E	m^2/s
Eddy viscosity	ν_E	m^2/s
Eddy thermal diffusivity	α_E	m^2/s
Heat transfer coefficient		
Individual	h	$W/(m^2 \cdot K)$
Overall	U	$W/(m^2 \cdot K)$
Mass transfer coefficient		
Individual	k	$kmol/(s \cdot m^2)$ (driving force)
Gas film	k_G	To define driving force use subscript:
Liquid film	k_L	c for $kmol/m^3$
Overall	K	p for bar
Gas film basis	K_G	x for mole fraction
Liquid film basis	K_L	
Stefan-Boltzmann constant	σ	$5.6703 \times 10^{-8}\ W/(m^2 \cdot K^4)$

Table 5. The Greek Alphabet

Greek Letter	Greek Name	English Equivalent
Αα	Alpha	a
Ββ	Beta	b
Γγ	Gamma	g
Δδ	Delta	d
Εϵ	Epsilon	e
Ζζ	Zeta	z
Ηη	Eta	a
Θθ	Theta	th
Ιι	Iota	ē
Κκ	Kappa	k
Λλ	Lambda	l
Μμ	Mu	m
Νν	Nu	n
Ξξ	Xi	ks
Οο	Omicron	o
Ππ	Pi	p
Ρρ	Rho	r
Σσ	Sigma	s
Ττ	Tau	t
Υυ	Upsilon	ü, $\overline{oo}$
Φφ	Phi	f
Χχ	Chi	H
Ψψ	Psi	ps
Ωω	Omega	ō

Appendix B contains tables of units and conversion factors prepared by the American Petroleum Institute (API), Washington, DC. Tables and conversion factors have been grouped into the following categories:

1. Space, Time
2. Mass, Amount of Substance
3. Heating Value, Entropy, Heat Capacity
4. Temperature, Pressure, Vacuum
5. Density, Specific Volume, Concentration, Dosage
6. Facility Throughput, Capacity
7. Flowrate
8. Energy, Work, Quantity of Heat, Power
9. Mechanics
10. Transport Properties
11. Electricity, Magnetism
12. Acoustics, Light, Radiation

Metric units recommended for general use are shown under the heading "API preferred metric unit."

CHAPTER 1.2

FLUID STATICS AND PRESSURE MEASUREMENT

This chapter deals with the principles of *hydrostatics* and methods of measuring pressure. Hydrostatics deals with the properties and characteristics of liquids that are not in motion. When dealing with stationary gases the subject is called *pneumatics*. In keeping with the general theme of the book, only incompressible fluids are discussed here. Basic definitions applicable to problem solving and flow measurement are presented.

PRESSURE DEFINITIONS

If a force F(N) is uniformly distributed over a flat surface of area A (m^2), then the pressure is $P = F/A$ (N/m^2 or lb_f/ft^2). As shown, the units of pressure are force per unit area. A *normal pressure* acts only in a direction perpendicular to the surface sustaining it. The term pressure can take on several meanings.

For fluids, *relative and absolute* pressures are defined in terms of a chosen reference. Relative pressure is the measured pressure with respect to the atmosphere. Absolute pressure is that measured with respect to a complete vacuum. In the former, relative pressure is dependent on location, temperature, humidity and other environmental factors. Since absolute pressure is referenced with respect to a complete vacuum, it is a fixed measurement or true measurement.

Barometric pressure refers to a measurement of the atmospheric pressure using a barometer.

Gauge pressure is that pressure expressed as a quantity measured from above some reference pressure, which is usually atmospheric. The relationship of gauge, barometric and absolute pressures is as follows:

$$\text{Gauge Pressure} + \text{Barometric Pressure} = \text{Absolute Pressure} \qquad (1)$$

Figure 2 defines the pressure relationships graphically along with common engineering units.

The terms *hydrostatic* or *static pressure* at a point in a fluid refer to the compressive stress at that point. Consider a fluid at rest in a containing vessel. Hydrostatic pressure is the stress normal to any surface exerted by the fluid. At any given point it has the same magnitude, regardless of the orientation of the surface. Hence, the pressure exerted on the containing walls of the vessel by a stationary fluid is normal and equal to, at each point, the static pressure of the adjacent fluid. In the case of a flowing fluid, the pressure on the containing surface is not wholly normal.

Consider now a differential cube of fluid. The variation of the static pressure in any direction in the fluid may be found from Newton's law:

$$F = (\text{mass} \times \text{acceleration})/g_c \tag{2}$$

If there is no relative motion, the only forces acting on the boundaries of the cube are gravity and the normal outward pressures. For forces acting in the x and y directions,

$$\rho X - \frac{\partial P}{\partial x} = \frac{\rho a_x}{g_c} \tag{3a}$$

$$\rho Y - \frac{\partial P}{\partial y} = \frac{\rho a_y}{g_c} \tag{3b}$$

where ρ = fluid density, lb_m/ft^3

X, Y = x- and y- component of extraneous forces (lb_f/lb_m), respectively

a_x, a_y = x- and y- component of the fluid's acceleration (ft/s^2), respectively

$g_c = 32.1740\ (lb_m)(ft)/(lb_f)(s^2)$.

Since y is measured vertically downward, $Y = g/g_c$; where g is the local acceleration due to gravity (ft/s^2). For the horizontal x direction, the force component X is zero.

Pressure will vary with depth, y, for a stationary fluid accordingly:

$$\frac{dP}{dy} = \rho g/g_c \tag{4}$$

If the density of the fluid is constant, then the pressure is the same everywhere at the same depth. These conclusions lead to Pascal's law,

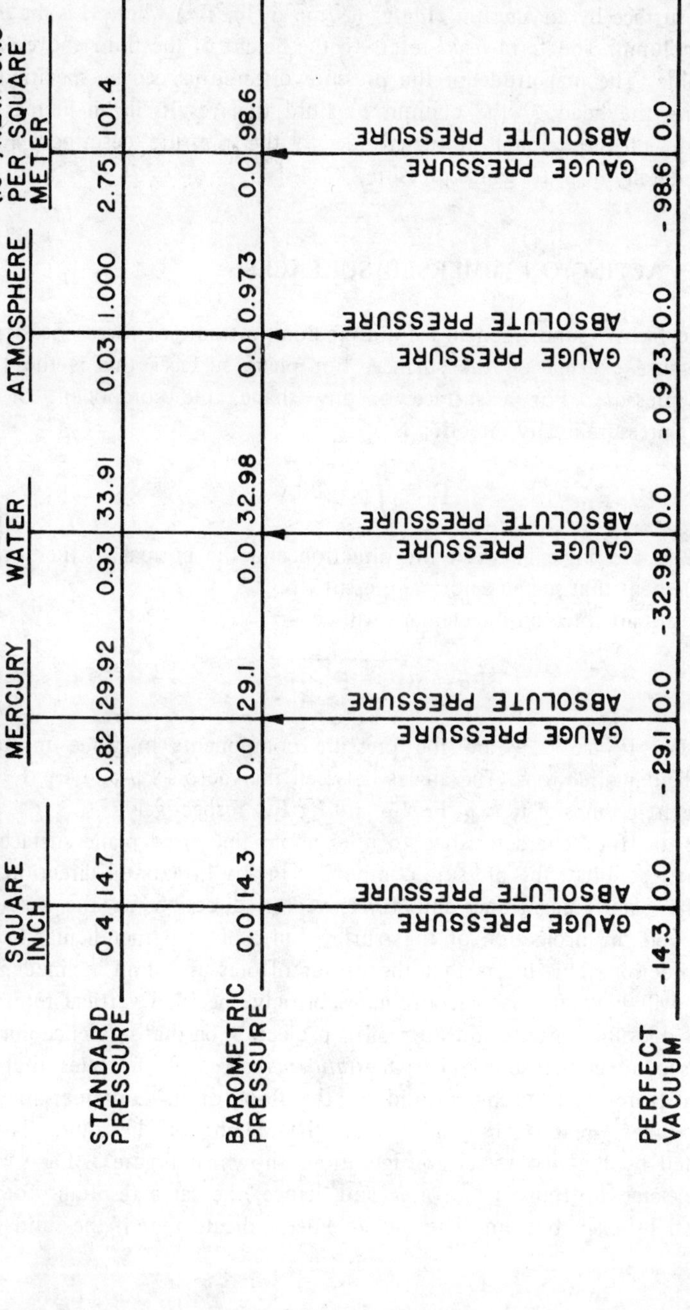

Figure 2. Shows the relationship of standard, barometric and vacuum pressure for different engineering units. Standard atmosphere is the pressure obtained in a standard gravitational field and is equivalent to 14.696 psi or 760 mm mercury at 0°C. Atmospheric pressure is a variable that must be measured with a barometer each time.

which for incompressible fluids states that any increase in the pressure at any point is accompanied by an equal increase at every point in the fluid.

If our differential fluid cubical is now extended to a column, then it can be said that the static pressure at the base of the column is greater than the top surface by an amount $H\rho g/g_c$ (N/m^2 or lb_f/ft^2) where H is the *head* of the column. The term *head* refers to the height of the fluid above some datum line. The magnitude of the pressure differential can be specified by measuring the head of the column of fluid that results in an increase in the static pressure equal in magnitude to the pressure described. Fluid heads generally measure gauge pressure.

FORCES ACTING ON IMMERSED SURFACES

If a surface is submerged in a fluid medium, a resultant force arises from the pressure exerted on the surface. For plane surfaces this is the total normal pressure. For a surface of any shape, the component of the resultant pressure in any direction is

$$F_a = \int p \cos\theta \, dA \qquad (5)$$

where θ is the angle between the direction and the normal to the surface at the element that sustains normal pressure, p.

The resultant force on the element will be

$$F_R = \sqrt{F_1^2 + F_2^2 + F_3^2} \qquad (6)$$

where F_1, F_2 and F_3 are the pressure components in three mutually perpendicular directions. The angles between the vectors F_R and F_1, F_2, F_3 have the arc-cosines of F_1/F_R, F_2/F_R and F_3/F_R, respectively.

There are four characteristics to note about immersed plane surfaces in fluids at rest. First, the pressure component in any horizontal direction will have the same magnitude and force vector direction as the resultant pressure on the projection of that surface on a plane perpendicular to any given direction. This infers that the center of pressure of an inclined plane surface will have the same coordinates on any specified vertical reference plane as does the center of pressure of its projection on that reference plane.

The second is the so-called *hydrostatic paradox*, which states that the resultant force exerted by a fluid on the floor of its container may be several times greater or smaller than the weight of the fluid. This is illustrated by the two vessel configurations shown in Figure 3. Each vessel has the same bottom surface area and, hence, the same resultant force is sustained by each bottom. The dashed lines indicate that if the fluid were

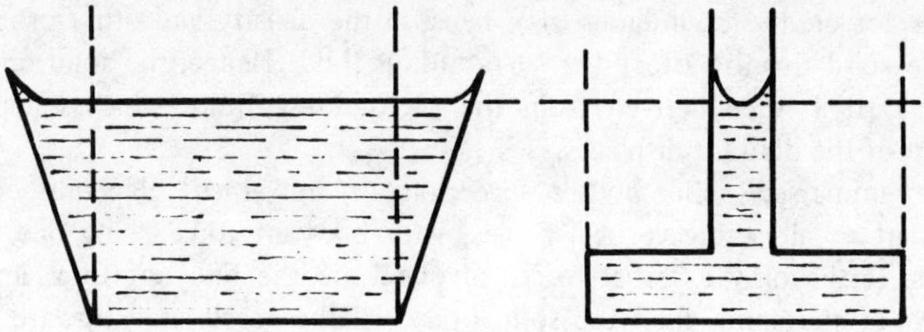

Figure 3. Illustrates the hydrostatic paradox.

filled to the same level, then the same volume or weight of fluid would exist in the dashed region. The fact that one vessel contains more fluid than the other has no bearing on the sustained resultant force.

Third, the resultant pressure on a closed surface is a vertical force that is equivalent to the weight of the fluid that could be contained within the surface under conditions such that the contents would be in equilibrium with the surrounding fluid if the surface were removed. The resultant force acts through the center of gravity of the enclosed fluid and acts in an upward or downward direction depending on whether the outer or inner side of the surface sustains the fluid pressure.

Finally, the pressure component in any given direction may be determined from the area of the projection of the given surface on a plane perpendicular to that direction times the pressure on the center of gravity of the surface.

FLOATING BODIES

Almost 1800 years prior to the dawning of Newtonian mechanics, Archimedes laid down one of the first laws of fluid statics. When a body is immersed partially or wholly in a fluid, it is acted on by an upward or buoyant force, which equals the magnitude of the weight of the fluid displaced by the body.

Imagine in irregularly shaped finite volume, V, which represents a fraction of the total volume of fluid contained in a vessel. The fluid occupying V is at rest because resultant forces from the surrounding fluid acting on it counterbalance the downward weight of the fluid contained in V. Consequently, fluid occupying V is subject to an upward force arising from the surrounding fluid, which matches the weight of the fluid within V. If the imaginary volume is replaced by a solid having the same configuration and hence occupies the same volume, the same exact forces act on the boundaries of V because the density and other properties of the solid do not affect the surrounding fluid. Hence, the solid object is subject to a force derived from the surrounding fluid, which equals the weight of the liquid it displaces.

The immersed solid body's direction of movement depends on the resultant of all acting vertical forces, with buoyant force being one of the forces. If the only external forces involved are the buoyant force and the weight of the solid, then the solid body will be accelerated upward if the buoyant force exceeds the weight. For a body wholly immersed in a fluid, acceleration in an upward direction will occur if the specific gravity of the body is less than that of the fluid; conversely, if the body's specific gravity exceeds that of the fluid, it will sink.

At equilibrium, buoyancy and weight are equal, and the centers of gravity and buoyancy of a floating body must lie in the same vertical line. A body that is floating in equilibrium has stability such that a small displacement brings into action forces that will restore the body to its original state of equilibrium.

The depth at which flotation occurs is the *draft*. If a body is floating at the interface between two fluids (e.g., gas and liquid or two immiscible liquids), the draft is defined as the vertical distance of the lowest point of the body below the interface.

A final note is that the specific gravity and volume of a solid body can be readily determined from the principles described above. By weighing any substance in a liquid in which it is insoluble, then weighing it in air, the difference in weight represents the weight of an equal volume of liquid:

$$M_w = m_a - m_f \tag{7}$$

where M_w = weight of an equal volume of liquid
m_a = weight of body in air
m_f = weight of body in fluid

The relative density is then as follows:

$$\text{Relative Density} = m_a/(m_a - m_f) \tag{8}$$

CAPILLARY ACTION

Flow through capillaries have a wide range of applications, the most common of which pertains to manometric techniques. In examining the interface between two immiscible fluids, it can be observed that a minimum area consistent with the volume of fluid enclosed and with the external forces acting on the fluids is attained. This condition is a consequence of the intermolecular attraction between the two fluids in the region of the interface. If we isothermally extend the area A of the interface, the work performed is

$$dW = \sigma dA \tag{9}$$

Parameter σ is the work performed per unit increase of area (sometimes called the "free energy" of the interface per unit area) and is a function of the state of the system. σ is more commonly referred to as the interfacial tension. Values for σ vary greatly with the composition of fluids and temperature. The effect of pressure is generally small on surface tensions. The dimensions on σ are in energy per unit area (same as force per unit area = dynes/cm^2). When one of the fluids is a gas, σ is referred to as the surface tension.

If the interfacial plane between the two fluids contacts a solid surface, the angle of contact lies between very specific limits, which depend on the properties of the two fluids. This is known as the contact angle, ϕ. Typical values for gas-liquid interfaces in contact with glass at room temperature are 0° for air-water and many organic liquids, and 140° for air-mercury.

A final note on capillarity concerns the phenomenon of capillary rise. If we have a small-diameter, vertical tube, as shown in Figure 4, a column

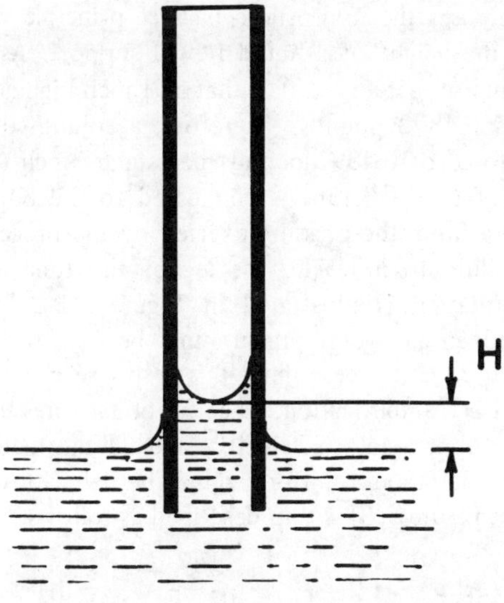

Figure 4. Illustrates capillary rise action.

of liquid will stand at a height that is greater, or less, than that corresponding to the static head at the base of the column by an amount:

$$H = \frac{4\sigma \cos\phi}{gD(\rho_\ell - \rho_g)} \tag{10}$$

where
D = capillary tube diameter (cm)
ρ_ℓ, ρ_g = densities of the liquid and gas, respectively (g/cm^3)
$g \simeq 981$ cm/s^2
H = head (cm)

Equation 10 is accurate for sufficiently small tubes where the meniscus is substantially spherical.

MANOMETRIC TECHNIQUES FOR PRESSURE MEASUREMENT

The manometer is the simplest device for measuring head or differential pressure. The basic manometer consists of a glass tube bent in the shape of the letter "U" and partially filled with some liquid. When both ends of the "U" tube are open to the atmosphere, as shown in Figure 5(A), the pressure on each side is equivalent and the column of liquid on one side exactly balances the column of liquid on the other side, so that the surfaces of the two liquids reach equilibrium at the same level.

The manometer uses the hydrostatic balance principle wherein a pressure is measured by the height of a liquid it will support. As an example, the weight of a column of mercury at 0°C, that is, 1 inch high and 1 inch in cross-sectional area, is 0.4892 pounds. Therefore, a column of mercury 1 inch high imposes a force of 0.4892 pounds per square inch (i.e., 0.4892 psi).

When one leg of the "U" tube is connected to the bottom section of a vessel containing a fluid, the pressure exerted by the process fluid will cause the manometer fluid to move up the leg of the tube exerting the least pressure. This situation is illustrated in Figure 5(B). The response of a manometer to sudden changes in pressure may be described as a second-order system. The basic equation describing this response can be derived from a force balance. For simplification, the frictional pressure drop can be assumed to be proportional to velocity, and all the manometer fluid can be assumed to accelerate at a uniform rate. The behavior of the manometer fluid illustrated in Figure 5(B) is thus described as follows:

$$\frac{AL\rho}{g_c}\frac{d^2H}{dt^2} = A\left(P_1 - 2H\rho\frac{g}{g_c}\right) - fA\frac{dH}{dt} \tag{11}$$

where A = the cross-sectional area of the tube
ρ = the density of the manometer fluid (it is assumed that the density of the gas above the fluid can be neglected)
g = gravitational acceleration constant
t = time
P_1 = the applied pressure
f = the frictional resistance to flow
L = the length of manometer tube filled with fluid
H = head of fluid (i.e., the height of manometer fluid above the reference mark)

Flow in a capillary tube or manometer is laminar; therefore, the pressure drop across a unit length of the "U" tube can be obtained from the Hagen-Poiseuille equation:

$$\frac{\Delta P}{L} = \frac{32u\mu}{D^2 g_c} \tag{12}$$

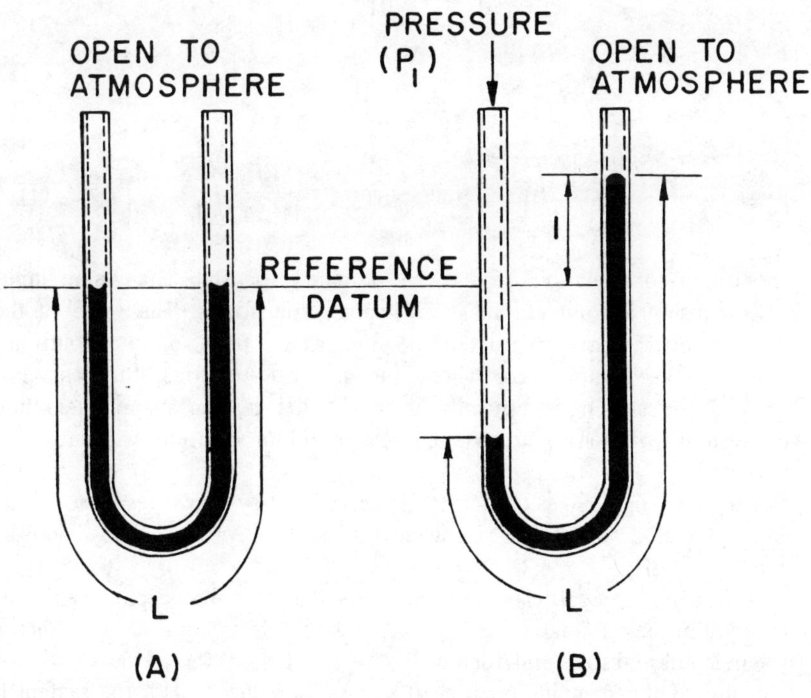

Figure 5. U-tube manometer.

whence the frictional resistance to flow is

$$f = \frac{32L\mu}{D^2 g_c} \tag{13}$$

where μ = viscosity of the manometer fluid
u = flow velocity
D = diameter of the tube

By substituting Equation 13 into 11 and rearranging,

$$\frac{L}{2g}\frac{d^2H}{dt^2} + \frac{16L\mu}{\rho g D^2}\frac{dH}{dt} + h = \frac{P_1 g_c}{2\rho g} + H_i \tag{14}$$

To describe the dynamic response of a second-order system, a characteristic damping coefficient is introduced. Defining ω_N as the natural frequency of the system in radians per second and ξ as the damping coefficient, Equation 14 becomes

$$\frac{1}{\omega_N^2}\frac{d^2H}{dt^2} + \frac{2\xi}{\omega_N}\frac{dH}{dt} + H = H_i \tag{15a}$$

where

$$\omega_N = \sqrt{\frac{2g}{L}}, \quad \text{rad/sec and } \xi = \frac{8L\mu}{D^2\rho g}\sqrt{\frac{2g}{L}} \tag{15b}$$

Solutions to Equation 15 can be obtained for step changes in input pressure using a complete table of Laplace transforms. The value of the damping coefficient will determine the exact form of the solution's response. The damping coefficient can give rise to three situations. For ξ = 1.0 the system is critically damped, that is, the manometer fluid (or system's response) arrives at its equilibrium position without overshooting.

When the damping coefficient is zero, the system's response is an undamped sine wave. The frequency of such a system is ω_N and its amplitude is $2H_i$.

For a damping coefficient less than 1.0, the system's output overshoots the equilibrium or final value and oscillates before coming to rest. Such a system is referred to as underdamped.

Finally, if the damping coefficient exceeds a value of 1.0, the system is overdamped. The fluid then comes to equilibrium slowly. The solutions to Equation 15 for each of these cases are summarized below:

Critically damped system – $\xi = 1.0$:

$$\frac{H}{H_i} = 1 - (1 + \omega_N t)\exp(-\omega_N t) \tag{16}$$

Underdamped system – $\xi < 1.0$:

$$\frac{H}{H_i} = 1 + \frac{\exp(-\xi\omega_N t)}{\sqrt{1-\xi^2}} \sin(\omega_N \sqrt{1-\xi^2}\, t - \phi) \tag{17}$$

where

$$\phi = \tan^{-1}\left[\frac{\sqrt{1-\xi^2}}{-\xi}\right]$$

Overdamped system – $\xi > 1.0$:

$$\frac{H}{H_i} = 1 + \frac{1}{\theta_b - \theta_a}\,(\theta_a \exp(-t/\theta_a) - \theta_b \exp(-t/\theta_b)) \tag{18}$$

where θ_a and θ_b are time constants obtained from two roots to the solution.

Water- or mercury-filled U-tube manometers are typically underdamped systems. Table 6 [1] gives theoretical values of the natural frequencies and damping coefficients for these manometers. Biery [2] noted that actual

Table 6. Theoretical Damping Coefficients for Water-Filled and Mercury-Filled Manometers [1]

	Fluid Properties					
Indicating Fluid	Density (g/cm^3)	Viscosity (cp)	Diameter (cm)	L (cm)	ω_N(rad/s)[a]	ξ [a]
Mercury	13.5	1.6	0.2	250	2.80	0.169
Mercury	13.5	1.6	0.3	250	2.80	0.075
Mercury	13.5	1.6	0.5	250	2.80	0.027
Water	1.0	1.0	0.2	250	2.80	1.429
Water	1.0	1.0	0.5	250	2.80	0.229

[a]Values computed from Equation 15b.

values for the damping coefficient are higher than predicted by Equation 18. The natural frequency of a manometer typically ranges between 0.5 and 1 cycle per second. Adjustable damping can be provided by including a valve between the two legs of the manometer. This approach is typically employed when measuring the pressure drop across an orifice with a manometer for flow measurement purposes.

Liquid manometers are widely used in both industrial and laboratory applications. Manometers can be used for both basic pressure measurement devices and standards of calibration of other instruments.

Only the height of fluid from the surface of one tube to the surface in the other is the actual height of the fluid opposing and balancing the pressure. This is so regardless of the geometry or size of the tubes. Even if the manometer tubes are unsymmetrical, all that will occur is that more or less fluid will be moved from one leg to another; however, the fluid height required to achieve equilibrium will only depend on the density of the manometer fluid and vertical height. Table 7 gives equivalent values of various common indicating fluids. The table demonstrates the versatility of the manometer. As an example, when water is used as an indicating fluid, a 10-inch fluid height measures 0.360 psi, whereas the same instrument

Table 7. Conversion Values for Common Manometer Fluids at 22°C [1]

1 inch of water	=	0.0360 lb/in.^{2}
		0.5760 oz/in.^{2}
		0.0737 in. mercury
		1.2131 in. red oil
1 foot of water	=	0.4320 lb/in.^{2}
		6.9120 oz/in.^{2}
		0.8844 in. mercury
		62.208 lb/ft.^{2}
		14.5572 in. red oil
1 inch of mercury	=	0.4892 lb/in.^{2}
		7.8272 oz/in.^{2}
		13.5712 in. water
		1.1309 ft water
		16.4636 in. red oil
1 ounce/in.^{2}	=	0.1228 in. mercury
		1.7336 in. water
		2.1034 in. red oil
1 pound/square inch	=	2.0441 in. mercury
		27.7417 in. water
		2.3118 ft water
		33.6542 in. red oil

utilizing mercury measures 4.892 psi. This represents a ratio of 13.57 to 1 (i.e., the ratio of the specific gravities of the two fluids).

There are three types of pressure measurements that can be made with manometers:

1. positive pressures or gauge pressures, which are those greater than atmospheric;
2. negative pressures or vacuums, which are pressures less than atmospheric; and
3. differential pressure, which is the difference between two pressures.

Connecting one leg of the U-tube to a source of gauge pressure causes the fluid in the connected leg to depress while the fluid rises in the vented leg. However, if the connection is made to a vacuum, the effect would be to reverse the fluid movement, causing it to rise in the connected leg and and recede in the open leg.

In obtaining differential pressures, both legs of the manometer are connected so that a pressure change between two points can be measured. The higher pressure depresses the fluid in one leg while the lower pressure allows the fluid to rise in the other. The true differential is measured by the difference in height of the fluid in the two legs.

Manometers are available in configurations other than the U-tube to provide greater convenience and to meet different service requirements. Figure 6 illustrates a well-type manometer. As shown, if one leg of the manometer is increased in area in comparison to the other, the volume of

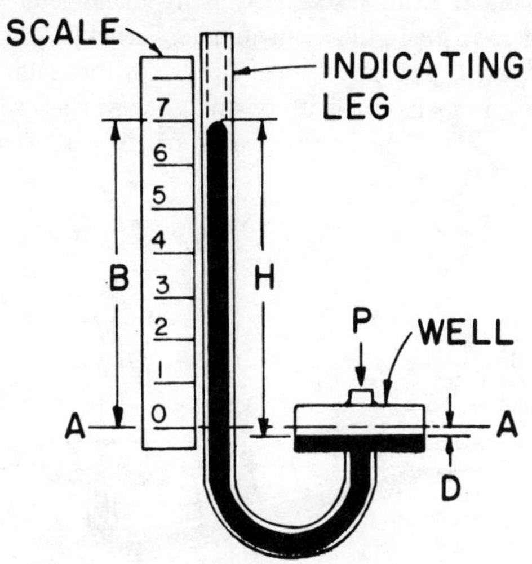

Figure 6. Well-type manometer.

fluid displaced represents little change in the vertical height of the large area leg compared to the change of height in the smaller area leg. It thus becomes only necessary to read the scale adjacent to the single tube rather than two, as in the U-type.

The well-type design lends itself to use of direct reading scales that are graduated in appropriate units for the process variable involved. The higher-pressure source to be measured must always be connected to the well leg, while the lower pressure source must be connected to the top of the tube.

In any measurement, therefore, the source of the higher pressure must be connected in such a manner to cause the manometer fluid to rise in the indicating tube.

The true pressure follows the same principles previously discussed, and the measurement obtained reflects the difference between the fluid surfaces. Obviously there must be some decrease in the well level; however, this can be compensated for by spacing the scale graduations in the proper amount needed to reflect and correct for the well level drop.

There are a variety of applications requiring accurate measurement of low pressures, such as drafts and low differentials in air and gas installations. A common arrangement for these cases is the use of an inclined-tube well manometer, shown in Figure 7. This design has the advantage of providing an expanded scale. For example, 12 cm of scale length can represent 1 cm of vertical length. Using scale divisions of 0.01 inches of liquid height, an equivalent pressure of 0.000360 psi per division can be detected with water as the manometer fluid.

To measure relatively high pressures, a longer indicating fluid tube is required. Rather than using excessively long manometer tubes for high-pressure readings, a dual tube manometer arrangement, as shown in Figure 8, can be used. This arrangement provides for reading the full range of the instrument in only one-half of the total vertical viewing distance. The system shown in Figure 8 consists of two separate manometers

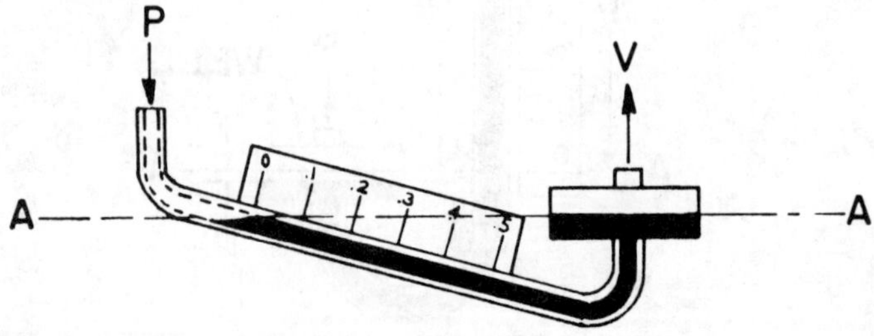

Figure 7. Inclined-tube well-type manometer for low-pressure measurements.

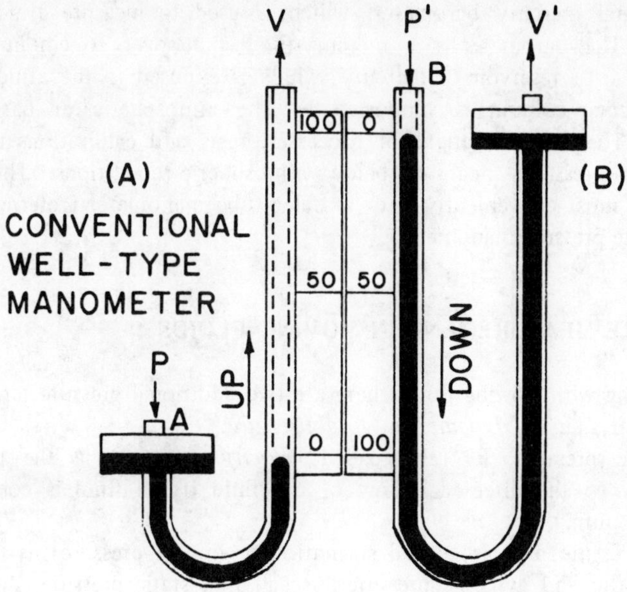

Figure 8. Dual-tube manometer system for measuring high pressures.

mounted on a single housing. Manometer (A) is a conventional well design having a zero scale at the bottom and graduated upward. Manometer (B) has a well and scale zero raised to the 100-inch level with the scale graduated downward. Connecting a positive pressure source at connections (A) and (B) causes the indicating fluid level to rise in tube (A) and fall in (B). Note, however, that these are essentially two separate manometers measuring the same pressure, with one indicating column rising and one falling. The fluid columns pass each other at the 50-inch mark and, as such, it becomes convenient to read the left scale on pressures from 0 to 50 inches progressing upscale and the scale on manometer (B) from 50 to 100 inches downscale. Only the lower 50 inches is thus required to read the entire 100 inches range. Gauge pressure connections must be made at connections A and B, while vacuum connections must be made at V and V′.

Another system frequently used is a sealed tube or absolute manometer. The term "absolute pressure" originates from the fact that in a perfect vacuum the complete absence of any gas is referred to as "absolute" zero. For an absolute pressure manometer, the pressure measured is compared to the vacuum or absolute pressure in a sealed tube above the indicating fluid column. The most common type of sealed tube manometer is the conventional mercury barometer, which is used to measure atmospheric pressure. The barometer is a mercury-filled tube over 76 cm in height, immersed in a reservoir of mercury, which is exposed to the atmosphere. The mercury column is supported by the atmospheric or barometric pressure. There are a variety of processes, tests and calibrations that are based on pressures near or below atmospheric conditions. These are measured most conveniently with a sealed-tube manometer (referred to as an absolute pressure manometer).

PRESSURE MEASUREMENT IN MOVING FLUIDS

In dealing with flowing fluids there are two additional pressure terms that are relevent, namely, *dynamic* and *total pressure.*

Dynamic pressure, P_v (also called *velocity pressure*), is the pressure equivalent to the directed energy of the fluid if the fluid is considered as a continuum.

Total pressure, P_t (also called stagnation or impact pressure), is the sum of the static and dynamic pressures. Recall that static pressure, P_s, is the actual pressure of the fluid, whether in motion or at rest. Total pressure can be sensed by a probe that is stationary with respect to the system boundaries when it locally stagnates the fluid isentropically.

Since $P_t = P_s + P_v$, the following relationship can be written from an energy balance and the first law of thermodynamics:

$$\int_{P}^{\dot{P}_t} \frac{dP}{d\rho} + \int_{\dot{u}}^{0} \frac{\dot{u}du}{g} = 0 \tag{19}$$

where the dot $(\cdot)$ signifies a point quantity. For a constant density, incompressible fluid, this relation becomes

$$(\dot{P}_t - P) = \dot{P}_v = \frac{\rho \dot{u}^2}{2g} \tag{20}$$

Equation 20 merely restates the relationship of total, static and dynamic pressures.

A similar analysis may be performed for compressible fluids, whereby for an ideal gas undergoing an isentropic process the following is obtained:

$$(\dot{P}_t - P) = P_v = \frac{\rho \dot{u}^2}{2g}\left(1 + \frac{N_{Ma}^2}{4} + (\gamma - 2)\frac{N_{Ma}^4}{24} + \ldots\right) \tag{21}$$

where γ is the ratio of specific heats and N_{Ma} is the Mach number defined as $u/(\gamma gRT)^{1/2}$. Note that as N_{Ma} approaches zero, compressibility effects become less important and Equation 21 reduces to Equation 20.

When a total pressure at a plane is needed, the quantity is more rigorously referred to as the *effective total pressure*. The point definition of total pressure given in Equation 20 must be modified by accounting for the flow-weighted average total pressure as follows [3]:

$$\int_A P_t dm = \int_A P dm + \int_A \left(\frac{\rho \dot{u}^2}{2g}\right)(\rho \dot{u} dA) \tag{22}$$

where m is the mass flowrate per unit time and the integrals are evaluated over the area, A, of the plane. Integrating Equation 22 gives the following:

$$P_t = P + \frac{\rho \int \dot{u}^3 dA}{2guA} \tag{23}$$

Benedict [3] notes that the term $\int \dot{u}^3 dA/2guA$ is really the actual kinetic energy per pound of fluid flowing. If we redefine point values as one-dimensional continuum quantities, then

$$\frac{K.E.}{W} = \frac{\int \frac{1}{2} u^2 (\rho u dA)}{\rho u A} = u^2/2g \tag{24}$$

and Equation 23 becomes

$$P_t = P + \rho\alpha \frac{u^2}{2g} \tag{25}$$

for $\alpha = 2.0$ for laminar flow

$1.02 < \alpha < 1.15$ for turbulent flow

Parameter α is the kinetic energy coefficient.

Equations 22–25 apply strictly to incompressible flows. The treatment of compressible fluids, is, unfortunately, not as straightforward.

As will be shown later, these pressure quantities are needed for evaluating flowrates. The simplest devices for measuring flowrates are head meters (examples are venturis, nozzles and orifices). These devices are discussed in Chapter 2.3 of Section 2. Flow measurement with these fluid meters is based on measurements of the static pressure. Static pressure can be sensed in a

Table 8. Summary of the Effect of Orifice Edge on Static Pressure Measurements, as Reported by Rayle [5]

Wall Tap Geometry	Measurement Error (%)	Wall Tap Geometry	Measurement Error (%)
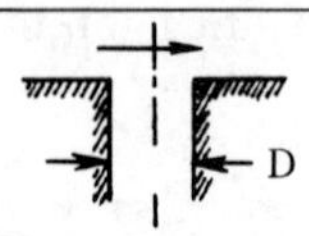	0	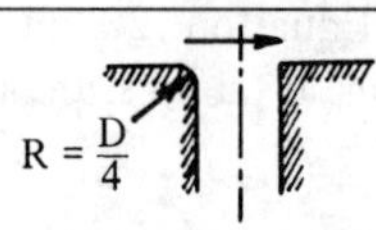	0.9
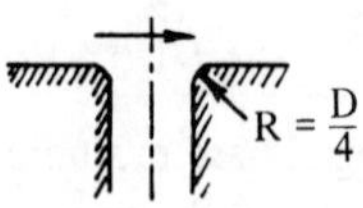	0.2		−0.5
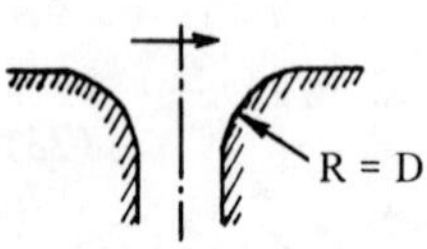	1.1	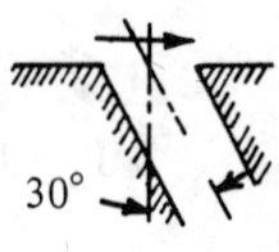	0.3
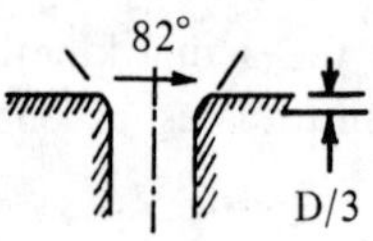	−0.1	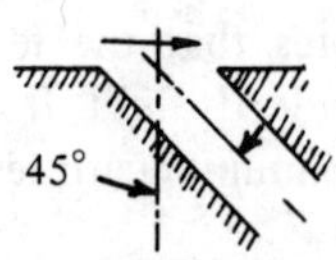	0.4
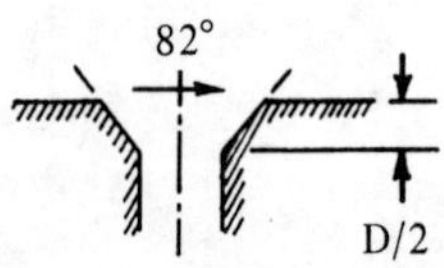	−0.3	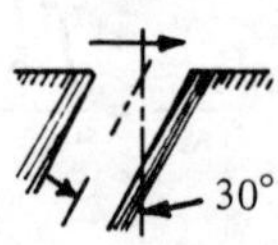	0

number of ways, the most common including wall taps, static tubes and critical taps on aerodynamic bodies. We shall describe only the first, since these are the most widely used. (The student is referred to Benedict [3] and Perry [4] for discussions on the others.)

The accuracy of static pressure measurements via wall taps depends on the shape and size of the hole, as well as on the actual measuring device. For the ideal case, infinitely square-edged holes installed normal to the flow provide the true static pressure. Rayle [5] has studied the effect of hole size and has determined absolute errors. His study showed that square-edged holes produce small overestimates of the static pressure; radius-edged holes introduce larger positive errors; and chamber-edged holes tend to underestimate the true static pressure. Table 8 summarizes some of the results reported by Rayle.

There are no hard and fast rules for the construction and performance of wall taps. There are, however, rough guidelines that can be applied. Shaw [6] recommends for turbulent incompressible flow in smooth pipe that the static pressure hole have a diameter 0.1 of the pipe diameter. Rayle [5] notes that a 0.762-mm-diameter hole with a 0.381-mm-deep countersunk should give nearly true static measurements.

Further discussions on static and total pressure measurements may be found in the literature [7-12].

CHAPTER 1.3

PROPERTIES OF FLUIDS

This chapter begins with a review of the general properties and behavior of all fluids. It is not the intention of this chapter to describe all the properties of fluids, but to emphasize those physical properties that characterize the flow resistance of most fluids. The principal property characterizing flow resistance is viscosity. This chapter is devoted mainly to discussing the viscosities of gases and liquids, with emphasis on the latter. With this format, the subject of rheology and characteristics of so-called non-Newtonian materials are introduced. Generalized effects of temperature and pressure on the viscosities of common fluids are described along with a description of viscosity in terms of molecular processes and the mechanisms of momentum transport. Examples are given illustrating how viscosity varies with conditions and the nature of the fluid. The information presented in this chapter will be needed throughout the book.

KINETIC THEORY OF GASES

Gases are nature's simplest substances, whose structures can be described quite readily by the kinetic theory of gases. The theory is based on the following assumptions:

1. Gases consist of a large number of atoms or molecules.
2. In the absence of external forces, atoms follow Newton's first law of motion (i.e., they move in straight lines).
3. Atoms/molecules interact (i.e., collide) with one another only infrequently.

In addition, the following condition is imposed—the total kinetic energy of the two molecules is the same before and after collision (i.e., an elastic collision).

Since gases consist of very large numbers of moving particles, then it follows that the motion of these particles is completely random. That is, gas molecules move in all directions and with a variety of speeds.

If a finite volume of gas is contained in a vessel, particles will collide with the vessel's walls and rebound. At the moment of collision a force is exerted on the wall, and this force divided by the area of the wall would be the momentary pressure exerted by the impact and rebound of the particle. Hence, the pressure exerted by the gas is the force exerted on the wall by the impacts of many molecules. That is, a pressure gauge provides a measurement of the average value of the force per unit exerted by the impacts of a tremendous number of molecules.

From Newton's second law of motion the force acting on the particle is equal to the change of momentum per unit time; therefore, the force acting on the wall is equal and opposite in sign to this value:

$$F = ma = m\frac{du}{dt} = \frac{d(mu)}{dt} \tag{26}$$

Since pressure is F/A, then

$$P_i = mu_1^2/V \tag{27}$$

where V is the volume of the vessel and u_1 is the velocity of the particle before and after its collision (since the collision is elastic). P_i is the pressure exerted by a single particle. If we consider all particles, each at different velocities, then

$$P = (m/V)(u_1^2 + u_2^2 + u_3^2 + \ldots) \tag{28}$$

or

$$P = Nm\bar{u}^2/V \tag{29}$$

where N is the number of molecules in the vessel and $\bar{u}^2$ is the average of the squares of the velocities ($\bar{u}^2 = (u_1^2 + u_2^2 + u_3^2 + \ldots)/N$).

Equation 29 assumes that all the molecules are moving in one direction. Since molecules move randomly it is logical to assume that on the average only one third of the particles move in any one of three directions. Defining the average total velocity vector, $\bar{C}^2$, in terms of the individual velocity components,

$$\bar{C}^2 = \bar{u}^2 + \bar{\nu}^2 + \bar{\omega}^2 \tag{30}$$

where u, ν, ω are the x-, y- and z-components of velocity.

Hence,

$$p = Nm\bar{c}^2/3V \tag{31}$$

The average kinetic energy of all the molecules is KE = ½ $m\bar{c}^2$; thus,

$$pV = \frac{2}{3} NKE \tag{32}$$

and from the ideal gas law (pV = nRT) it can be shown that

$$E = \frac{3}{2} RT \tag{33}$$

where E is the total kinetic energy associated with the random motion of molecules.

Equation 33 states simply that the kinetic energy of the random motion is proportional to the absolute temperature. Hence, for gases, temperature is a measure of the average kinetic energy of the chaotic motion.

GENERAL PROPERTIES OF LIQUIDS

Liquids are simply the *condensed phases* of gases (so are solids for that matter). The term "condensed phases" emphasizes the high density of the liquid as compared with the low density of gases. Consider a vessel designed to contain an amount of oxygen not to exceed a few hundred pounds mass: the weight of water required to fill the same vessel would be several hundred tons. Another way of saying this is that the volume per mole of gases is very large, whereas for liquids it is quite small. At conditions of standard temperature and pressure (STP) any gas occupies 22,400 cm^3/mol. Most liquids occupy between 10 and 100 cm^3/mol [13]. Hence, the molar volume of a liquid is 500–1000 times smaller than that of a gas.

Since the volume ratio of gas to liquid is as high as 1000, the ratio of distances between molecules in the gas as compared to the liquid is $\sqrt[3]{1000}$, or 10. That is, the distance between gas molecules is ten times farther apart than those of a liquid. In liquids, the distance between molecules is on the order of the molecular diameter. This large difference in molecular distances between gases and liquids accounts for the drastic difference in properties between the two states of matter. Intermolecular forces (i.e., van der Waals forces) have a relatively short field of influence such that the effect of these forces decreases dramatically with increasing distances between molecules. In fact, the magnitude of van der Waals forces in liquids is on the order of 10^6 times larger than it is in gases because of the shorter distances between molecules.

Because molecular distances are so small and intermolecular forces correspondingly so large for liquids, the properties of liquids depend greatly on the details of the forces acting between molecules. Hence, there are no simple relations or a universal equation of state to describe liquid behavior as there are for gases. Hence, properties of liquids vary greatly whereas generalities can be made about many of the physical properties of gases.

One example is the dependence of volume on temperature. For liquids, volume varies with temperature at constant pressure by an expression of the following form:

$$V = V_0(1 + \kappa T) \tag{34}$$

V_0 is the volume of the liquid at some reference temperature (usually 0°C) and κ is the coefficient of thermal expansion. Gases follow a general dependency similar to Equation 34; however, κ is approximately constant for all gases. For liquids, κ depends on the specific liquid.

V_0 in Equation 34 is a function of pressure and follows the relationship

$$V_0 = \hat{V}_0[1 - \beta(p - 1)] \tag{35}$$

where $\hat{V}_0$ is the volume at STP and β is the coefficient of compressibility. Parameter β is observed to be constant over a wide range of pressures for a particular material but is different for each substance and for the solid and liquid states of the same material.

Equation 35 states that the volume of a liquid decreases linearly with pressure. For gases, volume is inversely proportional to pressure. Even more dramatic a contrast is that β for liquids is typically 10^{-6} atm^{-1}, whereas for gases it is significant. As an example, if we subject water to a pressure change from 1 to 2 atm, Equation 35 predicts less than 10^{-3}% reduction in volume. If a sample of air were subjected to the same pressure change, its volume would be reduced by more than a factor of two. Because of this characteristic of insignificant volume changes from moderate pressure changes, liquids are referred to as *incompressible fluids*. That is, for practical purposes, values of β are taken to be zero for liquids.

If we go back to our vessel discussed earlier, evacuate the gas and place a known quantity of a pure liquid in it, then a portion of the liquid will evaporate so as to fill the remaining volume of the vessel with vapor. After equilibrium is established, the pressure exerted by the vapor in the vessel becomes dependent on the temperature of the system only. This pressure developed is the vapor pressure of the liquid and will increase rapidly with temperature. The normal boiling point of the liquid is the temperature at which the vapor pressure becomes equal to the atmospheric pressure.

The amount of heat absorbed during the transformation of liquid to vapor state is the *heat of vaporization*. For vaporization to occur, liquid molecules are pulled apart against intermolecular forces. The energy needed to accomplish this is measured by the heat of vaporization. From thermodynamics, the relationship between vapor pressure and the heat of vaporization is

$$\ln P_v = -\frac{\dot{Q}_{vap}}{RT} + \ln P_\infty \tag{36}$$

where $\dot{Q}_{vap}$ is the heat of vaporization, R is the gas law constant, T is absolute temperature, and P_∞ is a constant defined as

$$P_\infty = (\text{atmos. press.}) \exp(+\dot{Q}_{vap}/RT_b) \tag{37}$$

where T_b is the boiling point of the liquid.

Equation 36 is one form of the Clausius-Clapeyron equation. As an aside, a plot of log P_v versus 1/T produces a straight line whose slope is $-\dot{Q}_{vap}/R$. By obtaining measurements of the vapor pressure at various temperatures and preparing such a plot, the heat of vaporization can be readily obtained for the liquid.

There are two other general properties of interest to our forthcoming discussions. One is interfacial tension, already described in the previous chapter, and the other is the coefficient of viscosity. As we will see, viscosity is simply a property of the fluid, which results in the resistance to flow when the fluid is subjected to stress. Because the molecules of a liquid are packed more closely, a liquid is more viscous than a gas. Both the close molecular spacing and high intermolecular forces contribute to this resistance of flow.

MOLECULAR THEORIES OF VISCOSITY

Let us consider a fluid (either gas or liquid) sandwiched between two flat plates located at a distance, y, apart. If the bottom plate is fixed and a force,

F, is applied to the upper plate to set it in motion at a constant velocity, u, in the x-direction, the following sequence of events occurs:

1. At time t = 0, the fluid has a velocity profile that is flat and vertical (see Figure 9(A)).
2. At a small increment of time to t_1, the velocity of the fluid goes through a transient or unsteady flow where the momentum of the fluid builds up (refer to Figure 9(B)).
3. Finally, at some large time, t_2, steady flow is achieved and the velocity distribution appears as in Figure 9(C).

The force per unit area applied to the upper plate (assuming the flow is laminar) is proportional to the velocity increase in distance, y (taking y as positive in the upward direction). This is defined by the following relation, where the constant of proportionality, μ, is the viscosity of the fluid:

$$\frac{F}{A} = \mu \frac{du}{dy} \tag{38}$$

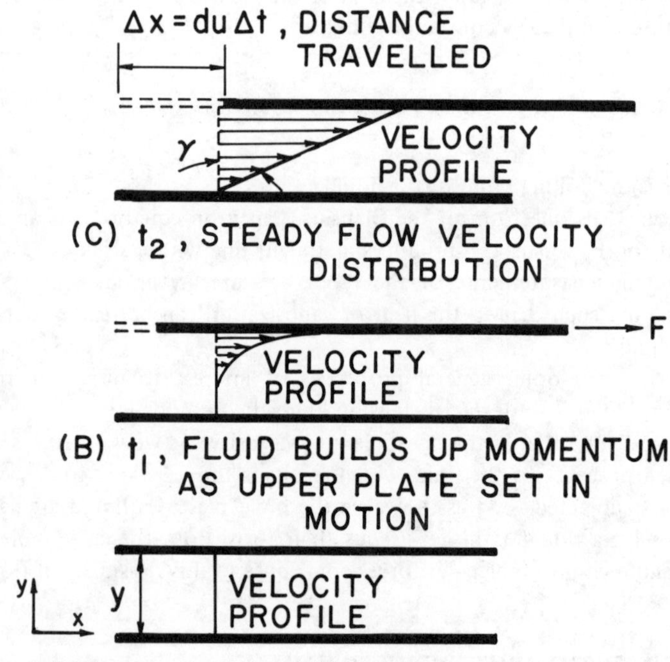

Figure 9. Illustrates the development of steady laminar flow.

The value of μ describes a physical property of the fluid that establishes the magnitude of resistance to flow. We shall first discuss the property of viscosity on a molecular scale and then relate this to macroscopic observations. Since properties of gases and liquids vary greatly, their molecular behavior is quite different and therefore should be discussed separately.

Gases

Bird et al. [14] describe a simple model for explaining the momentum transport mechanism of low-density gases based on the following assumptions:

1. Gas molecules are considered rigid, nonattracting spherical particles of the same size.
2. The number of molecules per unit volume is small such that the average distance between molecules is many times their diameter.

From the kinetic theory of gases it can be shown that at equilibrium the molecular velocities have an average fluid velocity as follows:

$$\bar{u} = \sqrt{\frac{8KT}{\pi m}} \tag{39}$$

and that the average distance traveled by a molecule between successive collisions is the mean free path:

$$\lambda = \frac{1}{\sqrt{2}\pi d_p^2 n} \tag{40}$$

where d_p is the molecular diameter and n is the number of molecules per unit volume. K is the Boltzmann constant.

From a momentum balance in the x-direction across any plane of constant, y, the following expression for the shear stress of the fluid is obtained:

$$\tau_{yx} = -\frac{1}{3} n m_p \bar{u} \lambda \frac{du_x}{dy}$$

$$= -\frac{1}{3} \rho \bar{u} \lambda \frac{du_x}{dy} \tag{41}$$

where m_p is the mass of a molecule ($\rho = nm_p$), and du_x/dy is the velocity gradient in the y-direction. Equation 41 is known as *Newton's law of viscosity*, which is more commonly expressed as

$$\tau_{yx} = -\mu \frac{du_x}{dy} \tag{42}$$

where in this model of low-density gases

$$\mu = \frac{1}{3} nm_p\bar{u}\lambda = \frac{1}{3} \rho\bar{u}\lambda \tag{43}$$

From Equation 39 then,

$$\mu = \frac{2}{3\pi^{3/2}} \frac{\sqrt{mKT}}{dp^2} \tag{44}$$

This last expression qualitatively describes the viscosity of a low-density gas composed of hard spheres. For practical calculations, Equation 44 is of limited value; however, the above derivation does provide insight into the momentum transfer of a low-density gas.

A more rigorous model based on the kinetic theory of monatomic gases at low density is based on the Chapman-Enskog theory [15]. This has also been extended by Curtiss et al. [16] to include multicomponent gas mixtures. Wilke [17] also recommends the following formula for multicomponent gas mixtures at low density:

$$\mu_{mix} = \sum_{i=1}^{n} \frac{x_i \mu_i}{\sum_{j=1}^{n} x_j \Phi_{ij}} \tag{45}$$

where n is the number of chemical species in the mixture; x_i, x_j are the mole fractions of species i and j; μ_i, μ_j are the viscosities of species i and j at the temperature and pressure of the system; and M_i, M_j are the corresponding molecular weights. Φ_{ij} is dimensionless and is defined as follows:

$$\Phi_{ij} = \frac{1}{\sqrt{8}} \left(1 + \frac{M_i}{M_j}\right)^{-1/2} \left[1 + \left(\frac{\mu_i}{\mu_j}\right)^{1/2} \left(\frac{M_j}{M_i}\right)^{1/4}\right]^2 \tag{46}$$

Liquids

Our understanding of the viscosity of liquids is based largely on empiricism, as the kinetic theory of liquids is not entirely developed. Detailed discussions are given by Eyring et al. [18] and Bird et al. [19]. Only the model of Eyring is described briefly here.

Because of the closeness of liquid molecules, their random motion is largely limited to vibration coupled to surrounding molecules. However, this vibration does produce random motion whereby molecules occasionally escape and enter interstices made available by other escaping molecules. If a fluid undergoes shearing (again as in Figure 9), the frequency of these molecular movements increases. The Eyring model explains this in terms of the potential energy barrier concept illustrated in Figure 10, whereby

$$-\Delta\widetilde{G} = -\Delta\widetilde{G}_0 \pm \left(\frac{\Delta x}{\delta}\right)\left(\frac{\tau_{yx}\widetilde{V}}{2}\right) \tag{47}$$

where $\Delta\widetilde{G}_0$ = molar free energy of activation in the stationary fluid
$\Delta\widetilde{G}$ = molar free energy of activation of the sheared fluid
$\widetilde{V}$ = molar volume of the liquid
Δx = width of the potential energy barrier
δ = distance between successive layers of molecules (it is assumed that molecules form a lattice structure of successive layers as

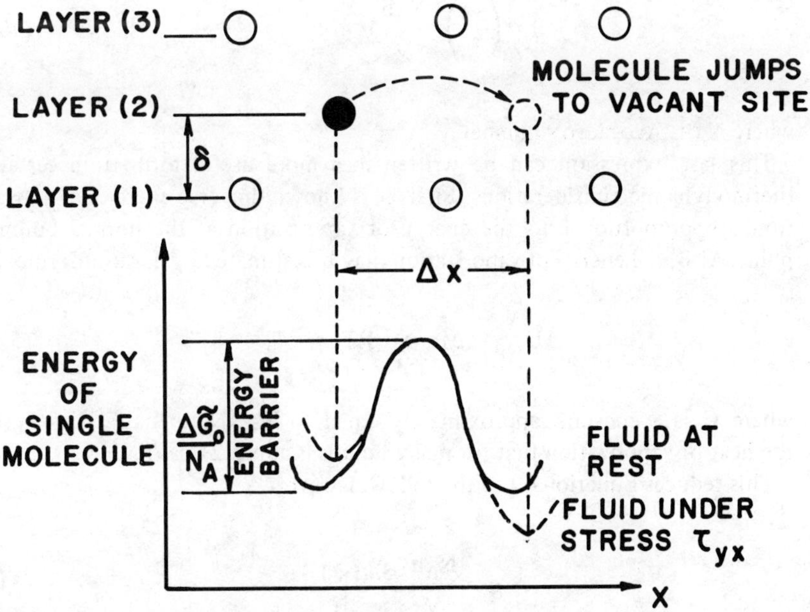

Figure 10. Illustrates the energy barrier concept.

shown in Figure 10, and presumably each molecule has its own energy barrier to overcome).

Note that $(\Delta x/\delta)(\tau_{yx}\widetilde{V}/2)$ represents the work done on the molecules as they move to the top of their energy barriers. The plus/minus sign indicates that the molecule is moving with the applied shear stress (plus sign) or against it (minus sign).

By considering the frequency by which molecules move or jump about, the following expression for the velocity profile is obtained:

$$-\frac{du_x}{dy} = \frac{\Delta x}{\delta}\left(\frac{KT}{h}e^{-\Delta\widetilde{G}_0/RT}\right)\left(e^{\Delta x\tau_{yx}\widetilde{V}/2\delta RT} - e^{-\Delta x\tau_{yx}\widetilde{V}/2\delta RT}\right)$$

$$= \frac{\Delta x}{\delta}\left(\frac{KT}{h}e^{-\Delta\widetilde{G}_0/RT}\right)\left(2\sinh\frac{\Delta x\tau_{yx}\widetilde{V}}{2\delta RT}\right) \quad (48)$$

where h is Planck's constant. (Note that some of the details to this derivation are given by Bird et al. [14]). Note that if the last group of terms in Equation 48 is small, the expression reduces to another form of Newton's law of viscosity:

$$\mu = \left(\frac{\Delta x}{a}\right)^2 \frac{N_A h}{\widetilde{V}} e^{\Delta\widetilde{G}_0/RT} \quad (49)$$

where N_A is Avogadro's number.

This last expression can be written in a more useful form from certain thermodynamic considerations. First, it is known that free energies of activation are proportional to the energy of vaporization at the normal boiling point. Also, the energy of vaporization may be estimated by Trouton's rule:

$$\Delta U_{vap} \simeq \Delta H_{vap} - RT_b = CT_b \quad (50)$$

where C is a constant approximately equal to 9.4. Note that $\Delta H_{vap}/T_b$ is the heat of vaporization that for many liquids is about 21 cal/mol °K.

This reduces Equation 49 to the following form:

$$\mu \simeq \frac{N_A h}{\widetilde{V}} e^{3.8T_b/T} \quad (51)$$

Equation 51 can serve as a rough guide for estimating viscosities from basic thermodynamic properties of a liquid and only for Newtonian liquids. However, there are more reliable empirical relations that are available for specific liquids.

PROPERTIES OF NEWTONIAN FLUIDS

Equation 38 is our governing expression for describing the laminar flow situation shown in Figure 9 and, from a macroscopic viewpoint, is Newton's law of viscosity, where $\tau_{yx} = F/A$.

$$\tau_{yx} = F/A = \mu \frac{du}{dy} \tag{52a}$$

or, if English units are used,

$$\tau_{yx} = \frac{\mu}{g_c} \frac{du}{dy} \tag{52b}$$

The conversion factor, g_c [value of 32.174 (lb_m/lb_f) $(ft\ s^{-2})$] is introduced because of the dual system of units we shall adhere to in our computations. Careful checks for dimensional consistency are needed in all practical calculations when dealing in English units. If SI units are employed the value of g_c is unity.

In writing Equation 52 we have assumed the ideal case of zero slip condition at the plate walls. The x-distance traveled by a layer of fluid immediately beneath the top plate during time Δt will be $du\Delta t$. Note the angle, γ, shown in Figure 9(C) provides an indication of the deformation of the fluid caused by the imposed stress, τ_{yx}. For small values of γ,

$$\gamma = \frac{du\Delta t}{dy} \tag{53}$$

The treatment of a fluid undergoing deformation is analogous to the deformations in elastic solids, where γ is referred to as the strain. In our case, the rate of deformation is the same as the velocity gradient:

$$\frac{du}{dy} = \frac{d\gamma}{dt} \tag{54}$$

The more exact name for du/dy used by rheologists is the *shear rate* of the fluid.

Many textbooks give an alternate interpretation of the system shown in Figure 9, where y is taken to be positive in the downward direction. Then Equations 52 and 54 become

$$\tau_{yx} = -\frac{\mu}{g_c}\frac{du}{dy} \tag{55}$$

and

$$\frac{d\gamma}{dt} = -\frac{du}{dy} \tag{56}$$

For an incompressible fluid, Equation 55 is more properly stated as

$$\tau_{yx} = -\frac{\mu}{\rho g_c}\frac{d(u\rho)}{dy} \tag{57}$$

The quantity of $u\rho$ is the *momentum concentration* (has units of momentum per unit volume). Hence, $d(u\rho)/dy$ is the momentum concentration gradient in the y-direction and is negative according to the sign convention for the y-direction. The quantity $\tau_{yx} g_c$ may be considered to be the flux of the x-directed momentum in the y-direction.

A Newtonian material is simply any substance whose shear rate is directly proportional to the stress or shearing force applied to the fluid. Equation 57 is a straight line, with the intercept at the origin and whose slope is μ. (A plot of τ_{yx} versus du/dy is referred to as a flow curve.) Hence, viscosity is a constant, independent of the shear rate, and only depends on the process temperature and pressure of the particular fluid. The energy dissipated by the deformation of these type fluids results from the collisions of relatively small molecules. Hence, virtually all gases, as well as liquids of low-molecular-weight, display Newtonian behavior.

PROPERTIES OF NON-NEWTONIAN FLUIDS

Non-Newtonian fluids are those materials showing nonlinearity in their shear rate-shear stress relationship. There are three general categories of non-Newtonians, namely, time-independent fluids, time-dependent fluids, and viscoelastic fluids.

Time-Independent Fluids

Time-independent fluids display the characteristic that the rate of shear at a given point is entirely dependent on the instantaneous shear stress at that point. These fluids are sometimes referred to as being "purely viscous."

There are two subgroups in this non-Newtonian category: "plastic flow" and "non-plastic" materials. The term "plastic flow" refers to fluids having a yield stress.

Yield stress refers to a minimum shear stress or force that must be applied to the fluid to initiate flow deformation. Typical flow curves for these type fluids are shown in Figure 11(A). Examples of materials having a yield stress are ketchup, mustard, toothpaste, mayonnaise, black liquors and various polymer solutions.

There are no rigorous models that describe fluids with yield stresses; however, there are several semiempirical relationships that appear adequate. The most commonly used and simplest is the Bingham plastic model, which attempts to approximate the nonlinear flow relationship by a straight line. The model is based on a yield stress, τ_y, and a parameter called the "plastic viscosity," or simply "plasticity," which is the slope of the equation:

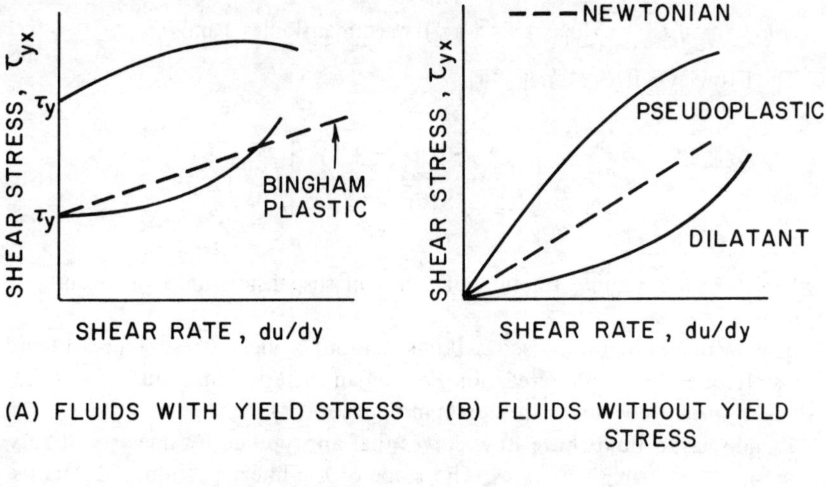

Figure 11. Flow curves for time-independent fluids.

$$\tau_{yx} = \frac{\eta}{g_c}\left(\frac{du}{dy}\right) + \tau_y \tag{58}$$

The Bingham plastic equation is also shown plotted in Figure 11(A). Although a Bingham plastic fluid is an idealized material, often it serves as a good approximation over a range of shear rate conditions.

Following the terminology used by Skelland [20], an "apparent viscosity" is defined analogous to Newton's law (i.e., $\mu = \tau_{yx} g_c/(du/dy)$); hence, for a Bingham plastic,

$$\mu_a = \eta + g_c \tau_y \Big/ \frac{du}{dy} \tag{59}$$

Equation 59 states that the apparent viscosity decreases with increasing shear rate (referred to as a "shear thinning" effect). It is therefore meaningless to report a value for viscosity without its corresponding shear rate for this type of material.

Other models that provide a more realistic approximation of the true flow curve include the following:

The Herschel-Bulkley Model [21-23]:

$$\tau_{yx} - \tau_y = \left[\frac{\eta'}{g_c}\left(\frac{du}{dy}\right)\right]^{1/m} \tag{60}$$

τ_y (lb_f/ft^2), η' (lb_f^{m-1}-lb_m-ft^{1-2m}-s^{-1}); m (dimensionless) and

The Crowley-Kitzes Model [24]:

$$\tau_{yx} = \frac{\mu_L}{g_c}\left[\frac{1.2 + S(C\tau_{yx}^{-0.2} + 1)^3}{1.2 - 2S(C\tau_{yx}^{-0.2} + 1)^3}\right]\left(\frac{du}{dy}\right) \tag{61}$$

where S is the volume fraction of solids in suspension and C is in units of $(lb_f/ft^2)^{0.2}$.

The second subgroup, viscous fluids without a yield stress, represent the most frequently encountered non-Newtonian materials in industrial applications. Typical flow curves are shown in Figure 11(B).

Pseudoplastic fluids have flow curves that are typified by linearity at very low and very high shear rates. The slope of the linear portion of the curve at high shear rates is called the "apparent viscosity at infinite shear" (μ_∞).

The slope of the linear portion at low shear is referred to as the "apparent viscosity at zero shear" (μ_0) (see Figure 12(A)). If the pseudoplastic flow curve is replotted on logarithmic coordinates, τ_{yx} values are often found to be linear over a wide range of shear rates. This convenient form of the plot enables many materials to be characterized by a simple power law expression:

$$\tau_{yx} = \frac{K}{g_c}\left|\frac{du}{dy}\right|^{n-1}\frac{du}{dy} \tag{62}$$

or

$$\mu_a = K\left|\frac{du}{dy}\right|^{n-1} \tag{63}$$

where parameter K is called the *consistency index* and is determined from the intercept on the τ_{yx} axis at shear rate of unity. n is called the *flow behavior index* and is the slope of the logarithmic plot (refer to Figure 12(B)). Values of n vary between 0 and 1, with increasing pseudoplasticity occurring as n approaches zero. Both parameters are temperature dependent, with K being the more sensitive to temperature.

As with the Bingham plastic model, a power law relation should always be tried first when attempting to correlate viscosity data for fluids without a yield stress. If that fails, some of the other models summarized in Table 9 or described by Skelland [20] should be attempted.

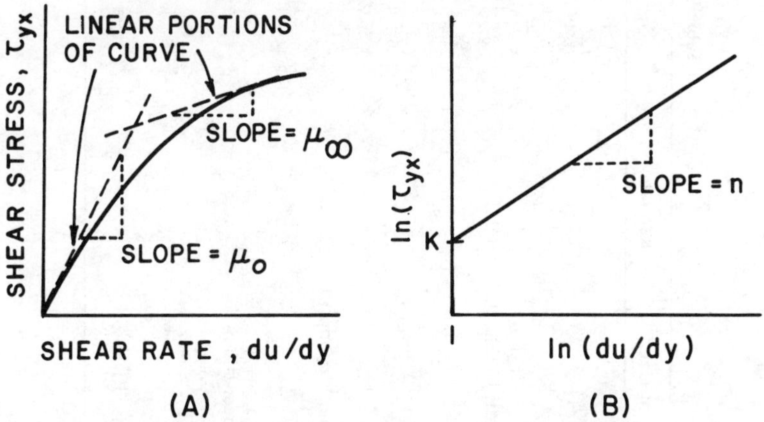

Figure 12. Details of flow curve for pseudoplastic fluids.

Table 9. Various Models for Fluids Without Yield Stresses

Model	Reference	Equation	Equation No.	Nomenclature
Ellis	25	$\tau_{yx} = \dfrac{1}{A + B\tau_{yx}^{\alpha-1}}\left(\dfrac{du}{dy}\right)$	64	A $ft^2\ s^{-1}\ lb_f$; B $ft^{2\alpha}\ s^{-1}\ lb_f^{-\alpha}$; α dimensionless
Prandtl-Eyring	26,27	$\tau_{yx} = A \sinh^{-1}\left[\dfrac{1}{B}\left(\dfrac{du}{dy}\right)\right]$	65	A lb_f/ft^2; B s^{-1}
Powell-Eyring	28	$\tau_{yx} = C\left(\dfrac{du}{dy}\right) + \dfrac{1}{B}\sinh^{-1}\left[\dfrac{1}{A}\left(\dfrac{du}{dy}\right)\right]$	66	A s^{-1}; B ft^2/lb_f; C $lb_f\text{-}s/ft^2$
Reiner-Philippoff	29	$\tau_{yx} = \dfrac{1}{g_c}\left[\mu_\infty + \dfrac{\mu_0 - \mu_\infty}{1 + (\tau_{yx}/A)^2}\right]\left(\dfrac{du}{dy}\right)$	67	μ_0, μ_∞ $lb_m/ft\text{-}s$; A lb_f/ft^2
Spencer-Dillon	30,31	$\tau_w = \dfrac{8u/D}{\dfrac{g_c}{\mu_{e0}}\left[1 + k\tau_w + \dfrac{(k\tau_w)^2}{2!} + \dfrac{(k\tau_w)^3}{3!} + \ldots\right]}$	68	τ_w = wall shear stress, lb_f/ft^2; μ_{e0} $lb_m/ft\text{-}s$; k ft^2/lb_f
Reiner-Rivlin	32	$\tau_w = \dfrac{1}{g_c}\left[(\mu_0 - \mu_\infty)^{-\tau_w F/4\mu_0} + \mu_\infty\right]\left(\dfrac{8u}{D}\right)$	69	F sec g_c; μ_0, μ_∞ $lb_m/ft\text{-}s$

Dilatancy is another characteristic displayed by some fluids without a yield stress (Figure 11(B)). Dilatant fluids are generally less common than pseudoplastics. There are two types of dilatancy: volumetric dilatancy and rheological dilatancy. Volumetric dilantancy refers to a "swelling" of the fluid; that is, an increase in the total volume of the fluid will occur when subjected to shearing. Rheological dilatancy refers to an increase in the apparent viscosity with increasing shear rate. The power law can be used to describe dilatant fluids, but with values of n greater than 1. Also, any of the models given in Table 9 are applicable.

Time-Dependent Fluids

With time-dependent fluids, shear rate is a function of both the magnitude and duration of deformation and the time lapse between consecutive applications of shearing. There are two subgroups in this category, thixotropic fluids and rheopectic fluids. The characteristic flow curves for both of these are shown in Figure 13.

Thixotropic fluids show a reversible decrease in shear stress with time at constant shear and temperature. The characteristic curve shown in Figure 13(A) is obtained by first applying an increasing shearing force to some maximum value and then immediately decreasing the force at a steady

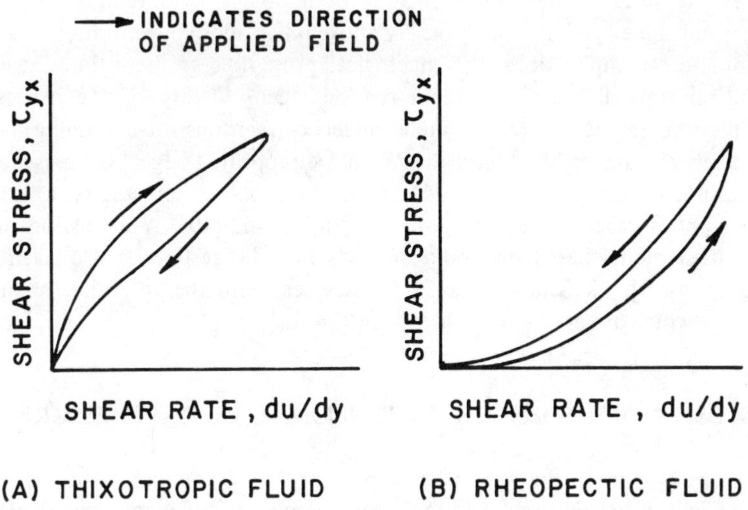

Figure 13. Typical flow curves for thixotropic and rheopectic fluids. The arrows indicate the chronological progress of applied shear in obtaining such curves.

uniform rate. The resultant curve is an envelope referred to as a hysteresis loop. The location of the loop and often the area bound within the curve are different for a different time history. Even varying the rate at which the shear rate is applied can significantly modify the hysteresis loop.

Rheopectic fluids are generally very rare. The concavity of the hysteresis loop shown in Figure 13(B) distinguishes these fluids from thixotropic materials. Rheopectics are characterized by a reversible increase in shear stress with time under a constant rate of shear and isothermal conditions. As with thixotropic fluids, the location of the hysteresis loop depends on the time history of the fluid and the rate at which du/dy is increased and decreased.

Although attempts have been made to relate the area bound by the hysteresis loop to an index of thixotropy or rheopectic index, no design procedures have been recommended. For isothermal, steady-flow situations, the calculation procedures applied to time-dependent fluids and outlined in Chapter 2.10 are applicable to thixotropic and rheopectic fluids.

Viscoelastic Fluids

Viscoelastic fluids display properties of both viscous and elastic materials. These fluids will flow when subjected to stress; however, a portion of their deformation gradually recovers on removal of the stress. In simpler terms, these are fluids typically described as having memories, in that once the shear is removed, the viscous properties return partially to their original state. Examples of these fluids include flour dough, nylon and other polymers, and some black liquors typically above 68% solids concentration.

For more complications, the rheological properties of these fluids at any point in time also are dependent on the recent history of the material. Needless to say there are no simple models or relations for describing these materials and many of us wish they would disappear. Lodge [33] presents a discussion of various theories and modeling approaches to these type fluids, and a more recent review is given by Bird et al. [19]. A discussion of a simplified model based on Hookean or elastic solid systems is also given by Skelland [20]. Skelland also briefly discusses a number of other models, most of which are noted in the literature [34-42].

VISCOSITY DEPENDENCIES ON TEMPERATURE AND PRESSURE

Extensive data on the viscosities of pure gases and common liquids can be found in the literature [4,43]. As a rule, common liquids will experience a decrease in viscosity with increasing temperature at constant pressure. With gases, viscosity increases with increasing temperature at constant pressure.

This follows from the kinetic theory of gases, which shows viscosity to be proportional to density, the molecules' average velocity and the mean free path of the molecules.

Newtonian viscosity usually is related to temperature through the Arrhenius relationship:

$$\mu = a' e^{E/RT} \tag{70}$$

where E = activation energy of flow per mole
R = gas constant per mole
T = absolute temperature
a′ = constant characteristic of the fluid

Through Newton's law, the rheological counterpart to this expression is

$$\tau_{yx} = \frac{a'}{g_c}\left(-\frac{du}{dy}\right)\exp\left(\frac{E}{RT}\right) \tag{71}$$

An analogous expression for power law fluids also can be developed (see Problem E.19 in Chapter 2.12, Section 2).

For many liquids, viscosity increases with increasing pressure at constant temperature. For gases, viscosity also increases with increasing pressure. Note that this contradicts the kinetic theory of gases, which predicts viscosity to be independent of pressure. For real gases at high reduced temperatures and low reduced pressures, however, the theory does apply to an extent.

Caution must be exercised in extrapolating viscosity data to other temperature and pressure conditions, especially with non-Newtonian fluids. With non-Newtonians, the best approach is to generate flow curve data with a viscometer close to the process temperature.

For common Newtonian fluids (gases and pure, low-density liquids), empirical methods are available for estimating viscosity at process conditions, but are recommended for use only when time does not permit experimental measurements to be made. One widely accepted method based on estimating viscosity from critical properties data is given in Figure 14.

Figure 14 is the empirical method of Uyehara and Watson [44], and consists of a plot of reduced viscosity versus reduced temperature. Reduced viscosity is defined as the ratio of the viscosity at a given temperature and pressure to the viscosity at the critical point. Similarly, the reduced temperature is defined as T/T_c. The parameter P_r is the reduced pressure (p/p_c). Figure 14 shows that the viscosity of a gas approaches a limit of low-density

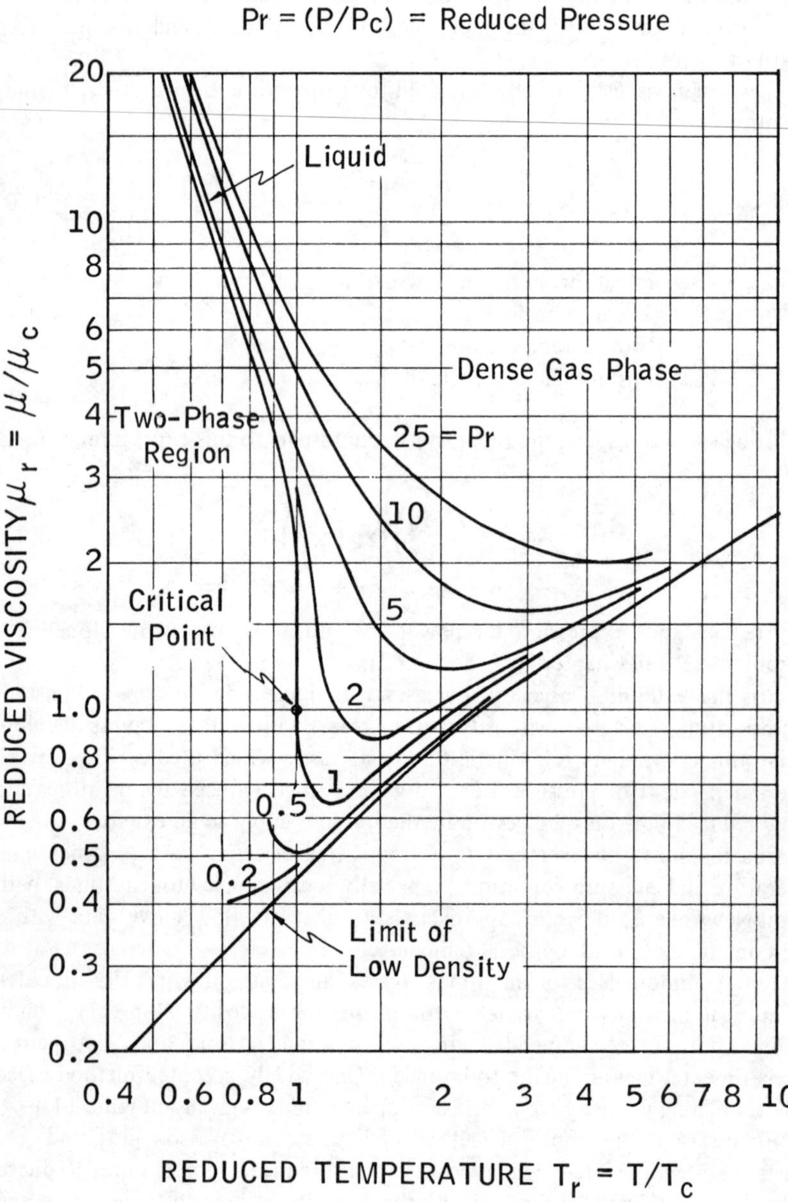

Figure 14. Uyehara and Watson's [44] generalized plot for estimating viscosity.

fluids (the limit is 1 atm pressure for most gases). As noted earlier, at low density the viscosity of a gas increases with increasing temperature, whereas the opposite is true for liquids. The critical viscosity can be estimated from critical properties data using the following formula:

$$\mu_c = \frac{61.6(MT_c)^{1/2}}{(\widetilde{V}_c)^{-2/3}} \tag{72}$$

where μ_c is in micropoise, M is the molecular weight of the fluid, T_c is in °K, and $\widetilde{V}_c$ is the critical molar volume in cc per gram mole.

To estimate the viscosity of an n-component gas mixture, so-called pseudo-critical properties are used in place of p_c, T_c and μ_c in Figure 14.

These are defined by the following formulas:

$$P_c' = \sum_{i=1}^{n} x_i P_{ci} \tag{73a}$$

$$T_c' = \sum_{i=1}^{n} x_i T_{ci} \tag{73b}$$

$$\mu_c' = \sum_{i=1}^{n} x_i \mu_{ci} \tag{73c}$$

where x_i is the mole fraction of the species.

The method is outlined by Hougen and Watson [45], and Hougen et al. [46] give a more detailed version of Figure 14.

A similar method to the one given by Uyehara and Watson is that of Carr et al. [47]. This method is based on a plot of reduced viscosity versus reduced pressure, with reduced temperature as the parameter. Both methods are good for rough approximations of the viscosities of pure gases and gas mixtures. More accurate methods are given by Comings [48].

CHAPTER 1.4

EQUATIONS OF MOTION AND ENERGY

This chapter reviews the general differential equations of conservation of mass and momentum. These two basic equations form the starting basis for describing all types of problems in viscous flow of purely isothermal fluids. For many flow systems a third equation is needed to define all the dependent variables of a flow system, that is, an equation based on the law of conservation of energy. This last differential equation, called the mechanical energy equation, provides a description of the energy interconversions that take place in the flowing fluid.

We shall start our discussion with some basic definitions and examples to illustrate the concept of momentum balances. This will acquaint students with the application of the principle of conservation of momentum to viscous flow problems.

The chapter is presented on an elementary level for the first-time students and as a review for practicing engineers who use these principles but may have forgotten their origin. Within this context only some short notes at the end of this chapter are provided on basic terminology of vector and tensor notation. We will not, however, use this form of mathematics in addressing flow problems. Vector and tensor mathematics are extremely useful when tackling complex flow configurations. The more advanced student may wish to refer to the literature [49, 50] for some of the details of vector and tensor manipulations in tackling flow problems.

MOMENTUM BALANCES

A momentum balance allows us to tabulate all the forces acting on a fluid element in the form of an equation or equations. By application of appropriate boundary conditions, these equations may be solved for the parameters of interest. The term "fluid element" refers to an imaginary finite volume or shell of fluid, which is conveniently chosen to evaluate the forces at play.

"Boundary conditions," in the text of fluid flow, refers to an approximation of the true state and/or physics of the flowing fluid at specified values of the system's independent variables. As an example, if a fluid is flowing over a solid boundary, the fluid at the solid-fluid interface is said to be stationary also. (Similarly, if the solid surface is moving, the velocity of the fluid at the solid-fluid interface is the same as the velocity with which the surface itself is moving.) This is referred to as the "no slip" condition and is one of the most often applied boundary conditions. Two other very common boundary conditions widely used for flowing systems are as follows:

1. At the gas-liquid interface the momentum flux in the liquid is assumed to be zero (i.e., $du/dy = 0$).
2. At liquid-liquid interfaces both the momentum flux and velocity gradient are continuous across the interface (i.e., they are the same on both sides of the liquid-liquid interface).

Momentum balances are an effective but limited tool that may be applied only to simple flow situations, namely laminar, rectilinear flows. By rectilinear, we mean that all fluid streamlines are straight and parallel to the solid boundaries of the flow system. Any deviation from this, such as nonlinear streamlines caused by entrance or exit flow conditions or obstructions in the flow stream causing intermixing of fluid layers, cannot be effectively described by a simple momentum balance. Therefore, we can readily use this tool to describe a very steady, uniform flow situation. The general momentum balance for a uniformly flowing fluid is

$$\left\{\begin{matrix}\text{Rate of Momentum}\\ \text{into a Fluid}\\ \text{Element}\end{matrix}\right\} - \left\{\begin{matrix}\text{Rate of Momentum}\\ \text{out of a Fluid}\\ \text{Element}\end{matrix}\right\} + \left\{\begin{matrix}\text{Pressure Forces and}\\ \text{Gravity Forces}\end{matrix}\right\}$$

$$= \left\{\begin{matrix}\text{Rate of Momentum}\\ \text{Accumulation}\end{matrix}\right\} = 0 \text{ at Steady State} \tag{74}$$

There are two approaches or interpretations applied when using Equation 74. One is a very fundamental approach in which we imagine a portion of the flowing fluid to be a shell of finite thickness; by allowing this thickness to approach zero and making use of the mathematical definition of the first derivative, a corresponding differential equation describing the momentum flux distribution is obtained. The other approach is to consider a finite volume of fluid and write a straight force balance. Both approaches enable the derivation of the necessary flow parameters of interest and are illustrated below.

Momentum Shell Balances

Let's consider an incompressible fluid flowing in the downward direction in a vertical tube. The system is illustrated in Figure 15 in which we define a cylindrical shell of fluid about the tube axis, having thickness Δr and length L.

$$\underbrace{\left\{2\pi r L\tau_{rx}|_r - 2\pi r L\tau_{rx}|_{r+\Delta r}\right\}}_{\substack{\text{Rate of Momentum Across}\\ \text{Cylindrical Surface of}\\ \text{Fluid}}} + \underbrace{\left\{2\pi r\Delta r\rho u_x^2|_{x=0} - 2\pi r\Delta r\rho u_x^2|_{x=L}\right\}}_{\substack{\text{Rate of Momentum in Across}\\ \text{Annular Surface}}}$$

$$\underbrace{2\pi r\Delta r L\rho g}_{\substack{\text{Gravity Force Acting}\\ \text{on Cylindrical Shell}}} + \underbrace{2\pi r\Delta r\,(P_1 - P_2)}_{\substack{\text{Rate of Momentum}\\ \text{in Across Annular}\\ \text{Surface}}} = 0 \tag{75}$$

Because the fluid is incompressible, and assuming ρ to be constant, the second term vanishes and Equation 75 is rearranged to the following:

$$\frac{r\tau_{rx}|_{r+\Delta r} - r\tau_{rx}|_r}{\Delta r} = \left(\frac{P_1 - P_2}{L} + \rho g\right)r \tag{76}$$

If we take the limit of this equation as Δr goes to zero, we recognize that the right-hand side (RHS) of this equation is the derivative $(d/dr)(r\tau_{rx})$, or

$$\frac{d}{dr}(r\tau_{rx}) = \left(\frac{P_1 - P_2 + L\rho g}{L}\right)r \tag{77}$$

and, on integration,

$$\tau_{rx} = \left(\frac{P_1 - P_2 + L\rho g}{2L}\right)r + \frac{C_1}{r} \tag{78}$$

The constant of integration C_1 must be zero or else the momentum will be infinite at $r = 0$ (this is a boundary condition).

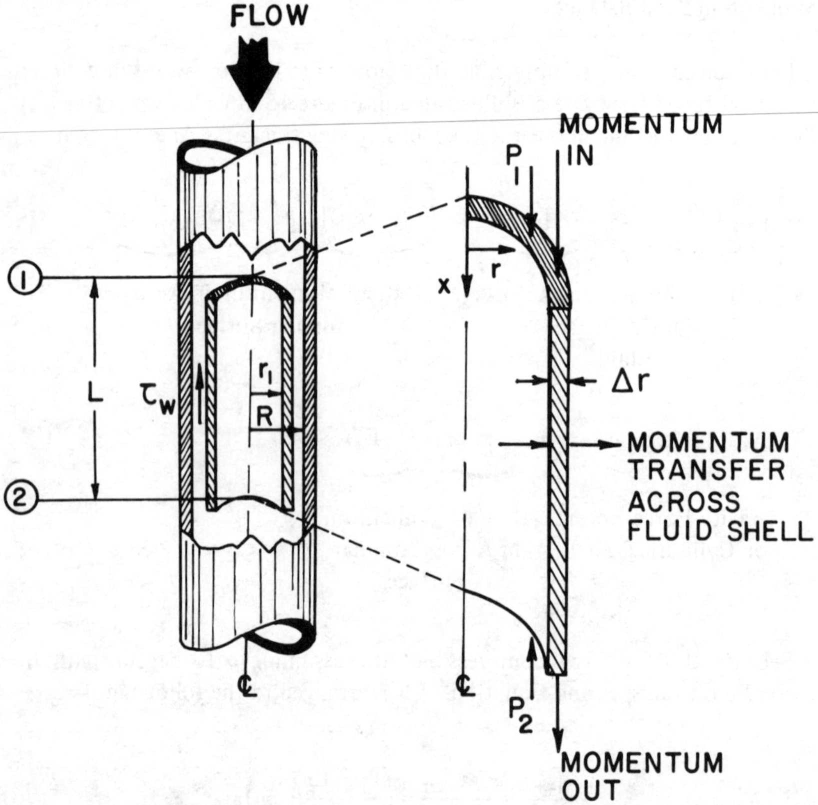

Figure 15. Incompressible fluid flowing downward in a vertical tube. A cylindrical shell of fluid is imagined so that a momentum balance may be developed.

Hence, the momentum flux is

$$\tau_{rx} = \left(\frac{P_1 - P_2 + L\rho g}{2L}\right) r \qquad (79a)$$

Bird et al. [14] write Equation 79a in a more convenient form, where the pressure term includes both the effect of static pressure and head, such that $P' = P + \rho g \Delta x$. Hence,

$$\tau_{rx} = \left(\frac{P_1' - P_2'}{2L}\right) r \qquad (79b)$$

Depending on the rheological property of the fluid, an appropriate expression for τ_{rx} can be introduced into Equation 79b to obtain the velocity profile across the tube. For a Newtonian fluid, as noted in the last chapter, $\tau_{rx} = -\mu du_x/dr$. Substituting this relation in Equation 79b produces an ordinary differential equation, which can be integrated to the following:

$$u_x = -\left(\frac{P_1' - P_2'}{4\mu L}\right) r^2 + C_2 \tag{80}$$

We can now apply one of our common boundary conditions, namely, $u_x = 0$ at $r = R$ (i.e., "no slip" at the wall) to solve for the constant of integration C_2. On doing this, $C_2 = (P_1' - P_2')R^2/4\mu L$ and the expression for the velocity distribution becomes

$$u_x = -\left(\frac{P_1' - P_2'}{4\mu L}\right) r^2 + \left(\frac{P_1' - P_2'}{4\mu L}\right) R^2$$

or

$$u_x = \frac{(P_1' - P_2')R^2}{4\mu L}\left[1 - \left(\frac{r}{R}\right)^2\right] \tag{81}$$

Simple Force Balances

By summing all the forces acting on the volume of fluid between points (1) and (2) in Figure 15, the following balance is obtained:

$$\pi R^2 P_1 - \pi R^2 P_2 + \pi R^2 L\rho g - 2\pi R L \tau_w = 0 \tag{82}$$

And by collecting terms,

$$\frac{P_1 - P_2}{\rho} + Lg = \frac{2L\tau_w}{\rho R} \tag{83}$$

If we define the sum of all the frictional losses to be $\Sigma F = \Delta P/\rho$, then $\Sigma F = \Delta P/\rho = (P_1 - P_2)/\rho - Lg$ where $L = x_2 - x_1$, and Equation 83 becomes

$$\tau_w = \frac{\Delta P}{\rho}\left(\frac{\rho R}{2L}\right) = \frac{R\Delta P}{2L} \tag{84}$$

Note that τ_w is really τ_{rx} at r = R. Hence, for any r (where $r \leqslant R$), the fluid shear stress is

$$\tau_{rx} = \frac{r\Delta P}{2L} \tag{85}$$

which is the same as Equation 79a. (In fact, the reader may wish to argue that the shell balance approach is really the same derivation, since the general momentum flux expression, Equation 79a, is developed by considering many hollow fluid shells from the tube axis to the pipe wall. All we have really done with the force balance is to consider the fluid as a cylindrical element composed of many shells). At this point we could substitute in Newton's law of viscosity and arrive at our parabolic velocity profile, Equation 81. Instead, we will carry the analysis a step farther and develop a relation between the volumetric flow rate and fractional losses for the system.

Combining Equations 84 and 85 through ΔP, an alternate expression for the shear stress at any point in the fluid is obtained:

$$\tau_{rx} = \tau_w\left(\frac{r}{R}\right) \tag{86}$$

Equation 86 states that the fluid shear stress distribution across the tube is linear and varies from zero at the tube centerline to τ_w at the tube wall.

If we go back to our shell concept, the volumetric flowrate through the differential annulus between r and r + Δr is $\Delta Q = u2\pi r\Delta r$, or, in differential form,

$$Q = \int_0^Q dQ = \pi \int_0^R u_x 2r dr = \pi \int_0^{R^2} u_x d(r^2) \tag{87}$$

The integral given on the RHS may be broken into two parts:

$$\pi \int_0^R u_x d(r^2) = \pi \left\{ u_x r^2 \Big|_0^{R^2} - \int_0^{R^2} r^2 du_x \right\}$$

and from the no-slip condition the term $u_x r^2 |_0^{R^2}$ vanishes.

Hence,

$$Q = -\pi \int_0^{R^2} r^2 du_x \tag{88}$$

Now substituting in for Newton's law; du = $(-\tau_{rx}/\mu)$ dr; and noting from Equation 86 that

$$r^2 = \frac{\tau_{rx}^2 R^2}{\tau_w^2}$$

and

$$dr = \frac{R}{\tau_w} d\tau_{rx}$$

Equation 88 becomes, for a Newtonian fluid,

$$Q = \frac{\pi R^3}{\mu \tau_w^3} \int_0^{\tau_w} \tau_{rx}^3 d\tau_{rx} \tag{89}$$

This equation can be integrated to give the following:

$$Q = \frac{\pi R^3 \tau_w}{4\mu} \tag{90}$$

where $Q = \pi R^2 U$, with U being the average linear velocity in the x-direction.

Substituting Equation 84 into the above expression for τ_w we obtain the famous Hagen-Poiseuille equation [51,52]:

$$Q = \frac{\pi R^4 \Delta P}{8\mu L} \tag{91}$$

Note that if we are dealing in English units, Equation 91 must be corrected by multiplying the RHS by g_c.

THE CONTINUITY EQUATION

The *continuity equation* is a mathematical statement of the law of conservation of mass. Consider a cubical element of fluid flowing in space with respect to the three coordinates shown in Figure 16. From a material balance,

$$\left\{\begin{matrix}\text{Mass}\\\text{Accumulation}\end{matrix}\right\} = \left\{\begin{matrix}\text{Mass into Cubical}\\\text{Element}\end{matrix}\right\} - \left\{\begin{matrix}\text{Mass out of Cubical}\\\text{Element}\end{matrix}\right\} \tag{92}$$

With this format, the mass input into the x face of the element and the mass out at the x + Δx face can be obtained over some differential time element, dt (from this point on we shall write small increments as derivatives).

Hence, following Equation 92 we obtain the following:

$$\frac{\partial \rho}{\partial t}\,dxdydzdt = \rho udydzdt - \left(\rho u + u\frac{\partial \rho}{\partial t}\,dx + \rho\frac{\partial u}{\partial x}\,dx\right) dydzdt \tag{93}$$

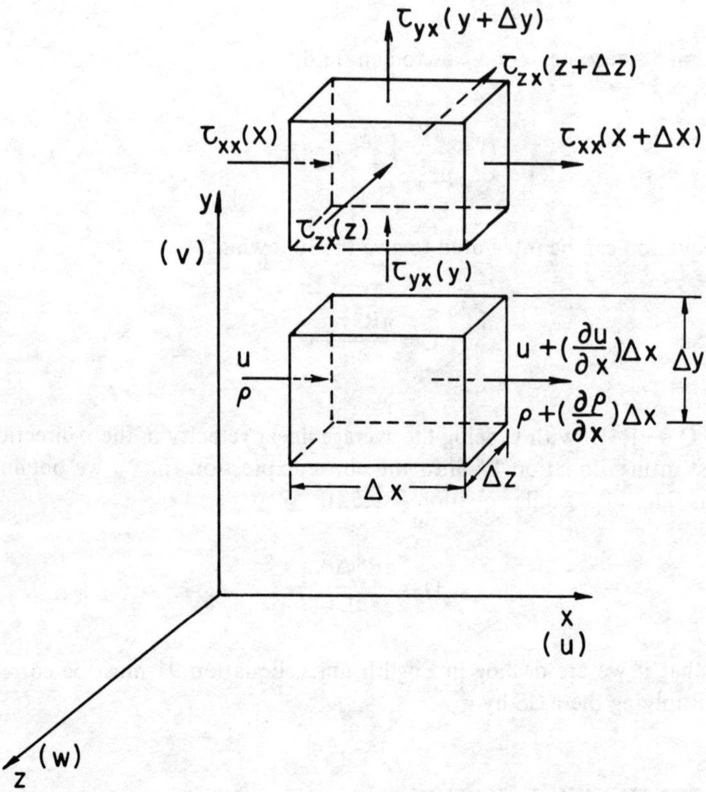

Figure 16. Illustrates one-dimensional flow of a cubical element of fluid. The top element shows the directions in which the x-component of momentum is transported through the surfaces.

which is rearranged to give

$$-u\frac{\partial \rho}{\partial x} - \rho\frac{\partial u}{\partial x} = \frac{\partial \rho}{\partial t} \tag{94a}$$

or

$$-\frac{\partial(\rho u)}{\partial x} = \frac{\partial \rho}{\partial t} \tag{94b}$$

Note that u refers to the fluid velocity in the x-direction.

This last differential equation is the law of conservation of mass, where both u and ρ are functions of x and t. Equation 94b only describes flow in the x-direction. The more general statement is

$$\frac{\partial \rho}{\partial t} = -\left(\frac{\partial}{\partial x}\rho u + \frac{\partial}{\partial y}\rho v + \frac{\partial}{\partial z}\rho w\right) \tag{94c}$$

Other forms of the continuity equation for two- and three-dimensional flows are given in Section 2. For the derivation details of the alternate forms see Knudsen and Katz [53].

EQUATIONS OF MOTION

The top fluid element in Figure 16 has arrows indicating the direction in which the x-component of momentum is transported through the fluid volume surfaces. Momentum transport into and out of the fluid element occurs because of (a) the bulk fluid motion, and (b) the velocity gradients that exist. The first mechanism is strictly convection, whereas the latter is molecular transfer.

Following the general momentum balance statement given by Equation 74, an expression describing the unsteady-state momentum transport for the x-component can be developed:

$$\underbrace{\frac{\partial}{\partial t}\rho u_x}_{①} = \underbrace{-\left(\frac{\partial}{\partial x}\rho uu + \frac{\partial}{\partial z}\rho wu + \frac{\partial}{\partial y}\rho vu\right)}_{②}$$

$$\underbrace{-\left(\frac{\partial}{\partial x}\tau_{xx} + \frac{\partial}{\partial z}\tau_{zx} + \frac{\partial}{\partial y}\tau_{yx}\right)}_{③} \underbrace{-\frac{\partial P}{\partial x} + \rho g_x}_{④} \tag{95}$$

where term (1) = rate of momentum accumulation
term (2) = rate of momentum in
term (3) = rate of momentum out
term (4) = sum of forces acting on fluid element

Equation 95 is the equation of motion for the x-component of momentum. For the y- and z-components, the equations of motion are

$$\frac{\partial}{\partial t}\rho v = -\left(\frac{\partial}{\partial x}\rho uv + \frac{\partial}{\partial y}\rho vv + \frac{\partial}{\partial z}\rho wv\right) - \left(\frac{\partial}{\partial x}\tau_{xy} + \frac{\partial}{\partial y}\tau_{yy} + \frac{\partial}{\partial z}\tau_{zy}\right) - \frac{\partial P}{\partial y} + \rho g_y \tag{96}$$

$$\frac{\partial}{\partial t}\rho w = -\left(\frac{\partial}{\partial x}\rho uw + \frac{\partial}{\partial y}\rho vw + \frac{\partial}{\partial z}\rho ww\right) - \left(\frac{\partial}{\partial x}\tau_{xz} + \frac{\partial}{\partial y}\tau_{yz} + \frac{\partial}{\partial z}\tau_{zz}\right) - \frac{\partial P}{\partial z} + \rho g_z \tag{97}$$

It is convenient to express the equation of motion in different coordinate systems. Appendix C also provides the equation of motion in cylindrical and spherical coordinates.

For an incompressible, Newtonian fluid, Equations 95 through 97 become the following:

x-component:

$$\rho\left(\frac{\partial u}{\partial t} + u\frac{\partial u}{\partial x} + v\frac{\partial u}{\partial y} + w\frac{\partial u}{\partial z}\right) = -\frac{\partial P}{\partial x} + \mu\left(\frac{\partial^2 u}{\partial x^2} + \frac{\partial^2 u}{\partial y^2} + \frac{\partial^2 u}{\partial z^2}\right) + \rho g_x \tag{98}$$

y-component:

$$\rho\left(\frac{\partial v}{\partial t} + u\frac{\partial v}{\partial x} + v\frac{\partial v}{\partial y} + w\frac{\partial v}{\partial z}\right) = -\frac{\partial P}{\partial x} + \mu\left(\frac{\partial^2 v}{\partial x^2} + \frac{\partial^2 v}{\partial y^2} + \frac{\partial^2 v}{\partial z^2}\right) + \rho g_y \tag{99}$$

z-component:

$$\rho\left(\frac{\partial w}{\partial t} + u\frac{\partial w}{\partial x} + v\frac{\partial w}{\partial y} + w\frac{\partial w}{\partial z}\right)$$

$$= -\frac{\partial P}{\partial z} + \mu\left(\frac{\partial^2 w}{\partial x^2} + \frac{\partial^2 w}{\partial y^2} + \frac{\partial^2 w}{\partial z^2}\right) + \rho g_z \tag{100}$$

Equations 98 through 100 represent the Navier-Stokes equation.

The equations of motion and continuity form the starting basis for describing any viscous flow problem. Rather than developing momentum balances, we can simply apply boundary conditions to the equation of motion and solve for the appropriate flow parameter of interest. As an example, let us again examine the tube flow problem. Using cylindrical coordinates (refer to Table C-2 in Appendix C) the equation of motion can be reduced to the form given by Equation 101 by ignoring end effects, such that $u_\theta = u_r = 0$.

$$\rho v\frac{\partial v}{\partial y} = -\frac{\partial P'}{\partial y} + \mu\left[\frac{1}{r}\frac{\partial}{\partial r}\left(r\frac{\partial v}{\partial r}\right) + \frac{\partial^2 v}{\partial y^2}\right] \tag{101}$$

and the continuity equation reduces to

$$\frac{\partial v}{\partial y} = 0 \tag{102}$$

which means that $\partial^2 v/\partial y^2 = 0$ also. Hence, Equation 101 becomes

$$0 = -\frac{dP'}{dy} + \mu\frac{1}{r}\frac{d}{dr}\left(r\frac{dv}{dr}\right) \tag{103}$$

Applying the boundary condition of v = 0 at r = R after integration, we obtain

$$v = \frac{\Delta P' R^2}{4\mu L}\left[1 - (r/R)^2\right] \tag{104}$$

Note that this is our familiar parabolic velocity profile derived back in Equation 81.

Flow problems involving the determination of fluid velocity as a function of time and space require simultaneous solution of the momentum and continuity equations. For many situations the flow is such that these equations may be simplified greatly. In the one-dimensional flow problem treated above we eliminated the two-equations in the y- and z-directions (or the θ- and y-directions in cylindrical coordinates). In two-dimensional flow problems, two momentum equations are needed. Naturally, terms describing the coordinate direction in which no flow exists vanish.

THE MECHANICAL ENERGY EQUATION

We shall not derive the mechanical energy equation, but merely present it in its most useful engineering form. The reader may refer to Chapters 3 and 7 of Bird et al. [14] for a discussion on the development of the equation of mechanical energy.

For steady incompressible flow of a unit mass of fluid between two points, (1) and (2), the total mechanical energy balance may be written as follows:

$$z_1 \frac{g}{g_c} + \frac{P_1}{\rho} + kE_1 - W_s' = z_2 \frac{g}{g_c} + \frac{P_2}{\rho} + kE_2 + \Sigma F \qquad (105)$$

where z = vertical height above an arbitrary horizontal datum plane
P = hydrostatic pressure
kE = average kinetic energy per unit mass
W_s' = shaft work done by unit mass of fluid (note the value of W_s' is negative when the work is done by an external source, e.g., a pump).

The last term, ΣF, is the mechanical energy converted into the thermal energy as a result of fluid friction. Note that $\Sigma F = \Delta P/\rho$ + frictional losses due to entrance/exit effects, flow through constrictions, etc. ΔP is the frictional pressure loss at fully developed flow situations.

Equation 105 is also referred to as the Bernoulli equation and, as we will see in Section 2, is directly applicable to analyzing numerous engineering flow problems.

NOTES ON VECTOR AND TENSOR NOTATION

Vector and tensor notation may be described as a form of shorthand mathematics, useful for representing and manipulating lengthy equations.

In our above discussions, we tabulated the nine components of the convective momentum flux (ρuv, ρuw, ρvw, etc.) in a rather cumbersome way. All these terms can be denoted simply as the *dyadic product* $(\rho\vec{u})(\vec{u})$, where $\vec{u}$ is a vector composed of u, v, w. Similary, $\vec{\tau}$, the stress tensor, can be used to denote the nine components of the shear vector $(\tau_{xx}, \tau_{xy}, \tau_{xz}, \tau_{yx}, \ldots)$.

The symbol ∇ (called "del") is a differential vector operator, which in rectangular coordinates is defined as follows:

$$\nabla = \vec{\gamma}_1 \frac{\partial}{\partial\zeta_1} + \vec{\gamma}_2 \frac{\partial}{\partial\zeta_2} + \vec{\gamma}_3 \frac{\partial}{\partial\zeta_3} = \sum_{i=1}^{n} \vec{\gamma}_i \frac{\partial}{\partial\zeta_i} \tag{106}$$

$\vec{\gamma}_i$ represents the unit vectors and ζ_i the variables associated with the x-, y- and z- axes.

A scalar product can be formed with the operator ∇ by the "dot" product (also called the divergence). For example, the continuity equation (Equation 94c), may be simply written as

$$\frac{\partial\rho}{\partial t} = -(\nabla\cdot\rho\vec{u}) \tag{107}$$

The reader may refer to the literature [14,19,49,50] for instructions on the use of vector and tensor mathematics in handling the Navier-Stokes equations.

CHAPTER 1.5
LAMINAR FLOW

In Section 2 calculation methods for handling flows through pipes and fittings will be presented. The frictional resistance of fluids as they flow through pipes and piping singularities (i.e., contractions, expansions, valves, tees, etc.), as well as the rates of heat and mass transfer between fluids and conduit walls, are critical parameters that establish design criteria for process operations. It is essential that we have as detailed an understanding as possible of how the fluid behaves in a flow situation. In this chapter we will examine briefly some of the predictions of the pipe flow equations developed from a momentum balance in the previous chapter. For the newcomer, we will also introduce the terms boundary layer and friction factor, and examine another common flow configuration given much analytical attention, namely, flow through an annulus.

VELOCITY AND SHEAR STRESS PROFILES

In the previous chapter an expression for the velocity profile of fluid flowing through a tube was derived

$$\left(\text{Equation 104, } u = \frac{(P'_1 - P'_2)R^2}{4\mu L}\,[1 - (r/R)^2]\right).$$

The maximum point velocity can be obtained by setting r = 0. Then

$$u_{max} = \frac{(P'_1 - P'_2)R^2}{4\mu L} \tag{108}$$

Or we can write this as

$$\frac{u_{max}}{R^2} = \frac{1}{4\mu}\frac{\Delta P'}{L} \tag{109}$$

Hence, Equation 104 also can be written in the following form:

$$u = u_{max}\ [1 - (r/R)^2] \tag{110}$$

Earlier we defined the volumetric flowrate as $Q = UA = U(\pi R^2)$, where U is the average fluid velocity across the tube and A is the cross-sectional area of the tube. Or, perhaps more descriptive, we may write

$$dQ = 2\pi rudr \tag{111a}$$

and Q is obtained by integrating the point velocity over the cross section from $0 \leqslant r \leqslant R$.

$$Q = UA = \int_0^Q dQ = 2\int_0^R 2\pi rudr \tag{111b}$$

By substituting the value of U (Equation 110) into Equation 111b and performing the integration, the following relation is obtained:

$$U = \frac{U_{max}}{2} \tag{112}$$

Hence, the velocity profile may be written as

$$\frac{u}{U} = 2\ [1 - (r/R)^2] \tag{113}$$

Equation 113 is that of a parabola, and the profile is shown plotted in Figure 17(A).

Shearing stresses exist in the boundary layer of the flowing fluid and are exerted in the direction opposite to the flow. As noted earlier, shear stresses represent the force of resistance to flow. Again, from our force balance analysis in the preceding chapter, a linear relationship between the shear stress in the fluid and the distance from the tube axis was obtained (Equation 86, $\tau = \tau_w(r/R)$). This linear distribution is shown in Figure 17(B).

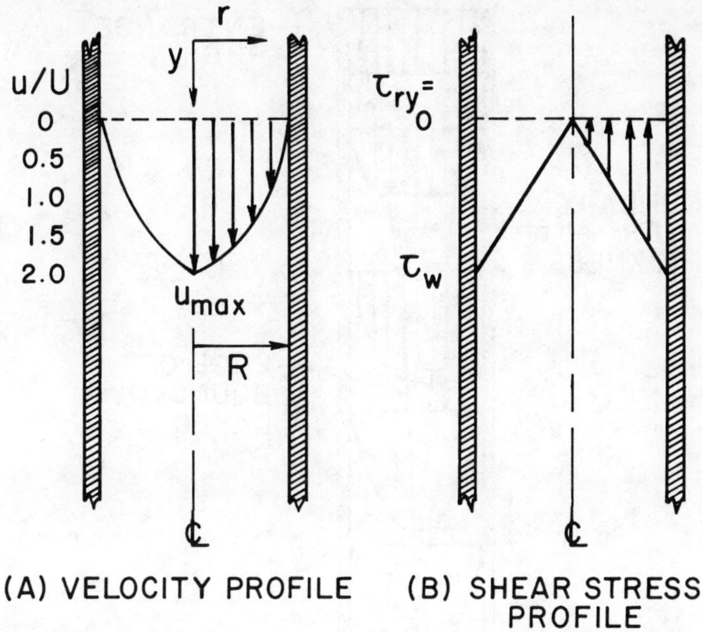

Figure 17. Velocity profile and shear stress distribution for laminar flow in a pipe.

FORMATION OF THE BOUNDARY LAYER

The boundary layer is defined by the region between the wall and a point in the viscous flowing fluid where the velocity is a maximum. The velocity gradient at the maximum velocity is zero.

Figure 18 illustrates how the boundary is formed. At the entrance of the conduit or pipe the boundary layer does not exist. As the flow continues along the pipe the boundary layer gradually increases in thickness until it eventually reaches a constant value at some distance from the entrance. Note that the dotted line in Figure 18 signifies the edge of the boundary layer. The last profile shown is that of the parabolic velocity distribution and signifies that fully developed flow exists.

RELATIONSHIP BETWEEN WALL SHEAR AND FRICTION FACTOR

Since shear stresses represent the force of resistance to flow, they are ultimately a measure of the friction losses. Friction losses in turbulent flow are known to be proportional to the kinetic energy of the fluid per

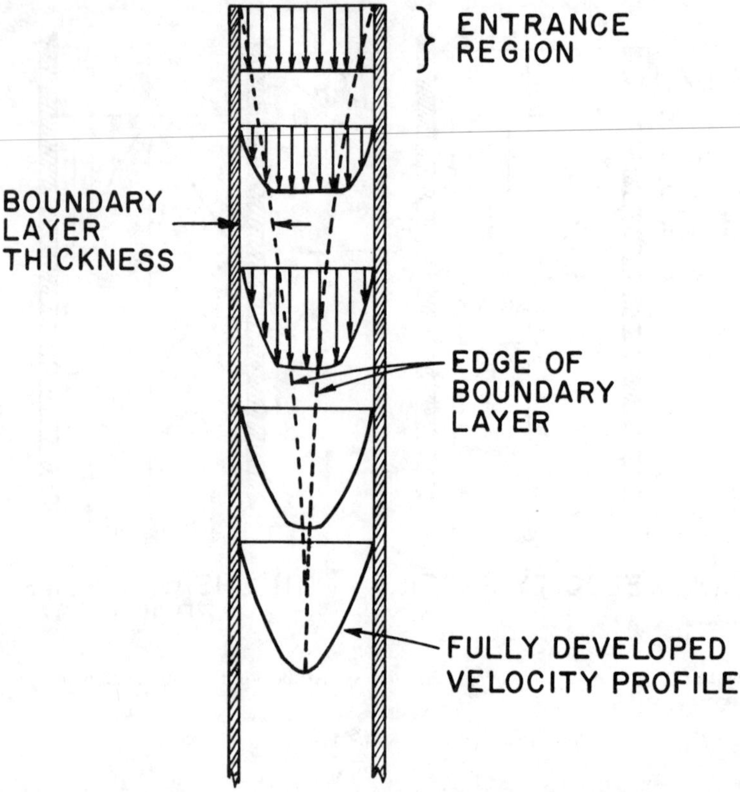

Figure 18. Illustrates the development of the boundary layer.

unit volume and the area of the solid surface contacting the flow. The force per unit area, A_w, acting on solid surface such as a pipe wall is

$$F/A_w \propto \frac{\rho U^2}{2g_c} = f\,\frac{\rho U^2}{2g_c} \tag{114}$$

where f is a proportionality constant.

We note that F/A_w is really the wall shear, τ_w; hence, $\tau_w = f(\rho U^2/2g_c)$.

In deriving the linear shear stress relation (Equation 86 in Chapter 1.4), it can be shown that wall shear may also be expressed as follows:

$$\tau_w = \left(\frac{R}{2}\right)\left(\frac{\Delta P}{L}\right) \tag{115}$$

By substituting for τ_w into the last expression, the proportionality constant f is solved. Factor f is commonly referred to as the friction factor, and Equation 116 defines the so-called "Fanning friction factor" for pipe flow:

$$f = \frac{g_c R}{\rho U^2} \frac{\Delta P}{L} \tag{116a}$$

or

$$f = \frac{g_c D}{2\rho U^2} \frac{\Delta P}{L} \tag{116b}$$

where D is the pipe diameter.

For laminar flow in a tube, Poiseuille's equation may be used to derive the following relation:

$$f = \frac{8\mu}{RU\rho} = \frac{16}{N_{Re}} \tag{117}$$

N_{Re} is a dimensionless parameter called the Reynolds number ($N_{Re} = DU\rho/\mu$). The friction factor definition in Equation 117 is good for Reynolds numbers up to 2000.

FLOW THROUGH AN ANNULUS

Incompressible fluid flow through an annulus is a very practical engineering flow configuration. It is most frequently encountered in double-pipe heat exchangers and condensers where one fluid flows through an inner tube and the other flows in the annular space, with the exchange of heat governed by conduction through the inner pipe wall. Figure 19 shows laminar flow of a viscous fluid through a simple annulus comprised of two smooth concentric tubes. In practice, the flow through the annular section may be laminar or turbulent, depending on the type of application.

By means of a momentum balance, identical to the approach used in Chapter 1.4 for the simple pipe flow problem, the following expressions may be derived for the velocity profile and shear stress distribution (i.e., momentum flux distribution) for steady incompressible flow in an annulus:

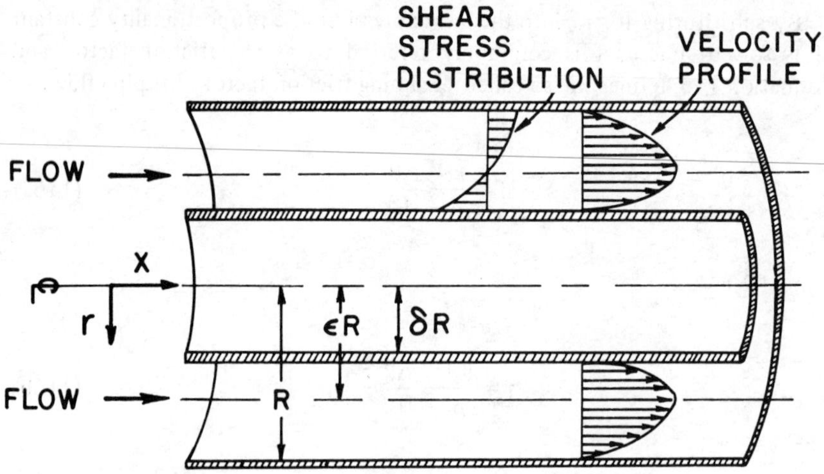

Figure 19. Viscous flow through an annulus.

$$u = \frac{(P_1 - P_2)R^2}{4\mu L}\left[1 - \left(\frac{r}{R}\right)^2 + \left(\frac{1 - \delta^2}{\ln(1/\delta)}\right)\ln(r/R)\right] \tag{118}$$

$$\tau_{rx} = \frac{(P_1 - P_2)R}{2L}\left[\left(\frac{r}{R}\right) - \left(\frac{1 - \delta^2}{2\ln(1/\delta)}\right)\frac{R}{r}\right] \tag{119}$$

Equations 118 and 119 describe the laminar flow case for a Newtonian fluid and were derived using the boundary conditions $u = 0$ at $r = \delta R$ and $u = 0$ $r = R$ (the ambitious reader may wish to perform the derivation).

Note that when δ is zero the second term in the brackets vanishes, and the expressions reduce to the pipe flow formulas developed earlier.

Expressions for the maximum velocity and the volumetric flowrate are

$$u_{max} = \frac{(P_1 - P_2)R^2}{4\mu L}\left\{1 - \left(\frac{1 - \delta^2}{2\ln(1/\delta)}\right)\left[1 - \ln\left(\frac{1 - \delta^2}{2\ln(1/\delta)}\right)\right]\right\} \tag{120}$$

$$Q = \frac{\pi(P_1 - P_2)R^4}{8\mu L}\left\{(1 - \delta^4) - \frac{(1 - \delta^2)^2}{\ln(1/\delta)}\right\} \tag{121}$$

Very often in heat exchanger applications fins are added to the outer wall of the inner tube. This has the effect of increasing the effective heat transfer area at the cost of pressure drop. This configuration is treated in the literature [4,53-55].

CHAPTER 1.6

TURBULENCE AND TURBULENCE MEASUREMENT

This chapter provides introductory material to turbulence theory. The classical turbulent pipe flow models developed by Prandtl, von Karman and Deissler are rederived, and physical significance is attached to the various aspects of these analogies. Descriptive background on other types of turbulence encountered in process operations handling fluids is also included, along with a generous listing of primary references recommended for further readings. The last subsection in this chapter presents the operational theory behind anemometry techniques for turbulence and flow properties studies of flowing systems. A list of nearly 90 references has been compiled covering specific applications of anemometry techniques for various applications in turbulence investigations.

THE NATURE OF TURBULENCE

The laminar flow pattern described consists of a steady advance of fluid layers such that a streamline flow is maintained. If the fluid's flowrate is increased sufficiently this pattern is no longer maintained and the flow becomes unsteady. This type flow is characterized by chaotic movements of portions of the fluid in different directions superimposed on the main flow of the fluid; the phenomenon is referred to as turbulence. The complex movement of fluid elements within the flow can only be described in terms of time-averaged values. Figure 20(A) shows typical velocity fluctuations in turbulent flow as measured with a hot-wire anemometer at the center of a pipe for air flow.

Momentum transfers between adjacent regions of flowing fluid play a dominant role in establishing turbulence. These momentum transfers are inertial effects, which cause velocity and fluid density to take on important roles. By contrast, in laminar flow only purely viscous effects determine the nature of the flow. Turbulent flow dominates when inertial effects (denoted by

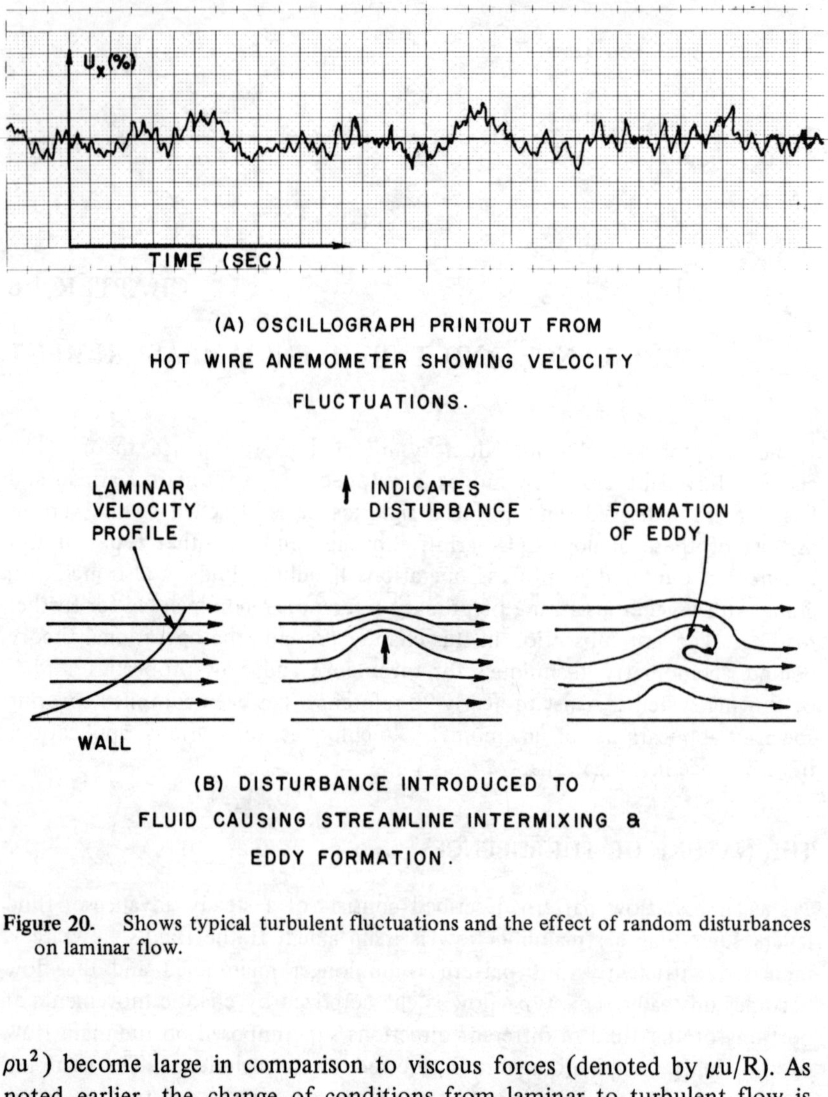

Figure 20. Shows typical turbulent fluctuations and the effect of random disturbances on laminar flow.

ρu^2) become large in comparison to viscous forces (denoted by $\mu u/R$). As noted earlier, the change of conditions from laminar to turbulent flow is characterized by a dimensionless group called the Reynolds number:

$$N_{Re} = \rho u D/\mu = uD/\nu \tag{122}$$

where D is a characteristic length (for the case of simple pipe flow, the diameter) and ν is kinematic viscosity (μ/ρ), having dimensions of (length)^2/time.

The transition from laminar to turbulent flow for a small portion of flowing fluid is illustrated in Figure 20(B). The figure at the left shows a steady laminar flow situation. By introducing a random disturbance (middle figure), density and velocity are in part damped out by viscous forces. In fact, if we look back at Equation 122 we see it can be written in the form $N_{Re} = (\rho u)/(\mu u/D)$, which is the ratio of inertial to viscous forces. As more momentum is transferred into the disturbance, inertial forces eventually become much greater than the viscous term, and so-called eddy currents are formed (right sketch, Figure 20(B)).

Numerous investigators [56-62] have studied the flow of fluids in pipes, confirming the flow regime dependency on Reynolds number. For flow in smooth pipes, the flow is always laminar for Reynolds numbers up to about 2000. Between Reynolds numbers of 2000 and 4000 the flow undergoes a transitional change (thus the name "transition region") from laminar to turbulent flow. Above Reynolds numbers of 4000 the flow is generally fully established turbulent flow.

The formation of turbulent eddy currents is best described by imagining ourselves to be moving within the fluid at the same velocity as the fluid's mean flowrate. If the flow is in a pipe, what we would see in the vicinity of the wall are patches of fluid torn into small eddies by the strong shearing forces within the fluid. These eddies tend to migrate toward the pipe centerline, some combining to form larger eddies, while many smaller ones simply diffusing into the "core" fluid with decaying intensity. We would further observe that these eddies are in fact superimposed on a much faster overall mean flow.

There are several situations that can lead to turbulent flow. Common cases are illustrated in Figure 21. One situation which we have just described is that of rapid flow of a fluid past a solid surface. This situation leads to unstable, self-amplifying velocity fluctuations, which form in the fluid in the vicinity of the wall and spread outward into the main fluid stream.

In a similar manner, turbulent eddies are formed from the velocity gradients established between a fast-moving fluid and a slower-moving fluid. A third general way in which turbulence is induced is by the relative movement of an object through the fluid streams. Examples of this last case are an impeller blade on an agitator and a falling sphere or cylinder. These situations cause eddies to form in the wake, resulting in an increase in the resistance of the movement of the object. This resistance is referred to as "form drag."

In the case of stirred vessels, turbulence can be very intense near the tips of the rotor blades. Most of the turbulence throughout the vessel arises from velocity gradients, whereby portions of high-velocity fluid are thrown from impeller blades onto slower-moving fluid. Some of this turbulence is, however, attributed to the high shearing over the blades themselves, which form separation behind each blade or neighboring baffles.

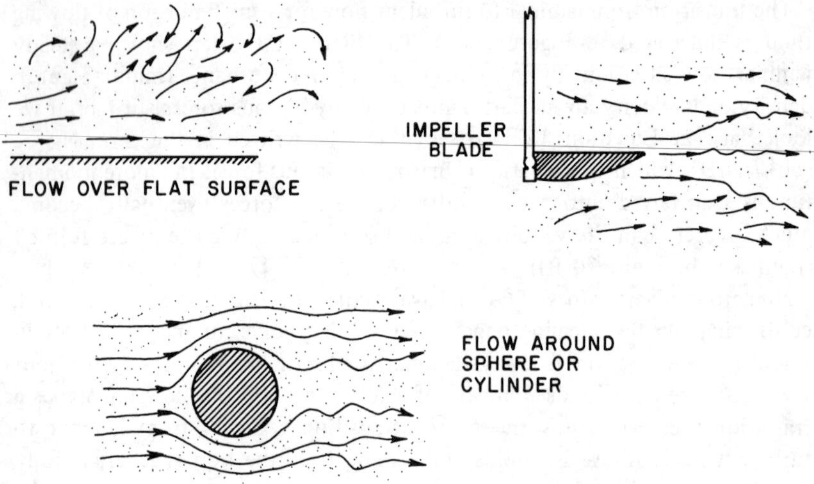

Figure 21. Examples of vortex shedding and eddy formation leading to turbulent flow.

Another mixing-related turbulence arises from submerged jets. In submerged jets (also referred to as "free turbulent jets") the fluid is expelled from a nozzle into a mass of miscible fluid, which is essentially at rest with respect to the discharging stream. The free jet expands outward in the form of a cone, entraining large amounts of the surrounding fluid at the periphery of the jet. The size of eddies formed are relatively uniform across any given section of such a jet.

When a submerged jet is introduced to an immiscible fluid the stream is referred to as a *restrained turbulent jet*. Turbulent eddies protrude from the sides of the free jet but are restrained from breaking away by surface tension forces. This type of turbulence decays quickly as fluctuations are damped by the elastic forces associated with surface tension effects. For discussions on turbulent jets refer to the literature [63,64].

It should be noted that the turbulent condition set by $N_{Re} > 4000$ is a general rule and one that is based mainly on experience with pipe flow. Turbulent conditions can, in fact, be generated at much lower Reynolds numbers. One example is flow in a corrugated conduit, where vortex shedding at the angularities can induce strong eddying at Reynolds numbers of only a few hundred. This principle is also applied to inline static mixers used for mixing two liquid streams or a chemical additive and a liquid. By means of fins, baffles or an arrangement of packing material through which fluids flow, vortex shedding occurs promoting eddies assisting in the necessary mass transfer (refer to Figure 22). Another example of this application is in heat exchangers used for viscous liquids. Here, only moderate flowrates are needed to promote eddy formation in narrow tubes. This promotes good heat transfer even though Reynolds numbers are only as high as perhaps 500 to 600. In these applications, the sharpness of the angularities is important since eddies shed much more readily from sharp corners than from smooth surfaces.

FLUCTUATIONS AND TANGENTIAL STRESSES

We can attempt to describe turbulence in a more quantitative manner by examining the instantaneous velocity fluctuations occurring in local patches of fluid. The instantaneous velocity in the x-direction of flow may be expressed as follows:

$$u_x = \bar{u}_x \pm u_x' \tag{123}$$

where $\bar{u}_x$ is the time-averaged velocity at any point in the flowing fluid and u_x' is the instantaneous velocity fluctuation. Fluctuation velocities may be positive or negative at times, with their time average being zero. That is,

$$\bar{u}_x' = 0 \tag{124}$$

Common practice is to express the amplitude of fluctuation velocities in the x-direction in terms of the mean of the squares of the fluctuating velocities, obviously making them positive. This is denoted by $\overline{(u_x^1)^2}$. A root mean square fluctuation velocity thus would be

Figure 22. Viscous flow in a static mixer unit. Flow division (resulting in a geometric progression) and radial mixing operate simultaneously in the static mixer unit, resulting in nearly plug flow characteristics. (courtesy of Kenics Division, Chemineer Inc.)

$$u'_{x_{rms}} = [\overline{(u'_x)^2}]^{1/2} \tag{125}$$

where $u'_{x_{rms}}$ is always positive. $\bar{u}_{x_{rms}}$ provides an indication of the *intensity of turbulence* in the x-direction.

The same definitions may be applied to the y-coordinate. That is,

$$u_y = \bar{u}_y \pm u'_y$$

$$\overline{u'_y} = 0 \tag{126}$$

and

$$u'_{y_{rms}} = \sqrt{\overline{(u'_y)^2}} \tag{127}$$

Note that the product $\overline{u'_x u'_y}$ for any point in the fluid is not necessarily zero, especially for pipe flow. The nomenclature used here implies that the product of u'_x and u'_y has been time-averaged and that these fluctuations are correlated with a particular eddy. These fluctuating velocities are responsible for establishing the degree of dissipation of energy, mass transfer and heat transfer in turbulent flows.

Visualize the fluid as consisting of individual parcels. In turbulent flow there is a continuous interchange of these parcels between adjacent regions of the fluid. This exchange of mass involves both gains and losses of momentum. That is, each parcel of fluid leaving a region with some velocity takes with it momentum to the next fluid region, which, in turn, may be moving at a different mean velocity. Fluid parcels transposed laterally from a faster-moving region tend to accelerate the slower-moving fluid region; conversely, a slower-moving parcel transposed to a faster-moving region has a retarding effect.

The rate of momentum transfer per unit area is $\rho u'_y u'_x$ and can be expressed in terms of a stress (i.e., equivalent force per unit area):

$$\tau = -\rho u'_x u'_y \tag{128}$$

The time-average of this product is called the Reynolds stress:

$$\tau = -\rho \overline{u'_x u'_y} \tag{129}$$

Continuing with our visualized model of parcel exchanges, let us assume that a particular parcel is carried from a distance, y_1, from the wall to new distance, y_2. Further assume that the parcel is moved from a slower-moving section to a faster-moving one. The instantaneous deficit in momentum occurring at distance y_2 may be expressed as follows:

$$\text{Momentum Deficit} = -(\Delta y)d(\rho\bar{u}_x)/dy \tag{130}$$

where $\Delta y = y_2 - y_1$.

If the momentum is conserved then we may write

$$\rho u'_x = -(\Delta y)d(\rho\bar{u}_x)/dy \tag{130a}$$

or

$$\tau = -\rho u'_x u'_y = u'_y(\Delta y)d(\rho\bar{u}_x)/dy \tag{130b}$$

Alternately, this last expression may be written as follows:

$$\tau = -\rho\overline{u'_x u'_y} = \rho \ell u'_{y_{rms}} d\bar{u}_x/dy \tag{131}$$

Equation 131 is obtained by representing the root mean squares of the displacement, Δy, and the velocity fluctuation, u'_y, by ℓ and $u'_{y_{rms}}$, respectively. The parameter ℓ is defined as the mean distance of travel of the parcels of fluid before they are mixed into the new fluid region. We can think of ℓ as a measure of the "scale of the turbulence eddies." It is commonly referred to as the *mixing length* or *Prandtl eddy length* and forms the basis for the famous Prandtl law model of the universal velocity profile for pipe flow (discussed shortly).

By analogy to Newton's law we can write

$$\epsilon' = \tau/(d\bar{u}_x/dy) = \rho \ell u'_{y_{rms}} \tag{132}$$

where ϵ' is the eddy viscosity. Equation 132 can be written in the alternate form:

$$\tau = \rho\epsilon'_M d\bar{u}_x/dy \tag{133}$$

where ϵ'_M is the eddy kinematic viscosity defined by $\epsilon'_M = \ell u'_{y_{rms}}$.

Some textbooks refer to ϵ' as the "eddy viscosity of momentum." This is somewhat more descriptive since it is essentially the drag on the fluid due to momentum effects. Prandtl hypothesized that it is proportional to the velocity fluctuations and to their distances (this is analogous to the product of the average molecular velocity and the mean free path in the kinetic theory of gases).

Since $u'_{y_{rms}}$ is difficult to obtain directly, Prandtl assumed u'_y to be related simply to the velocity fluctuation ($u'_y \propto \ell \, d\bar{u}_x/dy$) and simply wrote the Reynolds stress (Equation 129) in the following form:

$$\tau = \rho \ell^2 (d\bar{u}_x/dy)^2 \tag{134a}$$

or, alternately,

$$\tau = (\rho \ell^2 d\bar{u}_x/dy)(d\bar{u}_x/dy) \tag{134b}$$

Hence, Equation 132 may be expressed as

$$\epsilon' = \rho \ell^2 (d\bar{u}_x/dy) \tag{135}$$

With this definition the eddy kinematic viscosity may be written as $\epsilon'_M = \ell^2 (d\bar{u}_x/dy)$.

Finally, we can account for both viscous and turbulent contributions with an equation analogous to Newton's law of viscosity:

$$\tau = -(\mu + \epsilon') d\bar{u}_x/dy \tag{136}$$

In the last series of equations only time-averaged values for the velocity components are necessary. From this point on we will drop the bar (–), realizing that values are time-averaged.

VELOCITY PROFILES IN PIPE FLOW

The velocity profile for laminar flow in a circular pipe was shown back in Chapter 1.5 to be parabolic in shape (Figure 17). Recall, in laminar flow, that the shape of the profile is due to the effects of viscous forces between adjacent fluid layers. In turbulent flow, the shape of the velocity profile is very much different. The majority of the velocity gradient occurs in the fluid region close to the wall where some streamline flow continues to persist; however, as we enter into the fluid core the velocity profile is more blunt than in the laminar case and is nearly flat at the pipe centerline. The reason for this flatness is that in turbulent flow inertial forces are high and have influence over a large portion of the tube (recall the existence of large eddies in the fluid core). Figure 23 compares the velocity profiles for laminar and turbulent gas flow in a horizontal pipe. It is of interest to note that although the two profiles differ greatly, both flows have the same average velocity.

In both cases the maximum velocity occurs at the pipe axis. Recall again from the previous chapter that the ratio of the average (i.e., U_{avg}, the average velocity over the tube area) to maximum velocity at the center is a constant equal to 0.5 for laminar flow. For turbulent flows this ratio is experimentally determined to lie above 0.8. This ratio is shown as a function of Reynolds number in Figure 24 using the water data of Nikuradse [57], and the air and

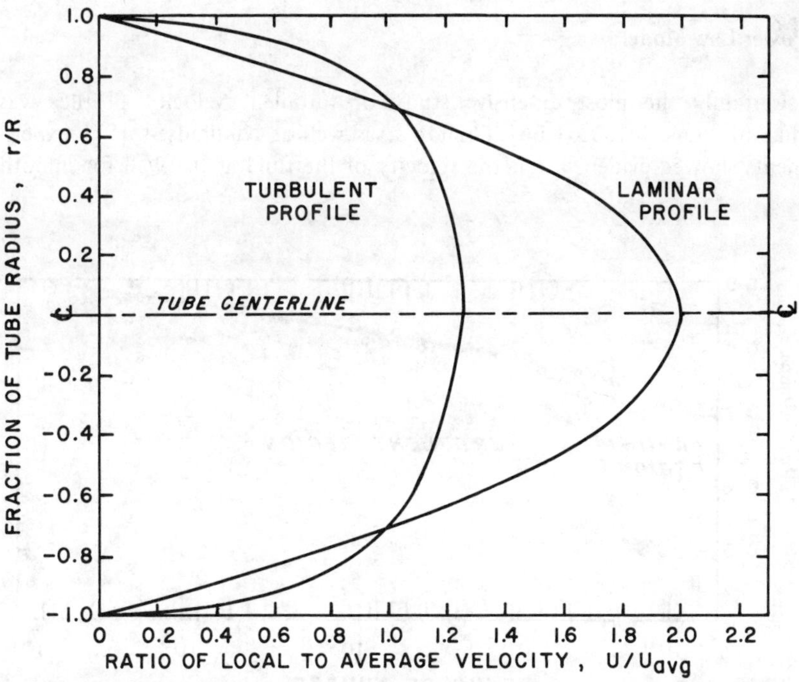

Figure 23. Compares the velocity profiles for turbulent and laminar flows in a tube. Note that the average fluid velocity in each case is the same.

water data of Stanton et al. [65]. As shown, in the laminar region (i.e., in the Reynolds number range below 2000) the ratio is nearly constant at 0.5.

In the transition region ($2000 < N_{Re} < 4000$), u_{avg}/u_{max} increases sharply, close to the value of 0.8. Above Reynolds number 4000, as the flow becomes fully turbulent, the correlation increases at a moderate slope.

Numerous investigations have been directed at studying the velocity profiles of turbulent flows. Nikuradse [57] extensively studied turbulent flow velocity distributions in smooth circular tubes, showing that the profile always takes on the same general shape shown in Figure 23. From these investigations a number of attempts have been made to develop from first principles an equation for describing the velocity distributions in tubes. In the discussions to follow we shall review the principal models developed for describing turbulent flow in smooth pipes. Much of the early work between the period 1920 to 1955 forms the basis for our current understanding of turbulence and has enabled sound engineering principles to be established.

PRANDTL'S MODELS

Power Law Model

Certainly the most extensive study of turbulent velocity profiles was that of Prandtl [60,61,66]. Prandtl's, as well as Nikuradse's [57] experiments showed evidence that the velocity of the fluid at the wall for smooth

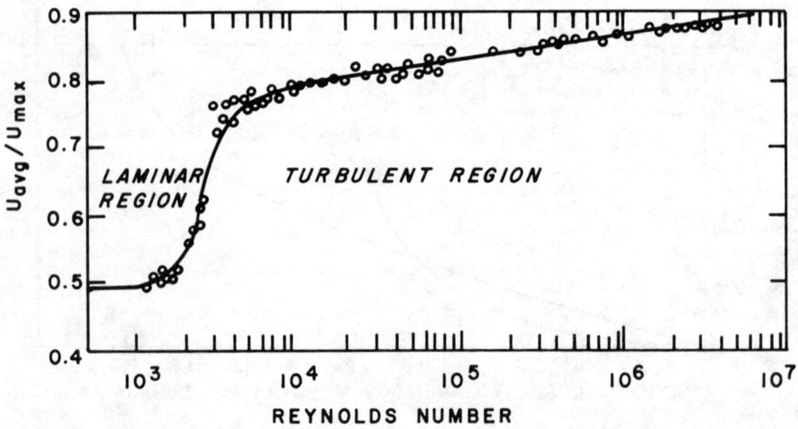

Figure 24. Correlation of air and water flows data showing the variation of U_{avg}/U_{max} with Reynolds number.

pipes is zero. Prandtl further postulated that there is a layer of fluid very close to the wall that is in laminar motion, and that only viscous forces contribute to shearing in this region. That is, at the wall $u \simeq 0$ for y close to 0. Hence, if τ_w is known (we can obtain τ_w from direct measurement or indirectly from $\Delta P/L$), then the velocity profile in the laminar layer can be computed. His earliest attempt to describe the flow in the turbulent region resulted in a power law relation:

$$\frac{u}{u_{max}} = C(y/R)^{1/7} \tag{137}$$

where C is a constant, y is the distance from the pipe wall, and R is the pipe radius. Equation 137 is known as the 1/7th power law velocity distribution and is related to the Blasius [67] friction factor:

$$\tau_w = \frac{f\rho u_{avg}^2}{2g_c} \tag{138}$$

The Blausius friction factor equation has been determined experimentally to be valid for turbulent Reynolds numbers up to 100,000. Hence, Equation 137 is also valid up to Reynolds numbers of 100,000.

Logarithmic Distribution

Prandtl further derived a logarithmic distribution from his mixing length theory (Equation 136):

$$\tau_w g_c = \underbrace{\mu \frac{du}{dy}}_{\substack{\text{viscous} \\ \text{shear}}} + \underbrace{\rho \left(\ell \frac{du}{dy}\right)^2}_{\substack{\text{turbulent} \\ \text{shear}}} \tag{139}$$

Again, y is defined as being the distance from the wall.

We know from our laminar flow analysis that $\tau = \tau_w(r/R)$ and, noting the following approximation,

$$(r/R) \simeq 1 - y/R \tag{140}$$

We then obtain

$$\tau_w g_c(1 - y/R) = \rho\left(\ell \frac{du}{dy}\right)^2 \tag{141}$$

or

$$\ell\left(\frac{du}{dy}\right) = \left(\frac{\tau_w g_c}{\rho}\right)^{1/2}(1 - y/R)^{1/2} \tag{142}$$

The group of terms $(\tau_w g_c/\rho)^{1/2}$ is given the special name *friction velocity* and is commonly denoted by u^*:

$$u^* = \sqrt{\frac{\tau_w g_c}{\rho}} \tag{143}$$

For conditions at the wall, $1 - y/R$ reduces to 1 and

$$u^* = \ell \frac{du}{dy} \tag{144}$$

Prandtl hypothesized that the eddy mixing length is proportional to the distance from the wall by some constant that he hoped to be universal. That is,

$$\ell = ky \tag{145}$$

Hence, Equation 144 may be restated as

$$\frac{du}{dy} = u^*/ky \tag{146}$$

Or, on integration,

$$u = \frac{u^*}{k}\,\ell n y + c \tag{147}$$

Using the eddy viscosity, Equation 135, which also may be written as $\epsilon' = k^2y^2(du/dy)$ from the above considerations, and solving for the constant of integration, C, in Equation 147, Prandtl developed the following velocity distribution equation:

$$u = u_{max} + 2.5\,u^*\ell n(y/R) \tag{148}$$

VON KARMAN ANALOGY

Von Karman [68, 69] attacked the velocity distribution problem through similarity theory, developing the following series of equations:

$$\epsilon'_M = \frac{k^2(du/dy)^3}{(d^2u/dy^2)^2} \tag{149}$$

$$\ell = \frac{k(du/dy)}{d^2u/dy^2} \tag{150}$$

$$\tau = \frac{\rho k^2(du/dy)^4}{g_c(d^2u/dy^2)^2} \tag{151}$$

By use of the boundary conditions,

$$\text{for } y = 0,\ du/dy = \infty \quad \text{and}$$

$$\text{for } y = R,\ u = u_{max}$$

the following velocity distribution was derived.

$$u = u_{max} + \frac{1}{k}\,u^*[\ell n(1 - \sqrt{1 - y/R}) + \sqrt{1 - y/R}] \tag{152}$$

CONSTRUCTING THE UNIVERSAL VELOCITY PROFILE

From Prandtl's theory and from the data of Nikuradse [57], Reichardt [70] and others, von Karman was able to construct a universal velocity profile through three equations. Each equation describes a different region in the flow: a laminar boundary layer, a buffer layer and the turbulent fluid core.

From the friction velocity (Equation 143), the following expression is obtained:

$$\tau_w = \frac{u^*\rho}{g_c} = \frac{\mu}{g_c}\frac{u_{\delta 1}}{\delta_1} \tag{153}$$

Equation 153 defines conditions at the edge of the boundary layer, where $y = \delta_1$ (i.e., the boundary layer thickness) and du/dy is approximated by $u_{\delta 1}/\delta_1$. $u_{\delta 1}$ is the velocity of the fluid at the edge of the boundary layer.

From this equation it is noted that

$$\frac{u_{\delta 1}}{u^*} = \frac{\delta_1 u^* \rho}{\mu} = \frac{\delta_1 u^*}{\nu} \tag{154}$$

which has the form of our familiar Reynolds number. Note that $u_{\delta 1}/u^*$ may be assumed to be a constant (since τ_w, g_c, ν are constants).

For conditions at the edge of the boundary layer, Prandtl's velocity distribution equation (Equation 148) may be written as

$$\frac{u_{\delta 1}}{u^*} = \frac{u_{max}}{u^*} - 2.5\,\ell n(R/\delta_1) \tag{155}$$

By writing the second term as $-2.5\,\ell n(R/\delta_1)$, the following manipulation can be made with the log term

$$\frac{R}{\delta_1}\left(\frac{u^*}{u^*}\right)$$

and since $\dfrac{u_{\delta 1}}{u^*}$ = constant = C_1, then

$$u^* = c_1\nu/\delta_1 \quad \text{(from Equation 154)}$$

Hence, the log term becomes

$$\frac{Ru^*}{\delta_1\left(\dfrac{C_1\nu}{\delta_1}\right)}$$

And Equation 155 now becomes

$$\frac{u_{\delta 1}}{u^*} = \frac{u_{max}}{u^*} - 2.5\,\ell n\left(\frac{Ru^*}{c_1\nu}\right) \qquad (156)$$

or

$$\frac{u_{max}}{u^*} = \frac{u_{\delta 1}}{u^*} + 2.5\left[\ell n\left(\frac{Ru^*}{\nu}\right) + \ell n\left(\frac{1}{c_1}\right)\right] \qquad (157)$$

The form of Equation 157 enables all the constants to be collected into one constant term:

$$\frac{u_{max}}{u^*} = c_2 + 2.5\,\ell n\left(\frac{Ru^*}{\nu}\right) \qquad (158)$$

where

$$c_2 = \frac{u_{\delta 1}}{u^*} + 2.5\,\ell n(1/C_1)$$

Substituting Equation 158 for u* back into Prandtl's equation (Equaion 148), the following is obtained:

$$\frac{u}{u^*} = c_2 + 2.5\left[\ell n\left(\frac{Ru^*}{\nu}\right) + \ell n(y/R)\right]$$

$$= c_2 + 2.5\left[\ell n\left(\frac{Ru^*}{\nu}\right)(y/R)\right] \qquad (159)$$

or

$$\frac{u}{u^*} = c_2 + 2.5\,\ell n\left(\frac{yu^*}{\nu}\right) \qquad (159)$$

Equation 159 may be written in its most general form by introducing the following dimensionless terms:

$$u^+ = u/u^*, \text{ velocity parameter} \tag{160}$$

$$y^+ = yu^*/\nu, \text{ friction distance parameter} \tag{161}$$

y^+ is the dimensionless distance from the wall and u^+ is the dimensionless velocity.

Hence, the universal velocity distribution equation for turbulent flow in circular pipes is

$$u^+ = c_2 + 2.5\,\ell n y^+ \tag{162}$$

The universal velocity profile is shown plotted in Figure 25. By empirically fitting Equation 162 to the dimensionless profile, von Karman [69] obtained three equations to describe separate regions of the flow:

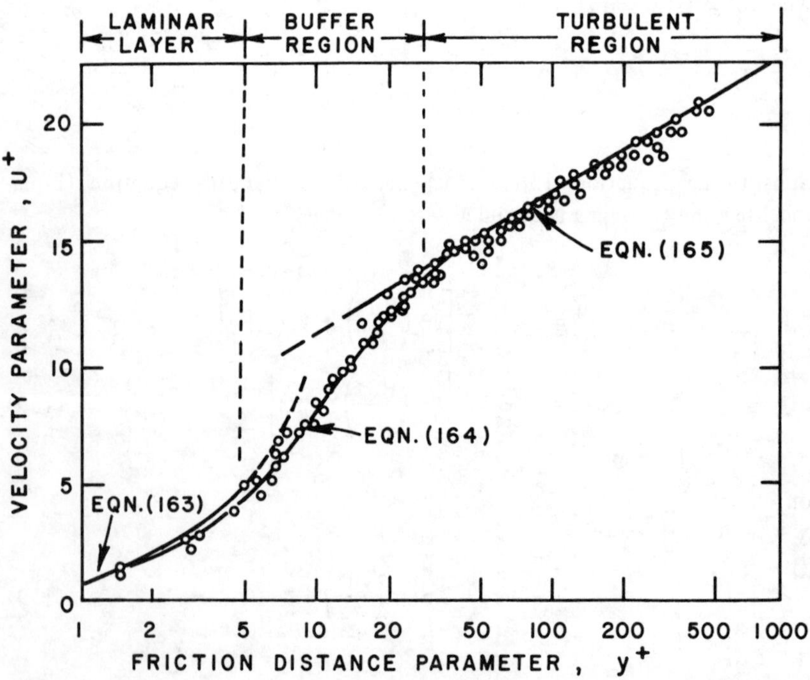

Figure 25. The universal velocity profile for turbulent pipe flow. Plotted are data of Nikuradse [57], Reichardt [70] and Deissler [59].

Laminar Sublayer . . . . $u^+ = y^+$, for $0 < y^+ < 5$ (163)

Buffer Region. . . . . . . $u^+ = -3.05 + 5.0 \ln y^+$, for $5 < y^+ < 30$ (164)

Turbulent Core . . . . . . $u^+ = 5.5 + 2.5 \ln y^+$, for $y^+ > 30$ (165)

The universal velocity profile has been used extensively in studying and developing useful analogies between momentum and heat transfer. The author's work [71-75] provides examples of how the above principles may be applied to analyzing flows involving heat transfer and two-phase (gas-liquid) flows.

DEISSLER'S MODEL

Deissler [59,76-80] approached the universal velocity profile problem differently by constructing a model for the laminar and buffer layers on the basis of the exponential decay of eddies penetrating these layers.

Using von Karman's model for τ, Equation 136, $\tau = -(\mu + \epsilon'_m \rho)(du/dy)$, in which the eddy viscosity could be defined for $y^+ > 30$ as

$$\epsilon' = k \frac{(du/dy)^3}{(d^2u/dy^2)^2} \tag{166}$$

For the region $0 < y^+ < 30$, the following relationship was proposed:

$$\epsilon' = n^2 uy(1 - e^{-\rho n^2 uy/\mu}) \tag{167}$$

where n is a constant approximately equal to 0.10.

Equation 167 predicts that the eddy viscosity approaches zero as the wall is neared.

In dimensionless form the velocity profile was derived in the following form:

$$y^+ = \frac{1}{n} \frac{\frac{1}{\sqrt{2\pi}} \int_0^{nu^+} e^{-[(nu^+)^2/2]} d(nu^+)}{(1/\sqrt{2\pi}) e^{-[(nu^+)^2/2]}} \tag{168}$$

This last expression applies to the laminar and buffer regions, for $0 < y^+ < 26$. Equation 168 must be numerically integrated to obtain u^+ values.

For the turbulent core, Deissler obtained the following expression:

$$u^+ = 3.8 + 2.78\,\ell n y^+ \tag{169}$$

ROUGHNESS AND TURBULENT FRICTION FACTORS

In the case of smooth pipe flow, the viscous sublayer near the wall is very thin at high Reynolds numbers. The introduction of roughness elements limits the thinning of the viscous sublayer.

Roughness elements are protrusions or interstices on the pipe walls. Standard pipe, even when new, has many imperfections and is considered quite rough. Drawn tubing when new, and plastic pipe generally are smooth-walled. Figure 26 conceptually shows the surface of a rough pipe wall. As shown, when fluid flows past these roughness elements eddies are formed behind them, causing so-called "form drag."

The height of a single unit of roughness is denoted by ϵ and is referred to as the *roughness parameter*. From dimensional analysis the friction factor, f, is both a function of the Reynolds number and the *relative roughness*, ϵ/D, where D is the pipe diameter. Experiments on artificially roughened pipe have shown that a different relationship between f and N_{Re} is obtained for each magnitude of the relative roughness [81,82].

Studies by Nikuradse [81,82], Rouse [83] and Moody [84] have led to the development of generalized friction factor charts. For laminar flow the friction factor is obtained from Equation 117.

For smooth pipes, the relationship between f and N_{Re} for turbulent flow may be approximated from the Blasius equation for $3000 < N_{Re} < 10^5$.

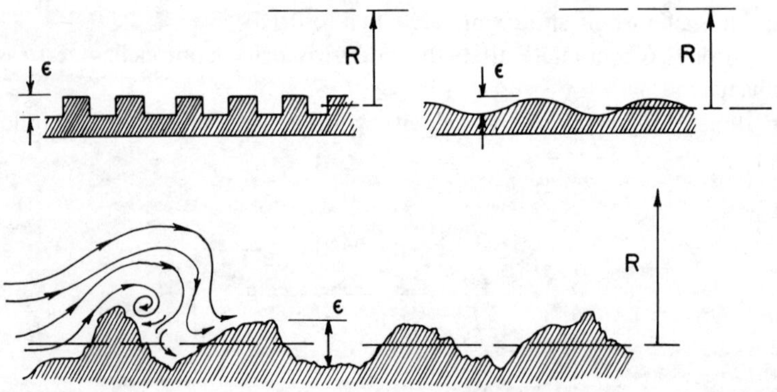

Figure 26. Blowup view of rough pipe wall. Also shown are several idealized types of roughness.

$$f = 0.04\ N_{Re}^{-1/4} \tag{170}$$

Moody [84] has given a complete friction factor plot for smooth and rough pipes (a log-log plot summarizing experimental data on friction factor as a function of Reynolds number and relative roughness), which is given in Section 2.

Moody also developed a chart giving the relative roughness of commercial pipe as a function of the diameter for different pipe materials. A portion of this plot is given in Figure 27.

As one would expect, velocity profiles obtained in rough pipes are distorted in comparison to smooth pipe flow. From turbulence theory, velocity distribution expressions have been developed. Nikuradse's velocity data for the turbulent core have been shown to be well correlated by the following expression:

$$u^+ = 2.5\ \ell n \frac{y}{\epsilon} + 8.5 \tag{171}$$

Rouse [83] developed a more general expression for the turbulent core applicable to both smooth and rough pipe flows. Further treatment is also given by Davies [85].

BOUNDARY LAYER SEPARATION

When fluid velocity is changed either in direction or magnitude due to a change in the direction or size of the conduit, additional friction over that formed in straight pipe flow is generated. This additional friction includes form friction or drag, which results from vortices that develop during boundary layer separation. Figure 28 shows several common situations when this occurs.

Figure 28(A) shows that when the cross section of a conduit is suddenly enlarged, the fluid stream separates from the wall and flows into the enlarged section as a jet. This fluid jet expands, filling the entire cross section of the larger conduit. The region between the expanding jet and the wall is filled with fluid in vortex motion, which is a characteristic of boundary layer separation. There is considerable friction that is generated in this turbulent region.

Figure 28(B) shows the effect of suddenly reducing the conduit cross section. The fluid stream cannot flow around the sharp edge, causing it

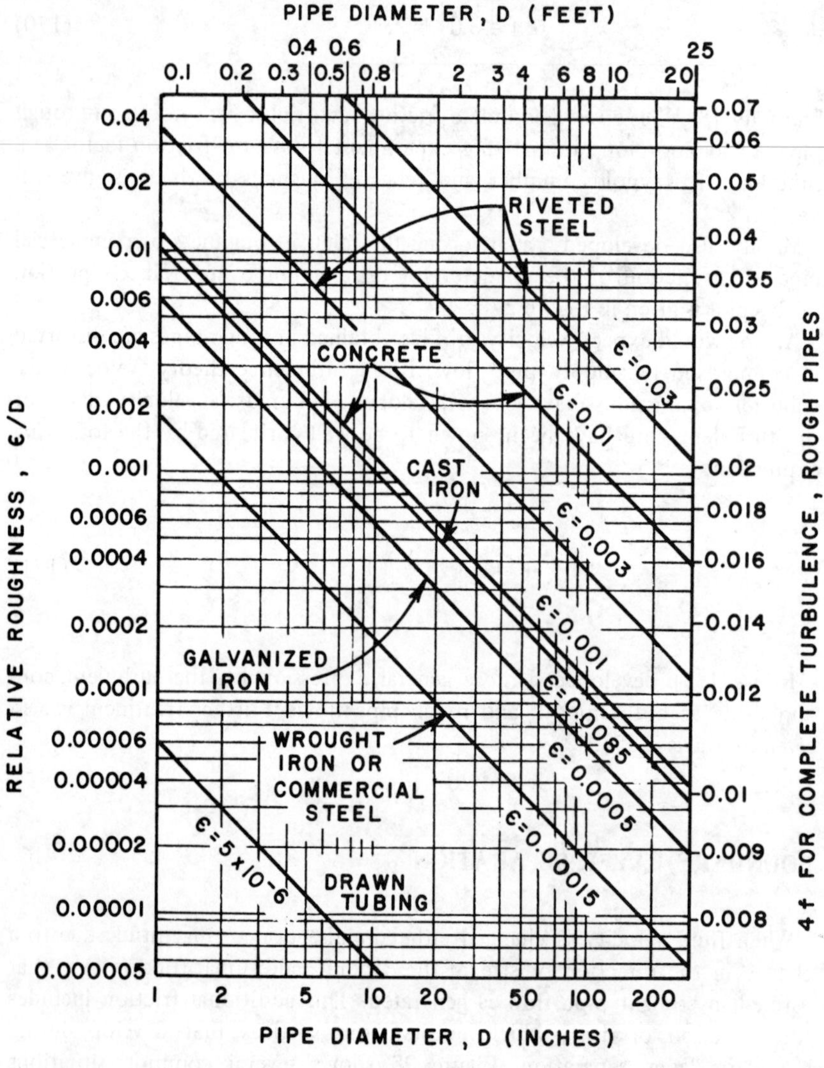

Figure 27. Moody plot of relative roughness of different pipe materials [84].

to break contact with the conduit wall. This situation also results in a jet, which flows into the stagnant fluid in the smaller section. The jet first undergoes contraction and then expands, filling the smaller cross section. Downstream from the point of contraction the normal velocity distribution is reestablished. The cross section of minimum flow area where the fluid jet

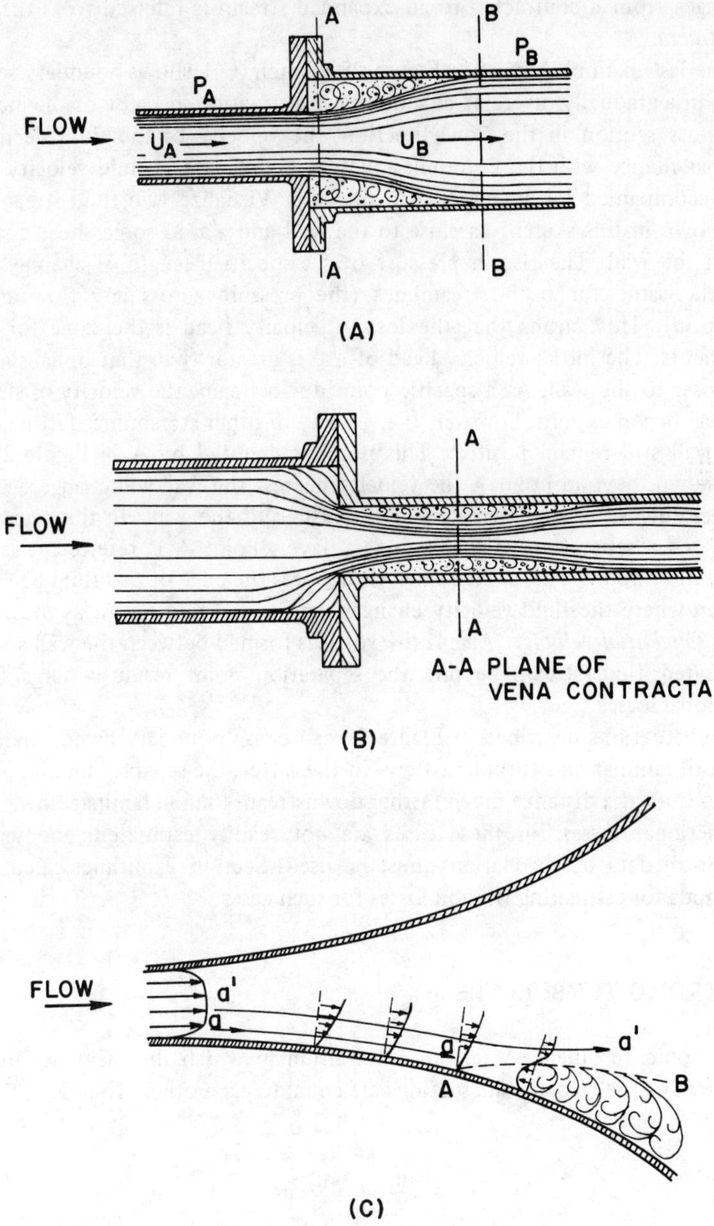

Figure 28. Illustrates flow situations resulting in boundary layer separation: (A) Sudden enlargement of conduit cross section. (B) Sudden contraction of cross section. (C) Gradual enlargement of cross section.

changes from a contracted to an expanded stream is referred to as the *vena contracta.*

The last example shown in Figure 28 (sketch (C)), shows boundary separation in a gradually diverging channel. In this situation, because of the increase of cross section in the flow direction, the velocity of the fluid decreases. In accordance with the Bernoulli equation, a decrease in fluid velocity must be accompanied by an increase in pressure. Visualize two fluid streamlines as shown in this sketch: aa close to the wall and a′a′ at some short distance from the wall. The rise in pressure over a specified length of channel must be the same for both streamlines (the pressure across any flow area is uniform). This means that the loss in velocity head is the same for both filaments. The initial velocity head of a′a′ is greater than that of aa, since aa is closer to the wall. At a specific point downstream, the velocity of streamline aa becomes zero; however, the velocity of other streamlines farther from the wall still remain positive. This point is denoted by A in Figure 28(C). As shown, beyond point A the velocity nearest the wall will change sign and a backflow situation is created between aa and the wall. In this region we then see a separation of the boundary layer. Point A is referred to as the *separation point.* The dashed line AB denotes the edge of this fluid backflow region where the fluid velocity changes sign, and is referred to as the *line of zero tangential velocity.* Again, the vortices formed between the wall and the separated fluid stream beyond the separation point result in large form-frictional losses.

The situations described in Figure 28 can cause boundary layer separation in both laminar and turbulent flow. In the latter, the separation point generally occurs at a distance much farther downstream than in laminar flow.

Frictional losses for these cases are not readily estimated, and usually empirical data or correlations must be used. Section 2 outlines calculation methods for estimating friction losses for such cases.

ISOTROPIC TURBULENCE

Isotropic turbulence refers to a condition in which the mean squares of all three of the velocity fluctuations are equal to each other. That is,

$$u_{x_{rms}} = u_{y_{rms}} = u_{z_{rms}} \tag{172}$$

In many practical applications of turbulence isotropy does not exist; however, there is a tendency for turbulence to approach isotropic conditions. As an example, at the edge of a boundary layer the mean velocity gradient locally is zero.

The degree of isotropic turbulence depends on the size spectrum of eddies. Qualitatively, we can think of large-, medium- and small-sized eddies existing in the flow. Large eddies have a long life expectancy and contain as much as one-fourth of the total kinetic energy of isotropic turbulence. Medium-sized eddies are associated with the largest contribution to kinetic energy. With small eddies, the dissipation by viscous effects increases as their size diminishes; although their total energy contribution is small, they undergo continuous reenergizing due to momentum transfer from larger eddies.

Davies [85] and Kolmogoroff [86] give relationships for estimating the frequencies of different size eddies based on the Blasius and Prandtl relations.

For the largest size eddies in pipe flow,

$$\phi \cong 0.4\, u_{avg}\, (N_{Re})^{-1/8}/D \tag{173}$$

where u_{avg} is the average or mean fluid velocity.

For medium-sized energy-containing eddies,

$$\phi = 4U_{avg}/D \tag{174}$$

And for the smallest size eddies,

$$\phi = 0.06U_{avg}(N_{Re})^{0.56}/D \tag{175a}$$

or

$$\phi = \tau_w [17/N_{Re}^{0.44}] \tag{175b}$$

Several properties of eddies found in pipe flow are given in Table 10.

TURBULENCE MEASUREMENT

Hot-wire and hot-film anemometry are the most reliable techniques developed to date for studying the microstructures of turbulent fluid flows. Due to its operating principle, anemometry techniques are applicable to determining other flow properties such as density, composition, temperature, pressure, etc.

Table 10. Typical Properties of Turbulent Eddies for Water Flow in Pipes[a]

Type Eddy	Length (mm)	Fluctuation Velocity (mm/s)	Eddy Reynolds Number	Eddy Frequency, ϕ (Hz)
Largest	25	90	2000	3
Prandtl	3.8	90	370	23
Energy-Dissipating	0.12	65	9	450
Energy-Containing	0.56	90	54	144
Kolmogoroff	0.025	37	1	1300

[a]Values are those reported by Davies [85].

Anemometry is based on the measurement of convective heat losses from a heated sensing element to the surrounding fluid. The rate of heat loss is a function of the sensor geometry and dimensions, and the temperature, pressure, density and thermal properties of the flowing fluid. If only one fluid property undergoes change, then the measured heat loss provides a direct measure of the quantity under study.

The sensing element (commonly referred to as the sensor) consists of either a fine metal wire suspended between two metal posts, or a thin metal film deposited on an insulating substrate. The sensor is positioned on the probe tip, from which a cable runs to the anemometer. Heating is supplied by an electric current from the anemometer. In simple terms, an anemometer consists of a Wheatstone bridge with one arm of the bridge comprised of the sensor and an amplifier. The current through the bridge heats the sensor while the amplifier output voltage provides a measure of the sensor's heat loss. Section 2 provides further information on sensor geometries.

The heat loss from the sensor can be expressed in terms of the Nusselt number:

$$N_{Nu} = \frac{(q/A)\ell'}{k_f \Delta T} \tag{176}$$

where ℓ' = length of the sensor body
q/A = heat flux
k_f = thermal conductivity of the fluid
ΔT = temperature difference between the sensor surface temperature and the fluid temperature

At relatively low fluid velocities the Nusselt number is a function of the Grashof number, N_{Gr}, Reynolds number, N_{Re}, and the Prandtl number, N_{Pr}. That is,

$$N_{Nu} = f(N_{Gr}, N_{Re}, N_{Pr}) \tag{177}$$

where N_{Gr} and N_{Pr} are defined as

$$N_{Gr} = \frac{g\beta(\Delta T)d^3}{\nu^2} \tag{178}$$

$$N_{Pr} = \nu/\alpha \tag{179}$$

where d = diameter of cylindrical body (i.e., the sensor)
g = gravitational acceleration
β = thermal expansion coefficient of the fluid
ν = kinematic viscosity of the fluid
α = thermal diffusivity of the fluid

Equation 177 applies to the case where natural convection is important in comparison to forced convection.

At high fluid velocities forced convection predominates. In this case, the Nusselt number is only a function of N_{Re} and N_{Pr}:

$$N_{Nu} = f(N_{Re}, N_{Pr}) \tag{180}$$

For compressible fluids at high velocities the Mach number also contributes to the heat transfer process:

$$N_{Nu} = f(N_{Re}, N_{Pr}, N_{Ma}) \ ,$$

$$\text{for } N_{Ma} > 0.3 \tag{181}$$

The equations of convective heat transfer are for the most part too complex to solve for the relationship between fluid velocity and heat transfer by direct means. One of the simplest cases is that of two-dimensional heat transfer from a cylinder (wire) immersed in an imcompressible flow stream. The relationship for heat transfer is given by the following:

$$q = k_f \ell' \left[1 + \sqrt{\frac{2\pi \rho C_p d u}{k_f}} \right] \Delta T \tag{182}$$

In dimensionless form this equation becomes

$$N_{Nu} = (T_m/T_0)^{0.17}(0.24 + 0.56 N_{Re}^{0.45}) \tag{183}$$

Equation 183 is applicable to air flow in the Reynolds range $0.02 < N_{Re} < 44$. T_m is the mean temperature between the fluid and wire, i.e., $T_m = \frac{1}{2}(T_s + T_0)$, where T_s, T_0 are the wire surface and surrounding fluid temperatures, respectively. When thermal equilibrium is reached, the rate of heat loss, q, from the sensor becomes equal to the power, $\mathcal{P}$, supplied to the sensor:

$$q = \mathcal{P} = I^2 \Omega_w \tag{184}$$

where Ω_w is the resistance of the sensor (in this case the wire).

The *overheat ratio* is defined as follows:

$$\psi = \frac{\Omega_w - \Omega_0}{\Omega_0} \tag{185}$$

where Ω_0 is the wire resistance at the fluid temperature.

In a given fluid of specific temperature and pressure, the overheat ratio is applied, defining the following relationship between supplied power and fluid velocity:

$$\psi I^2 = A' + B'(u)^{n'} \tag{186}$$

Parameters A′, B′ and exponent n′ are constants that must be determined through probe calibration. Typical calibration methods for probes employed in airflow studies are illustrated in Figure 29.

When the sensor is immersed in a flowing stream it gives response to fluctuations in the flow. In open-loop systems the current to the sensor is maintained constant, producing a response that is determined primarily by the probe itself. In the closed-loop system, the probe forms part of a negative feedback path that maintains a constant resistance. In this latter case the temperature of the sensor is maintained constant and the response is then

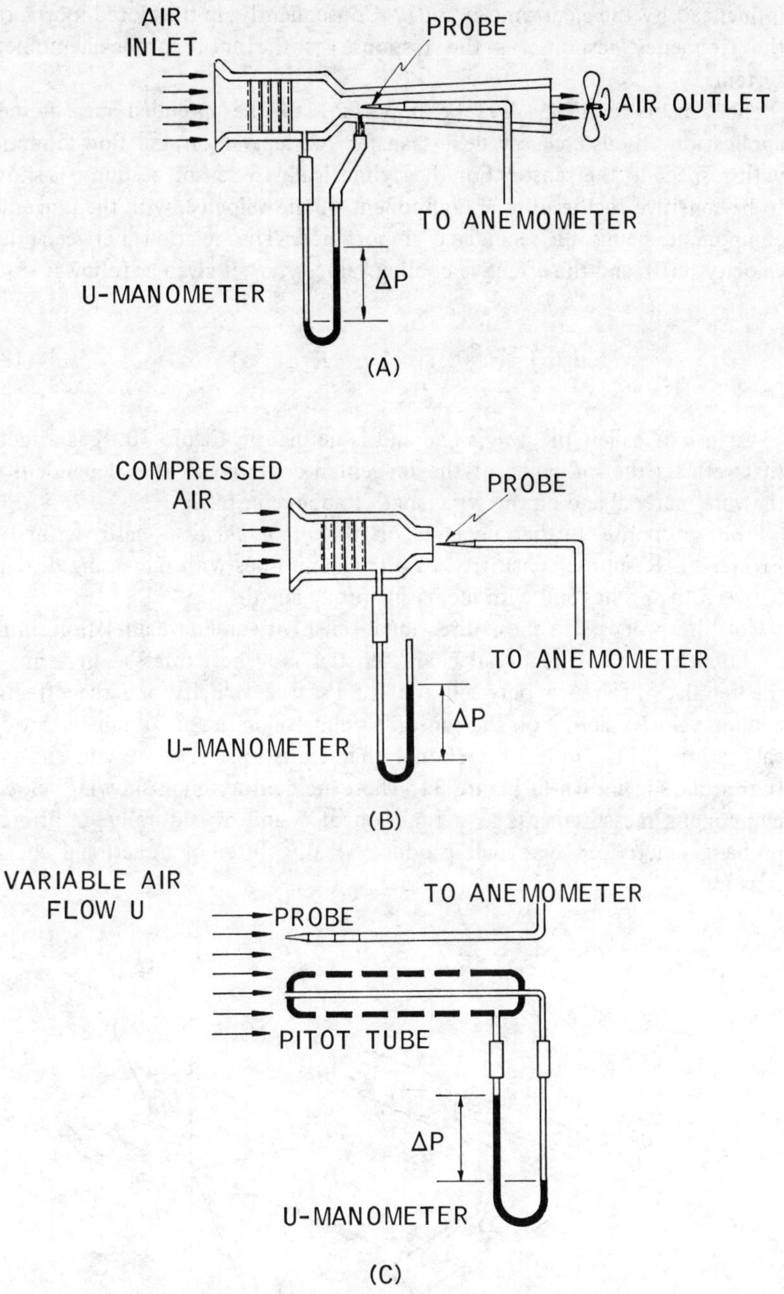

Figure 29. Various methods for calibrating hot-wire probes.

influenced by the electronic circuitry. Consequently, in the closed-loop mode the frequency response is the response of the entire probe-anemometer system.

The sensor geometry is very important to the intended measurement application. It also causes heat transfer to depend on the flow direction with respect to the sensor. For the cylindrical wire sensor, response is shown to be sensitive to the normal component of the velocity, with the tangential component being of secondary importance. The relation between fluid velocity, u(0), and the effective cooling velocity, u_c, is given as follows:

$$u_c^2(\alpha) = (u^2(0))\,(\cos^2\alpha + K^2 \sin^2\alpha) \tag{187}$$

Angle α is called the *yaw angle* and is defined in Figure 30. K is a factor that reflects the influence of the tangential component. It is dependent on the wire material and on the wire length to diameter ratio.

Probe response further depends on the *pitch angle,* θ, also defined in Figure 30. Response sensitivity to pitch diminishes with increasing distance between the prongs and with increasing prong length.

Hot-film probes also show directional sensitivity and have a relation similar to Equation 185. Directional characteristics may be expressed in terms of the relative difference between the fluid's true velocity and the effective cooling velocity acting on the sensor. Cooling velocities are obtained through calibration [87]. Typical directional characteristics for a hot-wire and hot-film probe are shown in Figure 31. These are contour maps showing velocity component measurements as a function of θ and α. Naturally, a different probe geometry or type will produce greatly different directional characteristics.

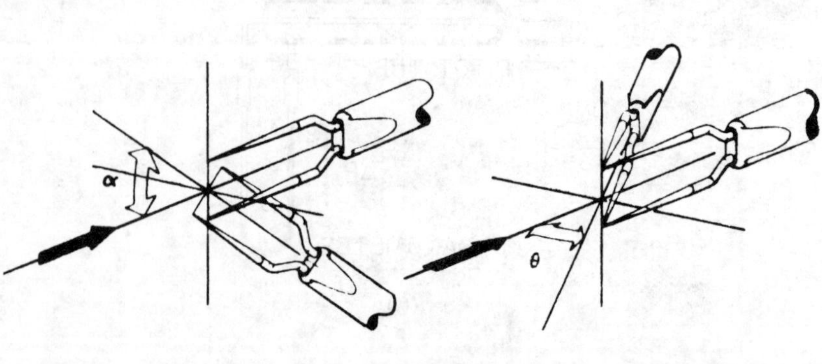

Figure 30. Defines the angles α and θ between the flow direction and probe sensor.

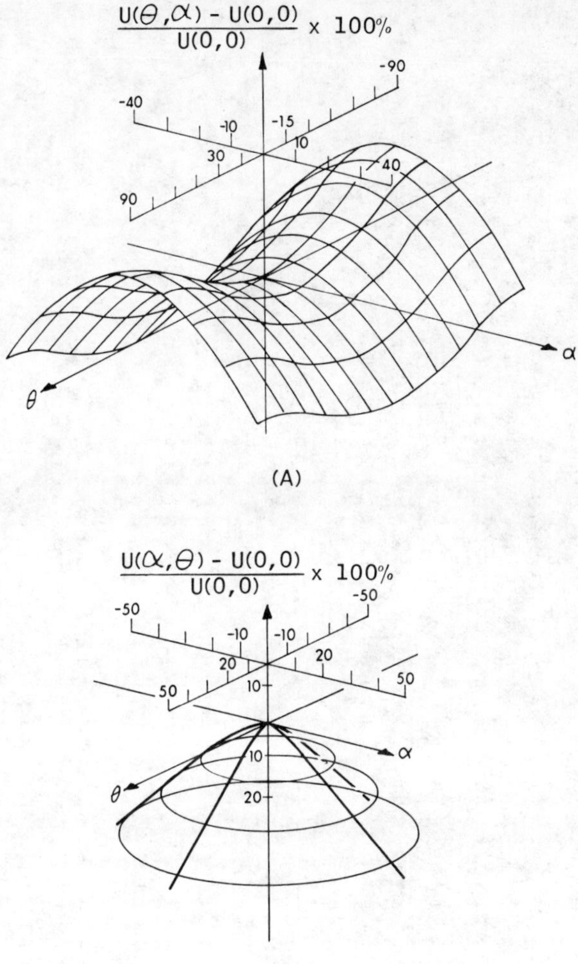

Figure 31. Typical directional characteristics for hot-wire and hot-film probes: (A) One type of hot-wire probe. (B) Contour for conical-shaped hot-film probe.

Hot-wire and hot-film probes have been employed for analyzing velocity fluctuations and, components, and for wall shear measurement in a variety of flow situations ranging from simple pipe flow to complex three-dimensional flow fields. Since turbulence measurement applications are specific, rather than describe some of these, references [88-171] have been compiled for the reader's use. Further discussions on anemometry techniques are given in Section 2.

CHAPTER 1.7

INTRODUCTION TO DIMENSIONAL ANALYSIS

For many engineering problems, theoretical equations describing a phenomenon are either not known or are in a form that cannot be solved readily. Often, since theoretical solutions are not possible, we must resort to empirical correlations for design and analysis. For problems involving fluid flow, these correlations may be developed in model studies in which flow conditions, fluid properties and size of the test systems may be different. This chapter provides an introduction to the theory of dimensional analysis. Covered are the use of dimensional analysis in the development of empirical correlations and its application to the design and analysis of engineering systems.

The principles of dimensional analysis may be applied to numerous engineering problems involving heat and/or mass transfer and are not limited to fluid flow applications.

PRINCIPLES OF DIMENSIONAL ANALYSIS

Dimensional analysis is based on the premise that the equation describing a phenomenon, even when the equation is unknown, must be dimensionally consistent. That is, different variables appear in an equation such that each term has the same dimensions. By dividing each term by any one term, the equation describing the phenomenon can be transformed into a generalized relation:

$$f(\Pi_1, \Pi_2, \Pi_3, \ldots \Pi_p) = 0 \tag{188}$$

where $\Pi_1, \Pi_2, \ldots \Pi_p$ represent dimensionless groups of variables.

The advantages of dimensional analysis are illustrated in the following example. In an earlier chapter we examined the drag force on a smooth spherical body submerged in a stream of incompressible fluid. Let us now determine the form of the drag force equation. From experimental data it is known that the drag force, F, depends on the sphere diameter, D, the fluid's viscosity, μ, density, ρ, and velocity, u:

$$F = f(D, \mu, \rho, u) \tag{189}$$

For the above function to be dimensionally consistent, Equation 189 must be reducible to the following form (the exact procedure for this reduction will be described shortly):

$$\frac{F}{\rho u^2 D^2} = f\left(\frac{\rho u D}{\mu}\right)$$

or

$$\Pi_1 = f(\Pi_2) \tag{190}$$

The groups $\Pi_1 (= F/\rho u^2 D^2)$ and $\Pi_2 (= \rho u D/\mu)$ are dimensionless. Observe that F has the dimensions of MLT^{-2}; similarly, $\rho \equiv ML^{-3}$, $u \equiv LT^{-1}$, $\mu \equiv ML^{-1}T^{-1}$ and $D \equiv L$, where M, L and T denote mass, length and time, respectively.

The drag force problem is greatly simplified by the above analysis. Note that a brute force approach to determining a drag force equation would involve correlating five variables in Equation 189. Instead, a correlation (Equation 190) was determined from a relation between only two dimensionless groups. For example, a plot of Π_1 vs Π_2 may be constructed from experimental data. From such a plot F is readily calculated for any sphere if D, μ, ρ and u are known. Such a plot is shown in Figure 181 of Section 2.

Dimensional analysis provides insight into how different variables in a problem relate to each other. By lumping the variables into groups, the number of parameters in the analysis is reduced. Dimensional analysis does not, however, provide a fundamental equation defining the mechanism of a phenomenon. It is strictly a means of developing engineering correlations. The dimensionless groups, which are always fewer in number than the original variables, still must be related to each other on the basis of additional theoretical knowledge or experimental observation.

Application of dimensional analysis is based on the assumption that the starting equation is dimensionally consistent. It may not always be assumed that any arbitrary equation is dimensionally consistent unless the equation is fundamental, that is, the equation contains all the variables required in its analytical derivation. Again, in the case of drag force on a sphere, if the fluid is water it can be argued that ρ and μ may be excluded since they are constant for water; then,

$$F = f(D,u) \tag{191}$$

Dimensional analysis cannot be applied to Equation 191 since it is not fundamental or dimensionally consistent. The variable F contains the dimensions of mass, whereas neither of the variables D and u does. Dimensional analysis alone cannot tell which variables influence the phenomenon and should be included in the formulation of the problem. It also fails to exclude variables that are irrelevant to the phenomenon. For drag force, experimental and/or theoretical knowledge is needed to state that F, D, u, μ and ρ form a complete set of variables. Extraneous dimensionless groups would appear in the solution if the fluid surface tension, which is known not to influence the drag force, was included in the analysis by mistake.

SYSTEMATIC DETERMINATION OF DIMENSIONLESS GROUPS

After establishing the variables affecting a phenomenon, the dimensionless groups required must be decided on. More dimensionless groups can be formed by combining the variables in different ways. However, only the requisite dimensionless groups, which form a complete set, are needed to describe the phenomenon. Requirements for a set of dimensionless groups to be complete are as follows:

1. The groups within the set should be independent of each other such that no group within the set is a product of powers of the other groups.
2. All possible dimensionless groups outside the complete set can be obtained as products of powers of the groups within the complete set.

Thus, many complete but different sets of dimensionless groups can be constructed from the large number of possible variables. The phenomenon may be described by any one of these complete sets. The selection of specific dimensionless groups to form a complete set depends on the ease with which selected groups can be regulated or varied in the experimental investigation. Often the chosen groups have a physical significance.

A rigorous method of computing the number of dimensionless groups in a complete set and determining a complete set of dimensionless groups is illustrated below. Basic knowledge of determinants and matrices is needed. Proof of the theorems involved in obtaining a solution are given by Langhaar [172], or may be found in books on advanced algebra.

Example: As noted in an earlier chapter, the pressure drop of an incompressible fluid flowing in a straight section of pipe has the following functional dependence:

$$\Delta P = f(L,\mu,\epsilon,\rho,D,u) \tag{192}$$

where L = pipe length
μ = fluid viscosity
ϵ = pipe roughness
ρ = fluid density
D = pipe diameter
u = fluid velocity

We wish to determine the number of dimensionless groups required to describe the phenomenon

$$f(\Delta P, L, \mu, \epsilon, \rho, D, u) = 0 \tag{193}$$

and determine a complete set of dimensionless groups.

Any dimensionless group formed from the variables has the form

$$\Pi = (\Delta P)^{k_1}(L)^{k_2}(\mu)^{k_3}(\epsilon)^{k_4}(\rho)^{k_5}(D)^{k_6}(u)^{k_7} \tag{194}$$

The dimensions of Π are, therefore,

$$\Pi \equiv (ML^{-1}T^{-2})^{k_1}(L)^{k_2}(ML^{-1}T^{-1})^{k_3}(L)^{k_4}(ML^{-3})^{k_5}(L)^{k_6}(LT^{-1})^{k_7}$$

The only restriction on the exponents $k_1 \ldots k_7$ is that their numerical values be such that Π is dimensionless.

Thus,

$$\left.\begin{aligned} k_1 + k_3 + k_5 &= 0 \\ -k_1 + k_2 - k_3 + k_4 - 3k_5 + k_6 + k_7 &= 0 \\ -2k_1 - k_3 - k_7 &= 0 \end{aligned}\right\} \tag{195}$$

The coefficients of the various k's in Equation 194 are the same as the elements of the following matrix created by writing the dimensions of the seven variables in different columns.

	k_1	k_2	k_3	k_4	k_5	k_6	k_7
	ΔP	L	μ	ϵ	ρ	D	u
M	1	0	1	0	1	0	0
L	−1	1	−1	1	−3	1	1
T	−2	0	−1	0	0	0	−1

The largest nonzero determinant of the above dimensional matrix is of order three. The matrix thus has a *rank 3*. A nonzero determinant of order three exists, as shown by constructing a 3 X 3 determinant from the last three columns of the matrix

$$\begin{vmatrix} 1 & 0 & 0 \\ -3 & 1 & 1 \\ 0 & 0 & -1 \end{vmatrix}$$

A rank 3 implies that all three equations represented by Equation 195 are linearly independent. That is, none of the three equations is redundant (i.e., no one equation can be obtained by a linear combination of the other two). A rank 2 implies that one of the equations is dependent and therefore should be ignored.

The number of dimensionless groups in a complete set is equal to the number of variables minus the rank of their dimensional matrix. In the above illustration, since there are seven variables and the rank of the matrix is 3, the number of dimensionless groups in a complete set is four.

The above conclusion could be reached without resorting to matrix algebra by the use of Buckingham's theorem [173], which states that the number of dimensionless groups is equal to the number of variables minus the number of dimensions in the problem. Normally, the rank of the matrix is the same as the number of dimensions in the problem. Buckingham's theorem is only a convenient guideline, however, and should be used with caution.

Equation 195 comprises three independent equations in seven unknowns. Since the number of unknowns is larger than the number of equations, the system is undetermined. Hence, we are left with the option of solving three unknowns in terms of the remaining four.

Solving unknowns k_5, k_6 and k_7 in terms of k_1, k_2, k_3 and k_4 from Equation 195, by elimination,

$$\left.\begin{aligned} k_5 &= -k_1 \qquad\qquad -k_3 \\ k_6 &= \qquad\quad -k_2 \quad -k_3 \quad -k_4 \\ k_7 &= -2k_1 \qquad\quad\; -k_3 \end{aligned}\right\} \tag{196}$$

Obviously, many values of k_5, k_6 and k_7 are possible depending on the values of k_1, k_2, k_3 and k_4. If we assign the value $k_1 = 1$ and $k_2 = k_3 = k_4 = 0$, then the solution is $k_5 = -1$, $k_6 = 0$ and $k_7 = -2$. If $k_2 = 1$ and $k_1 = k_3 = k_4 = 0$, then the solution is $k_5 = 0$, $k_6 = -1$, and $k_7 = 0$. If $k_3 = 1$ and $k_1 = k_2 =$

$k_4 = 0$, then $k_5 = -1$, $k_6 = -1$ and $k_7 = -1$. Similarly, if $k_4 = 1$ and $k_1 = k_2 = k_3 = 0$, then the solution is $k_5 = 0$, $k_6 = -1$ and $k_7 = 0$.

The four different solutions obtained above can be arranged in a matrix:

	ΔP	L	μ	ϵ	ρ	D	u
	k_1	k_2	k_3	k_4	k_5	k_6	k_7
Π_1	1	0	0	0	−1	0	−2
Π_2	0	1	0	0	0	−1	0
Π_3	0	0	1	0	−1	−1	−1
Π_4	0	0	0	1	0	−1	0

Each row of the matrix contains values of the seven exponents in a dimensionless group. Hence,

$$\Pi_1 = \frac{\Delta P}{\rho u^2}, \quad \Pi_2 = \frac{L}{D}, \quad \Pi_3 = \frac{\mu}{D\rho u}, \quad \text{and} \quad \Pi_4 = \frac{\epsilon}{D}$$

These four dimensionless groups form a complete set, since ΔP, L, μ and ϵ occur in only one of the groups and, hence, none of the groups can be a product of powers of the other three. This may be written as

$$\frac{\Delta P}{\rho u^2} = f\left(\frac{L}{D}, \frac{\mu}{\rho D u}, \frac{\epsilon}{D}\right) \tag{197}$$

Furthermore, since ΔP is expected to be a linear function of L, and since D is contained explicitly in both Π_3 and Π_4, L/D can be removed outside the functionality:

$$\frac{\Delta P}{\rho u^2} = \frac{L}{D} f\left(\frac{\mu}{\rho D u}, \frac{\epsilon}{D}\right) \tag{198}$$

Note that the solutions to Equation 196 were obtained by assigning numerical values to k_1, k_2, k_3 and k_4 in a manner that only one of the k's had a nonzero value at a time. This results in ΔP, L, μ and ϵ values occurring only once, and the four dimensionless groups become independent. The seven variables were deliberately arranged in a particular order and the solution of k_5, k_6 and

k_7 obtained in terms of k_1, k_2, k_3 and k_4 so as to obtain the four familiar dimensionless groups $\Delta P/\rho u^2$, L/D, ϵ/D and $\rho Du/\mu$. It was not essential that a solution be obtained in terms of k_1 to k_4. Any three unknowns could be solved for in terms of the remaining four. The only requirement is that the coefficients of the three k's being solved for constitute a nonzero determinant in the dimensional matrix. Thus, a solution could have been obtained for k_1, k_2 and k_3 in terms of k_4, k_5, k_6 and k_7 since the coefficients of k_1, k_2 and k_3 in the dimensional matrix also yield a nonzero determinant. As previously noted, a phenomenon may be described by alternate sets of dimensionless groups.

MODEL TESTING

Often, information about the performance of an engineering system, which may not yet be built, is obtained by studying the performance of a small-scale replica of the system. Model tests are of use when the necessary information about a process or phenomenon is not known or cannot be predicted from theory, and when tests on the actual system are either too expensive or not possible. An example is the design of a new airplane wing on the basis of tests on a model wing in a wind tunnel. Another example is in the scaleup of a chemical process unit such as a slurry reactor. Many times, the purpose of model tests is limited to determining certain important aspects of the phenomenon since it may be impractical to build a model that would represent the phenomenon in its entirety. For example, in the design of a new airplane wing, a key objective of a model study could be to predict the conditions at which flutter occurs.

For a model to be completely similar to the prototype, it is essential that dimensionless groups forming a complete set have the same numerical values for both the model and prototype. This is possible only if geometric similarity exists (i.e., the ratio of all corresponding dimensions between the model and the prototype are the same). To achieve similarity, physical properties and experimental conditions for the model test must be such that identical values of dimensionless groups are obtained between the model and the prototype. Very often it is not possible to establish physical properties and test conditions that provide identical dimensionless groups. In such cases, the dimensionless groups that have a greater influence on the phenomenon are maintained identical between the model and the prototype while the less important groups are allowed to differ, thus giving an approximate solution. If the model simulation is directed at some specific aspects and not concerned with the entire phenomenon, then only those dimensionless groups having greatest influence need be maintained identical.

INSPECTION ANALYSIS OF DIFFERENTIAL EQUATIONS AND ANALOGIES

A procedure was described for obtaining dimensionless groups from the variables involved in a phenomenon. If the differential equations governing the phenomenon are also known, the dimensional groups and the necessary conditions for similarity can be obtained more conveniently through inspection analysis.

Inspection analysis is a method of determining which conditions are necessary for similarity to exist in geometrically similar systems. It involves converting the governing differential equations into dimensionless form by the use of a characteristic length, time, velocity, temperature, etc., and determining the conditions under which the dimensionless differential equations yield identical solutions.

Again, consider flow around a sphere. The flow is governed by the equations of continuity and motion. If constant density and viscosity are assumed, these equations are

$$\frac{\partial u}{\partial x} + \frac{\partial v}{\partial y} + \frac{\partial w}{\partial z} = 0 \tag{199}$$

and

$$\rho\left(\frac{\partial u}{\partial t} + u\frac{\partial u}{\partial x} + v\frac{\partial u}{\partial y} + w\frac{\partial u}{\partial z}\right)$$

$$= -\frac{\partial P}{\partial x} + \mu\left(\frac{\partial^2 u}{\partial x^2} + \frac{\partial^2 u}{\partial y^2} + \frac{\partial^2 u}{\partial z^2}\right) + \rho g_x \tag{200}$$

plus two additional equations for the y- and z-components of velocity (refer back to Chapter 1.4). Recall that u, v and w are the x-, y- and z-components of fluid velocity, and ρ and μ are fluid density and viscosity, respectively. P represents the pressure and g_x is the x-component of acceleration due to gravity. Equations 199 and 200 can be made dimensionless by using quantities that are characteristic of the system. For flow around a sphere, convenient characteristic quantities are the sphere diameter, D, and the approach velocity of the fluid, u_0. The choice of characteristic quantities should be specified carefully. Using D and u_0 we may define the following dimensionless variables.

$$\tilde{u}_x = \frac{u_x}{u_0}, \quad \tilde{v} = \frac{v}{v_0}, \quad \tilde{w} = \frac{w}{w_0}$$

$$\tilde{x} = \frac{x}{D}, \quad \tilde{y} = \frac{y}{D}, \quad \tilde{z} = \frac{z}{D}, \quad \tilde{t} = \frac{tu_0}{D}$$

and

$$\tilde{P} = \frac{P - P_0}{\rho u_0^2} \tag{201}$$

where P_0 is a reference pressure. Substituting for u, v, w, x, y, z, t and P from Equation 201 into Equations 199 and 200, the following equations are obtained:

$$\frac{\partial \tilde{u}}{\partial \tilde{x}} + \frac{\partial \tilde{v}}{\partial \tilde{y}} + \frac{\partial \tilde{w}}{\partial \tilde{z}} = 0 \tag{202}$$

and

$$\frac{\rho u_0^2}{D}\left(\frac{\partial \tilde{u}}{\partial \tilde{t}} + \tilde{u}\frac{\partial \tilde{u}}{\partial \tilde{x}} + \tilde{v}\frac{\partial \tilde{u}}{\partial \tilde{y}} + \tilde{w}\frac{\partial \tilde{u}}{\partial \tilde{z}}\right)$$

$$= -\frac{\rho u_0^2}{D}\frac{\partial \tilde{P}}{\partial \tilde{x}} + \mu\frac{u_0}{D^2}\left(\frac{\partial^2 \tilde{u}}{\partial \tilde{x}^2} + \frac{\partial^2 \tilde{u}}{\partial \tilde{y}^2} + \frac{\partial^2 \tilde{u}}{\partial \tilde{z}^2}\right) + \rho g_x \tag{203}$$

Multiplying the last equation by $D/\rho u_0^2$ yields

$$\left(\frac{\partial \tilde{u}}{\partial \tilde{t}} + \tilde{u}\frac{\partial \tilde{u}}{\partial \tilde{x}} + \tilde{v}\frac{\partial \tilde{u}}{\partial \tilde{y}} + \tilde{w}\frac{\partial \tilde{u}}{\partial \tilde{z}}\right)$$

$$= -\frac{\partial \tilde{P}}{\partial \tilde{x}} + \frac{\mu}{\rho u_0 D}\left(\frac{\partial^2 \tilde{u}}{\partial \tilde{x}^2} + \frac{\partial^2 \tilde{u}}{\partial \tilde{y}^2} + \frac{\partial^2 \tilde{u}}{\partial \tilde{z}^2}\right) + \frac{Dg_x}{u_0^2} \tag{204}$$

Observe that each term in Equations 202 and 204 is dimensionless and the variables are lumped together into dimensionless groups $\rho u_0 D/\mu$ and $u_0^2/g_x D$. The first dimensionless group, $\rho u_0 D/\mu$, called the Reynolds number, is a ratio of the variables in the first and third terms of Equation 203, which represent the inertial and viscous forces, respectively, in the overall momentum balance. Thus, physically, the Reynolds number signifies the relative importance of inertial to viscous forces. A Reynolds number of less than unity, for example, means that the viscous forces are large in comparison to the inertial forces. The second dimensionless group, u_0^2/Dg_x, is called the Froude number and represents the relative importance of the inertial to gravity forces.

Now, if the variables ρ, u_0, D, μ and g in two systems (the model and prototype) are such that the Reynolds and the Froude numbers for the two systems are the same, then both systems are described by identical dimensionless Equations 202 and 204. If the dimensionless initial and boundary conditions for the two systems are also the same (which is feasible only if geometric similarity exists), then the solution of the two systems also must be the same. Inspection analysis thus shows that specification of the two dimensionless groups, the Reynolds and Froude numbers, completely defines the system. These two dimensionless groups therefore form a complete set. The procedure described above can be used to obtain a complete set of dimensionless groups for any phenomenon if the governing differential equations are known.

A knowledge of the governing differential equations also helps to reveal analogies between different phenomena and design correlations. The dimensionless differential equations describing a large number of problems in heat, mass and momentum transport are exactly of the same form. For example, the differential equations and the initial and boundary conditions for a specific heat transfer process may be obtained by changing the notation, and certain dimensionless groups in the equations for an analogous problem in mass or momentum transfer. If, for two different problems, the dimensionless differential equations and the initial and boundary conditions are identical in form, the solutions also must be the same. For this reason, many analogies exist among heat, mass and momentum transfer correlations. If the correlation for one process is known, the correlation for an analogous process may be obtained by an appropriate change in notation.

IMPORTANT DIMENSIONLESS GROUPS AND THEIR SIGNIFICANCE

The Reynolds and Froude numbers, obtained in the foregoing example, were shown to represent the ratios of different forces. Other physically meaningful dimensionless groups are obtained depending on the phenomenon under investigation. Catchpole and Fulford [174] and Boucher and Alves [175] provide a listing of the dimensionless groups that occur in the engineering problems. The most common of these groups and their significance are listed in Table 11. A more in-depth coverage of dimensional analysis is given in the literature [176-179].

Table 11. Important Dimensionless Groups and Their Significance

Dimensionless Group	Symbol	Definition	Significance; Ratio of
Reynolds Number	N_{Re}	$\frac{\rho uL}{\mu}$	$\frac{\text{Inertial force}}{\text{Viscous force}}$
		ρ = fluid density u = fluid velocity μ = fluid viscosity L = characteristic dimension	
Froude Number	N_{Fr}	$\frac{u^2}{Lg}$	$\frac{\text{Inertial force}}{\text{Gravitational force}}$
Euler Number	N_{Eu}	$\frac{p}{\rho u^2}$	$\frac{\text{Pressure}}{2 \times \text{velocity head}}$
		p = pressure	
Mach Number	N_{Ma}	$\frac{u}{u_c}$	$\frac{\text{Fluid velocity}}{\text{Velocity of sound}}$
Weber Number	N_{We}	$\frac{\rho Lu^2}{\sigma}$	$\frac{\text{Inertial force}}{\text{Surface tension force}}$
		σ = surface tension	
Drag Coefficient	C_D	$\frac{(\rho - \rho')Lg}{\rho u^2}$	$\frac{\text{Gravitational force}}{\text{Inertial force}}$
		ρ = density of object ρ' = density of surrounding fluid	

Table 11, continued

Dimensionless Group	Symbol	Definition	Significance; Ratio of
Fanning Friction Factor	f	$\frac{D}{L} \cdot \frac{\Delta P}{2\rho u^2}$ D = pipe diameter L = pipe length	$\frac{\text{Wall shear stress}}{\text{Velocity head}}$
Nusselt Number (heat transfer)	N_{Nu}	$\frac{hL}{k_f}$ h = heat transfer coefficient k_f = thermal conductivity	$\frac{\text{Total heat transfer}}{\text{Conductive heat transfer}}$
Prandtl Number	N_{Pr}	$\frac{C_p \mu}{k}$ C_p = heat capacity	$\frac{\text{Momentum diffusivity}}{\text{Thermal diffusivity}}$
Peclet Number (heat transfer)	N_{Pe}	$\frac{C_p \rho u L}{k}$ $= N_{Re} N_{Pr}$	$\frac{\text{Bulk heat transport}}{\text{Conductive heat transfer}}$
Grashof Number	N_{Gr}	$\frac{g b^3 \rho^2 \beta \Delta T}{\mu^2}$ β = coefficient of expansion ΔT = temperature difference b = height of surface	$N_{Re} \times \frac{\text{buoyancy force}}{\text{viscous force}}$

Stanton Number	N_{St}	$\frac{h}{\rho u C_p}$ $= N_{Nu}N_{Re}^{-1}N_{Pr}^{-1}$	$\frac{\text{Heat transferred}}{\text{Thermal capacity of fluid}}$
J Factor for Heat Transfer	j_H	$\frac{h}{\rho u C_p}\left(\frac{C_p\mu}{k}\right)^{2/3}$	Proportional to $N_{Nu}N_{Re}^{-1}N_{Pr}^{-1/3}$
Nusselt Number (mass transfer)	N_{Nu}	$\frac{k_c L}{\mathscr{D}}$ k_c = mass transfer coefficient $\mathscr{D}$ = molecular diffusivity	$\frac{\text{Total mass transfer}}{\text{Diffusive mass transfer}}$
Schmidt Number	N_{Sc}	$\frac{\mu}{\rho\mathscr{D}}$	$\frac{\text{Momentum diffusivity}}{\text{Molecular diffusivity}}$
Peclet Number (mass transfer)	N_{Pe}	$\frac{Lu}{\mathscr{D}} = N_{Re}N_{Sc}$	$\frac{\text{Bulk mass transport}}{\text{Diffusive mass transport}}$
J Factor for Mass Transfer	j_D	$\frac{k_c}{u}\left(\frac{\mu}{\rho\mathscr{D}}\right)^{2/3}$	Proportional to $N_{Nu}N_{Re}^{-1}N_{Sc}^{-1/3}$

NOMENCLATURE FOR SECTION 1

Chapter	Symbol	Definition	English Units	SI Units
1.2	a_x, a_y	x- and y-components of acceleration	ft/s^2	m/s^2
	A	Area	ft^2	m^2
	D	Diameter	ft	m
	f	Frictional resistance to flow; friction factor	–	–
	F_i	Force or pressure component	lb_f	N
	F_R	Resultant force, i.e., sum of all forces	lb_f	N
	g	Resultant acceleration of gravity	$32.2\ ft/s^2$	$9.81\ m/s^2$
	g_c	Conversion factor	$32.174\ ft\text{-}lb_m/(lb_f\text{-}s^2)$	1
	H	Fluid static head	ft	m
	m_a, m_f	Weight of body in air and fluid, respectively	lb_m	kg
	M_w	Weight of an equal volume of liquid	lb_m	kg
	p	Local pressure or normal stress	lb_f/ft^2	Pa
	P	Pressure	lb_f/ft^2	Pa
	t	Time	sec	sec
	W	Work	hp-hr	kJ
	X, Y	x- and y-components of force or acceleration	lb_f/lb_m	N/kg
	α	Kinetic energy correction term, see Equation 1	–	–
	θ	Angle	deg	rad
	θ_a, θ_b	Time constants	sec^{-1}	s^{-1}
	ρ	Density	lb_m/ft^3	kg/m^3

Nomenclature for Section 1, continued

Chapter	Symbol	Definition	English Units	SI Units
	σ	Interfacial or surface tension	dyne/cm	mN/m
	ϕ	Contact angle; also defined in Equation 17	deg	rad
1.3	a′	Constant in Equation 70	–	–
(new symbols introduced)	a	Constant in Equation 49; also acceleration (ft/s^2)	–	–
	A,B	Viscosity model constants, see Table 9 for definitions	Variable	–
	c	Constant in Equation 61	$(lb_f/ft^2)^{0.2}$	$Pa^{n'}$
	$\bar{C}$	Average velocity vector	ft/s	m/s
	dp	Molecular diameter	in.	microns
	D	Tube diameter	ft	m
	E	Activation energy per mole; also total kinetic energy associated with molecular movement	Btu/lb-mol	J/kg-mol
	$\tilde{G}$	Molar free energy of activation	Btu	J
	h	Planck's constant	6.625×10^{-27} erg-sec	same
	ΔH_{vap}	Enthalpy of vaporization	Btu/lb-mol	kJ/kg-mol
	K	Consistency index, see Equation 63	$lb_m\text{-}s^{n-2}\text{-}ft^{-1}$	$kg\text{-}s^{n-2}\text{-}m^{-1}$
	KE	Kinetic energy per unit mass	Btu/lb_m	J/kg
	m	Dimensionless constant in Equation 60	–	–
	m_p	Mass of particle	lb_m	kg
	M_{ij}	Molecular weights of species i and j	lb_m/lb-mol	kg/kg-mol
	n	Number molecules per unit volume; also, number of chemical species in mixture	–	–
	n	Flow behavior index, see Equation 63	–	–
	N	Number molecules	–	–
	N_A	Avogadro's number	–	–
	P_r	Reduced pressure	–	–
	P_v	Vapor pressure	$lb_f/in.^2$	Pa
	P_∞	Defined in Equation 37	–	–
	$\dot{Q}_{vap}$	Heat of vaporization	Btu/lb_m	kJ/kg

Nomenclature for Section 1, continued

Chapter	Symbol	Definition	English Units	SI Units
	R	Gas law constant	$(lb_m\text{-}ft^2)/$ $s^2\text{-}°R\text{-mol}$	$(kg\text{-}m^2)/$ $s^2\text{-}°K\text{-mol}$
	STP	Standard temperature and pressure	–	–
	T	Temperature, absolute	°R	°K
	T_r	Reduced temperature	–	–
	u	Velocity	ft/s	m/s
	U	Average velocity	ft/s	m/s
	U_{vap}	Energy of vaporization	Btu	J
	V	Volume	ft^3	m^3
	$V_0, \hat{V}_0$	Reference volumes	ft^3	m^3
	$\hat{V}$	Molar volume	ft^3/lb-mol	m^3/g-mol
	$\hat{V}_c$	Critical molar volume	ft^3/lb-mol	m^3/g-mol
	x_i	Mole fraction of species i	–	–
	Δx	Width of potential energy barrier	in.	cm
	β	Coefficient of compressibility	atm^{-1}	Pa^{-1}
	γ	Strain	sec^{-1}	s^{-1}
	δ	Distance between successive layers of molecules	in.	cm
	η	Plasticity coefficient	$lb_f\text{-}s/ft^2$ or cp	mPa-s
	η'	Rheological parameter	$lb_f^{m-1}\text{-}lb_m\text{-}$ $ft^{1-2m}\text{-}s^{-1}$	Variable
	κ	Coefficient of thermal expansion	in./(in.-°F)	mm/(mm-°K)
	K	Boltzmann constant	$lb_m/s^3\text{-}°R^4$	$kg/s^3\text{-}°K^4$
	λ	Mean free path	ft	m
	μ	Viscosity	$lb_f\text{-}s/ft^2$ or cp	mPa-s
	μ_c	Critical viscosity	$lb_f\text{-}s/ft^2$ or cp	mPa-s or micropoise
	μ_L	Parameter defined in Equation 61	–	–
	$\mu_{0,\infty}$	Apparent viscosities at zero and infinite shears	$lb_f\text{-}s/ft^2$ or cp	mPa-s
	τ	Shear stress	$lb_f/in.^2$	Pa
1.4	C_1, C_2	Constants of integration	Variable	
(new symbols introduced)	g_x, g_y, g_z	x-, y- and z-components of acceleration	ft/s^2	m/s^2
	L	Length	ft	m

Nomenclature for Section 1, continued

Chapter	Symbol	Definition	English Units	SI Units
	$\Delta P'$	Pressure drop including both static pressure and head	lb_f/ft^2	Pa
	Q	Volumetric flowrate	ft^3/s or gpm	m^3/s
	r	Distance from center of tube	ft	m
	R	Tube or pipe radius	ft	m
	u,v,w	x-, y- and z-components of velocity	ft/s	m/s
	U	Average linear velocity in x-direction	ft/s	m/s
1.5	f	Friction factor	–	–
(new	N_{Re}	Reynolds number	–	–
symbols introduced)	δ	Fraction of outer tube radius, see Figure 19	–	–
	ϵ'	Fraction of tube radius, see Figure 19	–	–
1.6	A′,B′	Constants in Equation 184	Variable	–
(new	C_1,C_2,C	Constants	Variable	–
symbols introduced)	C_p	Specific heat	Btu/(lb-°F)	kJ/(kg-°K)
	d	Wire diameter	in.	mm
	k	Universal constant in Equation 145	–	–
	k_f	Thermal conductivity of fluid	Btu/(hr-ft^2-°F/ft)	W/(m-°K)
	K	Constant in Equation 185	–	–
	ℓ	Prandtl mixing length, see Equations 145 or 150	ft	m
	ℓ'	Length of sensor body	ft	m
	n	Constant in Equation 167, equal to 0.10	–	–
	N_{Gr}	Grashof number, Equation 178	–	–
	N_{Ma}	Mach number	–	–
	N_{Nu}	Nusselt number, Equation 176	–	–
	N_{Pr}	Prandtl number, Equation 179	–	–
	$\mathcal{P}$	Power	kW or Btu/sec	kW
	q	Rate of heat transfer	Btu/hr	J/s
	T_m	Mean temperature	°R	°K
	T_0	Fluid temperature	°R	°K
	T_s	Surface temperature	°R	°K

Nomenclature for Section 1, continued

Chapter	Symbol	Definition	English Units	SI Units
	$\bar{u}$	Time-average velocity	ft/s	m/s
	u_x	Instantaneous velocity in x-direction	ft/s	m/s
	u'_x	Instantaneous velocity fluctuation	ft/s	m/s
	u^+	Dimensionless velocity parameter, see Equation 160	–	–
	u^*	Friction velocity, Equation 143	ft/s	m/s
	u_δ	Velocity at edge of boundary layer	ft/s	m/s
	y^+	Friction distance parameter, Equation 161	–	–
	α	Thermal diffusivity	ft^2/s	mm^2/s
	β	Coefficient of thermal expansion	in./(in.-°F)	mm/(mm-°K)
	δ_1	Boundary layer thickness	in.	mm
	ϵ	Roughness height	ft	m
	ϵ'	Eddy viscosity, Equation 135	cp	mPa-s
	ϵ'_M	Eddy kinematic viscosity, Equation 149	ft^2/s	mm^2/s
	θ	Pitch angle, see Figure 30	deg	deg
	ν	Kinematic viscosity	ft^2/s	mm^2/s
	ϕ	Frequency of eddies	Hz or cps	Hz or s^{-1}
	Ω	Resistance	ohms	ohms

REFERENCES FOR SECTION 1

1. Cheremisinoff, N. P. *Process Level Instrumentation and Control* (New York: Marcel Dekker, Inc., 1981).
2. Biery, J. C. "Numerical and Experimental Study of Damped Oscillating Manometers: I, Newtonian Fluids," *AIChE J.* 9 (1963).
3. Benedict, R. P. *Fundamentals of Temperature, Pressure, and Flow Measurement* (New York: John Wiley & Sons, Inc., 1969).
4. Perry, J. H., Ed. *Chemical Engineer's Handbook*, 3rd ed. (New York: McGraw-Hill Book Co., 1950).
5. Rayle, R. E. "Influence of Orifice Geometry on Static Pressure Measurements," ASME Paper 59-A-234 (December 1959).
6. Shaw, R. "The Influence of Hole Dimensions on Static Pressure Measurements," *J. Fluid Mech.* 7(Pt. 4):550 (April 1960).
7. Hurd, C. W., K. P. Chesky and A. H. Shapiro. "Influence of Viscous Effects on Impact Tube," *Trans. ASME, J. Appl. Mech.* 253 (June 1953).
8. Chambre, P. L., and S. A. Schaaf. "The Impact Tube," in *Physical Measurements in Gas Dynamics and Combustion* (Princeton, NJ: Princeton University Press, 1954).
9. Folson, R. G. "Review of the Pitot Tube," *Trans. ASME* 1447 (October 1956).
10. Glaser, A. H. "The Pitot Cylinder as a Static Pressure Probe in Turbulent Flow," *J. Sci. Instr.* 29:219 (July 1952).
11. Marris, R. E. "Multiple Head Instrument for Aerodynamic Measurements," *The Engineer* 315 (August 1961).
12. Schulze, W. M., G. C. Ashby, Jr. and J. R. Erwin. "Several Combination Probes for Surveying Static and Total Pressure and Flow Direction," *NACA-TN 2830* (August 1952).
13. Castellan, G. W. *Physical Chemistry*, 2nd ed. (Reading, MA: Addison-Wesley Publishing Co., Inc., 1971).
14. Bird, R. B., W. E. Stewart and E. N. Lightfoot. *Transport Phenomena* (New York: John Wiley & Sons, Inc., 1960).
15. Chapman, S., and T. G. Cowling. *Mathematical Theory of Non-Uniform Gases*, 2nd ed. (New York: Cambridge University Press, 1951).
16. Curtiss, C. F., and J. O. Hirschfelder. *J. Chem. Phys.* 17:550-555 (1949).
17. Wilke, C. R. *Chem. Phys.* 18:517-519 (1950).
18. Glasstone, S., K. J. Laidler and H. Eyring. *Theory of Rate Processes* (New York: McGraw-Hill Book Co., 1941), Chapter 9.
19. Bird, R. B., R. C. Armstrong and O. Hassager. *Dynamics of Polymeric Fluids–Fluid Mechanics Vol. I* (New York: John Wiley & Sons, Inc., 1977).

20. Skelland, A. H. P. *Non-Newtonian Flow and Heat Transfer* (New York: John Wiley & Sons, Inc., 1967).
21. Herschel, W. H., and R. Bulkley. *Proc. Am. Soc. Test. Mat.* xxvi:621 (1926).
22. Herschel, W. H., and R. Bulkley. *Kolloid-Z*, xxxix:291 (1926).
23. Scott-Blair, G. W. *An Introduction to Industrial Rheology* (Philadelphia: P. Blakiston's Son and Co., 1938), p. 76.
24. Crowley, P. R., and A. S. Kitzes. *Ind. Eng. Chem.* 49:888-892 (1957).
25. Reiner, M. *Deformation, Strain, and Flow* (New York: Wiley-Interscience, 1960).
26. Eyring, H. J. *J. Chem. Phys.* 4:283 (1936).
27. Prandtl, L. *Z. Angew. Math. Mech.* 8:85 (1928).
28. Christiansen, E. B., N. W. Ryan and W. E. Stevens. *AIChE J.* 1:544 (1955).
29. Philippoff, W. *Kolloid-Z.* 71:1-16 (1935).
30. Spencer, R. S. *J. Polymer Sci.* 5:599-600 (1950).
31. Spencer, R. S., and R. E. Dillon. *J. Colloid Sci.* 3:168-169 (1948).
32. Reiner, M., and R. S. Rivlin. *Kolloid-Z.* 44:9 (1928).
33. Lodge, A. S. *Elastic Liquids* (New York: Academic Press, Inc., 1964).
34. White, J. L., and A. B. Metzner. *J. Appl. Polymer Sci.* 7:1867-1889 (1963).
35. Spriggs, T. W., and R. B. Bird. *Ind. Eng. Chem.* (Fundamentals) 4: 182-186 (1965).
36. Spriggs, T. W. *Chem. Eng. Sci.* 20:931-940 (1965).
37. Williams, M. C., and R. B. Bird. *AIChE J.* 8:378-382 (1962).
38. Leslie, F. M. *Quart. J. Mech. Appl. Math.* 14:36 (1961).
39. Oldroyd, J. G. *Proc. Roy. Soc.* A218:122 (1953).
40. Frohlich, H., and R. Sack. *Proc. Roy. Soc.* A185:415-430 (1946).
41. Oldroyd, J. G., D. J. Strawbridge and B. A. Toms. *Proc. Phys. Soc.* B64:44-7 (1951).
42. Toms, B. A., and D. J. Strawbridge. *Trans. Faraday Soc.* 49:1225-1232 (1953).
43. Weast, R. C., Ed. *Handbook of Chemistry and Physics*, 49th ed. (Cleveland, OH: CRC Press, Inc., 1968).
44. Uyehara, O. A., and K. M. Watson. *Nat. Petrol.* Tech. Section 36, 764 (October 4, 1944).
45. Hougen, O. A., and K. M. Watson. *Chemical Process Principles*, Part III, (New York: John Wiley & Sons, Inc., 1947), p. 873.
46. Hougen, O. A., K. M. Watson and R. A. Ragatz. *C. P. P. Charts*, 2nd ed. (New York: John Wiley & Sons, Inc., 1960).
47. Carr, N. L., R. Kobayashi and D. B. Burroughs. *Am. Inst. Mining Metall. Eng. Petroleum Tech.* 6:47 (1954).
48. Comings, E. W. *High Pressure Technology* (New York: McGraw-Hill Book Co., 1956).
49. Thomas, G. B. *Calculus and Analytical Geometry*, 2nd ed. (Reading, MA: Addison-Wesley Publishing Co., Inc., 1953).
50. Curle, N., and H. J. Davies. *Modern Fluid Dynamics–Volume 1: Incompressible Flow* (New York: D. Van Nostrand Co., 1968).
51. Hagen, G. *Ann. Phys. Chem.* 46:423-442 (1839).
52. Poiseuille, J. L. *Comp. Rend.* 11:961,1047 (1840); 12:112 (1841).
53. Knudsen, J. G., and D. L. Katz. *Fluid Dynamics and Heat Transfer* (New York: McGraw-Hill Book Co., 1958), pp. 25-27.

54. Kern, D. Q. *Process Heat Transfer* (New York: McGraw-Hill Book Co., 1950).
55. Rohsenow, W. M., and J. P. Hartnett, Eds. *Handbook of Heat Transfer* (New York: McGraw-Hill Book Co., 1973).
56. Stanton, T. E., D. Marshall and C. N. Bryant. *Proc. Roy. Soc. (London)* 97A:413 (1920).
57. Nikuradse, J. *VDI-Forschungsheft* 356 (1932).
58. Reichardt, H. *NACA TN 1047* (1943).
59. Deissler, R. G. *NACA TN 2138* (1950).
60. Moody, L. F. *Trans. ASME* 66:671 (1944).
61. Prandtl, L. *Z. Ver. Deut. Ing.* 77:105 (1933); *NACA TN 720* (1933).
62. von Karman, T. *NACA TN 611* (1931).
63. Abramovich, G. N. *The Theory of Turbulent Jets* (Cambridge, MA: The M.I.T. Press, 1963).
64. Rajaratnam, N. *Turbulent Jets* (New York: Elsevier Scientific Publishing Co., 1976).
65. Stanton, T. E., and J. R. Pannell. *Trans. Roy. Soc. (London)* 214A:199 (1914).
66. Prandtl, L. *Z. Angew. Math. Mech.* 5:136 (1925); *NACA TN 1231* (1949).
67. Blasius, H., *Mitt. Forschungsarb.* 131:1-40 (1913).
68. von Karman, T. *NACA TN 611* (1931).
69. von Karman, T. *J. Aeronaut. Sci.* 1:1 (1934).
70. Reichardt, H. *NACA TN 1047* (1943).
71. Cheremisinoff, N. P., and E. J. Davis. "Heat and Momentum Transfer Analogies for Stratified and Annular Flows," *Proc. Two-Phase Flow and Heat Transfer Symp.* (1976).
72. Cheremisinoff, N. P., and E. J. Davis. "Stratified Turbulent-Turbulent Gas-Liquid Flow," *AIChE J.* 25(1) (1979).
73. Cheremisinoff, N. P., and E. J. Davis. "Pressure Drop and Holdup," *AIChE DIMP Report 4* (June 1977).
74. Cheremisinoff, N. P., and E. J. Davis. "Heat Transfer to Two-Phase Stratified Gas-Liquid Flows," *AIChE DIMP Report 5* (June 1977).
75. Cheremisinoff, N. P., et al. "Heat Transfer with Stratified Gas-Liquid Flow," *AIChE J.* (February 1980).
76. Deissler, R. G. *NACA TN 2242* (1950).
77. Deissler, R. G. *Trans. ASME* 73:101 (1951).
78. Deissler, R. G., and C. S. Eian. *NACA TN 2629* (1952).
79. Deissler, R. G. *NACA TN 3145* (1954).
80. Deissler, R. G., and M. F. Taylor. *NACA TN 3451* (1955).
81. Nikuradse, J. *Proc. Int. Cong. Appl. Mech.* 3rd Congress, Stockholm, 1:239 (1980).
82. Nikuradse, J. *VDI Forschungsheft* 361 (1933).
83. Rouse, H. *Elementary Mechanics of Fluids* (New York: John Wiley & Sons, Inc., 1948).
84. Moody, L. F. *Trans. ASME* 66:671 (1944).
85. Davies, J. T. *Turbulence Phenomena* (New York: Academic Press, Inc., 1972).
86. Kolmogoroff, A. N. *Comp. Rend. Acad. Sci. U.S.S.R.* 30:301; 32:16 (1941).
87. *DISA Probe Manual*, DISA Elektronik A/S, Herlev, Denmark (February 1970).

88. Andersen, O. K. "Time Resolution Power in Correlation Measurements with DISA Type 55A01 Hot-Wire and Hot-Film Anemometers," DISA Information No. 4 (December 1966).
89. Baldwin, L. V. "Heat Transfer from Yawed Cylinders in Rarefied-air Flows," an ASME Publication (1959).
90. Baldwin, L. V., and V. A. Sandborn. "Hot-Wire Calometry: Theory and Application to Ion Rocket Research," NASA Tech. Rep. R 98 (1961).
91. Bellhouse, B. J., D. L. Schultz and N. B. Karatzas. "The Measurement of Fluctuating Components of Velocity and Skin Friction with Thin-film Heated Elements with Application in Water, Air and Blood Flows," Oxford University Rep. No. 1003 (1966).
92. Bomelberg, H. J. "Miniature Hot-wire Pressure Gages for Wind Tunnel Work," Report No. 1095, Ballistic Research Lab., Maryland (1960).
93. Borden, A. "Time Constants and Frequency Response of Coated Hot Wires Used as Turbulence-sensing Elements," Navy Department, The David W. Taylor Model Basin, Report 952 (June 1957).
94. Bradshaw, P., and R. F. Johnson. "Turbulence Measurements with Hot-Wire Anemometers," National Physical Laboratory, Notes on Applied Science No. 33 (1963).
95. Breslin, J. P., and M. S. Macovsky. "Effects of Turbulence Stimulators on the Boundary Layer and Resistance of a Ship Model as Detected by Hot Wires," Navy Department, The David W. Taylor Model Basin, Report 724 (August 1950).
96. Chae, J. L., and V. A. Sandborn. "A Resistance Thermometer for Transient Temperature Measurements," Mechanics Paper No. 1, Colorado State University Fluid (1964).
97. Champagne, F. H. "Turbulence Measurements with Inclined Hotwires," Boeing Scientific Research Laboratories, Flight Science Laboratory, Washington, DC, Report No. 103 (December 1965).
98. Christensen, O. "New Trends in Hot-Film Probe Manufacturing," DISA Information No. 9 (February 1970).
99. Collis, D. C., and M. J. Williams. "Two-Dimensional Convection from Heated Wires at Low Reynolds Numbers," *J. Fluid Mech*. 6 (1959).
100. Cooper, R. D., and M. P. Tulin. "Turbulence Measurements with the Hot-Wire Anemometer," AGARDO-graph (August 1955).
101. Corssin, S. "Extended Applications of the Hot-Wire Anemometer," *NACA TN 1864* (1949).
102. Dahm, M., and C. G. Rasmussen. "Effect of Wire Mounting System on Hot-Wire Probe Characteristics," DISA Information No. 7 (January 1969).
103. Dell'Osso, L. "Turbulence Measurements in Water in an Open Channel with the Hot-Film Anemometer," Department of Chemical Engineering, College of Rice University (May 1966).
104. Dosanjh, D. S. "Studies of Transient Hot-Wire Response in a Shock Tube," Johns Hopkins University, Dept. Aeronautics, Rep. CM 725 (1955).
105. Dreyer, G. F. "Calibration of Hot-Film Sensors in a Towing Tank and Application to Quantitative Turbulence Measurements," U.S. Naval Academy, Annapolis, MD (March 1967).

106. English, J., B. Fletcher and W. Steckelmacher. "A Wide-Range Constant-Resistance Pirani Gauge with Ambient Temperature Compensation," *J. Sci. Instrum.* 42:77-80 (1965).
107. van der Hegge Zijnen, B. G. "On the Construction of Hot-Wire Anemometers for the Investigation of Turbulence," *Appl. Sci. Res.* A2 (1951).
108. Forbes Pearson, S. "Losses in Small Refrigerator Compressors," *Engineering* (October 1958).
109. Frenkiel, F. N. "The Influence of the Length of a Hot-Wire on the Measurement of Turbulence," *Phys. Rev.* 75 (April 1949).
110. Frenkiel, F. N. "On the Hot-Wire Length Correction," *Phys. Rev.* 76 (October 1949).
111. Friehe, C. A., and W. H. Schwarz. "Deviations from the Cosine Law for Yawed Cylindrical Anemometer Sensors," *J. Appl. Mech.*, Paper No. 68-WA/APM-16 (1968).
112. Fugita, H., and L. S. G. Kovasznay. "Measurement of Reynolds Stress by a Single Rotated Hot-Wire Anemomenter," paper presented at the Twentieth Anniversary Meeting of the Division of Fluid Dynamics, American Physical Society at Lehigh University, November, 1967.
113. Goldschmidt, V. W., and M. K. Householder. "Measurements of Aerosols by Hot-Wire Anemometry," *Adv. Hot-Wire Anemometry*, University of Maryland (1968), pp. 134-152.
114. Grant, H. P., and R. E. Kronauer. "Fundamentals of Hot-Wire Anemometry," *Proc. Symp. on Measurements in Unsteady Flow, Mass* (New York: American Society of Mechanical Engineers, 1962).
115. Grant, H. L., R. W. Stewart and A. Mailliet. "Turbulence Spectra from a Tidal Channel," *Fluid Mech.* No. 12 (1961).
116. Guitton, D., and R. P. Patel. "An Experimental Study of the Thermal Wake Interference Between Closely Spaced Wires of a X-Type Hot-Wire Probe," McGill University, M.E.R.L. Report 69-7 (1969).
117. Hinze, J. O. "Turbulence, An Introduction to its Mechanism and Theory" (New York: McGraw-Hill Book Co., 1959).
118. Hoole, B. J., and J. R. Calvert. "The Use of a Hot-Wire Anemometer in Turbulent Flow," *J. Roy. Aeronautical Soc.* (July 1967).
119. Hubbard, P. G. "Constant-Temperature Hot-Wire Anemometry with Application to Measurements in Water," Department of Mechanics and Hydraulics, Graduate College of the State University of Iowa (June 1954).
120. Humphrey, R. L., and F. M. Manion. "Low-Pass Filters for Pneumatic Amplifiers," DISA Information No. 1 (January 1965), p. 13.
121. Inoue, E. "Influence of the Length of Hot-Wire on the Measurement of Turbulent Flow," *J. Phys. Soc., Japan*, 7(5) (1952).
122. Janour, Z. "Abkühlung von Hitzdrahten endlicher Lange im Gasstrom," *Czechosl. J. Phys.* 9 (1959).
123. Kennelly, A. E., C. A. Wright and J. S. Blyevelt. "The Convection of Heat from Copper Wires," *Trans. Am. Inst. Electr. Eng.* 28:363-396 (1909).
124. Kidron, I. "Application of Modulated Electromagnetic Waves for Measurement of the Frequency Response of Heat-Transfer Transducers," DISA Information No. 4 (December 1966).

125. King, L. V. "On the Convection of Heat from Small Cylinders in a Stream of Fluid," *Phil. Trans. Roy. Soc., London, Ser. A* 214 (1914).
126. Kovasznay, L. S. G., and S. I. A. Tormark. "Heat Loss of Wires in Supersonic Flow," Bumblebee Series, Report No. 127 (1950).
127. Kovasznay, L. S. G. "The Hot-Wire Anemometer in Supersonic Flow," *J. Aeronautical Sci.* 17(9) (September 1950).
128. Kovasznay, L. S. G. "Development of Turbulence Measuring Equipment," *NACA TN 2839* (1953).
129. Kovasznay, L. S. G. "Turbulence in Supersonic Flow," *J. Aeronautical Sci.* 20(10) (October 1953).
130. Ladenburg, R. W., et al. Eds. "Physical Measurements in Gas Dynamics and Combustions," in *High Speed Aerodynamics and Jet Propulsion*, Vol. IX (Princeton, NJ: Princeton University Press, 1954).
131. Laufer, J., and R. McClellan. "Measurement of Heat Transfer from Fine Wires in Supersonic Flows," *J. Fluid. Mech.* 1:276 (1956).
132. Laurence, J. C., and L. G. Landes. "Auxiliary Equipment and Techniques for Adapting the Constant Temperature Hot-Wire Anemometer to Specific Problems in Airflow Measurements," *NACA TN 2841* (1952).
133. Laurence, J. C. "Applications of the Constant Temperature Hot-Wire Anemometer to the Study of Transient Air Flow Phenomena," *I.S.A. Proc. I*, 155-159 (1952).
134. Laurence, J. C., and V. A. Sandborn. "Heat Transfer from Cylinders," *Proc. Symp. on Measurement in Unsteady Flow, Mass.*, May 21-23 (New York: American Society of Mechanical Engineers, 1962).
135. Levey, H. C. "Heat Transfer in Slip Flow at Low Reynolds Numbers," *J. Fluid Mech.* 6:385 (1959).
136. Ling, S. C. "Measurements of Flow Characteristics by the Hot-Film Technique," Ph.D. Thesis, State University of Iowa (June 1955).
137. Ling, S. C., and P. G. Hubbard. "The Hot-Film Anemometer: A New Device for Fluid Mechanics Research," *J. Aeronautical Science* (September 1956).
138. Macovsky, M. S. "The Measurement of Turbulence in Water," a progress report prepared for presentation at the Seventh Underwater Ballistics Conference, Navy Department, The David W. Taylor Model Basin, Report No. 670 (October 1948).
139. Maye, J. P. "Error Due to Thermal Conduction Between the Sensing Wire and Its Supports when Measuring Temperatures with a Wire Anemometer Used as a Resistance Thermometer," DISA Information No. 9 (February 1970).
140. McLeod, M. G. "A Spinning Hot-Wire Anemometer for Simultaneous Measurement of u, v, and w," UTIAS Tech. Note No. 125 (March 1968).
141. Melnik, W. L., and J. R. Weske. "Advances in Hot-Wire Anemometry," International Symposium on Hot-Wire Anemometry, University of Maryland (July 1968).
142. Merceret, F. J. "An Experimental Study to Determine the Utility of Standard Commercial Hot-Wire and Coated Wedge-Shaped Hot-Film Probes for Measurement of Turbulence in Water Contaminated Air Flows," U.S. Government Research, Development Reports (January 1969).

143. Merceret, F. J. "On the Use of Hot-Film Anemometry to Measure Turbulence in the Presence of Heavy Rain," The Johns Hopkins University, Department of Mechanics (July 1969).
144. Morkovin, M. V. "Fluctuations and Hot-Wire Anemometry in Compressible Flows," AGARDOgraph (November 1956).
145. Morkovin, M. V., and R. E. Phinney. "Extended Application of Hot-Wire Anemometry to High-Speed Turbulent Boundary Layers," AFOSR TN 58-469, Johns Hopkins University, Department of Aeronautics (1958).
146. Moon, I. M. "Direction Sensitive Hot-Wire Anemometer for Two-Dimensional Flow Study Near a Wall," paper presented at the ASME Symposium on Measurements in Unstable Flow, 1962.
147. Nielsen, P. E., and C. G. Rasmussen. "Measurement of Amplitude and Phase Characteristics," DISA Information No. 4 (December 1966).
148. Norman, B. "Hot-Wire Anemometer Calibration at High Subsonic Speeds," DISA Information No. 5 (June 1967), pp. 5-19.
149. Patterson, A. M. "Turbulence Spectrum Studies in the Sea with Hot Wires," Pacific Naval Laboratory, Esquimalt, BC, Canada (1958).
150. Patterson, G. K., and J. L. Zakin. "Hot-Film Anemometry Measurements of Turbulence in Pipe Flow: I. Organic Solvents," University of Missouri at Rolla, Department of Chemical Engineering (December 1966).
151. Piret, E. L. "Heat Transmission from Fine Wires to Water," *Ind. Eng. Chem.* (1947).
152. Rasmussen, C. G. "CT Hot-Wire Anemometry–The Most Important Measuring Technique for Pure Fluid Research," DISA Information No. 1 (January 1965), p. 11.
153. Rasmussen, C. G. "The Air Jet Hot-Wire Microphone," DISA Information No. 2 (July 1965), p. 5.
154. Reichardt, H. "Uber das Messung Turbulenter Langs- und Querschwankungen," *Zamm* 18(6) (December 1938); available as *NACA TM 1313* (1951).
155. Rodriguez, J. M., G. K. Patterson and J. L. Zakin. "Effects of Probe Geometry on Turbulence Measurements in Liquids Using Hot-Film Constant Temperature Anemometry," University of Missouri, Rolla, MO (March 1969).
156. Roshko, A. "On the Development of Turbulent Wakes from Vortex Streets," *NACA TN 1191* (1954).
157. Rubatto, G. "Calibration of Probes for Flow Velocity Measurements in Liquids, in the Range 2–5 m/sec," DISA Information No. 9 (February 1970).
158. Sandborn, V. A. "A Hot-Wire Vacuum Gage for Transient Measurements," Air Force Ballistic Syst. Div., USAF (1963).
159. Schuh, H. "On Measuring Low Level Turbulence," *Trans. Instr. Measurement Conf.* (1952).
160. Serth, R. W., and K. M. Kiser. "The Effect of Turbulence on Hot-Film Anemometer Response in Viscoelastic Fluids," State University of New York at Buffalo, Department of Chemical Engineering (August 1968).
161. Spangenberg, W. G. "Heat-Loss Characteristics of Hot-Wire Anemometers at Various Densities in Transonic and Supersonic Flow," *NACA TN 3381* (1955).

162. Stevens, R. G., A. Borden and P. E. Strausser. "Summary Report on the Development of a Hot-Wire Turbulence-Sensing Element for Use in Water," Navy Dept., The David W. Taylor Model Basin, Report 953 (December 1956).
163. Staritz, R. F. "Die elektronische Messung der Stromungsgeschwindigkeit und der Turbulenz," *VDI-Zeitschrift*, 3 (1960).
164. Thun, R. E., G. F. Caudle and E. R. Pasciutti. "Rugged Film Resistor Thermometer for the Measurement of Surface Temperatures," *Rev. Sci. Instr.* 31(4) (April 1960).
165. Townsin, R. L. "Turbulence Detection–Results from the Use of an Unobstructive Technique During Ship Model Testing," *Quart. Trans. Inst. Nav. Arch.* (July 1955).
166. Vrebalovich, T. "Application of Hot-Wire Techniques in Unsteady Compressible Flows," *Proc. Symp. on Measurement of Unsteady Flow* (New York: American Society of Mechanical Engineers, 1962).
167. Wehrmann, O. "Methoden und Anwendungen der Hitzdraht-Mebtechnik für Stromungsvorgange," *Z. Konstruktion*, 8 (1958).
168. Weske, J. R. "Methods of Measurement of High Air Velocities by the Hot-Wire Method," *NACA TN No. 880* (February 1943).
169. Wills, J. A. B. "The Correction of Hot-Wire Readings for Proximity to a Solid Boundary," *J. Fluid Mech.* 12(3) (1962).
170. Wills, J. A. B. "A Traversing Orifice-Hot-Wire Probe Pair for Use in Wall Pressure Correlation Measurements," National Physical Laboratory (NPL) Aero Rep. 1155 (1965).
171. Wise, B., and D. L. Schultz. "Turbulence Measurements in Supersonic Flow with the Hot-Wire Anemometer," OUEL Report No. 83 (1959).
172. Langhaar, H. L. *Dimensional Analysis and Theory of Models* (New York: John Wiley & Sons, Inc. (1951).
173. Buckingham, E. "On Physically Similar Systems; Illustration of the Use of Dimensional Equations," *Phys. Rev.* 4(4):345 (1914).
174. Catchpole, J. P., and G. Fulford. "Dimensionless Groups," *Indus. Eng. Chem.* 58(3):46 (1966); 60(3):71 (1968).
175. Boucher, D. F., and G. E. Alves. "Dimensionless Numbers," *Chem. Eng. Prog.* 55(9):55 (1959).
176. Rayleigh, "The Principle of Similitude," *Nature* 95:66 (1915).
177. Johnstone, R. E., and M. W. Thring. *Pilot Plants, Models, and Scale-Up Methods in Chemical Engineering* (New York: McGraw-Hill Book Co., 1957).
178. Birkhoff, G. *Hydrodynamics* (Princeton, NJ: Princeton University Press, 1960).
179. Klinkerberg, A., and H. H. Mooy. "Dimensionless Groups in Fluid Friction, Heat, and Material Transfer," *Chem. Eng. Prog.* 44(1):17 (1948).

SECTION 2

CALCULATIONS AND DESIGN BASICS FOR INCOMPRESSIBLE FLOWS

SECTION OVERVIEW

As noted in the first section, the science of fluid mechanics is concerned with the motion of fluids and the effects that process conditions have on that motion. It is an exceedingly wide and complex subject, and, not surprisingly, one that has been tackled in numerous textbooks. To the chemical engineer, it is the backbone of his education.

The design of virtually all unit operations requires knowledge of the fluid mechanics. The principles outlined in Section 1 of this volume form the basis for describing/modeling flow phenomena for any type operation. In this section, these principles are applied to specific design practices for handling incompressible flows. Our discussions will be geared toward problem solving for industrial applications; consequently, we shall depart from using rigorous definitions and apply terminology that may be less exact but sufficient for process engineering. For the purposes of our discussions, then, the flow of incompressible fluids is taken to mean the flow of liquids.

There are a large number of unit operations that handle incompressible fluids, and it would be beyond the scope of this first volume to describe each in detail. Indeed, we would have to present a major handbook. Rather, we have elected to devote most of the discussions to the handling and transportation aspects.

There are twelve chapters in this section. Chapters 2.1 through 2.3 deal with piping systems. The first two outline calculation methods for pressure drop and flow parameters through piping. The third provides descriptive information and selection criteria for flow control and measurement devices.

Chapters 2.4 through 2.8 cover various aspects of pumping operations. Chapters 2.4 and 2.5 provide information on pump classifications, typical data on pump performances and characteristics, and recommendations for pump selection. Overall pumping system performances are described in these chapters. Chapters 2.6 and 2.7 are somewhat specialized, covering subjects not normally treated in standard textbooks, but useful for practising engineers. Chapter 2.7 is devoted entirely to oscillating displacement pumps, which have played an increasingly important role in a range of process operations in different industries. Chapter 2.7 covers types, selection criteria and maintenance practices for mechanical seals for pumps. The final chapter dealing with pumps, Chapter 2.8, covers a major problem for many pumping operations–cavitation.

Chapter 2.9 deals with flow past immersed bodies. Semiempirical correlations are presented for describing these kinds of flows. Nomrally, these flows are turbulent in nature. This information is of importance in process industries in equipment such as tubular heat exchangers, cooling towers, water-walled furnaces and in various reactor configurations where flow over or around baffle arrangements is encountered.

Chapter 2.10 describes simple flow configurations for non-Newtonian fluids. The majority of this chapter deals with laminar flow conditions, as it is most frequently encountered with fluids of this type. Because proper description of this type flow depends on the rheological properties of the fluid, a discussion of methods for characterizing non-Newtonian behavior is included.

In recent years chemical engineers have become more involved in environmental control, particularly in the area of wastewater treatment. The types of flows most often encountered in these applications are in the form of partially filled pipe flow or in open channels. Chapter 2.11 describes the characteristics of open channel flows and presents some useful correlations for estimating volumetric discharges. Flow control and measurement for these flows are also discussed.

Finally, since our aim is to solve engineering problems, Chapter 2.12 provides specific examples and solutions to selected problems. This chapter has been prepared with students in mind; however, practising engineers may wish to peruse some of the solved problems presented here. Additional calculation procedures not described in other chapters are included among the solved problems. Problems are grouped in accordance with the main subject titles noted above.

A final note–there are a number of important incompressible flows that we have elected not to cover in this first volume. Two major ones that will be treated in future volumes are liquid/liquid flows and slurry flows. It is felt that these topics are best covered under separate discussions pertaining to multicomponent flow mixtures.

CONTENTS

LIST OF FIGURES FOR SECTION 2

Chapter 2.8

Chapter 2.9

Chapter 2.10

Chapter 2.11

Chapter 2.12

LIST OF TABLES FOR SECTION 2

CHAPTER 2.1

PRINCIPLES OF FLUID FLOW

This chapter reviews the fundamental principles of fluids in motion. Outlined are the generalized equations for describing fluid flow behavior. These equations serve as the starting basis for deriving many design expressions for fluid flow operations.

THE BERNOULLI EQUATION

The principal equation for computing pressure drop for flowing liquids in pipes and fittings is the generalized Bernoulli equation:

$$\underbrace{-\frac{\Delta P}{\rho}}_{\text{pressure change}} = \underbrace{\frac{\alpha\Delta(u^2)}{2g_c}}_{\text{kinetic energy change}} + \underbrace{\frac{g\Delta Z}{g_c}}_{\text{elevation change}} + \underbrace{F}_{\text{frictional head loss}} \tag{1}$$

where		(English Units)	(cgs Units)
	F = frictional head loss	$ft\text{-}lb_f/lb_m$	$Pa\cdot m^3/kg$
	g = acceleration of gravity	ft/s^2	m/s^2
	g_c = conversion factor	$32.174\ ft\cdot lb_m/lb_f\text{-}s^2$	
	ΔP = pressure change	lb_f/ft^2	Pa
	u = fluid velocity	ft/s	m/s
	Z = elevation	ft	m
	ρ = fluid density	lb_m/ft^3	kg/m^3

α = kinetic energy correction term whose value depends on the velocity profile (for turbulent flow α = 1.1; for laminar flow α = 2.0)

Most design equations for single-phase flow can be derived from the generalized Bernoulli equation. The importance of each term in the equation varies with application (see Table 1).

For Newtonian liquid flow, constant properties (viscosity and density) can be assumed. Exceptions to this rule are non-Newtonian liquids and non-isothermal flow. In nonisothermal flow, properties change because of either heat exchange, or heat production or consumption in the fluid by chemical reaction or friction losses.

If isothermal conditions exist across the pipe cross section but not along the flow axis, pressure drop can be computed by dividing the pipe into a number of small lengths and calculating the pressure drop in each section.

REYNOLDS NUMBER AND FRICTION FACTOR

The Reynolds number is a dimensionless group defined as the ratio of inertial forces to viscous shearing forces acting on a fluid in motion. It is a function of a characteristic linear dimension of the fluid boundary surface, D, the relative fluid velocity, u, and fluid physical properties, namely, viscosity, μ, and density, ρ:

$$N_{Re} = \frac{Du\rho}{\mu} = \frac{Du}{\nu} \tag{2}$$

The parameter ν is kinematic viscosity, $\nu = \mu/\rho$. The characteristic linear dimension, D, describes the system under consideration: for example, D can be the diameter of the opening for orifices, the discharge diameter for a nozzle, the inside diameter of a pipe or round conduit, or the equivalent diameter of a rectangular conduit. The equivalent diameter is defined as

$$D = D_H = 4R_H \tag{3}$$

where R_H is the hydraulic radius.

$$R_H = \frac{\text{cross-sectional area of flow}}{\text{wetted perimeter}} \tag{4}$$

Table 1. Important Terms in the Bernoulli Equation

System	Important Term
Constant-Diameter Horizontal Pipes	Friction term, F
Vertical or Inclined Pipes	Include deviation term, $g\Delta z/g_c$
Piping with Cross Section Changes	Kinetic energy term, $\alpha\Delta(u^2/2g_c)$

Table 2 summarizes specific formulas of the hydraulic radius for various flow area geometries.

For a fluid in motion through a conduit of known geometry, the mechanical energy loss per unit mass of fluid flowing per unit length of conduit is expressed by an equation of the general form

$$\frac{Dg_c}{u^2}\left(\frac{dE_M}{dx}\right) = 2f = \text{function of } (N_{Re}) \tag{5}$$

where (dE_M/dx) is the mechanical energy loss term and f is the coefficient of friction (friction factor).

The friction factor is a dimensionless group that relates to two other dimensionless groups: the ratio of the loss of total head between two points to the velocity head, and the ratio of the distance between those two points and a characteristic length, which establishes fluid velocity.

Various flow regimes exist that are defined as laminar, transitional and completely turbulent. For laminar flow conditions, the friction factor is

$$f = 64/N_{Re} \tag{6}$$

Table 2. Formulas of Hydraulic Radius for Various Flow Area Geometries (see Equations 3 and 4)

Cross Section Geometry	Description	Hydraulic Radius, R_H
Circle	Diameter, D	$\frac{D}{4}$
Annulus	Inner diameter, d Outer diameter, D	$\frac{(D - d)}{4}$
Rectangle	Sides: a, b	$\frac{ab}{2(a + b)}$
Triangle, Equilateral	Sides: a	$\frac{1}{12}a\sqrt{3}$
Triangle, General	Sides: a, b, c	$\sqrt{\frac{s(s - a)(s - b)(s - c)}{(a + b + c)}}$, where $s = \frac{1}{2}(a + b + c)$
Ellipse[a]	Major axis: 2a Minor axis: 2b	$\frac{ab}{k(a + b)}$

[a]Values of k for $z = (a - b)/(a + b)$; z = 0.2, 0.4, 0.6, 0.8, 1.0; k = 1.010, 1.040, 1.092, 1.127, 1.273.

A critical Reynolds number region exists in which laminar flow exists only by virtue of the fluid's ability to damp oscillations or forces that otherwise would result in turbulence.

For partially turbulent flow, the Colebrook [1] equation has proven reliable:

$$\frac{1}{\sqrt{f}} = 2 \log_{10}\left(\frac{\epsilon/D}{3.7} + \frac{2.51}{N_{Re}\sqrt{f}}\right) \tag{7}$$

where ϵ/D is the relative roughness of the surface in a conduit.

For completely turbulent flow conditions, the following equation is often used:

$$f = \frac{1}{\left(\log_{10} \frac{3.7}{\epsilon/D}\right)^2} \tag{8}$$

The well-known Moody [2] chart shown in Figure 1 presents the graphical relationship between friction factor and Reynolds number for single-phase, full-conduit flow.

Table 3 summarizes various friction factor correlations and their range of applicability.

ROUGHNESS ELEMENTS

All surfaces have irregularities that can disturb the flow properties of a fluid passing over them. The mean distance between the maximum height and depth of these irregularities is referred to as the absolute roughness, ϵ. Values for absolute roughness for various surface conditions have been reported by Hutchinson [3].

The ratio ϵ/D in the Moody plot is referred to as the relative roughness of the surface. For a conduit, ϵ/D is the ratio of the absolute roughness to the effective diameter of the conduit. Any value of relative roughness can represent either a smooth or rough surface condition, depending on the Reynolds number.

THE CONTINUITY EQUATION

The continuity equation mathematically expresses the law of conservation of mass. Figure 2 represents a differential element of fluid flowing in a three-dimensional space. The fluid element has dimensions dx, dy, dz. With this orientation, the mass flow of fluid through the cubical element can be

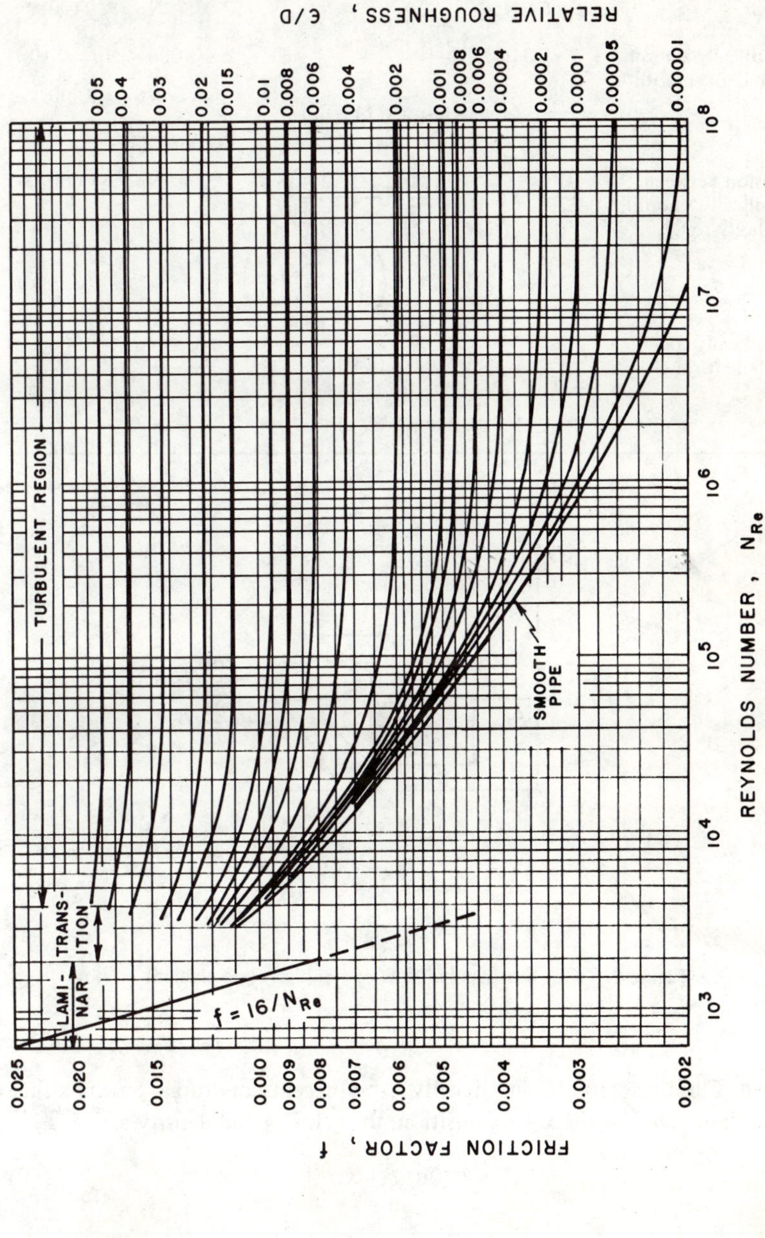

Figure 1. Moody chart for obtaining friction factors [2].

Table 3. Friction Factor Equations

Type of Flow	Equation Giving f	Range of Application
Laminar	$f = 64/N_{Re}$	$N_{Re} < 2100$
Hydraulically smooth or turbulent smooth	$f = 0.316/N_{Re}^{0.25}$	$4000 < N_{Re} < 10^5$
	$\dfrac{1}{\sqrt{f}} = 2 \log_{10} (N_{Re} \sqrt{f}) - 0.8$	$N_{Re} > 4000$
Transition between hydraulically smooth and wholly rough	$\dfrac{1}{\sqrt{f}} = 2 \log_{10} \left(\dfrac{\epsilon/D}{3.7} + \dfrac{2.52}{N_{Re} \sqrt{f}} \right)$	$N_{Re} > 4000$
	$= 1.14 - 2 \log_{10} \left(\dfrac{\epsilon}{D} + \dfrac{9.35}{N_{Re} \sqrt{f}} \right)$	
Hydraulically rough or turbulent rough	$\dfrac{1}{\sqrt{f}} = 1.14 - 2 \log_{10} (\epsilon/D)$	$N_{Re} > 4000$
	$= 1.14 + 2 \log_{10} (D/\epsilon)$	

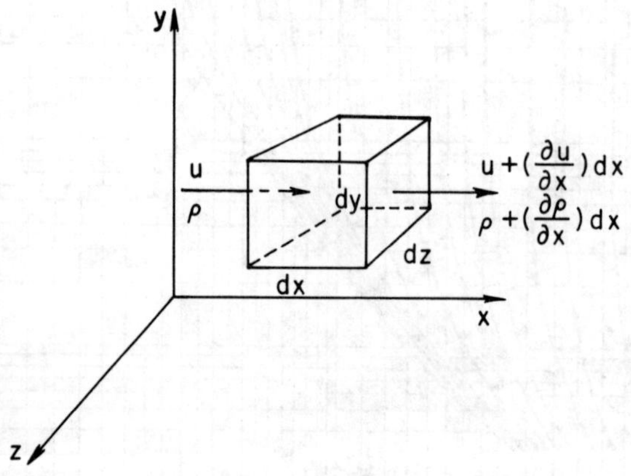

Figure 2. One-dimensional flow of a cubical fluid element.

derived. Fluid velocity, u, and density, ρ, represent conditions on the x face of the fluid cube. At the x + dx position, the velocity and density are

$$u + (\partial u/\partial x)dx \tag{9}$$

and

$$\rho + (\partial \rho/\partial x)dx \tag{10}$$

respectively. $\partial u/\partial x$ and $\partial \rho/\partial x$ represent the rate of change of velocity and density with respect to the x-axis. The variables u and ρ are dependent, while x as well as time are independent variables.

For a differential time, dt, a mass balance can be developed:

$$\text{Mass} - \text{Input to x-face} = \rho u\,dy\,dz\,dt \tag{11}$$

$$\text{Mass} - \text{Output from x + dx face} = \left(\rho + \frac{\partial \rho}{\partial x}dx\right)\left(u + \frac{\partial u}{\partial x}dx\right)dy\,dz\,dt \tag{12}$$

By neglecting the dx^2 term in Equation 12 we obtain

$$\left(\rho u + u\frac{\partial \rho}{\partial x}dx + \rho\frac{\partial u}{\partial x}dx\right)dy\,dz\,dt = \text{Mass} - \text{Output} \tag{13}$$

A simple mass balance expression states

$$\text{Input} = \text{Output} + \text{Accumulation} \tag{14}$$

where

$$\rho u\,dy\,dz\,dt = \left(\rho u + u\frac{\partial \rho}{\partial x}dx + \rho\frac{\partial u}{\partial x}dx\right)dy\,dz\,dt + \frac{\partial \rho}{\partial t}dx\,dy\,dz\,dt \tag{15}$$

where $(\partial\rho/\partial t)dx\,dy\,dz\,dt$ represents the accumulation of mass in the fluid element.

The general material balance can be simplified to

$$-u\frac{\partial \rho}{\partial x} - \rho\frac{\partial u}{\partial x} = \frac{\partial \rho}{\partial t} \tag{16}$$

The left-hand side (LHS) of this equation is an exact derivative. The continuity equation for one-dimensional flow is stated as follows:

$$-\frac{\partial(\rho u)}{\partial x} = \frac{\partial \rho}{\partial t} \tag{17}$$

This is the law of conservation of mass of a fluid flowing parallel to the x-axis. That is, Equation 17 describes the case of one-dimensional flow in which both the velocity and fluid density are functions of only x and t.

A similar analysis can be used to develop the equation of continuity for flow in three dimensions. Knudsen and Katz [4] and Bird et al. [5] perform this analysis in detail. Table 4 summarizes the various forms of the continuity equation for both compressible and incompressible fluids under steady- and unsteady-state flow conditions.

Table 4. Various Forms of the Continuity Equation[a]

	Unsteady State	Steady State
	Compressible Fluids	
One-Dimensional Flow (x-direction)	$\frac{\partial(\rho u)}{\partial x} = -\frac{\partial \rho}{\partial t}$	$\frac{\partial(\rho u)}{\partial x} = 0$
Two-Dimensional Flow (x and y direction)	$\frac{\partial(\rho u)}{\partial x} + \frac{\partial(\rho v)}{\partial y} = -\frac{\partial \rho}{\partial t}$	$\frac{\partial(\rho u)}{\partial x} + \frac{\partial(\rho v)}{\partial y} = 0$
Three-Dimensional Flow	$\frac{\partial(\rho u)}{\partial x} + \frac{\partial(\rho v)}{\partial y} + \frac{\partial(\rho w)}{\partial z} = -\frac{\partial \rho}{\partial t}$	$\frac{\partial(\rho u)}{\partial x} + \frac{\partial(\rho v)}{\partial y} + \frac{\partial(\rho w)}{\partial z} = 0$
	Incompressible Fluids	
One-Dimensional Flow (x-direction)	$\frac{\partial u}{\partial x} = 0$	$\frac{\partial u}{\partial x} = 0$
Two-Dimensional Flow (x and y direction)	$\frac{\partial u}{\partial x} + \frac{\partial v}{\partial y} = 0$	$\frac{\partial u}{\partial x} + \frac{\partial v}{\partial y} = 0$
Three-Dimensional Flow	$\frac{\partial u}{\partial x} + \frac{\partial v}{\partial y} + \frac{\partial w}{\partial z} = 0$	$\frac{\partial u}{\partial x} + \frac{\partial v}{\partial y} + \frac{\partial w}{\partial z} = 0$

[a]v represents fluid velocity in the y-direction; w represents fluid velocity in the z-direction.

MOMENTUM EQUATIONS

Newton's second law of motion states that the time rate of change of momentum is proportional to external forces. That is,

$$\text{Mass} \times \text{Acceleration} = \text{External Force} \tag{18}$$

or

$$m \frac{du}{dt} = Fg_c \tag{19}$$

The product on the LHS of Equation 19 is referred to as the inertial force. In fluid flow systems inertial forces act in three directions. The resultant inertial force can be related to the external forces acting on the fluid in motion. There are three major external forces that can produce fluid motion, namely, gravity forces, normal or pressure forces, and shear forces. Shear forces result from the resistance of the fluid to undergo deformation.

For the general case of three-dimensional flow where there are three component velocities, u, v and w, an expression for unsteady flow for the x-direction is

$$du = \frac{\partial u}{\partial x} dx + \frac{\partial u}{\partial y} dy + \frac{\partial u}{\partial z} dz + \frac{\partial u}{\partial t} dt \tag{20}$$

Dividing this expression by dt and noting that u = dx/dt, v = dy/dt and w = dz/dt,

$$\frac{du}{dt} = u\frac{\partial u}{\partial x} + v\frac{\partial u}{\partial y} + w\frac{\partial u}{\partial z} + \frac{\partial u}{\partial t} \tag{21}$$

The term du/dt represents the fluid's acceleration in the x-coordinate direction.

The inertial force in the x-direction is the product of mass and acceleration; then, defining mass as

$$m = \rho dxdydz \tag{22}$$

Using Equation 21, the inertial force in the x-direction becomes

$$F_x = \frac{\rho}{g_c}\left(u\frac{\partial u}{\partial x} + v\frac{\partial u}{\partial y} + w\frac{\partial u}{\partial z} + \frac{\partial u}{\partial t}\right) dxdydz \tag{23}$$

To apply Newton's second law, the inertial force of the fluid element is equated to the sum of the external forces. The overall momentum equation in the x-direction is

$$\frac{\rho}{g_c}\left(u\frac{\partial u}{\partial x} + v\frac{\partial u}{\partial y} + w\frac{\partial u}{\partial z} + \frac{\partial u}{\partial t}\right) dxdydz$$

$$= -\rho\frac{\partial \Omega}{\partial x} dxdydz + \left(\frac{\partial P_x}{\partial x} + \frac{\partial \tau_{yx}}{\partial y} + \frac{\partial \tau_{zx}}{\partial z}\right) dxdydz \tag{24}$$

where Ω = the force potential (for example, the potential energy of the fluid per unit mass)

Similar expressions can be written for the y- and z-directions.

DIMENSIONLESS NUMBERS

Dimensionless numbers have the advantage that the number of dimensionless ratios or groups needed to express a relationship is always less than the number of dimensionless variables involved. Dimensional analysis is a

Table 5. Dimensionless Numbers and Symbols

Dimensionless Number	Symbol	Definition
Brinkman Number	N_{Br}	$\frac{\mu(u/D)^2}{k(\Delta T)/D^2}$
Euler Number	N_{Eu}	$\frac{\Delta P}{\rho u^2}$
Fanning Friction Factor	f	$\frac{\rho D \Delta P}{2G^2 \Delta L}$
Fourier Number	N_{Fo}	$\frac{kt}{C_p L^2}$
Froude Number	N_{Fr}	$\frac{u^2}{gL}$
Graetz Number	N_{Gz}	$\frac{\pi}{4}\frac{D}{x} N_{Pe}$
Heat Transfer Factor	I_H	$(N_{St})(N_{Pr})^{2/3}$
Knudsen Number	N_{Kn}	λ_m/D
Lewis Number	N_{Lc}	$\frac{k}{C_p D_\nu}$
Mass Transfer Factor	$\bar{j}_M$	$\frac{k_c}{u}\left(\frac{\mu}{\rho D_\nu}\right)^{2/3}$
Mach Number	N_{Ma}	u/C
Nusselt Number	N_{Nu}	hD/k
Péclet Number	N_{Pe}	$Du\rho C_p/k$
Prandtl Number	N_{Pr}	$C_p\mu/k$
Reynolds Number	N_{Re}	$Du\rho/\mu$
Schmidt Number	N_{Sc}	$\frac{\mu}{\rho D_\nu}$
Sherwood Number	N_{Sh}	$\frac{k_c L}{D_\nu}$
Stanton Number	N_{St}	$h/C_p u\rho$
Weber Number	N_{We}	$Du^2\rho/\sigma$

method of grouping variables. Functional relationships between dimensionless parameters must be determined experimentally; however, mathematical methods for arriving at desired groups exist.

Equation 2 gives one dimensionless number, the Reynolds number, which can be interpreted as the ratio of inertial forces, $\rho u^2/D$, to viscous forces, $\mu u/D^2$, acting on a fluid. Other examples include

$$N_{Fr} = \frac{\rho u^2/D}{\rho g} = \frac{\text{inertial forces}}{\text{gravity forces}} \tag{25}$$

$$N_{Pr}N_{Re} = \frac{\rho C_p u \Delta T/D}{k\Delta T/D^2} = \frac{\text{heat transport by convection}}{\text{heat transport by conduction}} \tag{26}$$

$$N_{Br} = \frac{\mu(u/D)^2}{k\Delta T/D} = \frac{\text{heat production by viscous dissipation}}{\text{heat transport by conduction}} \tag{27}$$

Table 5 is a partial list of some of dimensionless groups encountered in fluid mechanics and heat and mass transfer applications. Also see Table 11, Section 1.

CHAPTER 2.2

GENERAL DESIGN PROCEDURES FOR SINGLE-PHASE FLOWS

This chapter reviews general design procedures for single-phase flow of incompressible fluids through pipes and pipe system components, such as valves and fittings. Calculation methods for determining pressure drop and pipe network analysis of head loss and flowrates are given.

PIPE SYSTEM EFFECTS ON PRESSURE DROP

Horizontal Straight Piping

The pressure drop in a horizontal length of straight pipe of constant diameter is due mainly to friction. It can be computed from the Fanning friction equation, in which the Fanning friction factor, f, is a function of the Reynolds number and relative pipe wall roughness (see Figure 1). The roughness is relatively independent of pipe diameter for a given class of pipe material. Thus, f can be expressed as a function of the Reynolds number and pipe diameter. For laminar flow conditions ($N_{Re} \leqslant 2000$), f is independent of pipe wall roughness and is a function only of the Reynolds number:

$$f = 16/N_{Re} \tag{28}$$

In the transition region ($2000 < N_{Re} < 4000$), the flow may be either laminar or turbulent, depending on various factors, such as changes in cross section, or the presence of restrictions in the piping, such as valves, fittings, etc. It is difficult to determine f as the true value lies between the limits of laminar and turbulent flows. The higher value of f is used generally since most applications with commercial pipe are turbulent.

The Fanning friction factor equation is accurate to within ±15% for smooth pipe and tubing, and ±10% for commercial steel pipe.

The following procedure can be used to compute the frictional pressure drop across a straight length of pipe:

Step 1. For noncircular conduits, calculate the equivalent hydraulic diameter from Equation 3.

Step 2. Compute the Reynolds number, N_{Re}, for d, given diameter and flow rate from Equation 2 or its variations given below:

$$N_{Re} = \frac{D_H u \rho}{\mu} = 50.6\left(\frac{Q\rho}{D_H \mu'}\right) \tag{29a}$$

$$= 6310\left(\frac{W}{D_H \mu'}\right) \tag{29b}$$

$$= 123.9\left(\frac{D\mu\rho}{\mu'}\right) \tag{29c}$$

where all units are in the English system except μ' in cP, and

where Q = volumetric flowrate, gpm
W = mass flowrate, thousands lb_m/hr

Step 3. Obtain the friction factor, f, from Figure 1, or if the flow is strictly laminar, use Equation 28.

Step 4. Using the Fanning equation (below), compute the frictional pressure drop:

$$(\Delta P)_f = \left(\frac{1}{144}\right)\left(\frac{4fL}{D_H}\right)\left(\frac{\rho u^2}{2g_c}\right) \tag{30a}$$

$$= 13.4\left(\frac{fLW^2}{\rho d_H^2}\right) \tag{30b}$$

where $(\Delta P)_f$ = frictional pressure drop (psi)
L = pipe length (ft)
D_H, d_H are in feet and inches, respectively

Step 5. For the case in which the pipe is not horizontal, compute the pressure drop due to the change in elevation:

$$(\Delta P)_e = \left(\frac{\rho g}{144 g_c}\right) \Delta z \tag{31}$$

where $(\Delta P)_e$ = pressure drop due to elevation changes (psi)
Δz = elevation (from beginning and end of pipe) (ft)

Step 6. Determine the total pressure drop by summing up the contributions:

$$(\Delta P)_t = (\Delta P)_f + (\Delta P)_e \tag{32}$$

Fittings and Pipe Components

Additional pressure drop in piping can be caused by bends, tees, valves, orifices and various flow restrictions. Fittings having the same nominal diameter as the pipe are handled in terms of an *equivalent length* of straight pipe. Equivalent lengths are calculated from resistance coefficients of the fittings. Equivalent lengths are then added to the actual pipe length and the sum is used in the Fanning equation to compute the total frictional pressure drop. Refer to Figure 3 for resistance coefficients for bends, ells and tees.

The following procedure can be used to estimate $(\Delta P)_f$:

Step 1. Use Figure 3 to obtain the resistance coefficient, K.

Step 2. Compute the frictional pressure drop:

$$(\Delta P)_f = \left(\frac{K}{144}\right)\left(\frac{\rho u^2}{2 g_c}\right) \tag{33a}$$

$$= 1.08 \times 10^{-4} (K \rho u^2) \tag{33b}$$

$$= 0.28 \left(\frac{K W^2}{\rho d_H^2}\right) \tag{33c}$$

Step 3. If there are long nonhorizontal bends, include the pressure drop due to elevation changes by using Equation 31.

Step 4. For piping systems using blanked-off tees and Y's, Equation 33 is still applicable with appropriate values of K.

Step 5. For tees that split or join streams, use the equations given in Table 6.

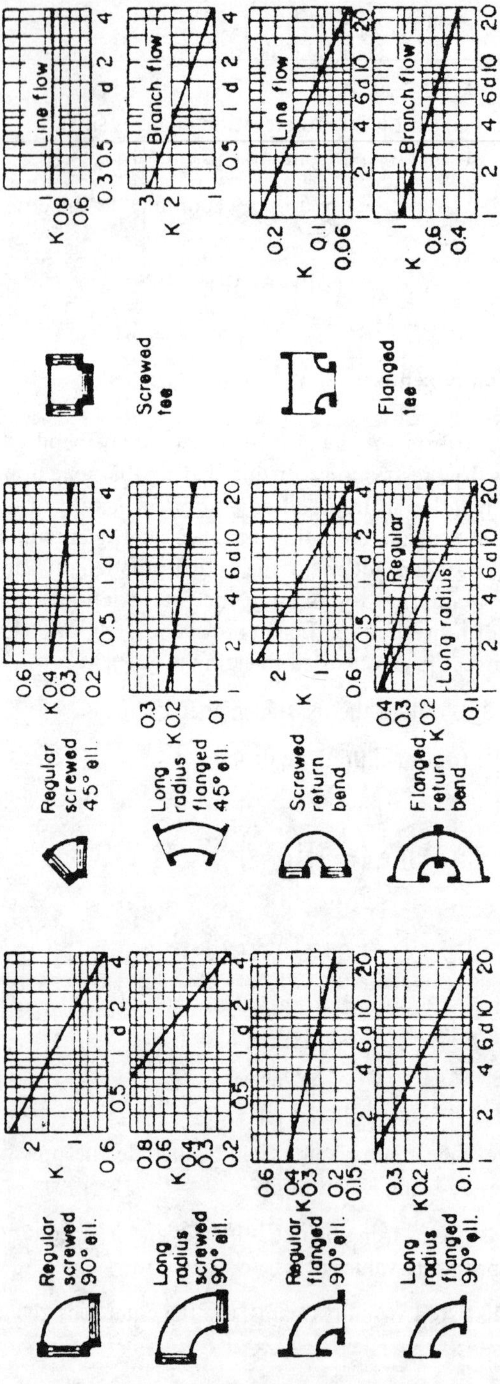

Figure 3. Resistance coefficients for bends, ells and tees [6].

Table 6. Pressure Drop Equations for Split and Join Pipe Flow Arrangements[a]

SPLIT FLOW

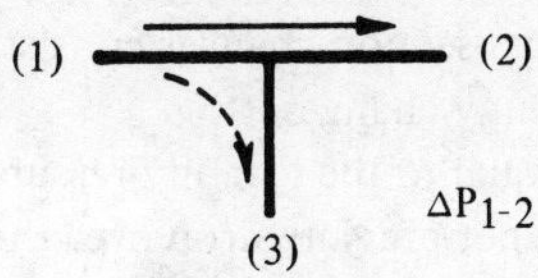

$$\Delta P_{1\text{-}2} = (1.08 \times 10^{-4})\rho(1.36U_2^2 - 0.64U_1^2 - 0.72U_1U_2) \quad (34)$$

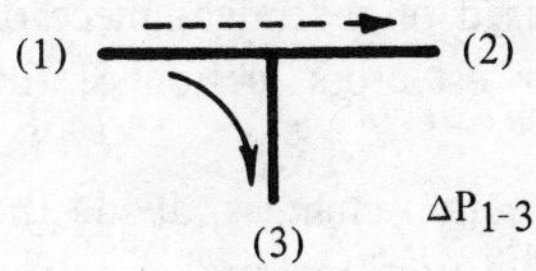

$$\Delta P_{1\text{-}3} = (1.08 \times 10^{-4})\rho(1.8U_3^2 - 0.37U_1U_3) \quad (35)$$

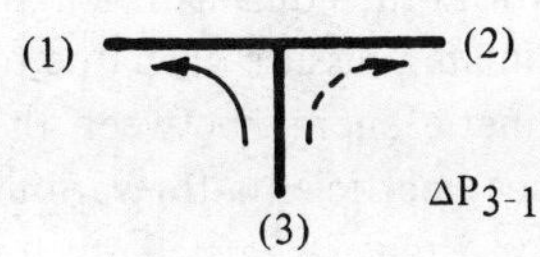

$$\Delta P_{3\text{-}1} = (1.08 \times 10^{-4})\rho(1.8U_1^2 - 0.37U_1U_3) \quad (36)$$

JOIN FLOW

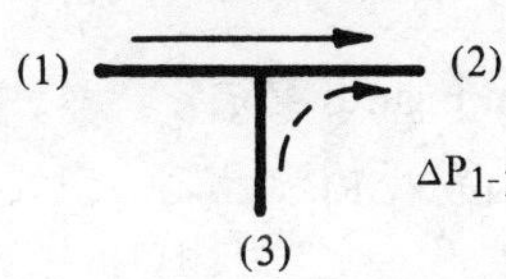

$$\Delta P_{1\text{-}2} = (1.08 \times 10^{-4})\rho\left\{2U_2^2 - 0.05U_1^2 - 2U_2\left(0.205U_3\frac{Q_3}{Q_2} + U_1\frac{Q_1}{Q_2}\right)\right\} \quad (37)$$

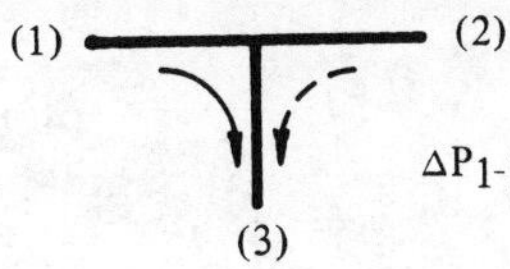

$$\Delta P_{1\text{-}3} = (1.08 \times 10^{-4})\rho\left\{2U_3^2 - 0.4U_1^2 - 0.41U_3\left(U_1\frac{Q_1}{Q_3} + U_2\frac{Q_2}{Q_3}\right)\right\} \quad (38)$$

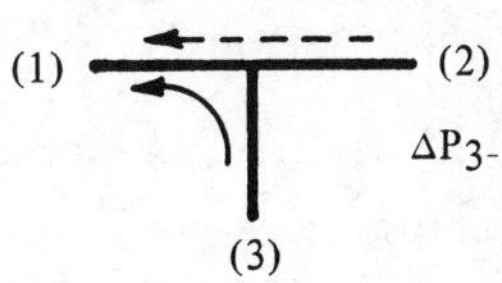

$$\Delta P_{3\text{-}1} = (1.08 \times 10^{-4})\rho\left\{2U_1^2 - 0.4U_3^2 - 2U_3\left(0.205U_3\frac{Q_3}{Q_1} + U_2\frac{Q_2}{Q_1}\right)\right\} \quad (39)$$

[a]Equations 34 through 39 account for both frictional pressure drop and pressure drop due to changes in kinetic energy. To allow for entrance and exit effects apply a multiplying factor of 1.3 to the computed pressure drop. Equations 34 through 39 are in English units.

Expansions and Contractions

Examples of cross section changes are exits and entrances of process vessels, reducers and diffusers. The pressure drop due to cross section changes consists of two components, namely friction and kinetic energy changes.

For pipe entrances into process vessels, $(\Delta P)_f$ is equal to the gain in pressure caused by the change in kinetic energy. Hence, the net pressure drop over the cross section change is zero.

For gradual contractions, $(\Delta P)_f$ is computed based on a straight piece of pipe with the inside diameter equal to the narrowest cross section of the contraction.

For lines containing fittings and various cross section changes, divide the line into sections of constant nominal diameter. The frictional pressure drop of each change in cross section can be accounted for in the equivalent length of the smaller diameter pipe attached to it. To estimate pressure drop due to changes in kinetic energy, compute the overall kinetic energy between the inlet and outlet of the line. Figure 4 gives resistance coefficients for various entrance and exit configurations. Figure 5 gives resistance coefficients for cross section changes.

The following procedure can be used to estimate frictional pressure drop when handling expansions and contractions:

Step 1. Obtain values for resistance coefficients from Figures 4 or 5.

Step 2. The following equation can be used to compute $(\Delta P)_f$:

$$(\Delta P)_f = \frac{K}{144}\left(\frac{\rho u^2}{2g_c}\right) \tag{40a}$$

$$= (1.08 \times 10^{-4})(K\rho u^2) \tag{40b}$$

$$= 0.28\left(\frac{KW^2}{\rho d^4}\right) \tag{40c}$$

where d = inside diameter (or equivalent hydraulic diameter) of the smaller pipe, in inches:

u = fluid velocity in smaller pipe, fps

Step 3. Compute the pressure drop due to kinetic energy changes:

$$(\Delta P)_K = \left(\frac{1}{144}\right)\left(\frac{\rho}{2g_c}\right)(u_2^2 - u_1^2) \tag{41a}$$

$$= (1.08 \times 10^{-4})\,\rho(u_2^2 - u_1^2) \tag{41b}$$

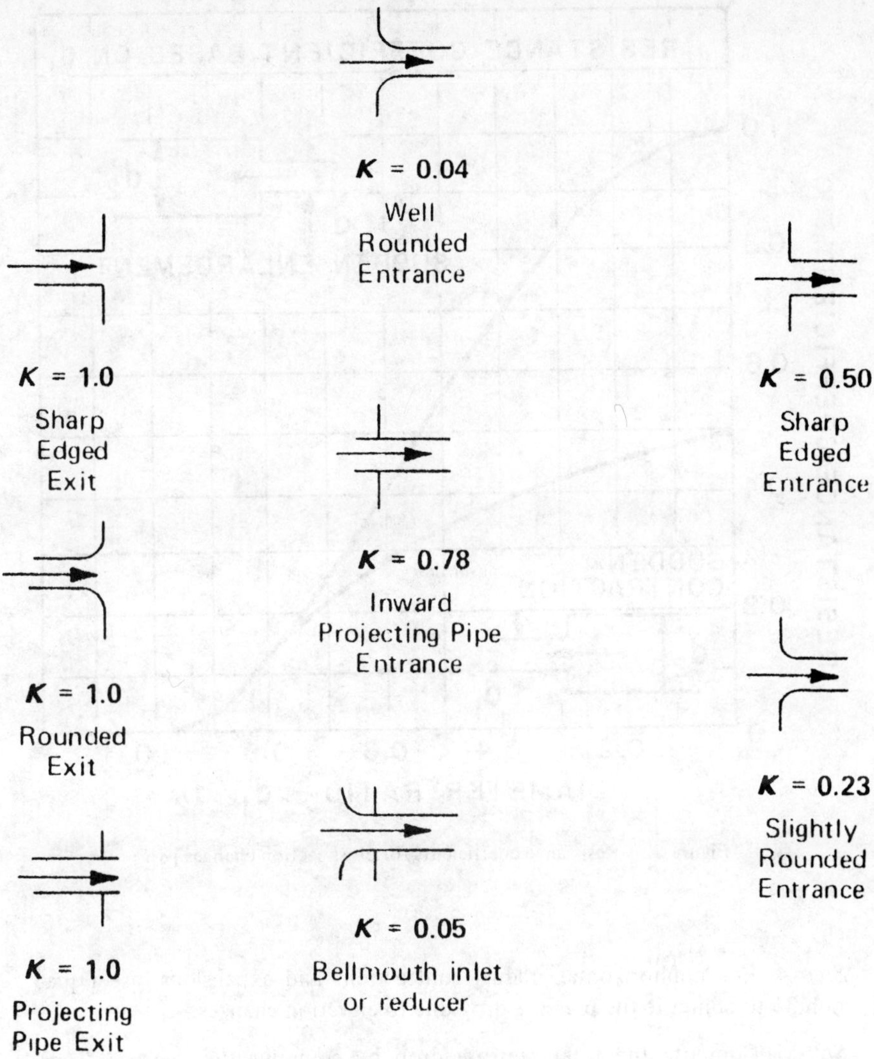

Figure 4. Resistance coefficients for various entrance and exit configurations [7].

$$= 0.28\left(\frac{W^2}{\rho}\right)\left(\frac{1}{d_2^4} - \frac{1}{d_1^4}\right) \qquad (41c)$$

where d_1, d_2 are the upstream and downstream diameters (or hydraulic diameters) in inches, respectively, and u_1, u_2 are the upstream and downstream velocities in feet per second, respectively.

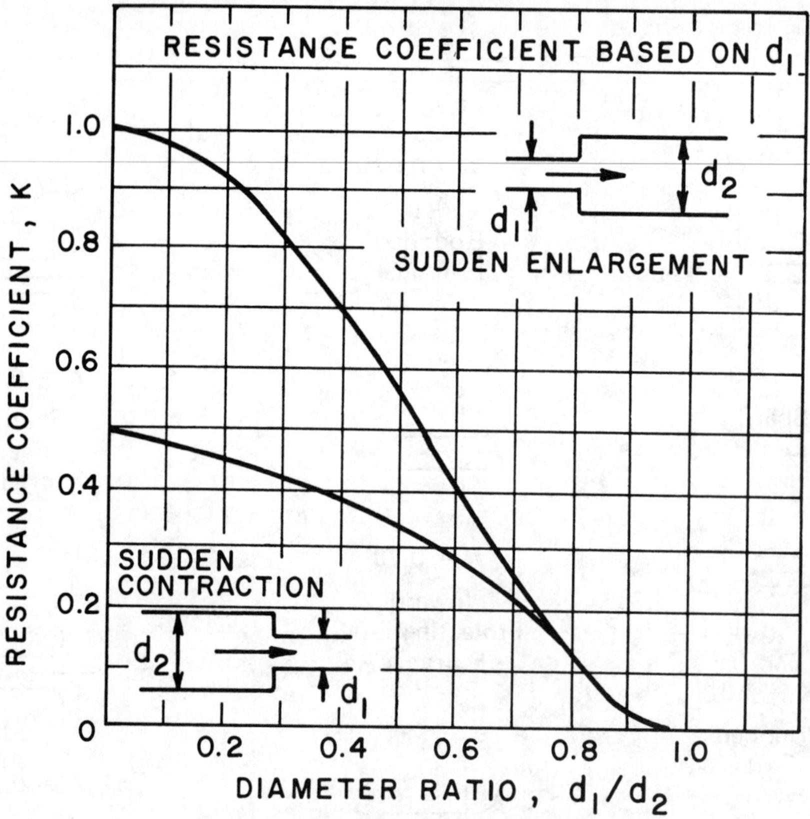

Figure 5. Resistance coefficients for cross section changes [8].

Step 4. For nonhorizontal gradual contractions and expansions, use Equation 31 to compute the pressure drop due to elevation changes.

Step 5. Compute the total pressure drop by summing the pressure drops obtained from Equations 31, 33 and 41.

OVERALL PRESSURE DROP CALCULATION PROCEDURE FOR PIPING SYSTEMS

For pipe systems having more than one piping component, divide the system up into sections of constant flowrate and constant nominal diameter. The following calculation procedure can be applied to each section.

Step 1. Compute the Reynolds number for each section, using equivalent hydraulic diameters for those sections having noncircular cross section.

Step 2. Using computed Reynolds numbers, obtain values for the friction factor from Figure 1 for each section.

Step 3. Determine all fittings in the system and find their total equivalent length:

$$L_{eq} = \left(\frac{d}{48f}\right) \Sigma K \tag{42}$$

where L_{eq} = equivalent length of all fittings (ft)
ΣK = sum of resistance coefficients of all fittings
d = nominal pipe diameter (in.)

For K values for orifices, nozzles and venturis, refer to the next chapter.

Step 4. Compute the frictional pressure drop using Equation 30, where L consists of the sum of the actual pipe lengths and the equivalent lengths of all fittings:

$$L = \text{length straight pipe} + L_{eq} \tag{43}$$

(For equivalent lengths of various valves see Chapter 2.3).

Step 5. Compute the overall kinetic energy change $(\Delta P)_k$. Check the flow cross section at the beginning and end of each constant flow section. If they are not equal, use Equation 41 to compute $(\Delta P)_k$. (Note that values of $(\Delta P)_k$ may be positive or negative.)

Step 7. Use Equations 34 through 39 in Table 6 to compute pressure drops for tees and Y's. Combine these with the pressure drops, $(\Delta P)_f$, across the various constant-flow sections.

Step 8. Compute the overall or total pressure drop for each constant flow section:

$$(\Delta P)_t = \Sigma(\Delta P)_f + (\Delta P)_k + (\Delta P)_e \tag{44}$$

where $(\Delta P)_t$ is the total pressure drop and $\Sigma(\Delta P)_f$ is the sum of the frictional pressure drops in all constant diameter sections, in psi.

GENERAL HEAD LOSS EQUATIONS

The following is a list of principal semitheoretical flow equations for estimating frictional head losses.

Darcy-Weisbach Equation

The Darcy-Weisbach equation can be used to estimate friction head losses:

$$h_L = f\frac{Lu^2}{2gD_H} = f\frac{Lu^2}{8gR_H} = \frac{\Delta P}{g\rho} \tag{45}$$

where h_L = friction head loss (ft of flowing fluid)
f = friction factor
L = length of pipe (ft)
D_H = hydraulic radius (ft)
R_H = hydraulic radius (ft)
u = fluid velocity (fps)
g = acceleration due to gravity $(32.2\ \text{ft/s})^2$

Hazen-Williams Equation

For fluid velocity, the Hazen-Williams equation is

$$u = 1.318\, CR_H^{0.63} S^{0.54} \tag{46}$$

where C = coefficient of roughness
S = slope of the energy line = h_L/L
u = fluid velocity (fps)

A more useful form of the equation is

$$h_L = \frac{4.73L}{C^{1.852}D^{4.87}}\, Q^{1.852} \tag{47}$$

where L and D are in ft and Q is the volumetric flowrate in ft^3/s. Typical values for the Hazen-Williams coefficient, C, are given in Table 7.

Manning Equation

The Manning equation is an empirical expression originally developed for open-channel flows; however, it can be applied to full-pipe flow:

Table 7. Typical Values of the Hazen-Williams Coefficient

Pipe Material	C
Very to Highly Smooth Pipes (all metals)	130-140
Smooth Wood	120
Smooth Masonry	120
Vitrified Clay	110
Cast Iron (old)	100
Iron (worn/pitted)	60-80
Polyvinyl Chloride (PVC)	150
Brick	100

$$u = \frac{1.486}{\beta} R_H^{2/3} S^{1/2} \tag{48}$$

or

$$h_L = \frac{4.666 \beta^2 L}{D^{5.333}} Q^2 \tag{49}$$

Values for the flow correction factor, β, given in Table 8, and the above equations are in English units.

PIPE NETWORK ANALYSIS: HARDY-CROSS METHOD

Pipe networks may include series pipes, parallel pipes and branching pipes, in addition to elbows, valves, meters and other devices that cause local disturbances and minor losses. There are several calculation techniques available to analyze flowrates and pressures or head losses throughout the pipe system. By dividing the pipe network into a series of loops, a system of equations can be prepared to describe flow conditions in each pipe section. For relatively simple pipe networks the following hand calculation procedure can be used. The method also can be readily written in the form of a computer program.

The Hardy-Cross method is an adaptation of the Newton-Raphson method. It solves one equation at a time before proceeding to the next equation during each iteration, instead of solving all equations simultaneously. For hand calculations of relative small pipe networks this method is acceptable. The starting equation is

$$\Delta Q_\ell^{(m+1)} = \Delta Q_\ell^{(m)} - \frac{F_\ell^{(m)}}{\dfrac{dF_\ell^{(m)}}{d(\Delta Q_\ell)}} \tag{50}$$

Table 8. Range of β Values for Various Pipe Materials

Pipe Material	
Cast Iron	0.011–0.015
Wrought Iron (black)	0.012–0.015
Wrought Iron (galvanized)	0.013–0.017
Smooth Brass	0.009–0.013
Glass	0.009–0.013
Riveted and Spiral Steel	0.013–0.017
Clay Drainage Tile	0.011–0.017
Concrete	0.012–0.016
Concrete Lined	0.012–0.018
Concrete-Rubble Surface	0.017–0.030
PVC	0.008–0.012
Wood	0.010–0.013

Usually the method applies only one iterative correction to each equation before proceeding to the next equation (even though several variations exist). After one iterative correction is applied to all equations the procedure is repeated until convergence.

In the procedure, the initially assumed flowrates in all pipes in the loop of that equation are adjusted on computing each ΔQ. Thus, each equation ($F_\ell = 0$) is evaluated with all ΔQ's summing to zero. Hence, $\Delta Q_\ell^{(m)} = 0$, so Equation 50 reduces to

$$\Delta Q = -\frac{F_\ell}{dF_\ell/d\Delta Q_\ell} \tag{51}$$

The head loss around the loop is

$$F_\ell = \Sigma K_i Q_i^{n_i} \tag{52}$$

and the derivative of F_1 is

$$\frac{dF_\ell}{d\Delta Q_\ell} = \Sigma | n_i K_i Q_i^{n_i - 1} | \tag{53}$$

By combining the above expressions the principal equation of the Hardy-Cross method is obtained, which allows computation of the flowrate for each loop in the pipe network:

$$\Delta Q = -\frac{\Sigma K_i Q_i^{n_i}}{\Sigma | n_i K_i Q_i^{n_i - 1} |} \tag{54}$$

Common practice is to use the Hazen-Williams equation to define the loss factor, K, and the exponential term, n:

$$\Delta Q = -\frac{\Sigma(K)_i Q_i^{1.852}}{1.852\ \Sigma|(K)_i Q_i^{0.852}|} \tag{55a}$$

$$= -\frac{\Sigma(h_f)_i}{1.852\ \Sigma|(h_f/Q)_i|} \tag{55b}$$

The Hardy-Cross calculation method is outlined in the following procedure:

Step 1. Assume an initial flowrate for each pipe such that all junction continuity equations are satisfied.

Step 2. Compute the sum of head losses around a loop of the network. Care must be taken to maintain proper signs. For example, if the direction of movement is opposite to the direction of flow in the loop, then h_f is negative. This step provides the numerator of Equation 55.

Step 3. Compute the denominator of Equation 55. Accumulate the absolute values of $n_i K_i Q_i^{n_i - 1}$ around the same loop.

Step 4. Evalute ΔQ from Equation 55.

Step 5. Repeat steps 2 through 4 for each network loop.

Step 6. Repeat the iterative procedure (steps 2 through 5) until all ΔQ's computed are sufficiently small to be neglected.

The calculation method is illustrated by the following example problem:

A pipe network can be divided into two flow loops shown in Figure 6, which shows line sizes, line lengths, and input and output flow streams. The system is a water distribution network constructed from new cast iron pipe. Compute the flowrate in each pipe section.

Solution: From Table 7 a value for the Hazen-Williams coefficient is obtained; K = C = 130.

An initial guess of the flowrates in each pipe are selected, obeying continuity at the junction of each pipe section. Based on these flowrates, the numerator and denominator of Equation 55 are computed. In this problem, head losses are computed in the clockwise direction so that we must be careful of signs. Table 9 tabulates values for the iterative procedure.

For the first iteration the difference for loop (I) is

$$\Delta Q_I = -\left\{\frac{1197}{1.852(1320.7)}\right\} = -0.49 \text{ cfs}$$

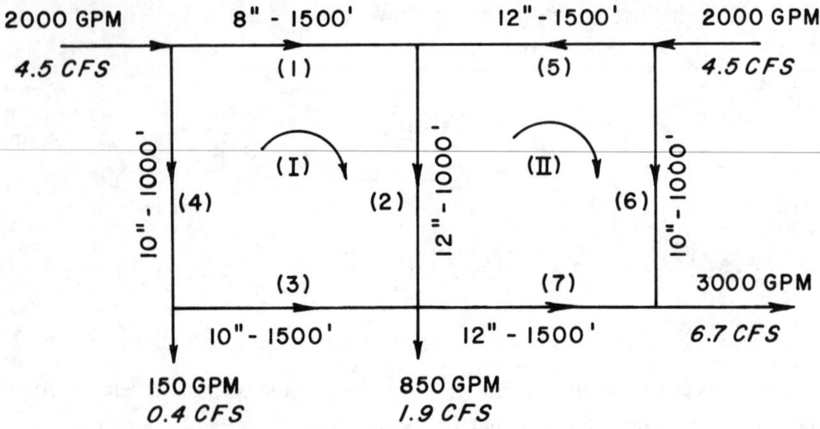

Figure 6. Pipe network for example problem.

For loop (II),

$$\Delta Q_{II} = -\left\{\frac{-5849}{1.852(1593)}\right\} = 1.98$$

These differences are unacceptable. Adding the ΔQ's to the initial flowrates provides estimates for the second iteration (refer again to Table 9).

For the second iteration,

$$\Delta Q_{I} = -\left\{\frac{-143}{1.852(1317)}\right\} = 0.06$$

$$\Delta Q_{II} = -\left\{\frac{-1740}{1.852(1242)}\right\} = 0.76$$

ΔQ_I is acceptable; ΔQ_{II} is not acceptable. A third trial is needed for loop (II), so ΔQ_{II} is added to the second iteration flowrates. Finally, for the third iteration convergence is achieved:

$$\Delta Q_{II} = -\left\{\frac{-100.6}{1.852(1087)}\right\} = 0.05$$

This procedure is also widely used with the Darcy-Weisbach equation to determine head losses; however, it is slightly more cumbersome via hand

Table 9. Calculations for Example Problem

Iteration	Loop (I) Pipe Section	Q (cfs)	$KQ_i^{1.852}$	$\lvert KQ_i^{0.852} \rvert$	Loop (II) Pipe Section	Q (cfs)	$KQ_i^{1.852}$	$\lvert KQ_i^{0.852} \rvert$
1	1	1.5	275.5	183.6	5	-3.5	-1323	378
	2	5.0	2561.1	512.2	6	1.0	130	130
	3	-2.6	-645.4	293.4	7	-5.7	-3265	573
	4	-3.0	-994.4	331.5	8	-5.0	-2561	512
	Totals		1197	1320.7			-5849	1593
2	1	1.0	130	130	5	-1.5	-275	184
	2	4.5	2107	468	6	3.0	995	331
	3	-3.1	-1057	341	7	-3.7	-1466	396
	4	-3.5	-1323	378	8	-3.0	-994	331
	Totals		-143	1317			-1740	1242
3	1	1.06			5	-0.74	-74.4	100.6
	2	4.56			6	3.76	1510.8	402
	3	-3.16			7	-2.94	-958	326
	4	-3.56			8	-2.24	-579	258
							-100.6	1087

calculations. Relatively simple computer programs can be written for the Hardy-Cross method.

If a computer with sufficient storage is available, a faster converging scheme based on the Newton-Raphson method should be used.

PIPE NETWORK ANALYSIS: NEWTON-RAPHSON METHOD

The Newton-Raphson method is an iterative scheme that starts with an estimate to the solution and iteratively computes better estimates. The method has quadratic convergence in comparison to most techniques, which have linear convergence. With quadratic convergence, fewer iterations usually are required to obtain a solution within a specified precision. The method is readily implemented in a computer algorithm. In this method, the equation containing the unknown is expressed as a function that equals zero when the correct solution is substituted into the equation. That is,

$$F(x) = 0 \tag{56}$$

It computes progressively better estimates of the unknown x by the following:

$$x^{(m+1)} = x^{(m)} - F(x^{(m)})/dF^{(m)}/dx \tag{57}$$

The superscripts in Equation 57 represent the number of iterations (not exponents).

Application of Equation 57 is illustrated by Figure 7. For the initial guess $x^{(0)}$ in Equation 57, $F(x) \neq 0$ (since $x^{(0)}$ is not the solution). The derivative $(dF(x^{(0)})/dx)$ gives the slope of the curve in Figure 7. By projecting a line with this slope to the x-axis, a closer approximation $(x^{(1)})$ to the solution is obtained. And from similar triangles,

$$\frac{x^{(1)} - x^{(0)}}{F(x^{(0)})} = -\frac{1}{\frac{dF^{(0)}}{dx}} \tag{58}$$

Solving for $x^{(1)}$,

$$x^{(1)} = x^{(0)} - \frac{F(x^{(0)})}{\frac{dF^{(0)}}{dx}} \tag{59}$$

The procedure is repeated (i.e., $x^{(1)}, x^{(2)}, x^{(3)}, \ldots x^{(m+1)}$) to produce the function $F(x^{m+1})$, which approaches zero.

The method may be used to solve any of the sets of equations describing flow in pipe networks, namely, the equations considering: (1) the flowrate in each pipe unknown; (2) the head at each junction unknown; or (3) the corrective flowrate around each loop unknown.

In vector notation the Newton-Raphson formula for a system of equations is

$$\vec{x}^{(m+1)} = \vec{x}^{(m)} - D_J^{-1}\vec{F}(x^m) \tag{60}$$

In Equation 60 the unknown vectors are $\vec{x}$ and $\vec{F}$, and D_J^{-1} is the inverse of the Jacobian (D_J^{-1} replaces $1/dF/dx$ in the Newton-Raphson formula for solving a single equation).

A system of equations can be represented in matrix form:

$$\vec{H} = \begin{vmatrix} H_1 \\ H_2 \\ \vdots \\ H_n \end{vmatrix} \quad \text{or} \quad \Delta\vec{Q} = \begin{vmatrix} \Delta Q_1 \\ \Delta Q_2 \\ \vdots \\ \Delta Q_n \end{vmatrix} \tag{61}$$

where H represents a system of equations for computing head losses and Q is a system for volumetric flowrates.

The Jacobian matrix is composed of derivative elements in which individual rows represent derivatives of that particular functional equation with respect to the variables making up the column headings. For a system of head equations the Jacobian is

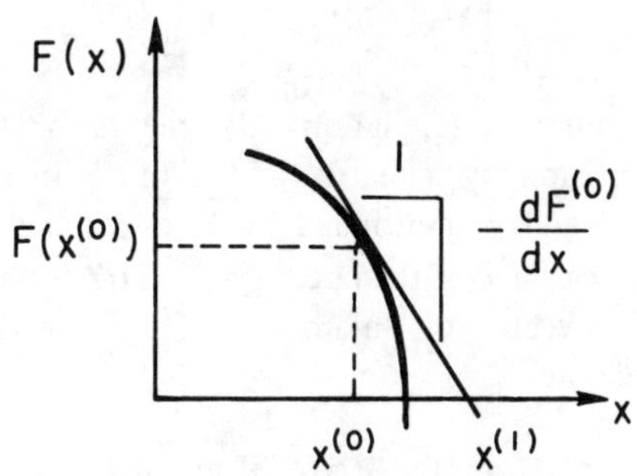

Figure 7. Illustrates application of the Newton-Raphson method to a single-equation solution.

Table 10. Pipe Capacities

Pipe Size (in.)	Water		70 SSU		100 SSU		150
	Gravity	Pressure	Gravity	Pressure	Gravity	Pressure	Gravity
3/4	1.42	4.70	0.58	4.21	0.39	4.01	0.25
1	2.65	9.20	1.04	8.27	0.692	7.85	0.432
1 1/4	5.30	18.9	3.24	16.9	2.02	16.1	1.30
1 1/2	8.10	28.4	5.90	25.5	3.74	24.2	2.38
2	15.60	54.7	11.95	49.0	10.22	46.7	6.55
2 1/2	25.10	87.4	23.04	78.3	19.50	77.0	13.32
3	44.50	154	33.2	138	29.5	131	27.1
4	91.00	317	80.5	284	77.0	270	72
5	164	573	139	514	131	489	123
6	267	930	212	834	202	794	187
8	550	1910	469	1710	436	1630	401
10	1010	3480	940	3120	885	2970	825
12	1610	5590	1398	5010	1305	4770	1220
14	2160	7250	1880	6500	1780	6190	1650
16	3020	10490	2610	9410	2470	8760	2300
18	4100	14500	3580	13200	3350	12400	3100
20	5500	19180	4850	17200	4600	16400	4300

$$D_J = \begin{bmatrix} \frac{\partial F_1}{\partial H_1} & \frac{\partial F_1}{\partial H_2} & \cdots & \frac{\partial F_1}{\partial H_n} \\ \frac{\partial F_2}{\partial H_1} & \frac{\partial F_2}{\partial H_2} & \cdots & \frac{\partial F_2}{\partial H_n} \\ \vdots \\ \frac{\partial F_m}{\partial H_1} & \frac{\partial F_m}{\partial H_2} & \cdots & \frac{\partial F_m}{\partial H_n} \end{bmatrix} \tag{62}$$

In the principal equation (Equation 60), the term $D^{-1}F$ contains the inverse Jacobian. Division by a matrix is undefined. In application D^{-1} is never really obtained and premultiplied by $\vec{F}$, as implied by Equation 60. Instead, the solution vector $\vec{Z}$ of the linear system $D_J\vec{Z} = \vec{F}$ is subtracted from the previous iterative vector of unknowns. In practice then, the Newton-Raphson formula becomes

$$\vec{H}^{(m+1)} = \vec{H}^{(m)} - \vec{Z}^{(m)} \tag{63}$$

Fewer computations are required to solve the linear system $D\vec{Z} = \vec{F}$ than to determine D^{-1}. The Newton-Raphson method obtains the solution to a

Flow Data (gpm)

SSU	200 SSU		300 SSU		500 SSU	
Pressure	Gravity	Pressure	Gravity	Pressure	Gravity	Pressure
3.79	0.18	3.64	0.12	3.45	0.068	3.21
7.43	0.324	7.14	0.216	6.75	0.130	5.66
15.3	0.943	14.7	0.634	13.9	0.389	12.9
22.9	1.73	20.9	1.17	19.8	0.706	19.4
44.2	4.86	42.4	3.24	40.1	1.94	37.4
70.5	9.65	67.8	6.55	64.2	3.92	59.8
124	23.2	119	15.7	113	9.22	105
256	69.8	246	54.0	233	32.3	217
463	116	445	99.10	421	71.0	392
751	176	722	159	683	143	570
1540	379	1480	350	1400	312	1310
2810	786	2700	735	2550	670	2380
4519	1106	4340	1085	4100	995	3820
5850	1580	5620	1470	5320	1350	4960
8480	2180	8150	2020	7700	1840	7180
13200	2920	12600	2720	11000	2500	10100
15500	4100	14900	3860	14100	3580	13100

system of nonlinear equations via iteratively solving a system of linear equations. A final point is that the method does require a reasonably accurate initialization, or else convergence may not be achieved.

FLOW CAPACITY AND PRESSURE DROP DATA

Data on flow capacities and pressure drop for water flowing through pipe are given below.

Table 10 gives pipe flow capacities in gpm for different viscosity fluids. Viscosities are given in units of Standard Saybolt Units (SSU). Values given are based on a head loss due to friction of fluids in a given pipe size for smooth pipe (C = 100) and are considered to be conservative. Pressure piping losses are based on a loss of 10 feet per 100 feet of pipe. For short runs and new fittings, use the next size smaller pipe. For long runs and/or many fittings, use the next larger size pipe.

Table 11 gives the pressure drop of water through Schedule 40 steel pipe. Values are based on the Saph and Schoder formula:

$$\Delta P = \frac{LQ^{1.86}}{1435\ D^5} \tag{64}$$

Table 11. Pressure Drop of Water Through Schedule 40 Steel Pipe

Flow		Pressure Drop of Water per 100 ft of Schedule 40 Steel Pipe (psi)													
gpm	ft^3/s	u fps	p psi	u fps	p psi	u fps	p psi	u fps	p psi	u fps	p psi	u fps	p psi	u fps	p psi
		1/8 in.		1/4 in.		3/8 in.		1/2 in.		3/4 in.					
0.1	0.00022	0.56	0.677												
0.2	0.00045	1.14	2.48	0.62	0.548										
0.3	0.00067	1.70	5.26	0.93	1.16	0.50	0.255								
0.5	0.00111	2.82	13.58	1.55	3.00	0.84	0.656	0.53	0.205	0.30	0.050				
0.6	0.00134	3.38	19.12	1.85	4.22	1.01	0.925	0.63	0.290	0.36	0.071				
												1 in.		1 1/4 in.	
0.8	0.00178	4.52	32.62	2.47	7.17	1.34	1.58	0.84	0.494	0.48	0.121	0.30	0.036		
1	0.00223			3.09	10.91	1.68	2.39	1.06	0.749	0.60	0.183	0.37	0.055	0.21	0.014
2	0.00446			6.18	39.60	3.36	8.68	2.11	2.72	1.20	0.665	0.74	0.199	0.43	0.051
3	0.00668					5.04	18.46	3.17	5.77	1.80	1.41	1.11	0.424	0.64	0.107
4	0.00891					6.72	31.55	4.22	9.86	2.40	2.42	1.49	0.724	0.86	0.183
		1 1/2 in.													
5	0.01114							5.28	14.92	3.01	3.64	1.86	1.09	1.07	0.276
6	0.01337							6.33	20.95	3.61	5.13	2.23	1.54	1.29	0.390
				2 in.											
8	0.01782	1.26	0.308							4.81	8.76	2.97	2.62	1.71	0.667
10	0.02228	1.58	0.466							6.01	13.28	3.713	3.97	2.142	1.01
15	0.03342	2.36	0.992	1.43	0.285							5.57	8.46	3.21	2.14
						2 1/2 in.									
20	0.04456	3.15	1.69	1.91	0.486							7.43	14.42	4.28	3.66
25	0.05570	3.94	2.54	2.39	0.736									5.36	5.54
30	0.06684	4.73	3.60	2.37	1.03	2.01	0.424							6.43	7.79
35	0.07798	5.51	4.79	3.35	1.37	2.35	0.566							7.50	10.38
40	0.08912	6.30	6.14	3.82	1.76	2.68	0.724							8.57	13.28
								3 in.							
50	0.1114	7.88	9.31	4.78	2.67	3.35	1.10	2.17	0.371						
60	0.1337	9.45	13.08	5.74	3.75	4.02	1.54	2.61	0.520						

										3½ in.					
70	0.1560			6.70	4.99	4.70	2.05	3.04	0.693	2.27	0.335				
80	0.1782			7.65	6.40	5.37	2.63	3.47	0.890	2.59	0.430				
90	0.2005			8.60	7.96	6.04	3.28	3.91	1.10	2.92	0.535				
												4 in.			
100	0.2228			9.56	9.69	6.71	3.98	4.34	1.34	3.24	0.650	2.52	0.346		
125	0.2785					8.38	6.03	5.43	2.01	4.05	0.984	3.15	0.523		
														5 in.	
150	0.3342					10.1	8.46	6.52	2.86	4.87	1.38	3.78	0.734		
175	0.3899					11.7	11.3	7.60	3.81	5.68	1.84	4.41	0.978	2.81	0.316
200	0.4456					13.4	14.4	8.69	4.89	6.49	2.36	5.04	1.25	3.21	0.405
		6 in.													
225	0.5013							9.77	6.09	7.30	2.94	5.67	1.56	3.61	0.505
250	0.5570	2.78	0.245					10.9	7.41	8.11	3.58	6.30	1.90	4.01	0.616
275	0.6127	3.06	0.292					11.9	8.84	8.92	4.27	6.93	2.27	4.41	0.734
300	0.6684	3.33	0.344					13.0	10.4	9.73	5.02	7.56	2.67	4.81	0.863
350	0.7798	3.89	0.457					15.2	13.8	11.4	6.87	8.82	3.55	5.62	1.15
				8 in.											
400	0.8912	4.44	0.587	2.57	0.149					13.0	8.58	10.1	4.56	6.41	1.47
450	1.003	5.00	0.731	2.89	0.185					14.6	10.7	11.3	5.66	7.22	1.83
500	1.114	5.55	0.887	3.21	0.225					16.2	13.0	12.6	6.89	8.02	2.23
550	1.225	6.11	1.07	3.53	0.270					17.8	15.5	13.9	8.25	8.82	2.67
600	1.337	6.66	1.25	3.85	0.316					19.5	18.2	15.1	9.68	9.62	3.13
						10 in.									
650	1.449	7.22	1.45	4.17	0.367	2.65	0.118					16.4	11.2	10.4	3.62
700	1.560	7.78	1.66	4.49	0.420	2.85	0.135					17.6	12.9	11.2	4.16
750	1.671	8.33	1.89	4.81	0.480	3.05	0.154					18.9	14.7	12.0	4.75
800	1.782	8.89	2.13	5.13	0.540	3.26	0.173					20.2	16.5	12.8	5.35
850	1.894	9.44	2.38	5.45	0.605	3.46	0.194					21.4	18.5	13.6	5.98
								12 in.							
900	2.005	10.0	2.66	5.77	0.627	3.66	0.216	2.58	0.090			22.7	20.6	14.4	6.65
950	2.117	10.6	2.93	6.09	0.744	3.87	0.238	2.72	0.099			23.9	22.8	15.2	7.36
1000	2.228	11.1	3.23	6.41	0.817	4.07	0.262	2.87	0.109					16.0	8.10
										14 in.					
1100	2.451	12.2	3.85	7.06	0.975	4.48	0.313	3.15	0.130					17.6	9.66
1200	2.674	13.3	4.53	7.70	1.15	4.88	0.368	3.44	0.153	2.85	0.096			19.2	11.4

Table 11, continued

Flow		Pressure Drop of Water per 100 ft of Schedule 40 Steel Pipe (psi)													
gpm	ft^3/s	u fps	p psi	u fps	p psi	u fps	p psi	u fps	p psi	u fps	p psi	u fps	p psi	u fps	p psi
1300	2.896	14.4	5.26	8.34	1.33	5.29	0.427	3.73	0.178	3.08	0.111			20.8	13.2
1400	3.119	15.6	6.01	8.98	1.53	5.70	0.490	4.01	0.204	3.32	0.127	16 in.		22.4	15.1
1500	3.342	16.7	6.84	9.62	1.74	6.10	0.556	4.30	0.232	3.56	0.145			24.1	17.2
1600	3.565	17.8	7.73	10.3	1.96	6.51	0.628	4.59	0.262	3.79	0.163	2.91	0.084		
1800	4.010	20.0	9.64	11.5	2.46	7.32	0.782	5.16	0.329	4.27	0.203	3.27	0.104		
2000	4.456	22.2	11.6	12.8	2.97	8.14	0.953	5.73	0.396	4.74	0.247	3.63	0.127	18 in.	
2500	5.570	27.8	17.6	16.0	4.49	10.2	1.44	7.17	0.601	5.93	0.374	4.54	0.192		
3000	6.684			19.2	6.30	12.2	2.02	8.60	0.842	7.11	0.525	5.45	0.270	4.30	0.149
3500	7.798			22.4	8.41	14.2	2.70	10.0	1.12	8.30	0.700	6.36	0.358	5.02	0.199
4000	8.912			25.7	10.8	16.3	3.46	11.5	1.44	9.48	0.896	7.26	0.459	5.74	0.255
4500	10.03			28.9	13.4	18.3	4.31	12.9	1.76	10.7	1.12	8.17	0.671	6.45	0.317
5000	11.14					20.4	5.20	14.3	2.18	11.9	1.36	9.08	0.695	7.17	0.386
6000	13.37					24.4	7.35	17.2	3.06	14.2	1.91	10.9	0.977	8.60	0.542
7000	15.60					28.5	9.80	20.1	4.08	16.6	2.54	12.7	1.30	10.0	0.723
8000	17.82							22.9	5.22	19.0	3.25	14.5	1.67	11.5	0.926
9000	20.05							25.8	6.51	21.3	4.06	16.3	2.08	12.9	1.15
10000	22.28							28.7	7.91	23.7	4.92	18.2	2.53	14.3	1.40
12000	26.74									28.5	6.92	21.8	3.55	17.2	1.97
14000	31.19											25.4	4.72	20.1	2.62
16000	35.65											29.1	6.06	22.9	3.36
18000	40.10											32.7	7.55	25.8	4.18
20000	44.56													28.7	5.08

CHAPTER 2.3

PIPING SYSTEM COMPONENTS

This chapter provides descriptive information on piping system components, which include control valves and flowmetering devices. Criteria and procedures are outlined for identifying, selecting, sizing and installing piping system components.

TYPES OF CONTROL VALVES

Control valves provide the normal mechanism for adjusting inputs of process control systems. Proper valve selection is essential for a well-designed piping system. For systems requiring relatively short time lags and having large capacities, fairly simple control systems are adequate, normally permitting the use of high-sensitivity controllers with quick-opening valves. For this type service, flow characteristics become secondary considerations, and valve characteristics such as size and material become primary.

As time lag increases and capacity decreases, required control schemes become more complex. In general, the larger the time lag, the smaller and more precise must be the variations in flowrate of the controlling medium for given changes in process conditions.

Common control valve types are illustrated in Figures 8 through 12. Table 12 provides brief descriptions of various control valves.

Knowledge of where the various parts are located in different valve types will help to maintain them and thus maintain long service. Preventive maintenance such as regular lubrication, tightening or replacing the packing, or replacing composition discs or resurfacing seats and discs on a regular basis will greatly extend the life of a valve. Figures 13 and 14 show the basic parts of common valves.

A brief description of the major valve types and guidelines for their installation and maintenance are given below:

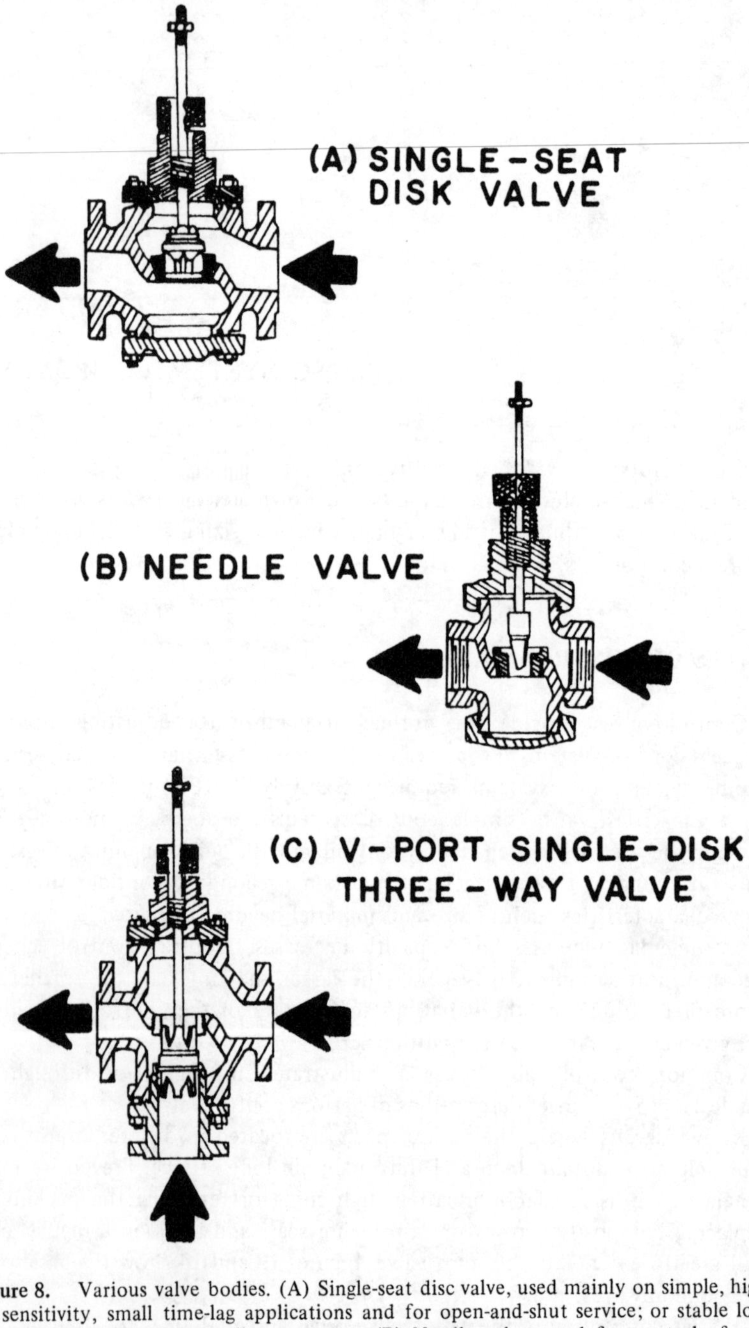

Figure 8. Various valve bodies. (A) Single-seat disc valve, used mainly on simple, high-sensitivity, small time-lag applications and for open-and-shut service; or stable load conditions and constant line pressures. (B) Needle valve, used for control of very small flows; employed generally where requirements are for less than 1/2-in. valve size. (C) V-port, single-disc three-way valve, suitable for mixing service.

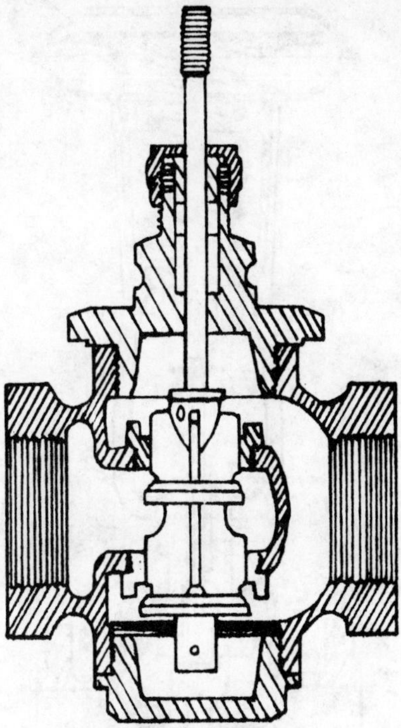

Figure 9. Double-seated, quick-opening inner valve with guided discs. These are used where close throttling flow is needed. Often used on low-sensitivity applications.

Gate Valves

As implied by the name, when the wheel is turned to lift the wedge of a gate valve it is literally like opening a flood gate. Fluid flows freely through a full port with little pressure drop. As with all standard valves, turning the handwheel counterclockwise opens it, while turning it clockwise closes it. Gate valves are neither meant to be throttling valves, nor are they designed for frequent operating cycles. Operation of a gate valve with the wedge partially open can cause vibration and rapid wear of the wedge and seat surfaces, leading to possible valve failure.

Nonrising stem gate valves are particularly useful when there are space limitations. The gate itself is threaded and rides up on the stem when the stem is rotated. Other variations include inside screw, resisting stem and outside screw and yoke (see Figure 13). "U" bolt gate valves were designed for quick disassembly and are recommended for use with viscous fluids where the fluid tends to collect in the bonnet structure.

Unless specially constructed, gate valves can be installed with either face to

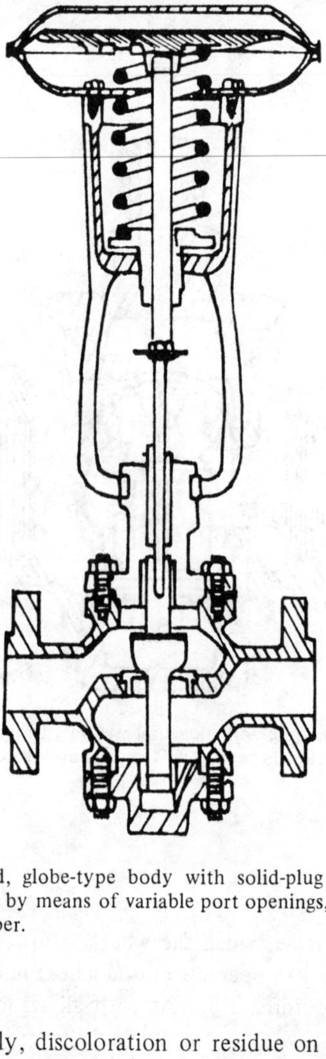

Figure. 10. Single-seated, globe-type body with solid-plug inner valve. The throttle plug controls flowrate by means of variable port openings, as determined by the curvature of the plug proper.

the inlet side. Normally, discoloration or residue on the bonnet indicates a packing leak. The packing nut should be tightened or the packing replaced. Packing material must be compatible with the fluid in service. Do not overtighten the packing nut. Tighten glad nuts evenly to prevent a cocked gland flange.

If valves are to be installed in a dirty or dusty environment, leather or metal stem jackets are recommended. A clean handwheel provides a firm grip when the valve is to be opened or closed.

Valves that require frequent use should be lubricated regularly. Most outside screw (OS) and yoke (Y) valves have a special grease fitting for this purpose.

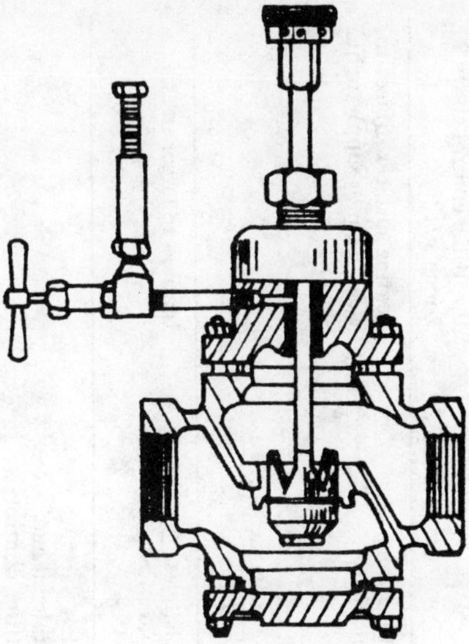

Figure 11. Illustrates a single-seat, V-port inner valve.

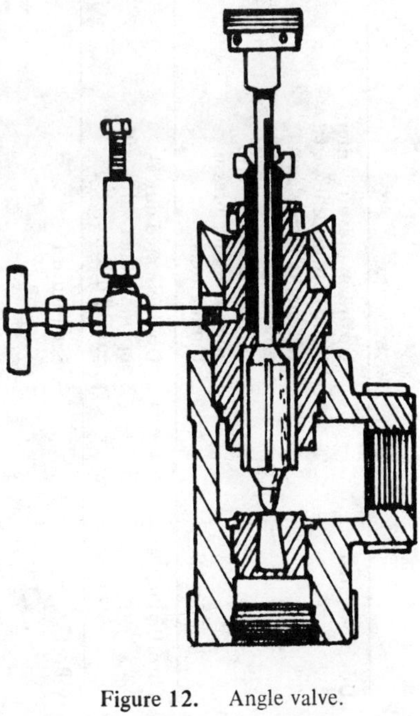

Figure 12. Angle valve.

Table 12. Descriptions of Common Control Valve Types

Valve Type	Description	Applications	Limitations/Disadvantages
Single-Seated	Quick-opening disc type.	Used on simple, high-sensitivity, small time-lag applications/open-and-shut service.	Impractical for throttling control if line pressures fluctuate widely, unless controller sensitivity is very high. Not recommended for throttling control above 2-inch size.
Double-Seated	Upstream pressure enters body between the seats, tending to create equal upward and downward forces. Balanced valve eliminates line pressure fluctuations.	Used for close throttling control, low-sensitivity applications.	Seating surfaces subject to wear. Cannot supply tight shutoff.
Throttle-Plug	Double-seated quick-opening inner valve with guided discs.	For large load changes. Handles large pressure drops well.	Limited to larger sizes.
V-Port	See Figure 11.	Used where widely varying flowrates encountered and where full throttling control over entire flow range is needed. Has greatest controllable range.	
Single-Seat Throttling	In sizes smaller than 1 inch they are generally preferred to double-seat type. Provide tight shutoff of controlling medium.	Same as single-seated.	Impractical for throttling control for lines with large pressure fluctuations.

Needle-Type	See Figure 8. Practically any required flow characteristic is possible.	For control of very small flows.	Limited to small sizes.
Three-Way Type	See Figure 8. Provide highly stable flows.	Service requirements include: • mixing fluids (e.g., mixing hot and cold water); • feeding two fluids to a vessel or pipeline, used as a throttling controller; • to divert fluid from one vessel or pipeline to two load demands.	
Angle Valves	Basically single-port valves.	Used to facilitate piping or where a self-draining piping system is needed. Used for handling fluids containing solids, slurries, flashing fluids.	Subject to shamming.
Butterfly Valves	Can be fitted with either manual or automatic valve positions. Main body may be diaphram-, cylinder-, float-, electric motor- or solenoid-operated.	For control of low-pressure, low-velocity fluids.	

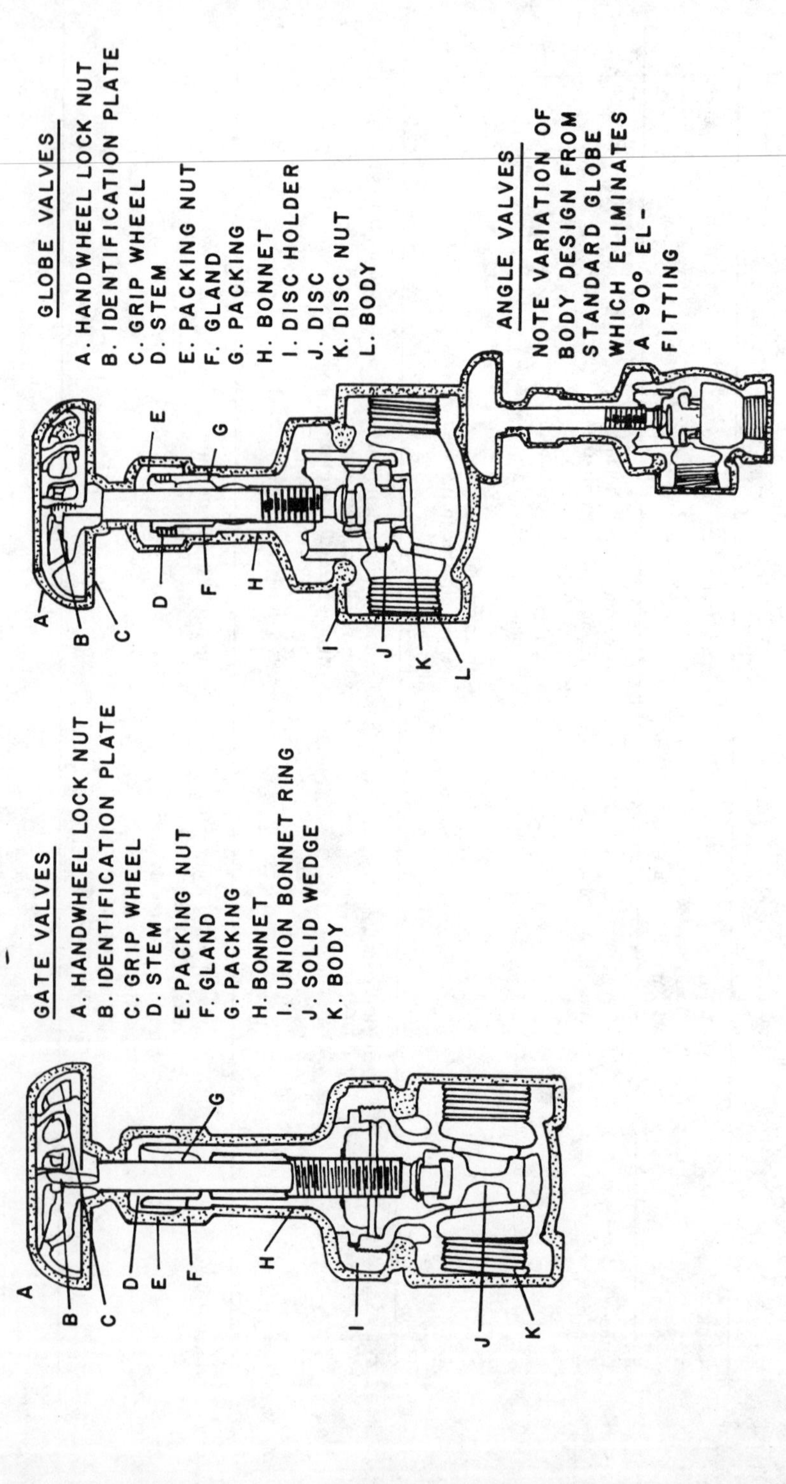
GATE VALVES
A. HANDWHEEL LOCK NUT
B. IDENTIFICATION PLATE
C. GRIP WHEEL
D. STEM
E. PACKING NUT
F. GLAND
G. PACKING
H. BONNET
I. UNION BONNET RING
J. SOLID WEDGE
K. BODY
GLOBE VALVES
A. HANDWHEEL LOCK NUT
B. IDENTIFICATION PLATE
C. GRIP WHEEL
D. STEM
E. PACKING NUT
F. GLAND
G. PACKING
H. BONNET
I. DISC HOLDER
J. DISC
K. DISC NUT
L. BODY
ANGLE VALVES
NOTE VARIATION OF BODY DESIGN FROM STANDARD GLOBE WHICH ELIMINATES A 90° EL-FITTING

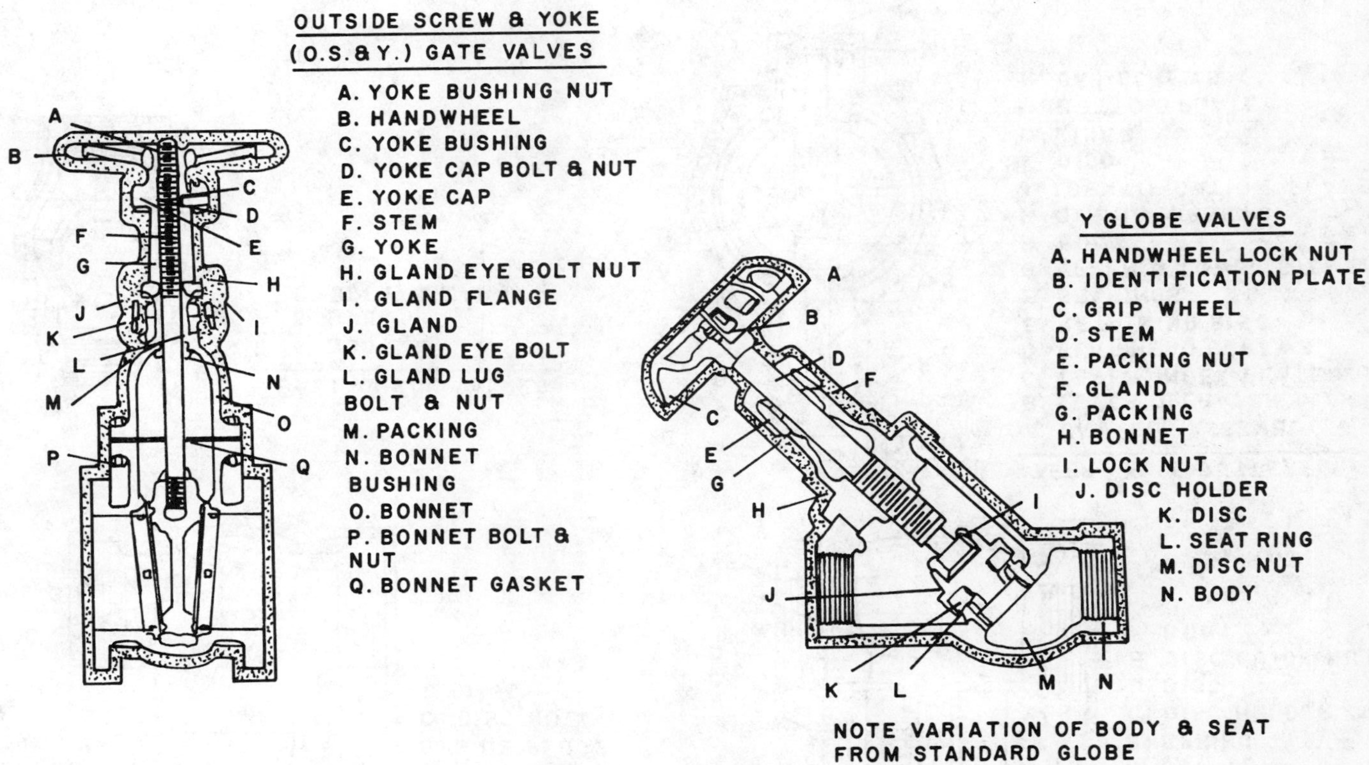

Figure 13. Basic parts of common valves. (courtesy of Jenkins Bros., New York, NY).

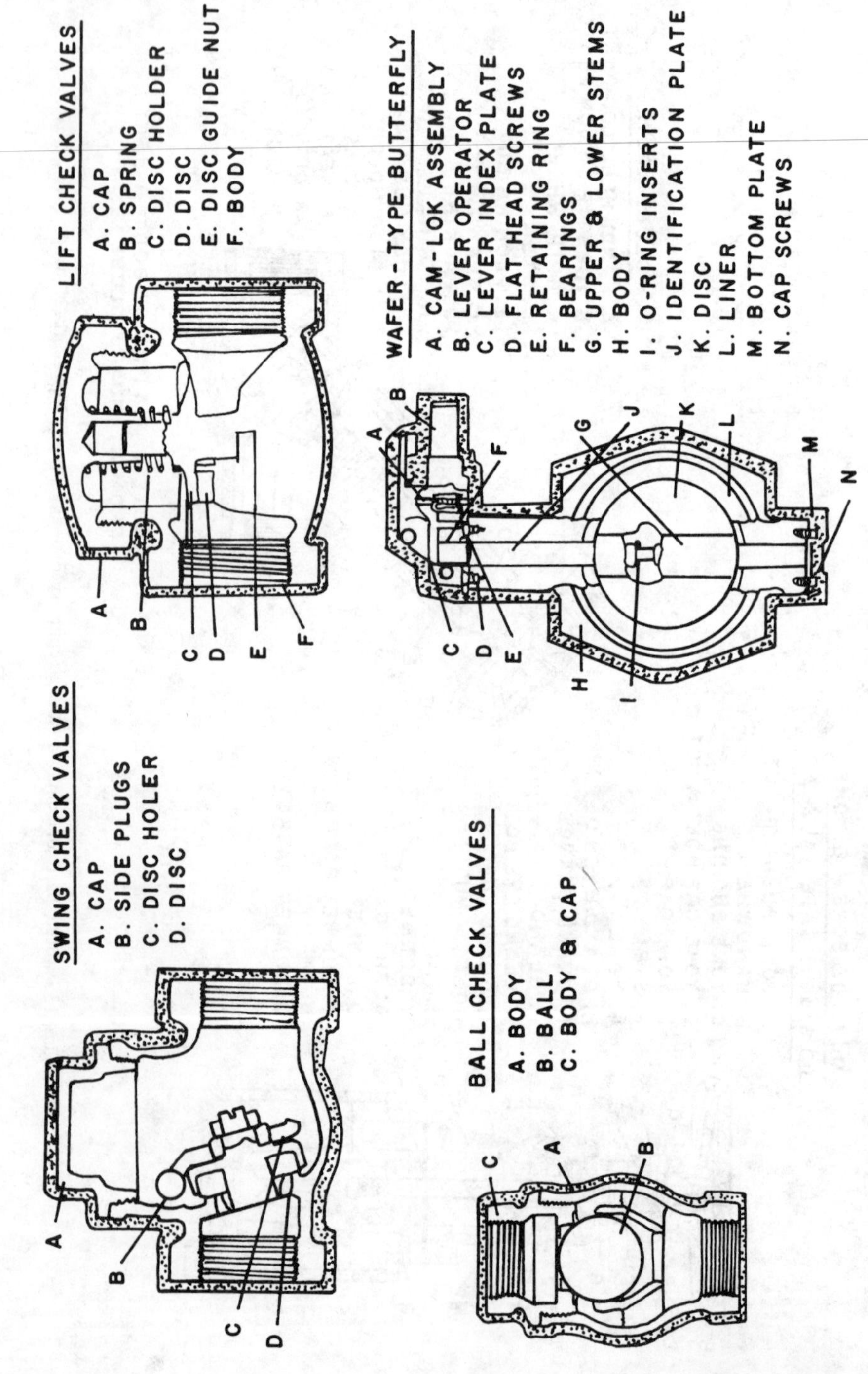
SWING CHECK VALVES
A. CAP
B. SIDE PLUGS
C. DISC HOLER
D. DISC
LIFT CHECK VALVES
A. CAP
B. SPRING
C. DISC HOLDER
D. DISC
E. DISC GUIDE NUT
F. BODY
BALL CHECK VALVES
A. BODY
B. BALL
C. BODY & CAP
WAFER-TYPE BUTTERFLY
A. CAM-LOK ASSEMBLY
B. LEVER OPERATOR
C. LEVER INDEX PLATE
D. FLAT HEAD SCREWS
E. RETAINING RING
F. BEARINGS
G. UPPER & LOWER STEMS
H. BODY
I. O-RING INSERTS
J. IDENTIFICATION PLATE
K. DISC
L. LINER
M. BOTTOM PLATE
N. CAP SCREWS

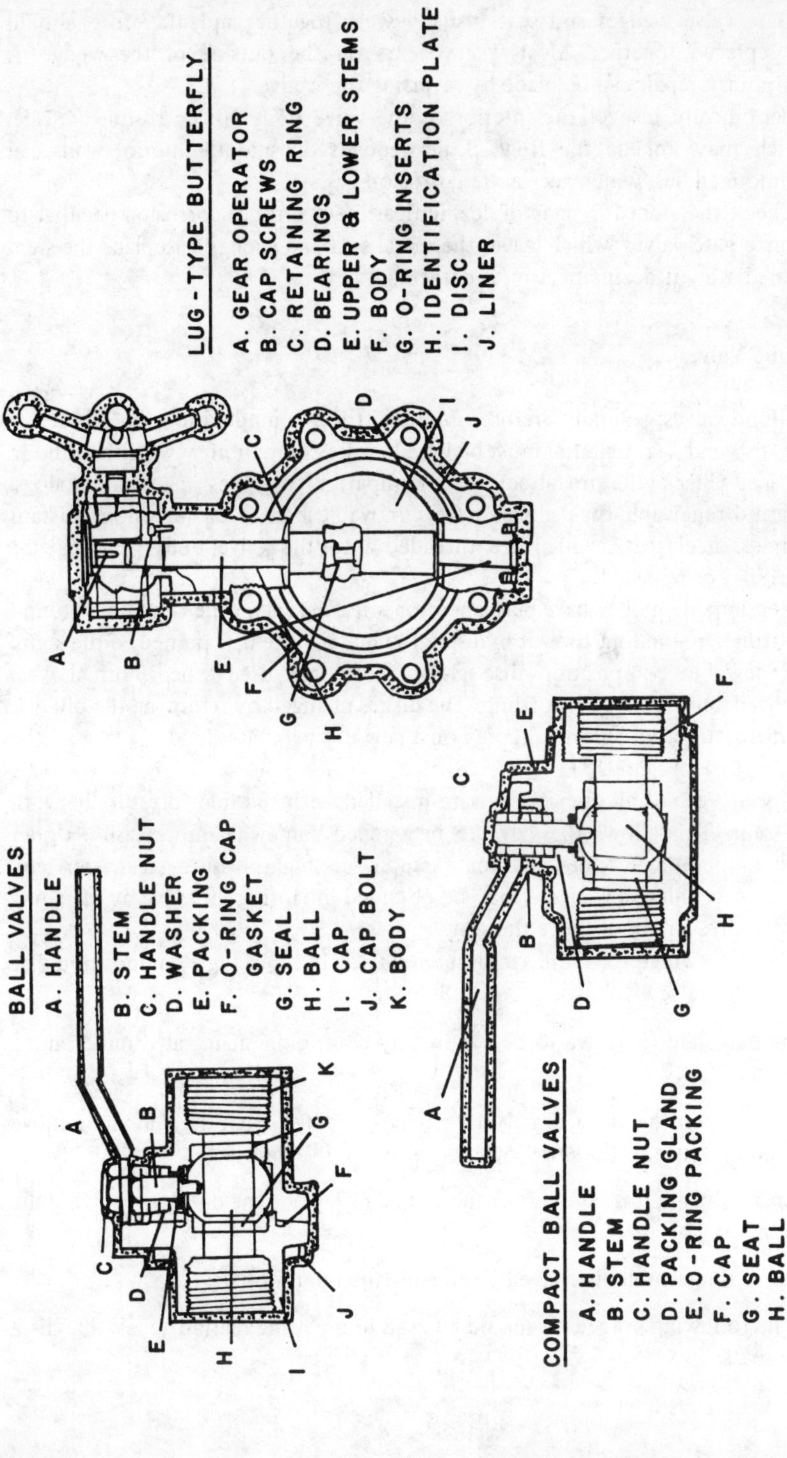

Figure. 14. Basic parts of common valves (courtesy of Jenkins Bros., New York, NY).

Gate valve wedges and seats usually wear together and, therefore, should be replaced together. Most wear occurs on the outside of the wedge. A temporary repair can be made by reversing the wedge.

Periodically inspect the interior of the valve body for corrosion or scale, which may impede the flow. Scale deposits on smooth interior walls can promote eddies, which accelerate corrosion.

Check the stem for signs of dezincification (a form of corrosion peculiar to bronze gate valves, which leaves the metal soft and spongy). Replace the stem immediately if dezincification is detected.

Globe Valves

Globe valves generally are used for throttling or modulation. The choice of the disc and seat material must be based on its compatibility with the fluid in service. Check the manufacturers' compatibility tables in their catalogs. Regrind/renewable-type globes have removable bronze or corrosion-resistant stainless-steel seats, which are threaded into the valve body and may be reground or replaced.

Regrindable globes have permanent or integral seats, which can be reground by using a regrinding tool or by using the metal valve disc pinned to the stem.

Renewable composition disc globe valves are used when normal wear results from frequent throttling. The disc is changed by removing the old disc from the disc holder and slipping on a new disc, which is held in place by the disc nut (see Figure 13).

About 90% of all globe valves are installed with the inlet pressure beneath the seat. This allows the valve to be opened somewhat more easily. Under severe operating conditions (for example, in high-pressure steam service), some mechanical advantage may be obtained in closing the valve by installing it with pressure flow above the seat.

The following procedure should be followed in replacing a composition disc in a globe valve.

Step 1. Isolate the valve to be repaired by closing the main valve and draining the line.

Step 2. Open the valve slightly to separate the disc from the seat and remove the bonnet using a flat-jawed wrench.

Step 3. Slip the disc holder off the stem and replace the disc by removing the disc nut.

Step 4. Reassemble the valve by reversing the operation.

The following procedure should be used in repairing regrind/renewable discs and seats:

Step 1. After the valve has been isolated and the line drained, remove the bonnet and place a small amount of grinding compound on the metal disc.

Step 2. Position a pin in the disc nut notch and stem hole, reinsert the trim and screw down the union ring. Then back off one full turn.

Step 3. Employ the stem as your grinding tool, exercising care not to overgrind. When finished, clean away the grinding compound, remove the pin, lubricate the threads and reassemble the valve.

Check Valves

There are several types of check valves. Swing check valves are like gate valves in the sense that they ensure almost free flow. Fluid velocities are generally low and nonpulsating, otherwise the swing disc will flap and pound on the seat causing excessive wear and possible failure.

Horizontal lift checks have a diaphragm setting that resembles that of globe valves. These are generally used to prevent backflow in steam, air, gas or water lines having higher flow velocities than swing check valves.

Ball checks are similar to lift checks; however, a metal ball replaces the metal or composition disc. These valves are used in heavy, viscous fluid service.

For all installations, flow through a check valve must be from under the seat, otherwise normal line pressure will force the disc against the seat causing a complete stoppage of flow. Swing checks can be installed in horizontal lines or in vertical lines where the flow is in the upward direction only. Lift checks should be installed only in horizontal lines.

All check valves usually trap fluid downstream of the valve. This must be drained when maintaining the line. A drain valve should be installed downstream or the valve body should be drilled, tapped and boss plugged.

Check valves should be installed as far from pumps as possible. Pulses will cause the disc to flap or pound on the seat, causing possible damage to the valve.

Many lift checks have replaceable seat rings. Discs are composition or metal and may be replaced or reground. The regrinding technique used is similar to that of globe valves, except that a screwdriver is used in a milled slot on top of the disc holder.

Most swing checks are provided with an access opening in line with the seat. This allows regrinding of the seat and disc with a screwdriver.

Ball Valves

Ball valves are widely used because of their very small pressure drop due to the straight-through hole. A quarter turn will fully open or close such a valve.

Elastomeric seats are used to provide a tight closure against a polished metal ball. The elastomeric material also serves as a lubricant, promoting easy operation.

Note that elastomeric seats, as well as the body and ball material, must be compatible with the fluid. In selecting a ball valve, features such as a blowout-proof stem and ample packing box should be included.

Ball valves can be operated in remote or dangerous locations through electric, pneumatic and hydraulic actuators. They are readily interfaced with automatic control systems.

When installing ball valves, pipe alignment is critical since strain on the valve itself may pinch or bind the ball. When soldering bronze ball valves some manufacturers recommend complete disassembly to avoid heat damage to the seats. A constant check on valve body temperature should be maintained during soldering operations.

Check for possible leaks at two locations: the packing or seats. If there is a packing nut under the handle, tighten or replace the packing. Care should be taken not to overtighten the packing or the seal may be ruined.

When installing threaded ball valves in a line, place a pipe wrench on the pipe and screw the body of the valve onto the pipe and tighten with a flat-jawed wrench on the body and hex flats. Then screw the other pipe into the cap while holding the cap hex flats stationary with a flat-jawed wrench. The pipe cap then should be tightened with a pipe wrench. Do not allow the cap to turn and cause excessive pressure on the ball and seats.

When reassembling a ball valve, torque up bolts or cap screws evenly. Rotate the ball during the final tightening to avoid binding. Check that the stem is squarely in the milled slot in the ball for proper alignment.

Butterfly Valves

Butterfly valves are best suited to relatively low-pressure service. Generally, the fluid friction coefficient is low. Because of the narrow design of the body sediment, buildup is usually minimum.

There are two basic types of butterfly valves: (1) the wafer design, which fits between pipe flanges and inside the circle of bolts; and (2) the lug type, which can be directly connected to a flange or flanges.

Butterfly valves operate with a quarter turn (like the ball valve). Manual operators may be lever type or geared. The position of the disc is noted by an indicator on the gearbox or from the position of the handle.

During installation, care should be taken not to damage the elastomeric liner with sharp tools. To install the wafer type, fit the bottom lug bolts loosely in the pipe flanges. Then drop the valve in place on top of the bolts, insert the remaining bolts and tighten (using the alternate, cross-over method to ensure even distribution of torque).

The valve should always be closed during installation or maintenance to maintain the disc within the body and to avoid possible damage. Gaskets are generally not necessary since the valve liner maintains a tight seal between the valve and pipe flanges.

If leakage is detected as a result of long and severe service, examine the liner, disc and "0" rings for signs of erosion.

GUIDELINES FOR VALVE SELECTION

As noted in the preceding discussions, there are five basic valve groups: gate, globe, check, ball and butterfly. Each valve type has its specific uses. The proper valve must be used in the right application. Even the highest-quality valve installed in the wrong service will require frequent attention or may result in valve failure, causing unnecessary downtime and lost production. The following notes summarize criteria useful to choosing the proper valve for a particular service.

On/Off Service

The simplest fluid control application is on or off service. This type of control is required on the main water supply to a building or feedwater control to a boiler system. In both cases, it is necessary to provide the fullest possible flow with a minimal pressure drop.

Gate valves are most often used for this type service. Gate valves provide almost full flow, minimal pressure drop, minimum turbulence and fluid trapped in the line. These valves are often used where operation is infrequent. They should never be used for throttling.

Modulation or Throttling

For this type operation, in addition to positive on or off service, the amount of fluid or gas in the service must be controlled or varied. Whenever the flow is reduced there is a corresponding drop in pressure across the valve.

Globe, "Y", Angle and Needle valves are best suited for this type service. The globe family of valves is designed specifically for throttling; hence, particular attention must be given to the several types of seating materials available to avoid unnecessary wear. Whether they consist of composition or metal, they must be compatible with the fluid in service.

Backflow Prevention

There are many services in which the reversal of flow could have disastrous results. One example is the failure in a boiler system, which could force live, superheated steam back into a city water system.

Check valves are designed to prevent backflow by automatically seating when the direction of flow is reversed. The three basic types are swing, check, lift check and ball check.

Fast or Remote Flow Control

Quarter-turn valves display features such as versatility and speed of operation, economy and lightweight service. Such valves are fully opened or closed by turning the hand 90° (hence, the name "quarter-turn valves").

Ball valves are probably the most versatile of all quarter-turn valves because of the variety of metals and different services in which they are used. In these designs a highly polished ball seats against elastomeric seals. They are easier to operate than their older counterpart, the plug valve.

Butterfly valves are often used to save weight and thus cost. They are used in larger, low-pressure services. The body is typically iron with a choice of disc metals to meet required service conditions. The tight closure is provided by a one-piece elastomeric liner seal.

VALVE MATERIAL SELECTION

Valves are manufactured primarily from bronze, iron and steel. Proper body material must be combined with the right seating and packing materials to resist wear and corrosion throughout the life of the valve.

Bronze

This is the oldest material used to make valves. It is used most often in valves for hot and cold water and low-pressure steam systems and for other noncorrosive services. Bronze is durable, noncorrosive and easily fabricated. It is often used as seating services in larger iron body valves to ensure tight closure by eliminating rust in this critical area.

Iron

Standard iron valves are generally used in medium-to-large pipelines to control noncorrosive fluids and gases where pressures do not exceed 250 psi at 450°F, or 500 psi cold water, oil or gas. Variations in trim combinations and seating elements make them suitable for long service in many mildly corrosive services.

Carbon Steel

Carbon steel is a strong, ductile material. Valves made from this metal are used in higher-pressure services, such as steam lines up to 600 psi at 850°F. Many steel valves are available with butt-weld ends for economy. They are generally used in high-pressure steam service, as well as other higher-pressure, noncorrosive services.

Stainless Steel

There are hundreds of stainless steel alloys developed for specific uses. Generally, 316SS can be used for over 90% of corrosive services.

Forged Steel

Valves forged from carbon steel are used in services up to 2000 psi and temperatures up to 1000°F in gate, globe and check valves. Forged steel ball valves may be used in high-pressure services but are limited in terms of temperature because of their elastomeric seats.

VALVE RATINGS

Valves generally have several identifying marks that provide information on the valve class, its type service, pressure and temperature ratings.

Valve Identification

Information identifying a valve and its class may be found in two places on a valve: cast into the body or stamped onto an identification plate, which is found under the handwheel nut on bronze, iron and forged steel valves or fastened to the yoke or yoke bonnet on steel valves.

Information cast on the body usually includes the valve size, pressure class (primary and secondary ratings).

Identification plates include the name of the manufacturer, figure number and other important data, such as size, body, stem, disc and seat material, pressure class, and secondary pressure rating and temperature rating, if it is different from the standard.

Pressure Ratings

The primary rating provides an indication of the safe operating pressure for a valve at some elevated temperature. This temperature is dependent on the materials used and the fabrication of the valve. When specific data are not available, a rule of thumb to follow is the temperature of saturated steam or the primary rating indicated on the valve body. For example, if a valve has a primary rating of 150 psi, as indicated by "150S," then it has a maximum operating temperature of 367°F (temperature of saturated steam at 150 psi).

The safe operating pressure of a valve is always relative to the working temperature. Basically, as the pressure increases there must be a proportional decrease in temperature to remain within the safe operating limits of the valve.

For example, a cast carbon steel valve rated at 150 psi may be operated safely up to 100°F at 275 psi. This same valve may sustain a working temperature of 800°F if the working pressure does not rise above 92 psi.

Manufacturers publish charts for a broad range of working pressures and temperatures in their comprehensive valve catalogs. Note that these are the maximum allowable nonshock pressures at the tabulated temperatures.

GUIDELINES FOR VALVE INSTALLATION

Threaded and Flanged Joints

Overly long pipe threads permit the pipe to enter the valve too deeply and press against the seats, preventing proper seating of the disc or wedge and damaging the valve. Pipe-sealing compound should be applied to the pipe-end only and not inside the valve body, where it can cause failure of the valve to the seat. The following steps should be followed when installing valves with threaded and flanged joints.

Step 1. Ream out all burrs that can impede flow. Then blow out welding scale and other foreign matter before making up the joints. Wipe away all chips and dirt from the pipe threads.

Step 2. Line up flanged joints carefully. Misaligned flanges can place unnecessary strain on a valve. Use a thread lubricant on the flange bolts and pull up the joint snugly. Pipe hangers are necessary to support valves and piping and to minimize pipeline stress, which can adversely affect the proper operation of valves.

Step 3. Use a pipe wrench on the pipe near the valve and a flat-jawed wrench on the valve itself on the end nearest the pipe. Keep the valve tightly closed during installation to prevent twisting and distorting. Place the pipe in a pipe vise (never the valve).

Step 4. On systems with flanged valves and piping, tighten bolts with the alternate-side cross-over method, which gives a much better bearing over the flange faces and reduces stress on the valve parts.

Step 5. Check for proper flow direction. In many cases you can mount a valve either way. But, particularly for globe and check valves, fluid flow must be correct.

Step 6. Whenever possible, install valves with the bonnet up to prevent dirt, condensate and sediment from collecting in the bonnet structure and causing unnecessary wear on the stem threads.

Step 7. Locate valves where they can be reached conveniently. If this is not possible, install chain-wheel operators extension stems or powered actuators. Be sure to allow clearance for rising-stem valves to open completely.

Step 8. Insulate piping systems containing hot fluids to prevent injury from contact and to prevent freezing of lines exposed to subfreezing temperatures.

Soldered Joints

The following steps should be followed when installing valves with soldered joints.

Step 1. Carefully measure the tubing run between fittings or valves and cut the tubing with a sharp hacksaw or round-wheel tubing cutter. The cut should be at a perfect right angle to the tubing.

Step 2. Place the end of the tubing into the valve or fitting to doublecheck your measurement. If the joint is to be made in an existing line, make sure it is completely drained and dry. Heat from the torch will turn any water present into steam, preventing a perfect solder joint.

Step 3. Remove all burrs from the outside and inside edges of the cut and clean it with a piece of fine sandpaper, emery cloth or steel wool. Then wipe clean both tubing and the inside of the valve or fitting.

Step 4. Coat the outside end of the tubing and inside end of the valve or fitting lightly with flux and twist the tubing or fitting at least 180° once or twice to distribute the flux evenly.

Step 5. Apply the flame to the fitting end and allow it to reach the proper temperature. The flame should not directly touch the solder or tubing. Fill the joint completely and wipe off the excess with a damp cloth to make a tight, good-looking, professional joint.

VALVE ACTUATORS

Valve actuators make it possible to operate valves in inaccessible or dangerous locations and can be interfaced with computers and other automated control systems. The operation of valves from a central location can be economical and provide more control since position indicators indicate at a glance whether the valve is open or closed.

In automatic control, output signals from the controller are transmitted to valves and their driving motors. Proper selection and specification of valves and valve actuators is an integral part of a process system design. Actuators are categorized on the basis of their power source, with the principal categories being pneumatically, hydraulically and electrically operated.

Pneumatically operated valve motors or actuators are among the most commonly used type in the process industries. Figure 15 illustrates a pneumatic valve motor whose operation is based on a diaphragm motor. The diaphragm is spring loaded in opposition to the driving air pressure such that the valve-stem position is proportional to the air pressure. The diaphragm is often a limp type, fabricated of rubber fabric or some resistant material, and is normally supported by a backup plate. Diaphragm motors typically have a maximum allowable stroke, or travel of the valve stem of 2–3 inches. Valve actuators with longer strokes are normally double-acting piston or a rotary pneumatic motor, which drives a rack or a worm gear. Such actuators are capable of driving very large valves. (For rotary motors, valve strokes can be as much as 5 feet.)

It should be noted that the application of the tank or system air pressure to the valve motor diaphragm may not always result in the desired valve stem position. For example, the valve stem could become frozen in its seal or stuffing box and, consequently, may not move at all or only sluggishly. This problem is overcome by incorporating a servomechanism, which supplies sufficient pressure via a pilot amplifier to the diaphragm to ensure that the valve stem achieves the proper position. An actuator of this type is referred to as a valve positioner. These devices normally use a higher source of air pressure than the normal instrument air (typically 100 psig).

The forcing signal transmitted to hydraulic actuators can be either a mechanical displacement or a pneumatic or electric signal, which is transformed to a mechanical displacement. The displacement in piston-type hydraulic actuators positions the cylinder valves and admits oil to either side of the piston. This action positions the piston shaft, which drives the valve stem.

One type of hydraulic actuator is the variable-delivery pump actuator, in which the input displacement fixes the stroke of a number of piston pumps connected in a parallel arrangement. The output-oil delivery of the pumps is

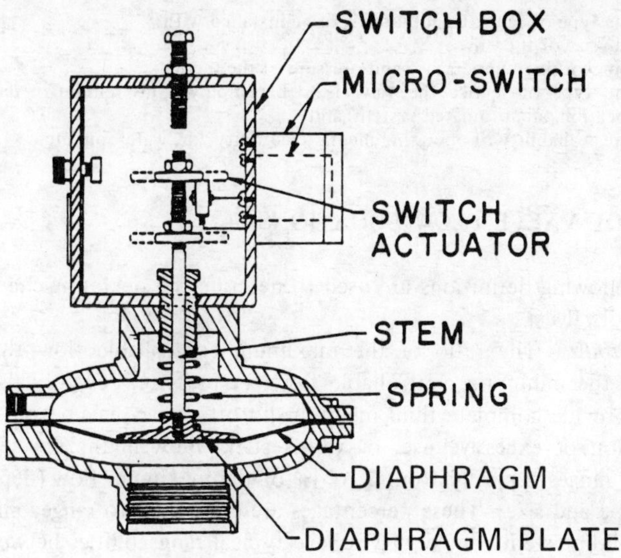

Figure 15. Shows a pneumatic actuator. This system controls liquid level in storage tanks and can be used to start and stop pumps, control valves, or for alarms. Static pressure from the fluid height changes the pressure supplied to the diaphragm. The pressure is opposed by spring tension. As pressure increases it causes the diaphragm to move. Diaphragm movement raises the activator stem. When the lower switch actuator contacts a microswitch lever it closes one set of contacts. These remain closed until downward movement due to decreasing liquid level causes the upper switch actuator to move the microswitch level in the opposite position. Switch closures may be field adjusted for the desired control points.

proportional to the forcing displacement and is fed to a positive-displacement motor to produce an output displacement. This output displacement is also proportional to the original forcing displacement. In some cases, the variable output of the piston pumps is fed to the positive-displacement motor in such a fashion that the rate of travel of the output displacement is proportional to the input displacement. This results in the actuator generating an integrating action. Hydraulic actuators are normally used in services in which high speeds and large forces are required.

Electric actuators are basically reversing motors that are operated by a circuit that is appropriate to the forcing signal. In general, these systems are slower than pneumatic and hydraulic actuators. Entirely electric control systems are often expensive.

There are five criteria to consider in selecting the proper valve actuator for a specific service:

1. the type of valve the actuator is to be installed with;
2. the size of the valve;
3. the working temperature and pressure of the service;
4. the type of service the valve is to be employed in: remote or dangerous location, an automated system; and
5. the availability of operating media: electricity, air, water, oil, etc.

CONTROL VALVE FLOW EQUATIONS

The following definitions are used to describe the operating characteristics of control valves:

Rangeability—The ratio of the maximum controllable flow through the valve to the minimum controllable flow. In practice, control valves do not always provide complete fluid medium shutoff. Valve seats become damaged by erosion or excessive use, or valves stick. Flow in the normally closed position ranges typically from 2 to 4% of the maximum flow (depending on valve type and size). These percentages are equivalent to rangeabilities of 50 to 25. Sliding stem control valves have typical rangeabilities between 20 and 70.

Turndown—The ratio of the normal maximum flow through the valve to minimum controllable flow. A rule of thumb applied to sizing control valves is that the maximum flow under operating conditions should provide approximately 70% of the maximum possible flow. The turndown is normally about 70% of the rangeability.

Lift—Refers to the extent to which a valve is open. Flow through a valve depends not only on the lift, but also on the pressure drop across the valve.

The relation between flow and pressure drop for a given lift can be expressed in the form of an orifice equation for incompressible fluids:

$$Q = C_V A \sqrt{\frac{\Delta P}{\rho}} \tag{65}$$

Flow control is achieved by moving the valve stem to vary the area of flow. The gain of a valve (i.e., the change of flow for a given change in stem position) is dependent on the change in area with stem position and also on the change in pressure drop with flow.

Control valve manufacturers have standardized on a flow coefficient (C_V in Equation 65) to provide the flow capacity of valves. The flow coefficient is defined as the flow of water in gpm for a pressure drop of 1 psi across the fully open valve. Flows for various valve stem positions are obtained from values of C_V and the valve characteristics. These can be plotted as percentage of maximum flow versus percentage lift or can be presented in the form of flowchart curves, as shown in Figure 16 for one manufacturer's two-way motor valve unit. An illustrative example on the use of this chart is given in the figure.

The inherent valve characteristic is the relationship between the stem position and the flow at constant pressure drop across a valve. Linear and equal percentage are the two common characteristics encountered. The characteristic of a parabolic plug valve is referred to as an equal-percentage characteristic. The term equal percentage refers to the fact that the flow increments resulting from a given change in lift are the same percentage as the actual flow, regardless of whether the valve is nearly open or nearly closed. Full flow is achieved with a relatively short stem movement, consequently, the terms "low-lift valve" and "quick-opening valve" are applied.

In sizing valves and valve capacities, the general formulas outlined below can be applied. The manufacturer should always be consulted for values of C_v and for more precise calculation procedures, particularly when sizing valves for service in viscous liquids, flashing liquids or gases at high-pressure drops. In establishing the proper valve size, all flow conditions must be known and C_v determined. Based on a computed C_v value, the valve size for the type valve under consideration can be selected from the manufacturer's values of C_v rating versus valve size.

For liquid service, apply Equation 65 to estimate C_v and the valve capacity.

For gases, the following equations can be applied for valve size and valve capacity:

$$C_v = \frac{\bar{\nu}\sqrt{S_g T_a}}{1360\sqrt{(\Delta P)P_2}} \tag{66}$$

$$\bar{\nu} = \frac{1360\, C_v \sqrt{(\Delta P)P_2}}{\sqrt{S_g T_a}} \tag{67}$$

where $\bar{\nu}$ is the flow quantity at 14.7 psia and 60°F (ft^3/hr), ΔP is the pressure drop at maximum flow (psi), P_2 is the outlet pressure at maximum flow conditions (psia), S_g is specific gravity (for air = 1.0), T_a is the flowing temperature absolute (i.e., °R = 460 + °F), and C_v is the valve-flow coefficient. Note that when the outlet pressure (under maximum flow, P_2) is less than one-half the inlet pressure at maximum flow, then the term $\sqrt{(\Delta P)P_2}$ in Equations 66 and 67 can be approximated by $P_1/2$, where P_1 is the inlet pressure.

For steam flow, use the following equations:

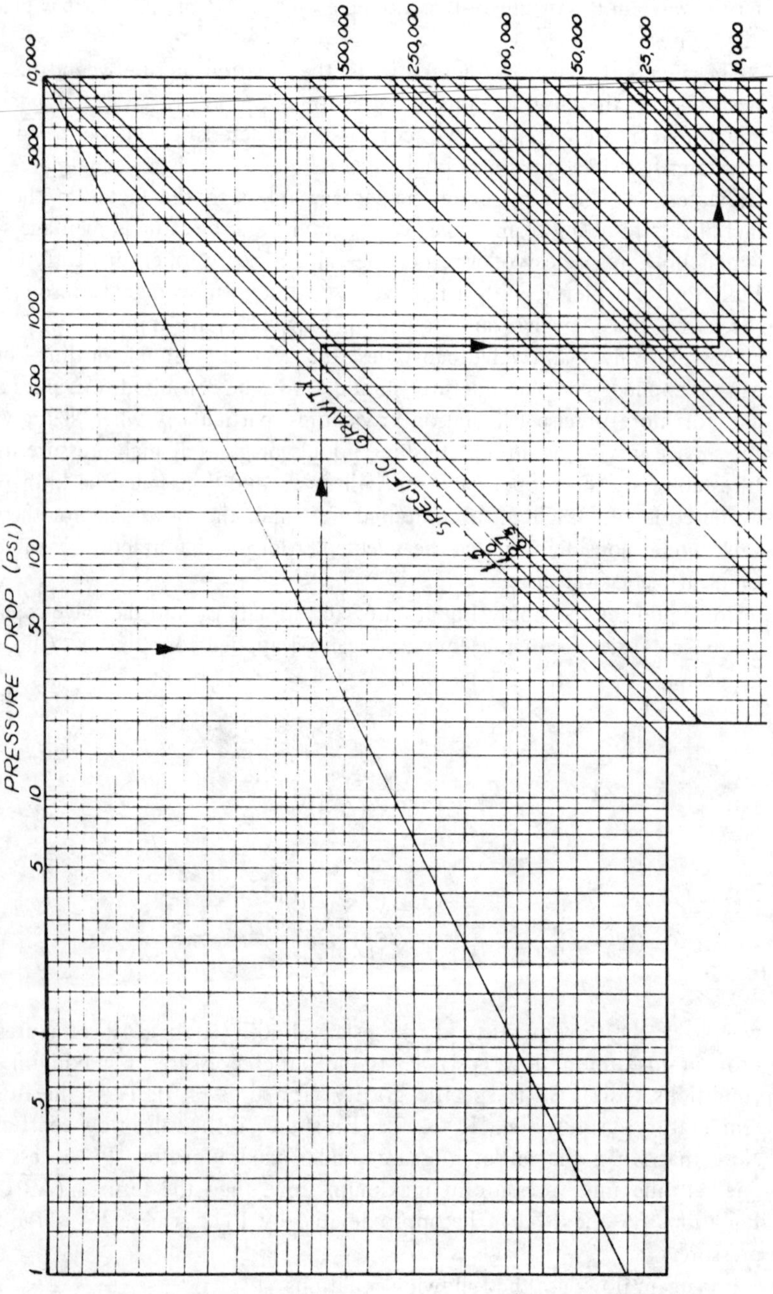
PRESSURE DROP (PSI)
10,000
5000
1000
500
100
50
10
5
1
500,000
250,000
100,000
50,000
25,000
10,000
SPECIFIC GRAVITY

EXAMPLE

A VALVE WITH A FLOW FACTOR $C_V = 45$ IS USED UNDER THE FOLLOWING CONDITIONS :

PRESSURE DROP ACROSS VALVE = 40 PSI
LIQUID SPECIFIC GRAVITY = 0.70

SOLUTION – ENTER CHART AT PRESSURE DROP OF 40 PSI – FOLLOW VERTICAL LINE TO TURN LINE – THEN FOLLOW HORIZONTAL LINE TO SPECIFIC GRAVITY 0.7 AND VERTICAL LINE TO FLOW FACTOR 45, READ FLOW OF 11,500 BARRELS / DAY.

CHART BASED ON FORMULA –

$$Q = 34.3\, C_V \sqrt{\frac{\Delta P}{SG}}$$

Q = FLOW (BARRELS/DAY)
ΔP = PRESSURE DROP ACROSS VALVE (PSI)
SG = SPECIFIC GRAVITY (FOR WATER = 1)
C_V = VALVE FLOW FACTOR

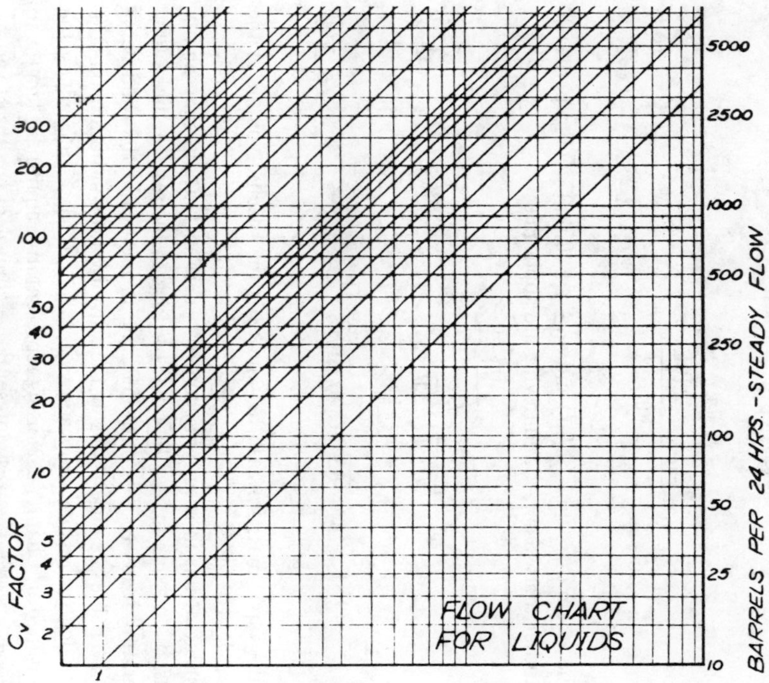

Figure 16. Liquid flowchart for a two-way motor valve system (courtesy of Liquid Level Lectronics, Inc., Houston, TX).

$$C_V = \frac{W\eta_0}{3\sqrt{(\Delta P)P_2}} \tag{68}$$

$$W = \frac{3\ C_V\sqrt{(\Delta P)P_2}}{\eta_0} \tag{69}$$

where W is the steam mass flow (lb/hr) and $\eta_0 = 1 + (0.0007 \times {}^\circ F \text{ superheat})$.

For vapors other than steam, the following are applicable:

$$C_V = \frac{W}{63.4}\sqrt{\frac{\nu_2}{\Delta P}} \tag{70}$$

$$W = 63.4\ C_V\sqrt{\Delta P/\nu_2} \tag{71}$$

where W is the vapor mass flow (lb/hr) and ν_2 is the specific volume at the outlet pressure, P_2 (ft^3/lb).

Further considerations in specifying control valves include the rangeability of the process, the maximum specific range of flows in the service, the normal range of operating loads to be controlled, the properties of the fluid being handled, and available pressure drops at the valve location at both maximum and minimum flows. Pressure drop across a valve can be estimated by the procedures outlined in Chapter 2.2, using representative equivalent lengths to represent the valve. Table 13 gives representative equivalent lengths of various valves in pipe diameters, L/D.

FLOW MEASURING DEVICES

Flow measurement is necessary to a wide range of applications. In the laboratory, high precision is often necessary. In industry, accurate measurements are needed for the testing and acceptance of materials-handling equipment. Less precise measurements may be used for material balances around various unit operations such as distillation, extraction, polymerization, wastewater handling processes, etc. Proper selection, installation and use for intended flow applications require working knowledge of the operating principles behind the various flow measuring instrumentation. Table 14 summarizes the major categories of flow measuring devices employed in closed channel flow. The operating principles behind these instruments and useful formulas that can be applied to their usage are described below by

Table 13. Representative Equivalent Lengths of Different Valves in Pipe Diameters (L/D) [7]

Type Valve	Description	Valve Stem Position	Equivalent Length In Pipe Diameters (L/D)
Globe Valves			
Stem Perpendicular to Run	With no obstruction in flat, bevel or plug-type seat	Fully open	340
	With wing or pin guided disc	Fully open	450
Y-Pattern	(No obstruction in flat, bevel or plug-type seat)		
	With stem 60 degrees from run of pipe line	Fully open	175
	With stem 45 degrees from run of pipe line	Fully open	145
Angle Valves	With no obstruction in flat, bevel or plug-type seat	Fully open	145
	With wing or pin guided disc	Fully open	200
Gate Valves			
Wedge, Disc, Double Disc or Plug Disc		Fully open	13
		Three-quarters open	35
		One-half open	160
		One-quarter open	900
Pulp Stock		Fully open	17
		Three-quarters open	50
		One-half open	260
		One-quarter open	1200
Check Valves			
Conventional Swing		0.5[a] Fully open	135

Table 13, continued

Type Valve	Description	Valve Stem Position	Equivalent Length In Pipe Diameters (L/D)
Clearway Swing		0.5[a] Fully open	50
Globe Lift or Stop; Stem Perpendicular to Run or Y-Pattern		2.0[a] Fully open	Same as globe
Angle Lift or Stop		2.0[a] Fully open	Same as angle
In-Line Ball		2.5 vertical and 0.25 horizontal[a]	
		Fully open	150
Butterfly Valves (8-inch and larger)		Fully open	
Cocks			
Straight-Through	Rectangular plug port area equal to 100% of pipe area	Fully open	18
Three-Way	Rectangular plug port area equal to 80% of pipe area (fully open)	Flow straight through	44
		Flow through branch	140

[a]Minimum calculated pressure drop in psi across valve to provide sufficient flow to lift disc fully. Note L/D valves are based on turbulent flow.

Table 14. Major Categories of Flow Measuring Devices for Closed-Channel Flows

A. Differential Pressure Meters
- 1. Venturi
- 2. Flow nozzles
  - 2.1 Critical flow nozzle
- 3. Orifice
- 4. Pitot tube
  - 4.1 Impact tube
  - 4.2 Pitot-static tube
  - 4.3 Pitot-venturi

B. Positive Displacement
- 1. Reciprocating piston
- 2. Nutating disc
- 3. Rotary piston
- 4. Rotary vane

C. Mechanical
- 1. Rotameter
  - 1.1 Glass
  - 1.2 Metal
- 2. Turbine meter
  - 2.1 Electromagnetic

D. Acoustic
- 1. Travel time difference method
- 2. Beam deflection method
- 3. Doppler method

E. Electrically Heated
- 1. Hot-wire
- 2. Hot-film

category. The treatment of this subject is somewhat superficial. Refer to the book by Cheremisinoff [9] for detailed discussions of each of these devices.

DIFFERENTIAL PRESSURE DEVICES FOR FLOW MEASUREMENT

Differential pressure flowmeters, also called head or rate meters, indirectly measure flow velocity by measuring a differential head. The main devices in this class are venturi meters, orifice meters, flow nozzles and pitot tubes.

Venturi Meter

The operation of the venturi meter is illustrated in Figure 17(A). As fluid passes through the venturi throat its velocity increases, resulting in a pressure

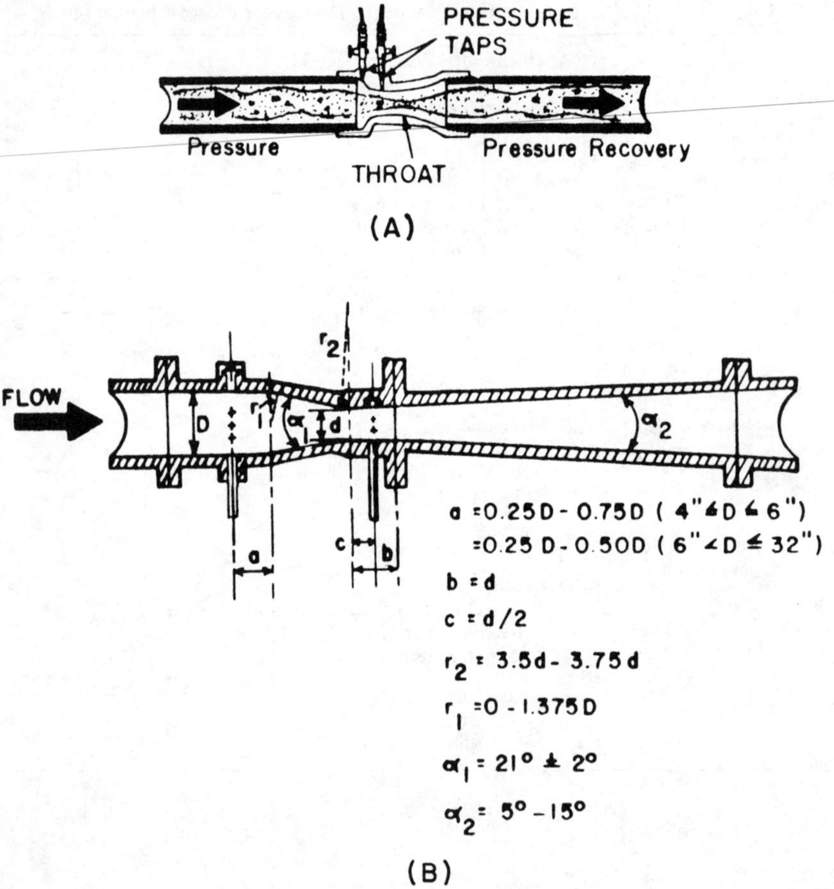

Figure 17. (A) Illustrates the operation of the venturi meter. (B) Recommended specifications for venturi meters.

differential between the inlet and throat regions. The section after the throat gradually increases in flow area; consequently, the fluid's velocity decreases, causing pressure recovery. The pressure drop across the throat can be directly recorded or translated into actual flow units by using differential pressure meters and capacity curves. The recommended specifications for a standard venturi, as designated by the American Society of Mechanical Engineers (ASME) [10,11] is given in Figure 17(B).

For an incompressible fluid under adiabatic conditions, an expression for pressure drop can be derived from the Bernoulli equation by ignoring frictional effects:

$$\Delta P = \frac{u_2^2 \rho}{2g_c}\left[1 - \left(\frac{A_2}{A_1}\right)^2\right] \tag{72}$$

The volumetric flowrate is

$$Q = CMA_2\sqrt{\frac{2g_c}{\rho}(P_1 - P_2)} \tag{73}$$

where M is the velocity of approach factor,

$$M = \left[1 - \left(\frac{A_2}{A_1}\right)^2\right]^{-1/2} \tag{74}$$

and C is the flow coefficient.

For compressible flow, the mass flowrate is

$$m = YCMA_2\sqrt{2g_c\rho(P_1 - P_2)} \tag{75}$$

when Y is the expansion factor defined by the following equation:

$$Y = \left[\left(\frac{P_2}{P_1}\right)^{2/\gamma}\left(\frac{\gamma}{\gamma - 1}\right)\frac{1 - (P_2/P_1)^{(\gamma-1)/\gamma}}{1 - (P_2/P_1)} \quad \frac{1 - \beta^4}{1 - \beta^4(P_2/P_1)^{2/\gamma}}\right]^{1/2} \tag{76}$$

β is the ratio of the throat diameter to the venturi's inlet diameter. Subscripts 1, 2 refer to upstream and throat conditions, respectively.

In specifying a venturi for a particular application, determine the flow requirements and establish line sizes. The above formulas, along with manufacturer's capacity curves, can be used to size units. For throat Reynolds numbers in excess of 200,000, a discharge coefficient of 0.99 can be used.

Installing a venturi involves setting the unit into the line as another section of pipe. The shorter cone should form the inlet or upstream end. Venturi meters can be installed in the horizontal, vertical or inclined position. They should be installed as far downstream as possible from sources of flow disturbances such as reducers, valves or fittings. Figure 18 shows the proper arrangement for venturis in liquid and gas services.

For vertical line installations (liquid service), pressure connections can be made to any part of the tube. In horizontal or inclined arrangements, make sure that pressure connections are made according to the manufacturer's recommendations, or faulty measurements will result.

For gas service, locate pressure connections at the top of the meter.

For liquid service, make connections on the side of the meter.

For steam service where the meter is above the line, connections should be made at the top of the meter; and when the meter is below the line, to the side of the meter.

A modified version of the standard venturi meter is the Dall flow tube. Although similar in design, it is considerably shorter than the venturi. Figure 19 illustrates the Dall tube.

The Dall tube has a lower permanent pressure loss for a given flowrate and differential than the venturi. Its differential pressure drop is nearly twice that of the venturi at the same β and flowrate.

In general, venturi and Dall meters provide accurate measurements within ±0.5%. They are self-cleaning devices in most applications.

Orifice Meter

The orifice plate is illustrated in Figure 20. The differential pressures between the upstream and downstream sides are measured with pressure taps located on either side of the orifice plate. Common designs include circular, eccentric and segmental metal diaphragms.

Flange-mounted orifice plates offer flexibility. They can be inserted and removed from a line without disturbing existing piping or the differential

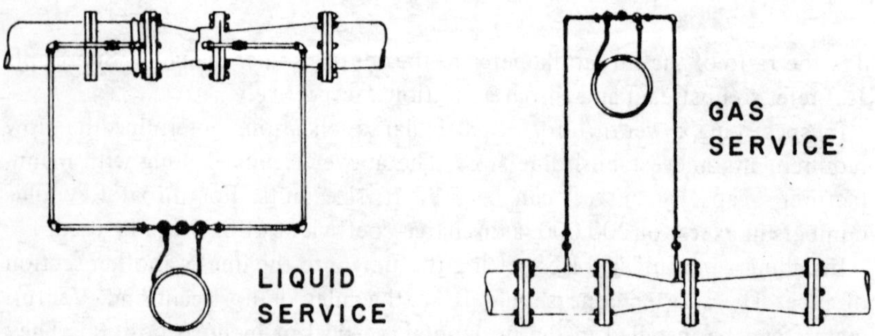

Figure 18. Illustrates proper installation arrangements for venturi meters in liquid and gas services.

pressure connections. The orifice plate's upstream edge should be sharp and and squared. These devices cannot be used for two-directional flow.

For incompressible flow, the volumetric flowrate is

$$Q = kA_2 \sqrt{\frac{2g_c}{\rho}(P_1 - P_2)} \tag{77}$$

where k is the flow coefficient (k = CM) and subscripts 1, 2 refer to upstream and downstream conditions, respectively.

For compressible flow, the mass flowrate is

$$m = YkA_2 \sqrt{2g_c\rho(P_1 - P_2)} \tag{78}$$

The expansion factor for an orifice employing flange taps or vena contracta taps is

$$Y = 1 - \left[0.41 + 0.35\left(\frac{A_2}{A_1}\right)^2\right]\left(\frac{P_1 - P_2}{g\rho P_1}\right) \tag{79}$$

where A_2 is the area of the orifice opening.

For an orifice with pipe taps,

$$Y = 1 - [0.333 + 1.145(\beta^2 + 0.7\beta^3 + 12\beta^{13})]\frac{P_1 - P_2}{g\rho P_1} \tag{80}$$

The discharge coefficient, k, is a function of the Reynolds number and β. For an orifice plate, an approximation can be made by assuming a flow coefficient of 0.62.

Table 15 gives pressure drop and flow capacity data for water flow through circular, straight-edge orifices.

Flow Nozzles

The standard flow nozzle is shown in Figure 21. The rounded approach has a curvature that is equivalent to the quadrant of an ellipse. The distance between the front face and tip is about one-half the pipe diameter. The flow equations for orifice plates are applicable to nozzles. For approximations, a discharge coefficient of 0.98 can be used.

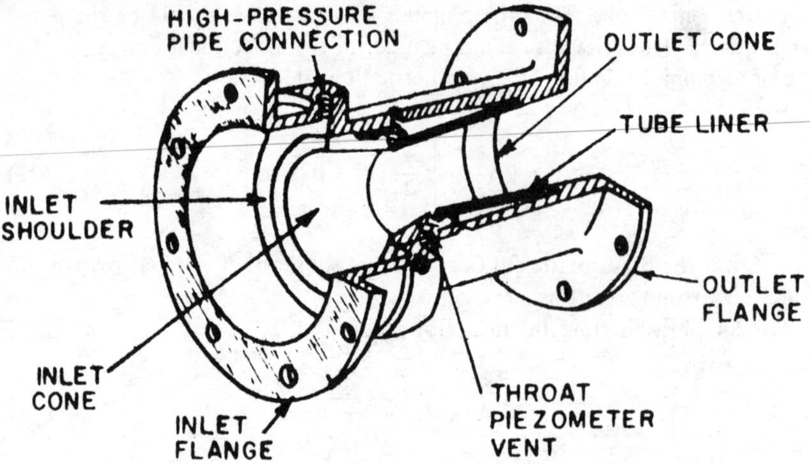

Figure 19. The Dall flow tube.

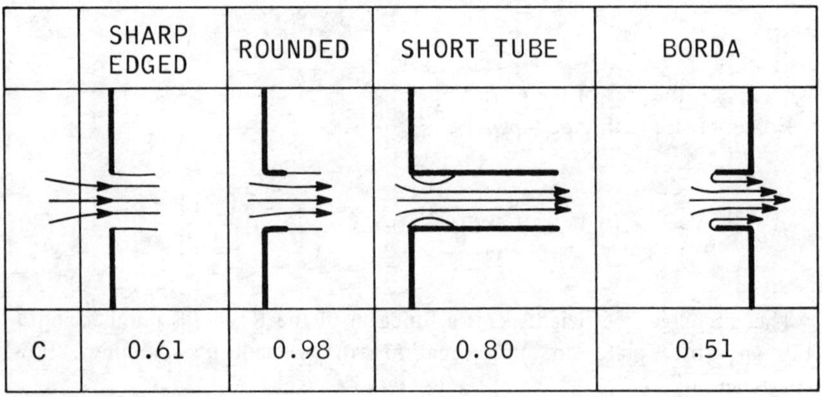

Figure 20. Basic design and operation of orifice plate meters.

Pitot Tubes

Pitot tubes (Figure 22) are available in a variety of designs. The single-tip type incorporates static pressure connections. Pitot tubes consist of two concentric tubes: the inner tube transmits the impact pressure, while the annular space between the two tubes transmits static pressure. Static pressure

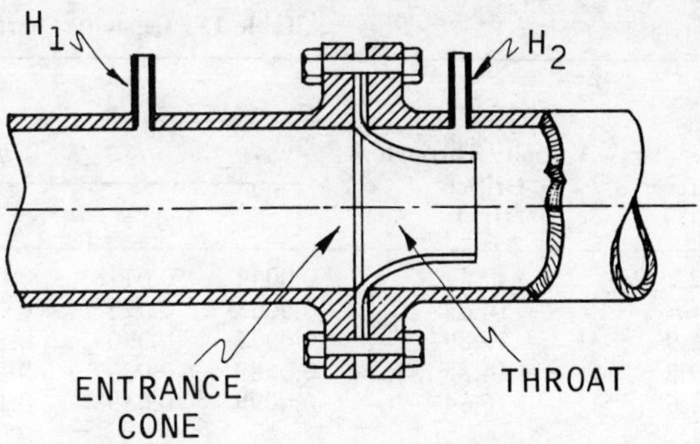

Figure. 21. Plan view of standard flow nozzle.

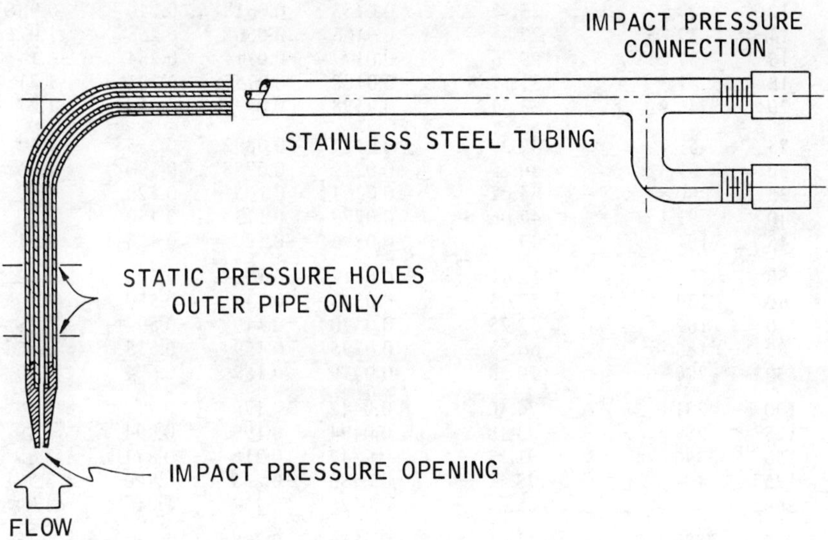

Figure 22. L-type pitot tube design.

taps are located well back from the tube's front face to prevent interference from eddy currents. Common designs are sharp- or round-nose.

Pitot tubes provide point measurements of the velocity head, ΔH. For ΔH, expressed in millimeters of fluid flowing, the point velocity of the fluid is

$$u_0 = C\sqrt{2g_c \Delta H} = C\sqrt{2g_c(\Delta P/\rho)} \qquad (81)$$

Table 15. Capacity Chart for Water Flow

			Diameter of			
			1/64	1/32	1/16	1/8
psi	Water (ft)	Velocity Through Orifice (fps)	Flow Through			
1	2.3	7.44	0.0044	0.0178	0.0711	0.285
2	4.6	10.52	0.0063	0.0252	0.101	0.404
3	6.9	12.89	0.0077	0.0309	0.123	0.494
4	9.2	14.88	0.0089	0.0357	0.142	0.570
5	11.6	16.64	0.0099	0.0398	0.159	0.637
6	13.9	18.22	0.0108	0.0437	0.174	0.699
7	16.2	19.69	0.0117	0.0472	0.188	0.754
8	18.5	21.04	0.0125	0.0505	0.201	0.806
9	20.8	22.32	0.0133	0.0535	0.213	0.855
10	23.1	23.53	0.0140	0.0564	0.224	0.900
12	27.7	25.77	0.0153	0.0618	0.246	0.988
14	32.3	27.84	0.0165	0.0668	0.266	1.07
16	37.03	29.76	0.0177	0.0714	0.284	1.14
18	41.6	31.57	0.0188	0.0757	0.301	1.21
20	46.2	33.27	0.0198	0.0798	0.318	1.27
25	57.7	37.20	0.0221	0.0892	0.355	1.42
30	69.3	40.75	0.0242	0.0978	0.390	1.56
35	80.8	44.02	0.0261	0.105	0.420	1.69
40	92.4	47.06	0.0279	0.113	0.450	1.80
45	104	49.91	0.0296	0.120	0.476	1.91
50	116	52.61	0.0312	0.126	0.503	2.02
60	139	57.63	0.0343	0.138	0.550	2.20
70	162	62.25	0.0370	0.149	0.594	2.38
80	185	66.55	0.0395	0.159	0.635	2.55
90	208	70.58	0.0419	0.169	0.674	2.70
100	231	74.40	0.0442	0.178	0.712	2.85
125	289	83.18	0.0494	0.199	0.794	3.19
150	346	91.13	0.0542	0.218	0.871	3.49
175	404	98.42	0.0586	0.236	0.940	3.77
200	462	105.2	0.0626	0.252	1.01	4.04
225	520	111.6	0.0664	0.268	1.08	4.28
250	577	117.6	0.0700	0.282	1.13	4.50
300	693	128.9	0.0767	0.309	1.23	4.94
350	808	139.2	0.0828	0.334	1.33	5.32
400	924	148.8	0.0886	0.357	1.42	5.70
450	1039	157.8	0.0940	0.379	1.51	6.05
500	1155	166.4	0.0990	0.399	1.59	6.37
600	1386	182.2	0.108	0.437	1.74	6.99
700	1617	196.9	0.117	0.472	1.88	7.54
800	1848	210.4	0.125	0.505	2.01	8.06
900	2079	223.2	0.133	0.535	2.13	8.55
1000	2310	235.3	0.140	0.564	2.24	9.00
1200	2771	257.7	0.153	0.618	2.46	9.88
1400	3233	278.4	0.166	0.667	2.66	10.7
1600	3695	297.6	0.177	0.714	2.84	11.4

Through Circular, Straight-Edge Orifices

Orifice (in.)							
3/15	1/4	3/8	1/2	5/8	3/4	7/8	1
Orifice (gpm)							
0.641	1.14	2.56	4.56	7.12	10.3	14.0	18.22
0.909	1.61	3.62	6.45	10.1	14.5	19.7	25.8
1.111	1.97	4.44	7.90	12.3	17.8	24.2	31.6
1.28	2.28	5.12	9.12	14.2	20.5	27.9	36.5
1.43	2.54	5.73	10.2	15.9	22.9	31.2	40.7
1.57	2.79	6.28	11.2	17.4	25.1	34.2	44.6
1.70	3.01	6.78	12.1	18.8	27.1	36.9	48.2
1.81	3.22	7.24	12.9	20.1	28.9	39.4	51.5
1.92	3.42	7.69	13.7	21.3	30.8	41.9	54.7
2.03	3.60	8.11	14.4	22.5	32.4	44.1	57.6
2.22	3.94	8.89	15.8	24.7	35.5	48.4	63.1
2.40	4.26	9.59	17.1	26.6	38.4	52.2	68.2
2.55	4.56	10.3	18.3	28.5	41.0	55.9	73.0
2.72	4.83	10.9	19.4	30.2	43.5	59.1	77.4
2.87	5.10	11.5	20.4	31.8	45.9	62.4	81.6
3.21	5.70	12.8	22.8	35.6	51.3	69.8	91.1
3.51	6.24	14.0	24.9	39.0	56.2	76.5	99.8
3.80	6.74	15.2	27.0	42.4	60.7	82.3	108
4.05	7.20	16.2	28.8	45.0	64.9	88.3	115
4.30	7.65	17.2	30.6	47.8	68.7	93.5	122
4.54	8.06	18.1	32.2	50.4	72.5	98.7	129
4.95	8.83	19.8	35.3	55.1	79.4	108	141
5.36	9.54	21.4	38.2	59.6	85.8	117	153
5.73	10.2	22.9	40.7	63.6	91.7	125	163
6.08	10.8	24.3	43.3	67.5	97.3	132	173
6.41	11.4	25.6	45.6	71.2	103	140	183
7.16	12.7	28.6	50.9	79.6	115	156	204
7.86	14.0	31.4	55.9	87.3	126	171	223
8.50	15.1	33.9	60.2	94.3	136	185	241
9.09	16.1	36.2	64.5	101	145	197	258
9.62	17.1	38.4	68.4	107	154	210	274
10.1	18.0	40.5	72.1	112	162	221	288
11.1	19.7	44.4	79.0	123	178	242	316
12.0	21.3	48.0	85.3	133	193	261	341
12.8	22.8	51.2	91.2	142	205	279	365
13.6	24.2	54.4	96.8	151	216	296	387
14.3	25.4	57.2	102	159	229	312	407
15.7	27.9	62.8	112	174	251	342	446
17.0	30.1	67.8	121	188	271	369	482
18.1	32.2	72.4	129	201	289	394	515
19.2	34.2	76.9	137	213	308	419	547
20.3	36.0	81.1	144	225	324	441	576
22.2	39.4	88.9	158	247	355	484	631
24.0	42.6	95.9	171	266	384	522	682
25.5	45.6	103	183	285	410	559	730

Table 15,

psi	Water (ft)	Velocity Through Orifice (fps)	Diameter of ¹⁄₆₄	¹⁄₃₂	¹⁄₁₆	¹⁄₈
			Flow Through			
1800	4157	315.7	0.188	0.756	3.01	12.1
2000	4619	332.7	0.198	0.798	3.18	12.7
2100	4850	341.0	0.203	0.817	3.25	13.1
2200	5081	349.0	0.208	0.837	3.34	14
2300	5312	356.8	0.212	0.856	3.42	13.7
2400	5543	364.5	0.217	0.875	3.49	14.0
2500	5774	372.0	0.221	0.892	3.55	14.2
3000	6929	407.5	0.242	0.978	3.90	15.6
3500	8083	440.2	0.262	1.05	4.20	16.9
4000	9238	470.6	0.280	1.13	4.50	18.0
4500	10393	499.1	0.297	1.20	4.76	19.1
5000	11548	526.1	0.212	1.26	5.03	20.2
6000	13857	576.3	0.342	1.38	5.50	22.0
7000	16167	622.5	0.370	1.49	5.94	23.8
8000	18476	665.5	0.395	1.59	6.35	25.5
9000	20786	705.8	0.419	1.69	6.74	27.0
10000	23095	744.0	0.442	1.78	7.12	28.5
12000	27714	815.0	0.485	1.95	7.79	28.5
14000	32333	880.3	0.524	2.11	8.42	33.7
16000	36952	941.1	0.560	2.26	9.00	36.0
18000	41571	998.2	0.594	2.40	9.55	38.2
20000	46190	1052	0.626	2.52	10.1	40.4

where ΔP is the indicated dynamic pressure (i.e., the difference between total pressure and static pressure).

To obtain an indication of the fluid's average or mean velocity, the device must be traversed over the channel's flow area and measurements numerically integrated. The technique is described at the end of this chapter.

POSITIVE DISPLACEMENT FLOWMETERS

Positive displacement-type flowmeters are used when a high degree of flow accuracy is required under steady flow conditions. They are characterized by one or more moving parts positioned in the flow stream. The main devices in

continued

Orifice (in.)							
3/15	1/4	3/8	1/2	5/8	3/4	7/8	1
Orifice (gpm)							
27.2	48.3	109	194	302	435	591	774
28.7	51.0	115	204	318	459	624	816
29.4	52.2	117	209	326	470	639	836
30.1	53.4	120	214	334	481	655	855
30.8	54.6	123	219	341	491	670	875
31.4	55.8	126	223	348	502	684	894
32.1	57.0	128	228	356	513	698	911
35.1	62.4	140	249	390	562	765	998
38.0	67.4	152	270	422	607	823	1078
40.5	72.0	162	288	450	649	883	1152
43.0	76.5	172	306	478	687	935	1222
45.4	80.6	181	322	504	725	987	1290
49.5	88.3	198	353	551	794	1078	1410
53.6	95.4	214	382	596	858	1168	1528
57.3	102	229	407	636	917	1247	1629
60.8	108	243	433	675	973	1321	1730
64.1	114	256	456	712	1025	1395	1822
70.3	125	281	499	780	1123	1529	1998
75.8	135	303	539	843	1212	1650	2158
81.1	144	324	576	901	1298	1765	2305
86.1	153	344	611	956	1376	1870	2445
90.9	161	362	645	1008	1450	1973	2578

this class are reciprocating piston, rotary piston, rotary vane meters and nutating disc.

Reciprocating Piston

Reciprocating piston meters are of the single and multipiston types (Figure 23(A)). Design choice depends on flow requirements and intended applications. They can be used to meter practically all liquids within ±1% accuracy, although extreme accuracy to 0.1% is achievable.

A piston meter is a backward-operating piston pump. The piston undergoes back and forth motion from the fluid under pressure. As the piston moves, reaching the end of its stroke, it shifts the intake and discharge valve. This,

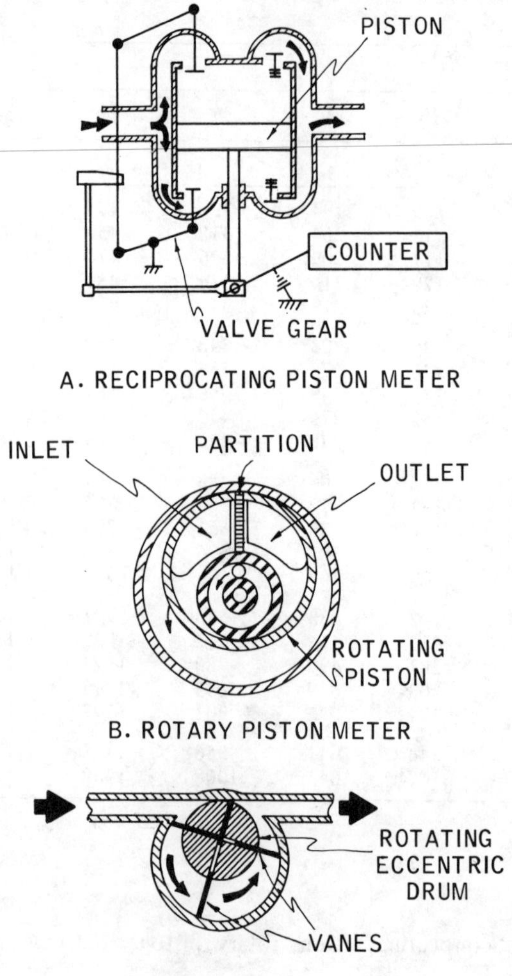

Figure 23. Positive displacement flowmeters.

in turn, operates a counter, which adds a fixed volume increment of fluid with each piston cycle.

Rotary Piston

The internals of a rotary piston meter are illustrated in Figure 23(B). A cylindrical working chamber houses a hollow cylindrical piston of equal length. The central hub of the piston is guided in a circular motion by two short inner cylinders. The piston and cylinder are alternately filled and emptied by the fluid flowing through the meter. A slot in the sidewall of the piston is removed such that a partition extending inward from the bore of the working chamber can be inserted. This restricts the piston to a sliding motion along the partition.

Rotary Vane

A rotary vane meter is shown in Figure 23(C). The vanes are spring loaded to ensure that they are in continual contact with the meter casing. As the eccentric drum rotates, a fixed quantity of liquid is swept through each section to the outlet. These are lightweight units. Flow accuracy is established by machined castings that form the measurement compartments. They can be used in gas and liquid applications.

Nutating Disc

A nutating disc meter consists of a movable disc mounted on a concentric sphere. The disc is contained in a working chamber with spherical sidewalls and top and bottom surfaces that extend conically inward. It is restricted from rotating about its own axis by a radial partition, which extends across the entire height of the working chamber. The disc is slotted to fit over its partition. Figure 24 shows a cutaway view of the nutating disc meter.

Nutating disk meters are limited to liquid service; however, they can handle a wide range of viscosities. Figure 25 is a nomograph for estimating the pressure loss across such a meter. The following example illustrates the use of Figure 25.

Example:

A nutating disk meter installed in a 1-in. nominal size line is used for water service. The maximum pressure drop across the meter is not to exceed 23 psi. Determine the minimum flowrate.

Solution:

Step 1. Locate the meter size on the first column of the nomograph.

Step 2. Locate the pressure loss value on the right-hand column.

Step 3. Using a straight-edge, draw an imaginary straight line through the two points and read the flowrate off of the middle column. For this example the answer is 29 gpm.

Figure 25 is good for water flow only.

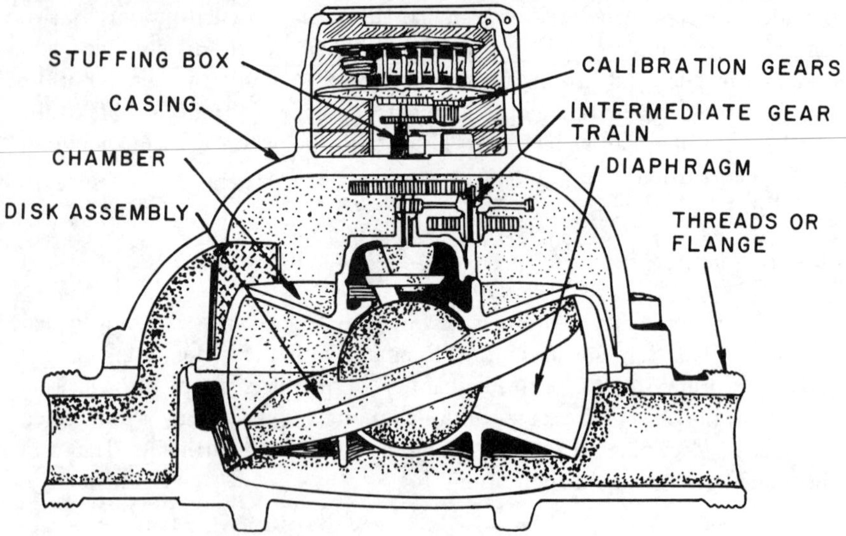

Figure 24. Cutaway view of a nutating disc meter.

MECHANICAL FLOWMETERS

Mechanical flowmeters measure flow by drag effects. The drag coefficient is a function of the Reynolds number and fluid properties. Consequently, these instruments must be recalibrated for each new fluid handled and calibrations periodically checked for fluid properties changes. The main devices in this class are the rotameter and the turbine meter.

Rotameter

The rotameter is a variable-area meter (Figure 26). It consists of a tapered vertical tube in which the fluid enters the bottom causing a float to ascend. The float is used to indicate the flowrate and is heavier than the metered fluid. As fluid flows through the tube, the float continues to rise until the drag forces are balanced by the float's weight and buoyancy forces.

From a force balance about the float an expression for the mean fluid velocity is obtained:

$$u_m = \left[\frac{2g\phi}{C_D A_B}\left(\frac{\rho_B - \rho_f}{\rho_f}\right)\right]^{1/2} \tag{82}$$

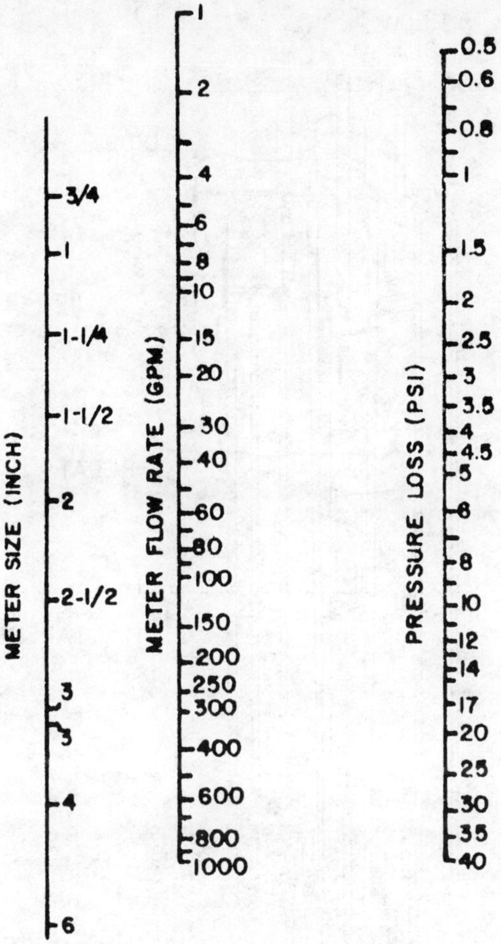

Figure 25. Nomograph for determining pressure loss or flowrate through a nutating disc meter. (courtesy of Hersey Products Inc., Spartanburg, SC).

where ϕ is the volume of the float, C_D the drag coefficient, and A_B the area of the front face of the float.

From this expression the volumetric flowrate for the annular flow area around the float is obtained (i.e., $Q = U_m A$), where

$$A = \frac{\pi}{4} [(D + \xi z)^2 - d^2] \tag{83}$$

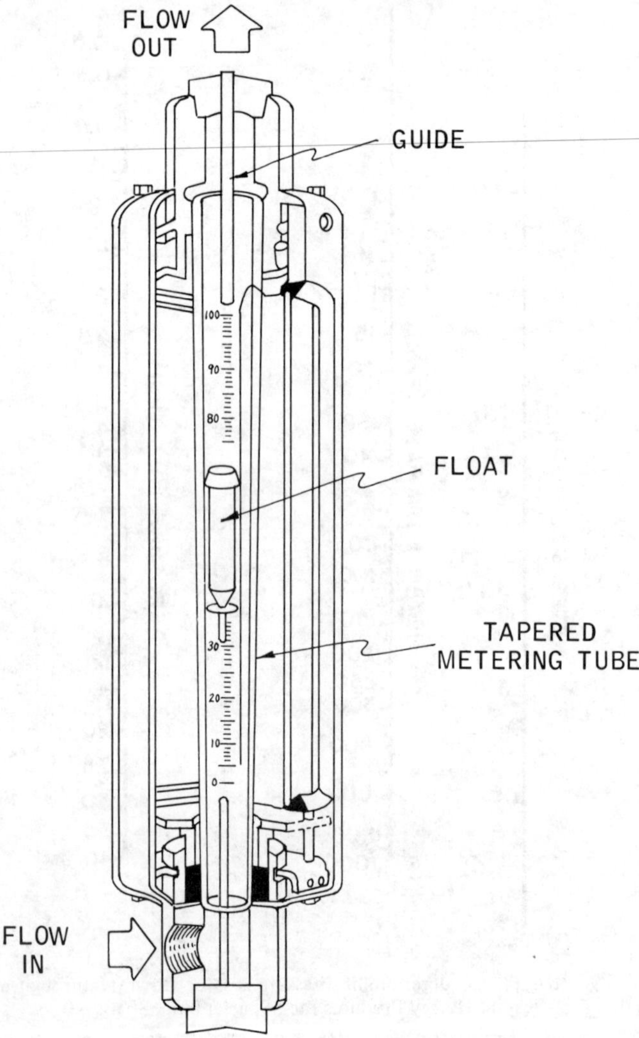

Figure 26. Illustrates the rotameter.

D is the inlet tube diameter, d the maximum float diameter, z the float's vertical distance from the entrance, and ξ a constant related to the tube taper.

For approximations, the area relationship given by Equation 82 can be assumed to be linear for the actual tube and float dimensions. With this assumption, a simple expression for the mass flowrate can be written as follows:

$$W = C'z[(\rho_B - \rho_f)\rho_f]^{1/2} \tag{84}$$

where C′ is an appropriate meter constant that must be obtained by calibration for the fluid in service.

In addition to being viscosity dependent, rotameters may be sensitive to fluid density variations. Special float construction can permit rotameters to be independent of fluid density changes. The mass flowrate expression then becomes

$$W = \frac{C'z\rho_f}{2} \tag{85}$$

where the float's density is specified as twice that of the fluid density.

There are a variety of float configurations, which are broken into two categories: extension type and spool type. Each float configuration is read differently. Figure 27 provides a guide for the proper manner in which to reach different float configurations.

Turbine Meter

The turbine meter (Figure 28) operates in a similar fashion to a hydraulic turbine. The turbine is positioned within a chamber through which the fluid passes. The fluid's impact on the blade surfaces causes rotation. The rotor can

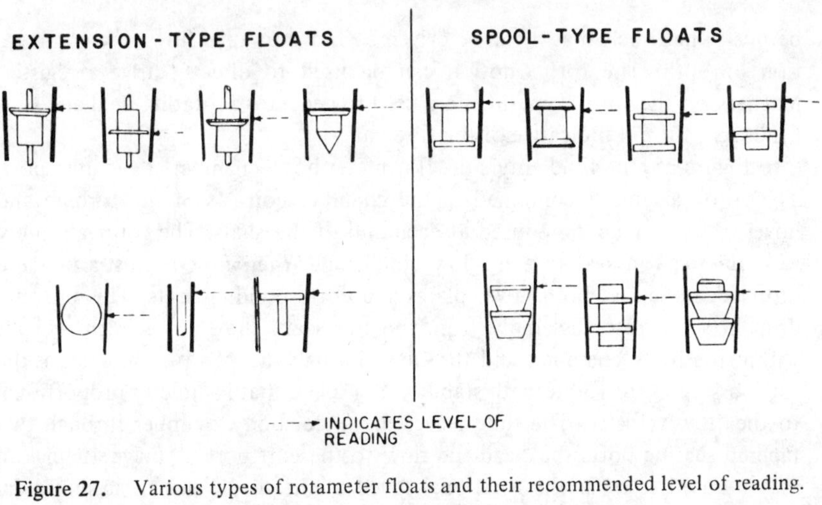

Figure 27. Various types of rotameter floats and their recommended level of reading.

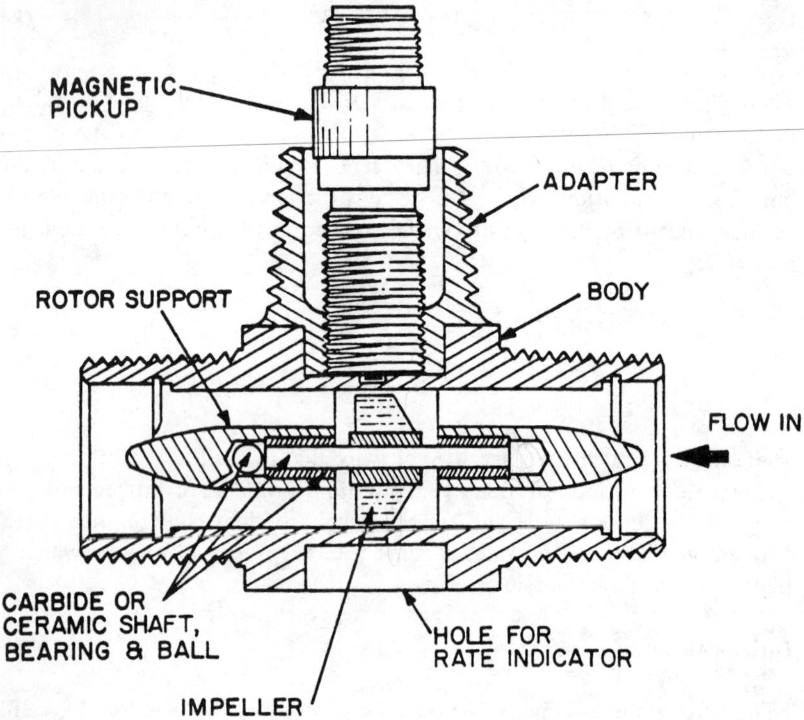

Figure 28. Shows the major components of the turbine meter.

be positioned such that it can be driven by radial or axial flow, or a combination of both. The rotor motion can be used to directly drive a register. Magnets on the rotor sometimes are used to generate a rotating field and thus produce a current indicative of the flow rate.

In Figure 28, the fluid enters the flowmeter body chamber at the inlet port. The rotor assembly contained in the chamber consists of the turbine and hover plate elements mounted at each end of the shaft. The rotor assembly rests on the top seat in a no-flow condition. When flow begins, a pressure drop occurs across both hover plates and corresponding seats. This pressure drop lifts the rotor assembly from the top seat. The fluid element is split within the body chamber, and the flow through the two turbines causes the rotor assembly to rotate with stability at a speed that is directly proportional to the flow velocity. The fluid leaves the meter body chamber through two tapered seating ports that lead the flow to the exit port. A magnetic pickup may be used to sense the rotational speed of the rotor, and an ac sinusoidal signal is generated. The frequency of these signals is directly proportional to the flowrate. The ac signal is transmitted by means of a conductor-shielded cable to an appropriate readout instrument.

MAGNETIC FLOWMETERS

Magnetic flowmeters (Figure 29) consist of a primary device (the magnetic flowtube) and a secondary device (the magnetic flow transmitter). The

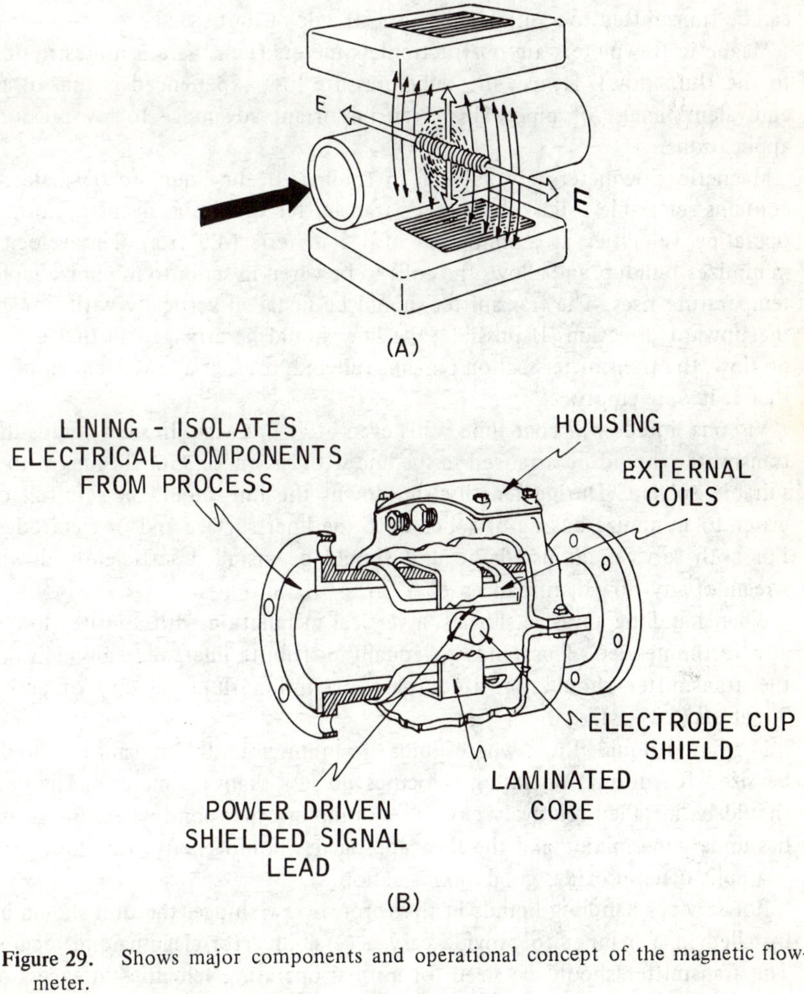

Figure 29. Shows major components and operational concept of the magnetic flowmeter.

magnetic flowtube is mounted in the process pipeline, and the secondary device can be remotely stationed.

The flowtube consists of an electrically insulated lined metal or unlined fiberglass-reinforced plastic metering tube with an opposed pair of metal electrodes in the tube wall. Electromagnetic coils are mounted external to the metering tube in a protective casing.

Electric current (ac) is used to excite the coils and generate a magnetic field at right angles to the axis of the fluid, passing through the metering tube. The fluid (liquid) must be conductive. As the fluid passes through the magnetic field an electrical potential develops across the electrode, which is proportional to the volumetric flowrate. The voltage signal generated is converted in the magnetic flow transmitter to either a standarized dc current signal, which can be transmitted to a suitable receiver, or a dc output signal.

Magnetic flowmeters are obstructionless meters (i.e., there is no restriction to the fluid flow). Hence, the only pressure loss experienced is that of an equivalent length of pipe. This is an important advantage to low-pressure applications.

Magnetic flowmeters are subject to fouling. If the liquid to be metered contains settleable solids, the meter's transmitter should be sized for normal operating velocities at a minimum of 1.5 meter/s (4.9 ft/s). This velocity minimizes buildup and allows the coils to be wired in series to minimize rapid temperature rises. The transmitter should be installed vertically with flow in the upward direction. If possible, the line should be arranged so that under no flow, the transmitter section remains full and, thus, at a lower temperature than if it were empty.

Viscous liquids will coat tube walls even after drainage. In such liquids the transmitter should be arranged in the line with easy access for flushing with a suitable solvent. During lengthy shutdowns the line should be left full of water to minimize baking of solids onto the liner surface and/or electrodes. For both services discussed the unit should be installed sufficiently downstream of any flow disturbances.

When handling abrasive slurries, a vertical installation with upward flow is also recommended. This tends to equally distribute liner wear. In addition, the transmitter should be sized for a maximum fluid velocity of about 3 meter/s (10 ft/s).

For liquid/liquid flows, where fluids are immiscible, the transmitter should be sized for normal operating velocities no less than 1.5 meter/s. The unit should be installed relatively close to a pump or other point where the steam has undergone mixing and the flow approaches homogeneity. An elbow, for example, often provides good mixing action.

For services handling liquids in the form of raw sludge, the unit should be installed in a manner to provide easy accessibility for cleaning and repairs. The transmitter should be sized for normal operating velocities in excess of 1.8 meter/s.

ULTRANSONIC FLOW MEASUREMENT

These devices use the Doppler frequency shift of ultrasonic signals reflected from discontinuities in the fluid stream to obtain flow measurements. These discontinuities can be suspended solids, bubbles or interfaces generated by turbulent eddies in the flow stream.

The sensor is mounted on the outside of the pipe and an ultrasonic beam from a piezoelectric crystal is transmitted through the pipe wall into the fluid at an angle to the flow stream (Figure 30). Signals reflected off flow disturbances are detected by a second piezoelectric crystal located in the same sensor. Transmitted and reflected signals are compared in an electrical circuit and the corresponding frequency shift is proportional to the flow velocity.

Doppler systems are applicable to a variety of liquid applications. Specific examples include slurries, acids, viscous fluids and various liquids containing suspended solids.

The advantages of Doppler flowmeters are as follows:

1. They allow noncontact flow measurements by means of an externally mounted transducer.
2. They are easy to install, requiring minimum piping alterations and downtime.
3. They do not undergo pressure loss.
4. There is no upper size limit on pipe diameters; hence, capital costs are the same regardless of application.

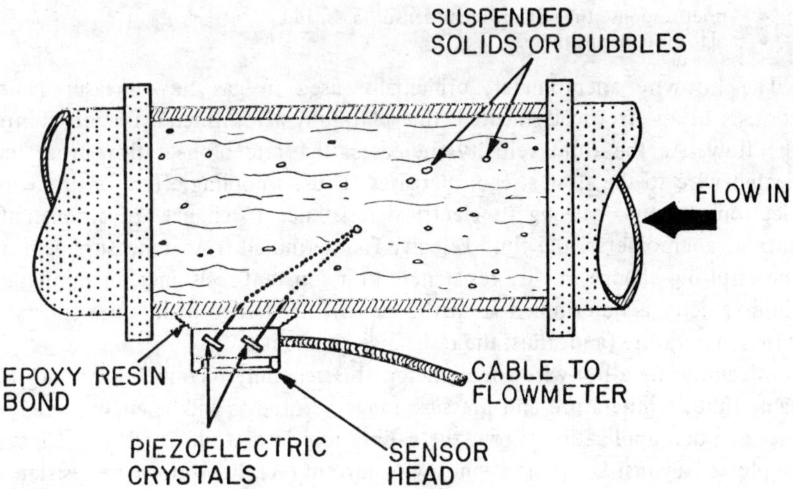

Figure 30. Illustrates application of the Doppler flowmeter to metering liquids in pipe flow.

The principal limitations with Doppler flowmeters are as follows:

1. Liquids to be metered must have an excess of approximately 2% suspended solids by volume.
2. Liquid linear velocities must exceed 0.15 meter/s.
3. The transducer head must be maintained at a temperature below 83°C.
4. Piping material must be of a homogeneous composition (composite materials such as concrete, reinforced concrete, vitreous clay and asbestos cement cannot be used with this device).
5. The pipe wall thickness cannot exceed 1.9 cm (0.75 in.).

The device can only be used for full pipe flow of liquids. Multiphase and some types of multicomponent flows cannot be metered by this method.

ELECTRICALLY HEATED TECHNIQUES

Electrically heated techniques consist of hot-wire or hot-film anemometry. These techniques are now oriented more toward research than industrial applications. Presently, they are the most widely used measuring systems for analyzing the microstructure of velocities in turbulent gas or liquid flows. The principle of operation makes it equally suited for determining other flow parameters such as density, pressure, temperature and composition. The hot-wire or hot-film anemometer possesses the following features:

- small sensing-element dimensions and, hence, high spatial resolution and little interference to flow,
- brief response time due to small sensor mass, and
- high system sensitivity.

The hot-wire anemometer, principally used in gas flow measurement, consists of an electrically heated, fine platinum wire, which is immersed into the flow. As the fluid velocity increases, the rate of heat flow from the heated wire to the flow stream increases. Thus, a cooling effect on the wire electrode occurs, causing its electrical resistance to change. In a constant-current anemometer, the fluid velocity is determined from a measurement of the resulting change in wire resistance. In a constant-resistance anemometer, fluid velocity is determined from the current needed to maintain a constant wire temperature (and, thus, the resistance constant).

In calibrating a hot-wire anemometer, it is generally recommended that the same fluid, temperature and pressure range be used as will be encountered in the intended application. From the calibration data, a plot of $I^2R_w/\Delta T$ can be plotted against $\overline{U}$; where I = hot-wire current (A), R_w = hot-wire resistance (Ω), ΔT = difference between the wire temperature and the fluid bulk temperature (°F), and $\overline{U}$ = mean local velocity of the fluid (fps).

Complete anemometry systems are commercially available for a range of applications. The electrical circuitry does not vary greatly from manufacturer to manufacturer. Indeed, the main difficulty in specifying a system for an intended application is in the selection of the proper probe or sensor. To be able to handle the great variety of measuring problems occurring in the field of anemometry, it is necessary to employ several different probe types. While wire sensors are always cylindrical, film sensors comprise two types: cylindrical and noncylindrical. Tables 16 and 17 give one anemometry sensor manufacturer's data on different probe configurations. These tables can serve as a guide to proper sensor configuration selection. In general, wire probes should be used wherever possible because they cost less and can be repaired in the event of wire breakage. Also, as a rule they have better frequency response than film probes.

For gaseous media, wire probes are normally employed, except for gases that are chemically aggressive or so heavily contaminated that the heat transfer from the wire is noticeably reduced by particles adhering to it, or so that damage occurs to the sensor wire from airborne particles striking it. Wire probes are also applicable for measurements in nonconducting liquids at low velocities.

For nonconducting fluids at high velocities, thinly coated film probes are recommended. Thinly quartz-coated film probes are preferred at high velocities in gaseous media because of their high resistance to mechanical influences.

DETERMINING MEAN VELOCITY

To evaluate the mean fluid velocity in a duct from local measurements use the following procedure:

Step 1. Divide the duct's cross section into a number of equal areas.

Step 2. Measure the local velocity at a representative point in each small area.

Step 3. Determine the mean value by averaging the results.

For rectangular ducts, the cross section can be divided into small squares or rectangles and the local velocity measured at the center of each.

For circular pipes, divide the cross section into several equal annular areas and a central circle. Local velocity measurements are made at the intersections of a diameter and the set of circles that bisect the annuli and the central circle.

For an N-point traverse on a circular cross section, readings should be made on each side of the cross section at

Table 16. Probe-Type Selection Chart (courtesy of DISA Electronics, Franklin Lakes, NJ)

Probe Selection Chart R = Recommended A = Applicable I = Inapplicable	Wire Probes										Film Probes									
	Gold-Plated Wire Probes	Miniature Probes	Subminiature Probes	High-Temperature Probe	Resistance Thermometer	Temperature-Compensated Probes	Gold-Plated X-Probes	Miniature X-Probes	Parallel-Sensor Probes	Triple-Sensor Probes	Fiber-Film Probes	Wedge-Shaped Probes	Conical Probes	Flush-Mounting Probes	Glue-On Probes	Fiber-Film X-Probes	V-Probes	Fiber-Film Triple-Sensor Probes	55M System with HF Bridge	55D80/81 Low-Velocity Anemometer
Medium																				
Gases and nonconducting liquids	R	R	R	R	R	R	R	R	R	R	R	R	R	R	R	R	R	R	R	R
Conducting liquids	I	I	I	I	I	I	I	I	I	I	R	R	R	R	I	R	R	I	I	I
State of Medium																				
Low and medium temperature up to 150°C	R	R	R	A	R	R	R	R	R	R	R	R	R	R	R	R	R	R	R	R
High temperatures up to 800°C	I	I	I	R	I	I	I	I	I	I	I	I	I	I	I	I	I	I	I	I
Contaminated flow	I	I	I	I	I	I	I	I	I	I	I	A	A	A	A	I	A	I	I	I

Flow conditions																				
Extremely low velocities	I	I	I	I	R	I	I	I	I	I	I	I	I	I	I	I	I	I	I	R
Low and medium velocities	R	R	R	R	R	R	R	R	R	R	R	R	R	R	R	R	R	R	R	I
High velocities	A	A	A	A	I	A	A	A	A	A	R	R	R	R	R	R	R	R	A	I
Large velocity gradients	R	A	A	A	I	A	R	A	I	R	R	A	A	I	I	R	A	R	A	I
Varying temperature	I	I	I	I	R	R	I	I	A	I	I	I	I	I	I	I	I	I	I	I
Space Conditions																				
Sufficient space	R	R	R	R	R	R	R	R	R	R	R	R	R	R	R	R	R	R	R	R
Little space	I	R	R	I	R	I	I	R	R	I	I	I	I	R	I	I	I	I	I	I
Very little space	I	I	R	I	I	I	I	I	I	I	I	I	I	I	I	I	I	I	I	I
Quantity to be Measured																				
Mean velocity	R	R	R	R	I	R	R	R	I	R	R	R	R	I	I	R	R	R	R	R
Instantaneous flow direction	I	I	I	I	I	I	R	R	I	R	I	I	I	I	I	R	R	R	I	I
Velocity fluctuations	R	R	R	R	I	R	R	R	R	R	R	R	R	I	I	R	R	R	R	R
Extremely low turbulence intensities	I	I	I	I	I	I	I	I	R	I	I	I	I	I	I	I	I	I	I	I
Low- and medium-turbulence intensities	R	R	R	R	I	R	R	R	I	R	R	R	R	I	I	R	R	R	I	I
High-turbulence intensities	R	A	A	A	I	A	R	A	I	R	R	A	A	I	I	R	A	R	I	I
Extremely high-frequency fluctuations	I	I	I	I	I	I	I	I	I	I	I	I	I	I	I	I	I	I	R	I
Turbulent shear stress	A	A	A	I	I	I	R	R	I	R	A	I	I	I	I	R	R	R	I	I
Spatial turbulence components	A	A	A	I	I	I	R	R	I	R	A	I	I	I	I	R	R	R	I	I
Wall shear stress	I	I	I	I	I	I	I	I	I	I	I	I	I	R	R	I	I	I	I	I
Temperatures and temperature fluctuations	A	A	A	A	R	I	I	I	I	I	A	A	A	A	A	I	I	I	I	I

Table 17. Characteristics of Hot-Wire and Hot-Film Probes

Sensor Type	Sensor Material	Sensor Dimensions	Thickness of Quartz Coating	Sensor Resistance R_{20} (approx.)	Temperature Coefficient of Resistance (TCR) α_{20} (approx.)
Gold-Plated Wire Sensors	Platinum-Plated Tungsten	5-μm diameter 1.25 mm long[a]	–	3.5 Ω	0.36%/°C
Miniature Wire Sensors	Platinum-Plated Tungsten	5-μm diameter 1.25 mm long	–	3.5 Ω	0.36%/°C
Subminiature Wire Sensors	Platinum-Plated Tungsten	5-μm diameter 0.45 mm long	–	1.3 Ω	0.36%/°C
Wire Sensors for High Temperatures	Platinum/Rhodium (90/10)	10-μm diameter 2.2 mm long	–	6 Ω	0.16%/°C
Wire Sensors for Temperature Measurements[c]	Platinum	1-μm diameter 0.4 mm long	–	50 Ω	0.35%/°C
Fiber-Film Sensors	Nickel	70-μm diameter 1.25 mm long[d]	0.5 μm 2 μm	6 Ω 6 Ω	0.40%/°C 0.40%/°C
Wedge-Shaped Film Sensors	Nickel	1 mm × 0.2 mm	0.5 μm 2 μm	15 Ω 15 Ω	0.30%/°C 0.30%/°C
Conical Film Sensors	Nickel	1.4 mm × 0.1 mm	0.5 μm 2 μm	15 Ω 15 Ω	0.35%/°C 0.35%/°C
Flush-Mounting Film Sensors	Nickel	0.75 mm × 0.2 mm	0.5 μm 2 μm	15 Ω 15 Ω	0.35%/°C 0.35%/°C
Glue-On Film Sensors	Nickel	1 mm × 0.2 mm	0.5 μm –	15 Ω –	0.4%/°C –

[a]Overall tungsten wire length is 3 mm. Wire ends are gold-plated to 25-30 μm diameter
[b]Depending on type of mounting.
[c]Sensor is specially suited for temperature measurements. Maximum sensor current:
[d]Overall fiber length is 3 mm. Fiber ends are gold-plated, limiting sensor length.
[e]At 20°C, decreasing with increasing temperature, 1 bar at 80°C (applies to probe design

(courtesy of DISA Electronics, Franklin Lakes, NJ)

Maximum Sensor Temperature	Maximum Ambient Temperature	Maximum Ambient Pressure	Minimum Velocity	Maximum Velocity	Frequency Limit (f_{cpo})	Frequency Limit (f_{cmax})	Medium
300°C	150°C	–[b]	0.2 m/s	200 m/s	90 Hz	400 kHz	Air
			0.01 m/s				Dist. water
300°C	150°C	–[b]	0.2 m/s	500 m/s	90 Hz	400 kHz	Air
			0.01 m/s	5 m/s			Dist. water
300°C	150°C	–[b]	0.2 m/s	500 m/s	300 Hz	150 kHz	Air
			0.01 m/s	5 m/s			Dist. water
800°C	750°C	100 bar	0.2 m/s	350 m/s	25 Hz	150 kHz	Air
–	150°C	–[b]	–	60 m/s	2 kHz		Air
350°C	150°C	–	0.2 m/s	350 m/s		175 kHz	Air
150°C	100°C	–	0.01 m/s	10 m/s		30 kHz	Water
350°C	150°C	70 bar	0.1 m/s	500 m/s	0.05 Hz	450 kHz	Air
150°C	100°C	70 bar[e]	0.01 m/s	25 m/s	0.4 Hz	30 kHz	Water
350°C	150°C	70 bar	0.1 m/s	500 m/s			Air
150°C	100°C	70 bar[e]	0.01 m/s	25 m/s		30 kHz	Water
200°C	120°C	70 bar	–	–	0.2 Hz		Air
150°C	100°C	70 bar[e]	–	–	0.5 Hz	30 kHz	Water
200°C	120°C	–	–	–			Air
–	–	–	–	–			

and limit sensor length.

1 mA.

listed).

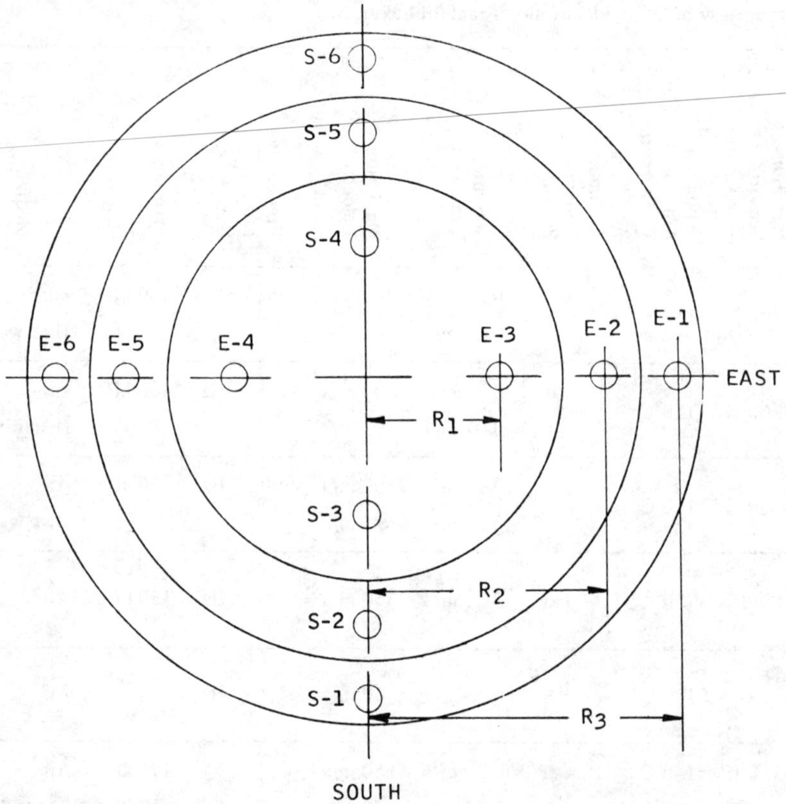

Figure 31. Recommended traversing positions for evaluating

$$100 \times \sqrt{\frac{2n-1}{N}} \text{ percent } \left(n = 1, 2, 3, \ldots \frac{N}{2}\right) \tag{86}$$

of the pipe radius from the center.

If the velocity distribution is unsymmetrical, traverse several diameters spaced at equal angles about the pipe. For a normal velocity distribution in a circular pipe, a 10-point traverse gives a mean velocity about 0.3% overestimate. A 20-point traverse gives a 0.1% overestimate.

Figure 31 illustrates proper traversing positions for rectangular and circular ducts.

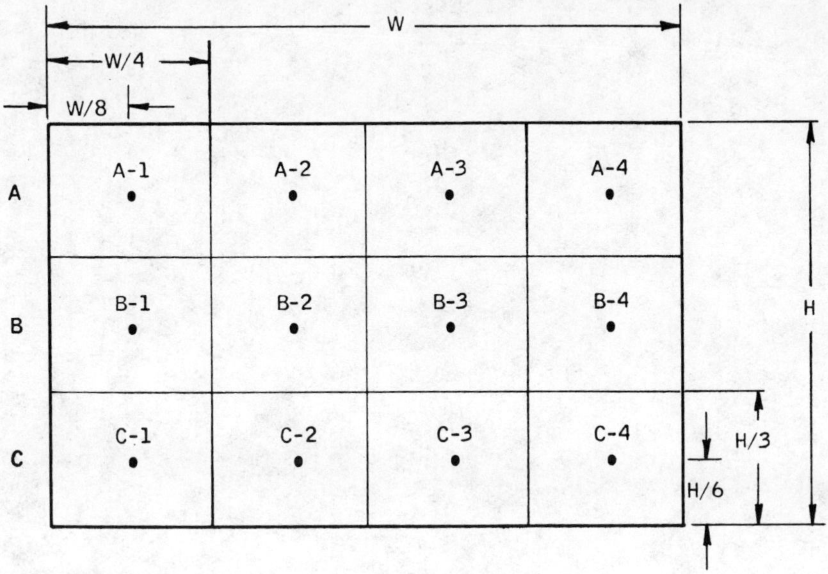

The minimum number of test points for uniform flow may be determined as follows:

Cross-Sectional Area (ft^2)	Number Of Test Points
<2	4
2–25	12
>25	20

mean fluid velocities in rectangular and circular ducts.

CHAPTER 2.4

PUMPS AND PUMPING SERVICES

This chapter presents general information and procedures for specifying pumps and process plant pumping services. Details on pump application technology are presented in subsequent discussions in this chapter. This chapter only provides supplemental information to acquaint the plant engineer with pumping technology and nomenclature. Emphasis is given to centrifugal pumps as these are most widely encountered in the chemical process industries. For more detailed discussions on pump types, designs and performance, the reader is referred to the literature [12,13]. Note also that data presented here represent typical conditions and should therefore only be applied for approximations. The manufacturer's actual data should always be used when specifying a specific pump style for an intended service.

PUMP CLASSIFICATIONS

The pump types most often employed in process plant applications fall into the following classes: centrifugal, axial, regenerative turbine type, reciprocating, metering and rotary. There are two categories of pumps under which these classes are grouped: dynamic pumps and positive displacement pumps.

Dynamic pumps include the classes centrifugal and axial. These pumps operate by developing a high liquid velocity and converting the velocity to pressure in a diffusing flow passage. In general, they tend to have lower efficiency than positive displacement pumps. However, they do operate at relatively high speeds to permit high flow rate in relation to the physical size of the pump. Also, dynamic pumps usually have significantly lower maintenance requirements than positive displacement pumps.

Positive displacement pumps operate by forcing a fixed volume of fluid from the inlet pressure section of the pump into the discharge zone of the pump. With reciprocating pumps this is performed intermittently. In the case of rotary screw and gear pumps it is done continuously. This category of pumps operates at lower rotative speeds than dynamic pumps. Positive displacement pumps also tend to be physically larger than equal-capacity dynamic pumps.

Table 18 gives a breakdown of pump types, construction styles and descriptive information. Table 19 reports various performance parameters. Note that data presented in Table 19 are typical and descriptive in nature, so should be applied cautiously to specific problems.

GENERAL FEATURES OF CENTRIFUGAL PUMPS

There are a variety of design features applied to the different construction styles of centrifugal pumps. A summary of the main features is given below.

Flange and Mount Bolts

The American Voluntary Standard (AVS) for centrifugal pumps for chemical industry use is applied by many manufacturers of small- to medium-sized pumps of light- to medium-duty ratings for flange, mounting bolt and shaft center locations. AVS recommendations cover dimensional aspects to allow replacement of one manufacturer's pump in an existing installation site with that of another manufacturer. Roughly half of the centrifugal pumps manufactured in the United States follow AVS standards.

Volutes and Diffusers

Many commercial pumps have diverging channels called volutes, which are cast into the discharge zone of the casing. Flow through the volute causes the velocity to decrease from near the tip speed (around 200 fps) to the discharge line velocity (16 fps). This, in turn, causes the pressure to increase.

Single-volute passages are utilized in most pumps because of simplicity in design. However, this type design causes an unbalanced load on the impeller because of the variation in pressure around the periphery. Double-volute configurations are employed when the unbalanced force level threatens to cause significant shaft deflection. This is typically above 500 feet per stage. Twin-discharge channels tend to balance the radial forces, thus reducing cycle stress in the shaft.

Some process pumps employ vaned diffusers for pressure conversion rather than volutes. Included are some inline pump styles, some axial pumps, and a few pumps designed for high head, low flow performance. Vaned diffusers offer the advantages of balanced radial force, compact size and peak efficiency at high head and low flow. Disadvantages include that they are more difficult to fabricate and repair than volute pumps.

Impeller Configurations

Common impellers are enclosed with full discs and shrouds (these are termed "closed"). Semiopen impellers are sometimes employed. These have a full back disc but no shroud. Fully open impellers have vanes but little or no disc material. These are employed occasionally in low head, solids handling service.

Many impellers employed are arranged from one side only (called single-suction designs). High flow pumps employ impellers that accept suction from both sides (called double-suction designs). Very high flowrate units employ "mixed flow" impellers, which involve an axial component in the direction of fluid flow. Many vertical, multistaged pumps (particularly deep well pumps) employ mixed flow impellers to maintain small diameters for installation convenience. Low diameter and high specific speed combinations limit the head per stage to relatively low values (typically 115–150 ft) and thus tend to require many stages. Figure 32 summarizes nomenclature for centrifugal pump impellers.

Wearing Rings

Many pumps are designed with close running clearance at the suction side to separate the discharge pressure zone of the pump from the suction pressure areas and to minimize back leakage. Wearing rings are generally fitted at the close clearance position in the pump casing and at an opposing position on the impeller, which allows easy restoration of the design clearance.

Many high-pressure centrifugal pumps are fitted with wearing rings at the back side (disc) of the impeller. This reduces the thrust force on the impeller, at the same time minimizing the pressure to which the stuffing box is exposed.

Nozzles

Many horizontal pumps are provided with suction and discharge nozzles on the top of the casing. In some horizontal pump styles, suction nozzles

Table 18. General Characteristics

Pump Type	Construction Style	Construction Characteristics
		Dynamic-
Centrifugal (horizontal)	Single-stage overhung, process type	Impeller cantilevered beyond bearings.
	Two stage overhung	Two impellers cantilevered beyond bearings.
	Single-stage impeller between bearings	Impeller between bearings; casing radially or axially split.
	Chemical	Casting patterns designed with thin sections for high-cost alloys.
	Slurry	Designed with large flow passages.
	Canned	No stuffing box; pump and motor enclosed in a pressure shell.
	Multistage, horizontally split casing	Nozzles located in bottom half of casing.
	Multistage, barrel type	Outer casing contains inner stack of diaphragms.
Centrifugal (vertical)	Single-stage, process type	Vertical orientation.
	Multistage	Many stages with low head per stage.
	Inline	Inline installation, similar to a valve.
	High speed	Speeds to 380 rps, heads to 5800 ft (1770 m).
	Sump	Casing immersed in sump for easy priming and installation.
	Multistage, deep well	Long shafts.
Axial	Propeller	Propeller-shaped impeller.
Turbine	Regenerative	Fluted impeller. Flow path resembles screw around periphery.
		Positive
Reciprocating	Piston, plunger	Slow speeds.
	Metering	Consists of small units with precision flow control system.
	Diaphragm	No stuffing box.

of Different Pumps

Normal Number of Stages	Maintenance[a]	Solids Tolerance	Notes
Type Pumps			
1	L	M-H	Capacity varies with head.
2	L	M-H	Used for heads above single-stage capability.
1	L	M-H	Used for high flows to 1083 ft (330 m) head.
1	M	M-H	Have low pressure and temperature ratings.
1	H	H	Low speed and adjustable axial clearance. Has erosion control features.
1	L-M	L	Low head capacity limits when used in chemical services.
Multi	L	M	Have moderate temperature-pressure ranges.
Multi	L	M	Used for high temperature-pressure ratings.
1	L	M	Used to exploit low net positive section head (NPSH) requirements.
1	L	M	Low-cost installation.
1	L	M	Low-cost installation.
1	M	L	High head/low flow. Moderate costs.
1	L	M-H	Low cost.
Multi	M-H	M	Used for water well service.
1	L	H	Vertical orientation.
1,2	H	M	Capacity independent of head. Low flow/high head performance.
Displacement Pumps			
1	H	M	Driven by steam engine cylinders or motors through crankcases.
1	M-H	L	Diaphragm and packed plunger types.
1	H	L	Used for chemical slurries. Can be pneumatically or hydraulically actuated.

Table 18,

Pump Type	Construction Style	Construction Characteristics
		Positive
Rotary	Screw	1, 2 or 3 screw rotors.
	Gear	Intermeshing gear wheels.

[a]L = low; M = medium; H = high.

are located at the end of the pump, coaxial with the shaft centerline. End suction pumps are most often used for services with moderate temperatures and flowrates.

Side suction connections are often used on horizontally split horizontal pumps. This arrangement permits positioning of the nozzle in the bottom half of the casing so that the bottom half does not have to be disturbed during maintenance.

Casing Construction and Orientation

Many centrifugal pumps have single casings (i.e., a single wall between the liquid under discharge pressure and the atmosphere). Double casings are employed in horizontal, multistage, high-pressure pumps and in vertical pumps. In the former, a heavy barrel-shaped casing surrounds the stack of stage diaphragms. The stack of diaphragms comprises the inner casing, while the barrel is the outer casing. Boiler feed pumps normally employ this arrangement.

In the case of vertical pumps, a vertical cylinder buried in the ground houses the pumping element. Suction liquid will enter the outer cylinder, flow to the bottom and then up through the pumping element stages. Again, the diaphragms of the stages in the pumping element comprise the inner casing.

Casings can be joined in the same plane as the shaft axis (axially split) or perpendicular to the shaft (radially split). Axially split horizontal pumps are often called "horizontally split," and radially split horizontal pumps are called "vertically split."

Radial jointing is employed on horizontal overhung pumps to allow easy removal of the rotor and bearing bracket assembly for maintenance purposes. It is also employed on high-pressure multistage pumps because of structural

continued

Normal Number of Stages	Maintenance[a]	Solids Tolerance	Notes
Displacement Pumps			
1	M	M	For high-viscosity, high-flow-high-pressure services.
1	M	M	For high-viscosity, moderate-pressure/moderate-flow services.

problems associated with bolting together the halves of axially split casings exposed to high internal pressure.

With horizontal pumps, their casings are supported by the base plate and the bearing brackets are, in turn, supported from the casing. Many casing-supported pumps have extensions along the sides of the casings at the centerline, which rests on pedestals. These are referred to as "centerline supported." This arrangement minimizes movement of the centerline as casing temperature increases. Casings for ambient and moderate temperature service are often supported at the bottom of the casing (these are said to be "foot supported").

Stuffing Boxes

Conventional stuffing boxes are filled with either packing material, such as braided rope or metallic foil rings, or with a mechanical seal. Pump casings designed for mounting only a mechanical shaft seal are said to have "built-in" seals (i.e., they cannot be converted to packing). This arrangement reduces shaft overhang and is generally less expensive than the conventional stuffing box. It is normally employed in clean, moderate suction pressure and temperature services. Disadvantages of this style include high sensitivity to dirt in service fluid, and fewer corrective measures can be applied to remedy chronic mechanical seal service problems.

Shaft Couplings

Pump shafts can be coupled to the drive shaft by a separate, removable, flexible-drive coupling, or they can be integral with the drive shaft. Most process plant pumps are coupled separately. The integral shaft style is referred to as "close coupled."

Table 19. Typical Performance

Pump Type	Construction Style	Capacity (dm³/s)	Capacity (gpm)	Maximum Head (m)	Maximum Head (ft)
					Dynamic-Type
Centrifugal (horizontal)	Single-stage overhung	1–320	15–5,000	150	492
	Two-stage overhung	1–75	15–1,200	425	1394
	Single-stage impeller-between bearings	1–2500	15–40,000	335	1099
	Chemical	65	1000	73	239
	Slurry	65	1000	120	394
	Canned	0.1–1250	1–20,000	1500	4921
	Multistage horizontal split	1–700	20–11,000	1675	5495
	Multistage, barrel type	1–550	20–9,000	1675	5495
Centrifugal (vertical)	Single stage	1–650	20–10,000	245	804
	Multistage	1–5000	20–80,000	1830	6004
	Inline	1–750	20–12,000	215	705
	High speed	0.3–25	5–400	1770	5807
	Sump	1–45	10–700	60	197
	Multistage deep well	0.3–25	5–400	1830	6004
Axial	Propeller	1–6500	20–100,000	12	39
Turbine	Regenerative	0.1–125	1–2000	760	2493
					Positive
Reciprocating	Piston, plunger	1–650	10–10,000	345,000	1.13×10^6
	Metering	0–1	0–10	51,700	1.70×10^5
	Diaphragm	0.1–6	4–100	34,500	1.13×10^5
Rotary	Screw	0.1–125	1–2,000	20,700	6.79×10^4
	Gear	0.1–320	1–5,000	3,400	11,155

[a]Depends on strength of materials. Can be in excess of 50,000 psi.

Parameters of Different Pumps

Maximum Discharge Pressure		Typical NPSH/Requirements		Efficiency (%)	Maximum Pumping Temperature	
(kPa)	(psi)	(m)	(ft)		(°C)	(°F)
Pumps						
4,100	600	2–6	6.56–19.7	20–80	455	851
4,100	600	2–6.7	6.56–22.0	20–75	455	851
6,800	980	2–7.6	6.56–24.9	30–90	205–455	401–851
1,400	200	1.2–6	3.94–19.7	20–75	205	401
4,100	600	1.5–7.6	4.92–24.9	20–80	455	851
68,900	10,000	2–6	6.56–19.7	20–70	540	1004
20,100	3,000	2–6	6.56–19.7	65–90	205–260	401–500
41,400	6,000	2–6	6.56–19.7	40–75	455	851
4,100	600	0.3–6	0.98–19.7	20–85	345	653
4,800	700	0.3–6	0.98–19.7	25–90	260	500
3,400	500	2–6	6.56–19.7	20–80	260	500
13,800	2,000	2.4–12	7.87–39.4	10–50	260	500
1,380	200	0.3–6.7	0.98–22.0	45–75		
13,800	2,000	0.3–6	0.98–19.7	30–75	205	401
1,030	150	2	6.56	65–85	65	149
10,300	1,500	2–2.5	6.56–8.20	55–85	120	248
Displacement Pumps						
a	a	3.7	12.1	65–85	290	554
345,000	50,000	4.6	15.1	20	300	572
24,100	3,500	3.7	12.1	20	260	500
20,700	3,000	3	9.84	50–80	260	500
3,400	500	3	9.84	50–80	345	653

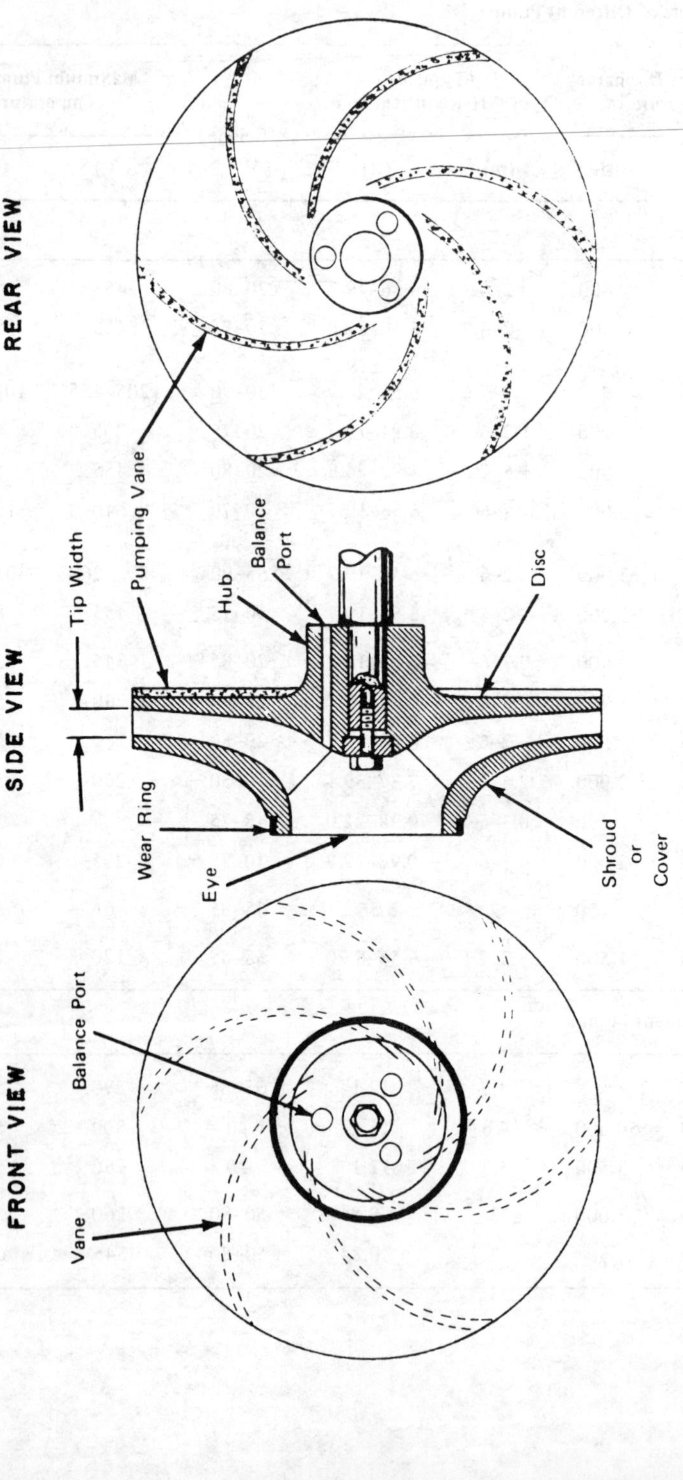

Figure 32. Major components of centrifugal pump impellers.

Cooling Arrangements

Centrifugal pumps normally have several cooling features that are essential to avoid bearing overheating, to maintain alignment and to assure proper mechanical seal performance.

There are three specific areas in which cooling is necessary in high-temperature service, namely, bearing oil cooling, stuffing box jacket cooling and seal flushing liquid cooling.

Cooling water is circulated through a bearing housing jacket, cooling coil or external oil cooler to maintain oil temperatures below 180°F (82°C). (This is the maximum allowable temperature for proper bearing lubrication.)

Cooling water is also circulated through the jacket to remove friction heat generated by the mechanical seal. It also keeps the seal flushing liquid from flashing. Stuffing box cooling helps to isolate the hot pump from the bearing housings.

For services between 400 and 600°F (200–315°C), the seal flushing liquid is cooled by an external cooler. This maintains temperature within the recommended limitations of the mechanical seal materials.

In addition to the above cooling water applications, low-pressure steam is used occasionally to cool seal plates and stuffing boxes in very high-temperature services (above 600°F).

CENTRIFUGAL PUMPS

Centrifugal pumps accomplish the generation of pressure by the conversion of velocity head into static head. The rotary motion of impellers adds energy to the service fluid in the form of a velocity increase. This velocity increase is converted into static head in the diffusing section of the casing. Centrifugal-type pumps have no valves. The flow is uniform and free of low-frequency pulsations. A pump operating at a fixed speed will develop the same theoretical head in feet of flowing fluid, regardless of density. However, the pressure corresponding to the developed head (in psi) depends on the fluid density.

The parameters that establish the maximum head (in feet of fluid) that a centrifugal pump can develop are the pump speed (rps), impeller diameter and the number of impellers in series combination. Impeller design and blade angle mainly affect the slope and shape of the head-capacity curve and normally have minor effects on the developed head. Conventional centrifugal pump impellers have a maximum tip speed around 200 fps (60 meter/s). In slurry pumps, impellers are limited to about half this tip speed to limit erosion problems.

Normal and maximum viscosity ranges are a major consideration in pump selection because of possible deterioration in performance with increasing viscosity. Deterioration can be both continuous and gradual. Table 20 can serve as a guide in selecting the proper centrifugal pump type for an application.

Most centrifugal pump types are not self-priming, meaning that they are not capable of evacuating vapor from the suction line so that liquid can flow into the line and into the pump casing without external assistance. The impellers on centrifugal pumps are designed specifically for efficient liquid pumping and are not operated at high enough tip speeds to convert them into vapor compressors. The differential head that the pump impeller can deliver is the same on vapor as on liquid. However, the equivalent differential pressure rise capability is much lower with vapor. Thus, centrifugal pump impellers are not capable of generating a significant reduction in the pressure exerted by the vapor in the suction line to permit liquid flow.

To prime a centrifugal pump before starting, the suction line and pump casing must be filled with liquid. When the suction source is at positive pressure or is positioned above the pump, priming is done by opening the suction valve and releasing (venting) the trapped vapor from a valved connection on the pump casing or discharge line (located inside the discharge block valve). Liquid then will flow into the suction line and pump casing to displace the venting vapor.

Specially designed centrifugal pumps are self-priming. One type accomplishes evacuation of the suction line vapor by entrainment of vapor bubbles from the suction side of the impeller in a charge of liquid held in the pump casing (or sometimes held in a holdup/separation bottled attached to it).

Table 20. Viscosity Sensitivity of Different Pump Types

Type Pump	Kinematic Viscosity $\nu = \mu/\rho$ (mm^2/sec)	Guidelines
Rotary	7	Nominal minimum viscosity for rotary pumps. Efficiency begins to decrease as viscosity increases. Viscosity should be specified for services when it exceeds this level.
Centrifugal	<30	Centrifugal pumps preferred over rotary when conditions allow application of either type.
Centrifugal	30	The head-capacity capability of a centrifugal pump begins to deteriorate.
Centrifugal	30–110	Centrifugal pump preferred over rotary despite some efficiency drop.
Positive Displacement	110–220	These are almost always used if the expected viscosity exceeds this level.
Rotary	110–650	Rotary preferred if viscosity falls in this range.

The liquid charge is recirculated to the suction side after separating the entrained vapor. There are other designs available; however, in general, self-priming centrifugal pumps have been applied on a limited basis in continuous process service.

Most centrifugal pumps have a self-venting feature. Small amounts of vapor trapped within the casing at startup (after suction priming is complete) are swept out into the discharge line when the pump is started. Horizontally split casings are not arranged to be self-venting, however, and are equipped with specially designed valved vent connections requiring manual operation. Single-stage centrifugal pumps with top discharge connections have good self-venting performance, even though the casing shape places a small high point vapor pocket in the top of the discharge volute.

A brief description of the major types of centrifugal pumps is given below.

Single-Stage Overhung

This design has a single-stage overhung impeller. Its casing is supported at the centerline. Two shaft bearings are mounted close together in the same bearing bracket, with the impeller cantilevered or overhung beyond them. This design is illustrated in Figure 33. This type design usually has top suction and discharge flanges; wearing rings both on the front and back of the impeller and casing; a single suction, closed impeller; and a single stuffing box fitted with a mechanical seal. There are also water cooling options on the pedestal, stuffing box and bearings. This pump type is well suited for high-temperature operation and can be used for handling flammable liquids.

Two-Stage Overhung

This is simply a modified version of the single-stage process pump (Figure 34). It is capable of higher head than its single-stage counterpart. In this style, the stuffing box pressure is roughly halfway between suction and discharge pressures.

Single-Stage (Impeller-Between-Bearings)

These designs have their impellers mounted between the bearings and thus have two stuffing boxes. Single-stage versions are capable of developing heads up to 1080 feet (330 meters). Casings can be axially split for temperatures up to 500°F (260°C) and radially split for temperatures to 850°F (455°C).

Inline Pumps

These are vertical pumps with casings designed to be bolted directly to piping (similar to a valve). There are two basic configurations: coupled and

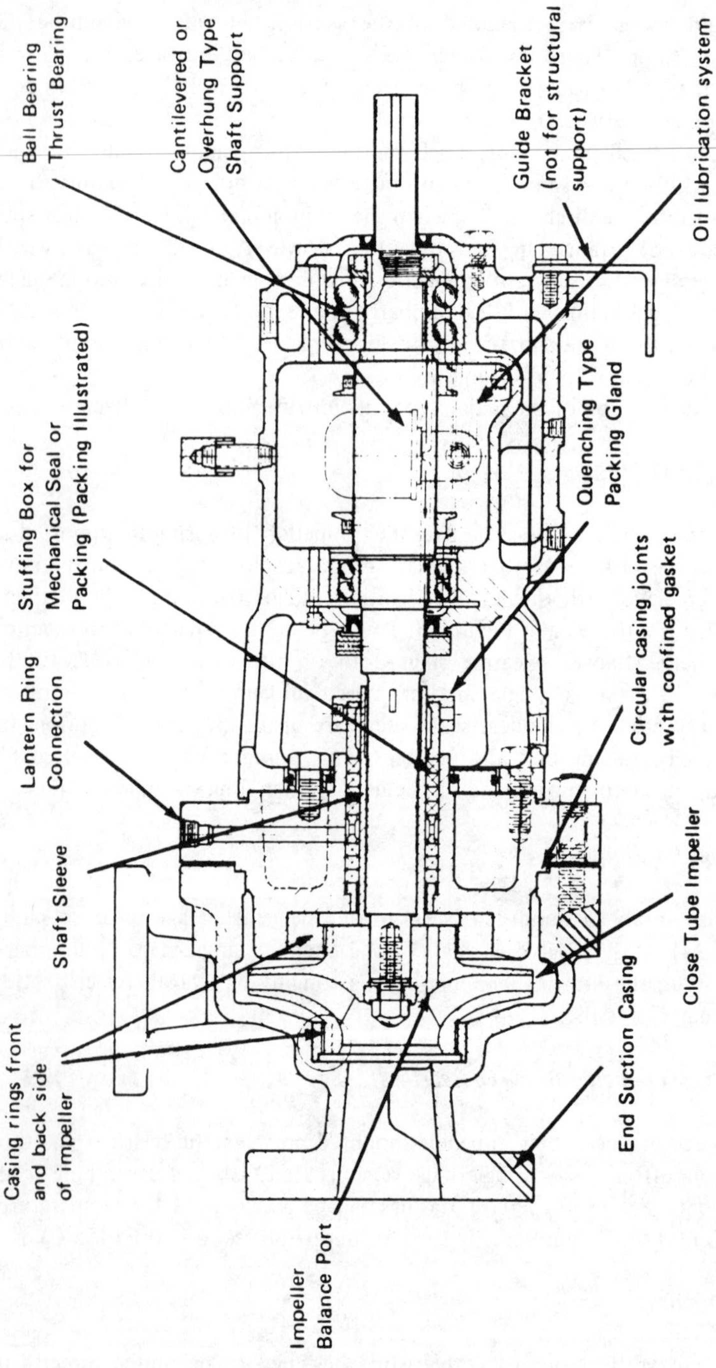

Figure 33. Illustrates a single-stage end suction-overhung centrifugal pump.

close-coupled. Service life and maintenance requirements for both styles are about the same. Generally, this type of pump is preferred if the cost of piping associated with installation can be reduced over the conventional horizontal styles. Chemical industry standards for inline pumps are summarized by Vetter and Fritsch [14].

High-Speed Centrifugal Pumps

These pumps are single-impeller models designed for speeds typically in the range of 170 to 280 rps and as high as 400 rps. They are capable of heads up to 5300 feet (1600 meters). Pumping temperatures are limited to about 500°F (260°C). High-speed pumps tend to have high NPSH requirement due to the sudden velocity increase as the liquid enters the impeller. Maintenance requirements for these pumps tend to be higher than for conventional-speed, single-stage pumps; however, it is about the same as multistage models, with which they compete for high head services.

Chemical Pumps

This class of pumps is designed with casing shapes that are cast in high-cost alloys. Casings are more frequently foot supported or bearing bracket supported than centerline supported. These pumps are limited to relatively low temperature, pressure and flowrates.

Slurry Pumps

Slurry pumps are used in services that have severe conditions of slurry pumping. They have a number of special design features that make them well-suited for this type service:

1. wide flow passages to avoid clogging;
2. open/semiopen impellers, which are less sensitive than closed impellers to clogging;
3. arrangements that break up large particles;
4. low fluid velocities, as low rotating and peripheral speed are possible;
5. adjustable rotor position to restore axial clearance without dismantling the pump; and
6. replaceable wearing plates and pumping plates on back of impeller, instead of wearing rings, which are subject to erosion.

Canned Pumps

These are motor pump units with the rotating rotor and impeller housed entirely within a pressure casing. This type design eliminates the need for a stuffing box. The pumped fluid serves both as a lubricant for bearings and as

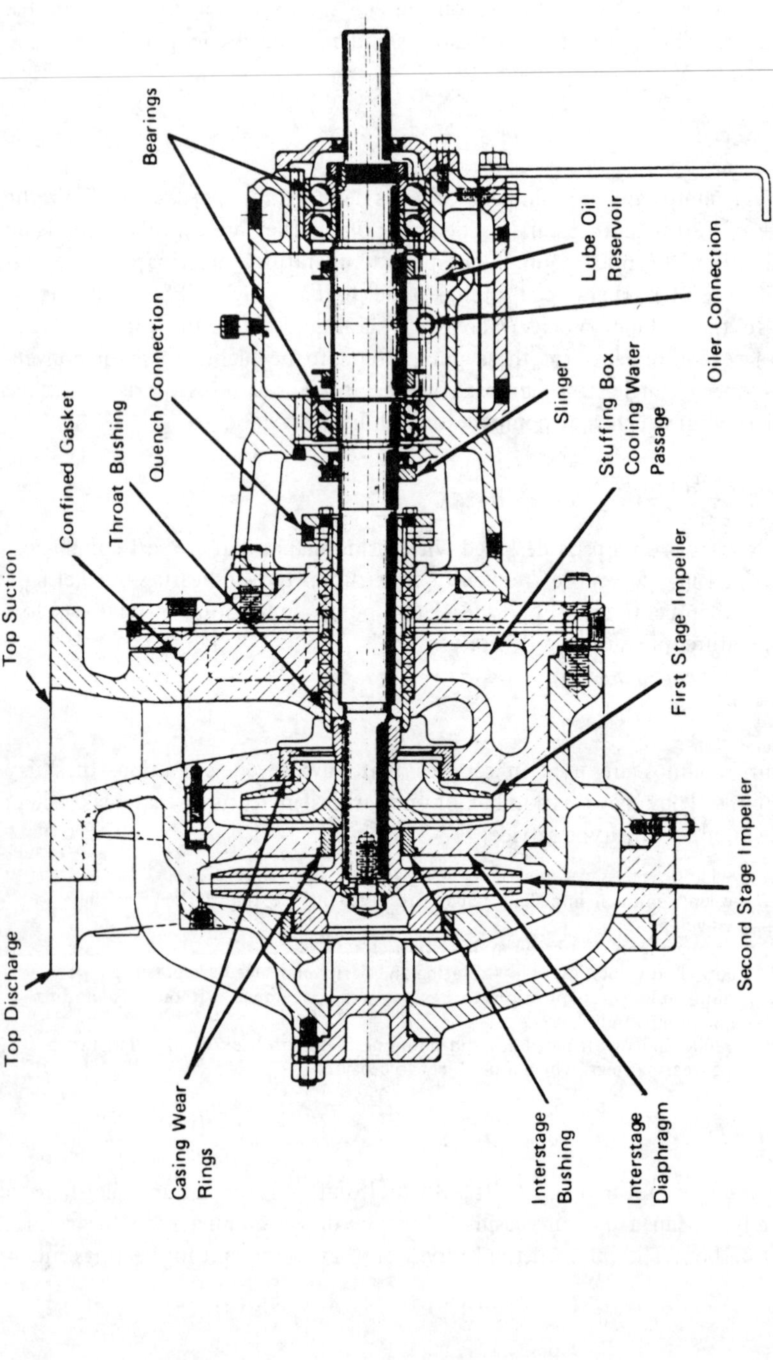

Figure 34. Illustrates a two-stage overhung centrifugal pump.

a coolant for the motor. Designs are limited to low-flow, low-pressure and low-temperature service.

Horizontal and Vertical Multistage Pumps

Horizontal multistage pumps are limited to about 12 stages due to the difficulty in limiting deflection over the long span between bearings. They have NPSH requirements that match those of single-stage pumps of the same capacity. These pumps are well suited for corrosive fluids handling.

Vertical multistage pumps may have as many as 24 (sometimes more) stages. Some high differential pressure models utilize opposed thrust arrangements. Below roughly 1200 feet (370 meters) head, pumps for as low as 1 foot (0.3 meter) NPSH at the suction flange are available. High specific speed impellers are frequently used. The first stage is normally at the bottom of the assembly, below grade. Vertical multistage pumps need a large number of close running clearances. Hence, these pumps are sensitive to damage by solids ingestion and by dry or two-phase generating conditions. In general, maintenance requirements for vertical multistage pumps are greater than for their horizontal counterparts.

Regenerative Turbine Pumps

Some manufacturers prefer to place regenerative turbine pumps in a separate class. In general, this type pump greatly resembles a conventional centrifugal pump but has the distinction of a much steeper head-capacity curve. The impeller consists of a solid disc with fluted vanes on each side of the perimeter, which impart energy to the liquid by multiple excursions from the impeller to the stator and back to the impeller (tracing dual screw shaped paths along the stator annulus).

Turbine pumps have shutoff pressures typically two or three times the design level. The steepness of the head curve causes the power requirement curve to rise as flow decreases, thus peaking at shutoff. Drivers for turbine pumps must be sized for minimum flow, rather than normal flow. Usually, a safety valve is needed inside the discharge block valve.

This type pump is extremely sensitive to dirt, temperature shocks, and to piping forces and moments on the pump flanges.

POSITIVE DISPLACEMENT PUMPS

Positive displacement pumps operate by forcing a fixed volume of liquid from the inlet pressure zone of the pump into the discharge zone of the pump. A brief description of each of the major types of positive displacement pumps is given below.

Reciprocating Pumps

Reciprocating pumps produce pulsating flow, develop high shutoff or stalling pressure, display constant capacity when motor driven, and are subject to vapor binding at low NPSH conditions. There are several construction styles. One type, the direct acting steam pump, consists of a steam cylinder end in line with a liquid cylinder end, with a straight rod connection between the steam piston and the pump piston or plunger.

Direct-acting steam pumps are available as simplex (one steam and liquid cylinder) and duplex (dual side by side) units. Duplex units are employed in larger-capacity services and to reduce the flow pulsations below that of the simplex. Dual pumps are designed with an interconnecting steam valve linkage arrangement so that one side pumps when the other side reaches the end of its stroke. Steam pumps consist of rod and piston design and are double acting (that is, each side pumps on every stroke). Consequently, a duplex pump will have four pumping strokes per cycle.

Another construction style is the power pump. Power pumps convert rotary motion to low-speed reciprocating motion via speed reduction gearing, a crankshaft, connecting rods and crossheads. Plungers or pistons are driven by the crosshead drives. Rod and piston construction, similar to duplex double-acting steam pumps, are used by the liquid ends of the low-pressure, higher-capacity units. The higher-pressure units are normally single-acting plungers. This latter style generally employs three (triplex) plungers. Three or more plungers substantially reduce flow pulsation relative to simplex and even duplex pumps.

In general, power pumps have high efficiency and are capable of developing very high pressures. They can be driven either by electric motors or turbines. They are relatively expensive pumps and rarely can be justified on the basis of efficiency over centrifugal pumps. However, they are frequently justified over steam reciprocating pumps where continuous-duty service is needed due to the high steam requirements of direct-acting steam pumps.

In general, the effective flowrate of reciprocating pumps decreases as viscosity increases because the speed must be reduced. High viscosity also leads to a reduction in pump efficiency. By contrast to centrifugal pumps, the differential pressure generated by reciprocating pumps is independent of fluid density. It is entirely dependent on the amount of force exerted on the piston.

Reciprocating pumps are most often used for sludge and slurry services, particularly where other types are inoperable or troublesome. Maintenance in such services tends to be high because of valve, cylinder, rod and packing wear.

Metering Pumps

Metering pumps are positive displacement pumps that provide precision control of very low flowrates. Flowrates can range from 1.6×10^{-3} to 0.16 gpm; however, there are some higher-capacity models that provide flows up to 0.66 gpm. Flow accuracy is typically within ±1%. Control schemes for metering pumps are used for controlling the proportioning of additives injected into the main flow stream. Other names for these type pumps are "proportioning pumps" and "controlled volume" pumps.

Metering pumps are available in two construction styles: diaphragm and packed plunger. The diaphragm design employs a hydraulic oil barrier between the reciprocating plunger and an impervious diaphragm, which, in turn, contacts the pumped liquid. The stuffing box works in lube oil in this design, so there is no process liquid leakage. With the packed plunger arrangement, the design resembles a small version of a conventional larger plunger pump, in which the stuffing box is exposed to the service liquid.

Usually the driver consists of an electric motor. Basically, the same design criteria applied to larger motor-driven reciprocating pumps can be applied to proportioning pumps.

Capacity variations normally are handled by manual resetting of the stroke adjustment. Controls are available for automatic stroke resetting and for remote, manual stroke resetting. The typical efficiency of this type pump is around 20%. Generally, viscosity effects on power requirements are negligible.

Finally, it should be noted that these pumps are designed for clean service. Nozzle connections and valves of metering pumps are small and, thus, are subject to plugging and/or valve sticking when handling dirty liquids. Figure 35 shows a cutaway view of a metering pump.

Diaphragm Pumps

This type of positive displacement pump operates by the periodic movement of a flexible diaphragm. It has the advantages of no stuffing boxes and high tolerance to abrasive slurries.

The diaphragm is flexed by pulsating fluid pressure on the drive side. Compressed air normally is used; however, steam and hydraulic oil systems are also available. Drive pressures normally pulsate between 0 and 15 psi (0–105 kPA) above the average discharge pressure level in the process stream.

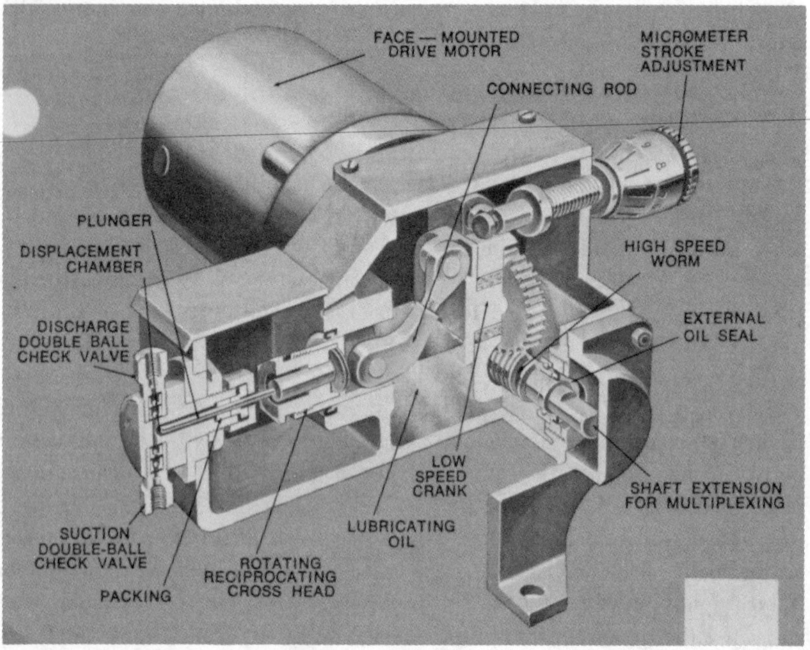

Figure 35. Cutaway view of a controlled-volume pump (courtesy of Milton Roy Co., Flow Control Div., Ivyland, PA).

Rotary Pumps

There are a wide variety of rotary pumps on the market; however, liquid services are limited primarily to external gear pumps and screw pumps. Sliding vane and internal gear pumps find limited application to process plant services.

The single screw pump is a special type of screw pump for handling slurries with relatively large particles. This design allows little fracturing of particles and very little abrasion damage to the pump. Single-screw pumps are employed extensively in the food processing and chemical industries for handling solid/liquid mixtures that are either abrasive or require gentle handling of the solid particles.

The working principle of the screw pump is that of Archimede's screw, invented by the Greek mathematician about 200 BC. Screw pump applications include fuel, lube and crude oil service; navy and marine cargo; oil burners; slurry handling; and a variety of high-viscosity materials such as polymers, copolymers and elastomers, cellulosics, syrups, fats and greases, soaps and solvents.

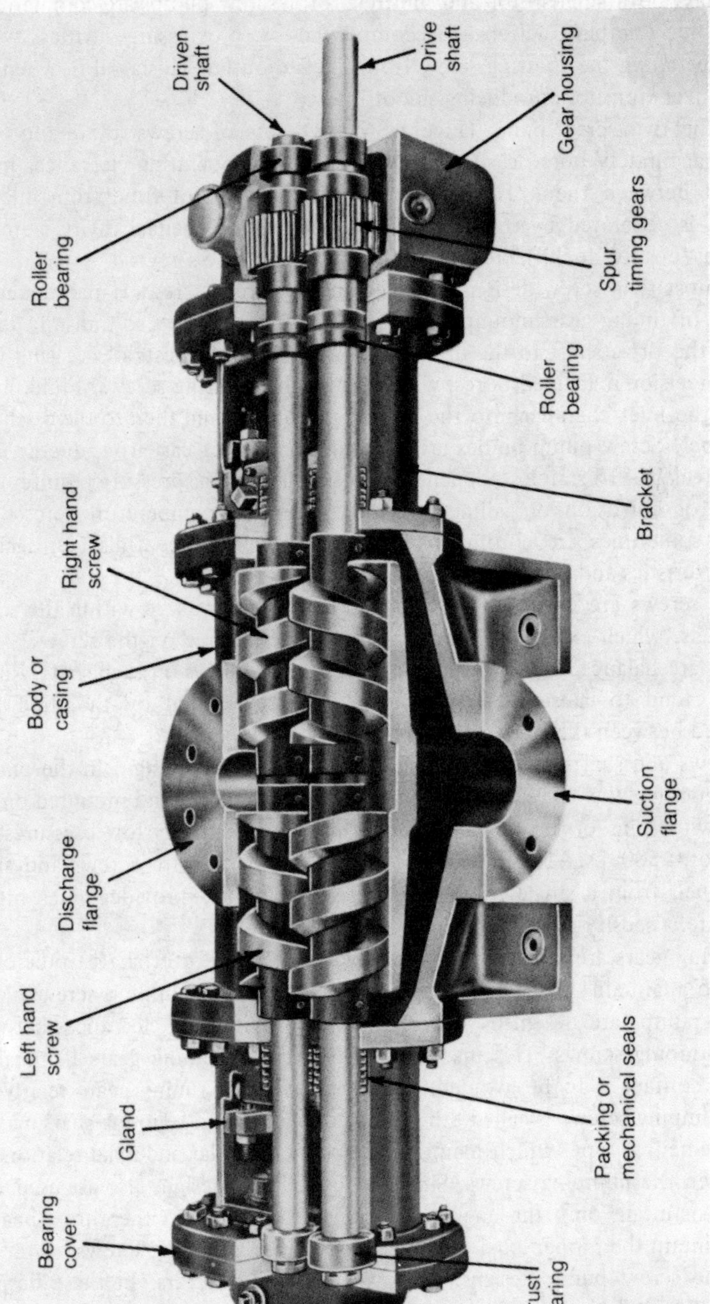

Figure 36. Shows features of the twin-screw pump. (courtesy of Worthington Pump Inc., Mountainside, NJ).

Today, both single-rotor and multiple-rotor screw pumps are commercially available. Pumping action is accomplished by progressing cavities, which advance along the rotating screw from inlet to outlet. This axial flow pattern minimizes vibration, producing smooth flow.

In the twin-screw pump (Figure 36), two sets of screws rotate and mesh in an accurately bored casing, having rather tight operating clearances maintained between them. The mechanical displacement of fluid from inlet to outlet is generated by trapping a slug of fluid in the helical cavity (referred to as a "positive lock"), created by the meshing of the screws.

In most twin screw designs, the clearance between screws is maintained by a pair of timing gears mounted on the shafts. These gears also transmit power from the drive shaft to the driven shaft. The body consists of a casing with two precision-machined bores, which house the rotating screws. Fluid passes from the inlet chamber into the pumping chamber and then to the discharge chamber. Screw pump bodies are commonly made of cast iron, ductile iron, cast steel or 316 stainless, depending on such factors as pressure requirement, need for corrosion or galling resistance, pumping temperature, etc. Body bores sometimes are chrome plated to improve surface finish, antigalling characteristics and hardness.

The screws are designed to withstand hydraulic pressure within the screw channels, which exerts both a radial and an axial force on the screws. Axial forces are balanced by using the matched pair of screws on each shaft. Radial forces tend to cause shaft deflection, and are resisted by the fluid film trapped between the screws and by the body bore diameter.

Screws and shafts may be of "pinned" or "integral" design. In the pinned screw design, the screws are machined as separate pieces and mounted on the shaft with pins or keys. This design is used for relatively low pressures, up to about 500 psi. The integral screw arrangement, with screws and shaft machined from a single forging or piece of barstock, provides much higher pressure, viscosity and shaft torque capabilities.

Timing gears are used to transmit power from one rotor shaft to the other and to maintain the proper clearance between the pumping screws. Most screw pumps are designed with relatively small internal clearances between the pumping screws. This mandates high-precision timing gears if metal-to-metal contact is to be avoided. The positioning of timing gears relative to the pumping screws is called "the timing of the screws." Timing gears may be double-helical type, which maintain the proper angular and axial relationship between the pumping screws. Single-helical or spur designs also are used. The latter maintain only the angular relationship, relying on the thrust bearing to maintain the proper axial relationships between pumping screws.

Some screw pump designs work without timing gears. In these designs, there is a drive screw that is used to drive the driven screws. The design of twin-screw pump bodies and the arrangement of screw assemblies are interrelated. Figure 37 illustrates typical body designs.

There are a variety of novel rotary pump types commercially available. The more conventional styles are compared in Table 21.

The main reason for selecting rotary pumps over centrifugals is to take advantage of their high viscosity capability. In addition, rotaries are simple in design and efficient in handling flow conditions that are generally considered too low for economic application of centrifugals. The importance of viscosity in rotary pump design is summarized in Table 22. Rotary pumps designed to handle high-viscosity liquids must be operated at reduced speeds (and, thus, at reduced flowrates).

Note that proper pump selection is not only dependent on the fluid viscosity level, but also on the rheological properties of liquid. For non-Newtonian fluid behavior, pump design and selection will depend also on how viscosity changes with shear rate. Shear rate/shear stress data (see chapters in Section 1 covering non-Newtonian behavior for a discussion of viscosity-shear rate dependency) should be included when preparing design specifications.

BASIC HYDRAULICS AND DEFINITIONS IN PUMPING SERVICE

The principle of conservation of energy states that the total energy input to a closed system (Figure 38) is equal to the total energy output from that system. In addition to the energy in the fluid at Point 1 (Figure 38), there is energy added by pumps. The output energy in the fluid plus the friction loss in the system is at Point 2. Bernoulli's equation is restated here in its more general form (total mechanical energy balance):

$$\frac{P_1}{\rho} + Z_1 + \frac{U_1^2}{2g} + E_p = \frac{P_2}{\rho} + Z_2 + \frac{U_2^2}{2g} + H_f \qquad (87)$$

The LHS of the equation refers to Point 1. The first energy term is pressure energy, P_1/ρ.

The second term is another energy form, derived through the fluid's static elevation (i.e., potential energy).

The third term, $U_1^2/2g$, is an expression for kinetic energy or velocity head.

The fourth term in the equation, E_p, represents pump energy. This is the amount of energy that is added to the fluid between Points 1 and 2.

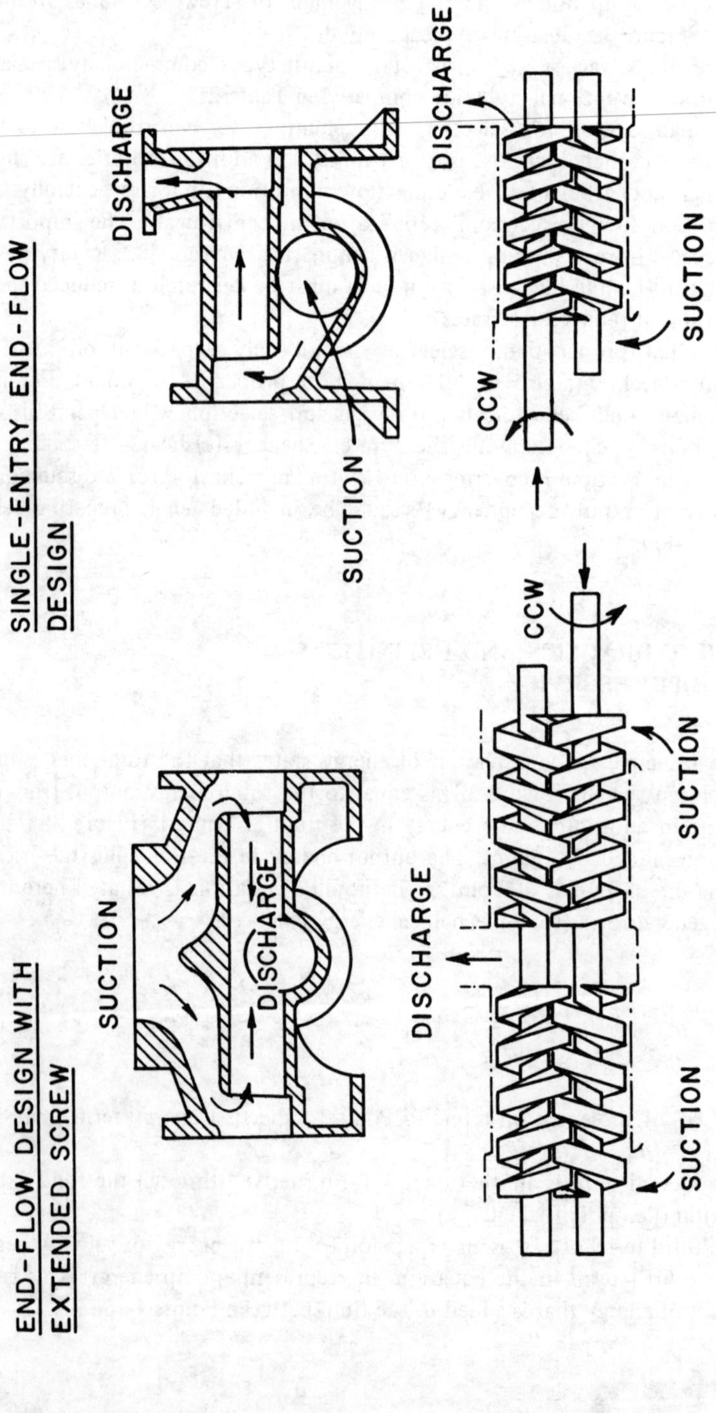
SINGLE-ENTRY END-FLOW DESIGN
DISCHARGE
SUCTION
DISCHARGE
SUCTION
CCW
CCW
END-FLOW DESIGN WITH EXTENDED SCREW
SUCTION
DISCHARGE
DISCHARGE
SUCTION
SUCTION

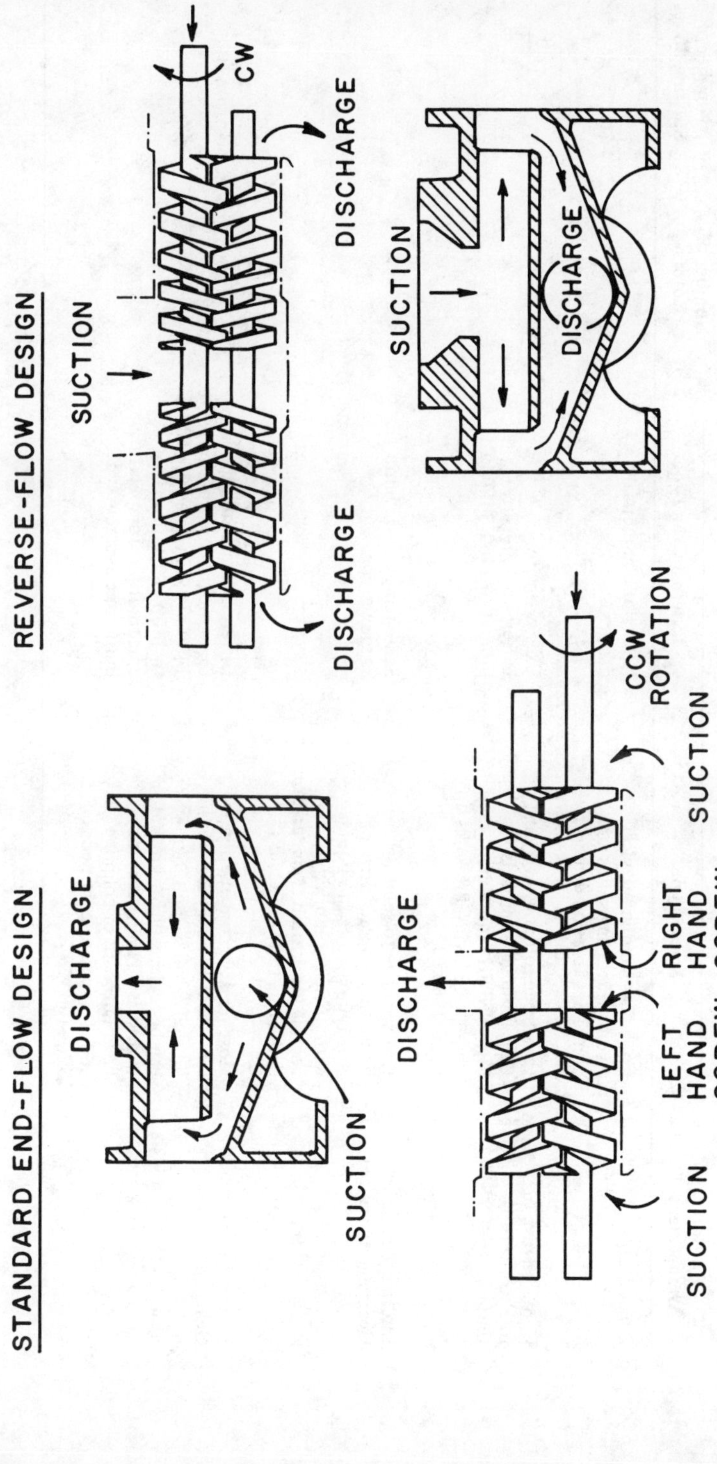

Figure 37. Typical twin-screw pump bodies and flow patterns.

Table 21. Comparison of Rotary Pump Styles

Rotor Arrangement	Description: Screw Type	Description: External Gear Type	Number of Rotors	Number of Stuffing Boxes
Single Rotor	For solids handling services. Avoids fracturing solids and minimizes abrasion damage.	No major commercial designs.	1	1
Contact Drive of Idle Rotor (no timing gears)	Uses three rotors. Lower cost than timing gear models. Has higher NPSH requirement than timing gear styles. Pressures up to 20,700 kPa.	Uses two rotors. Differential pressure limited to 2400/3450 kPa.	2–3	1
Internal Timing Gears and Bearings	Less sensitive to solids in liquid than contact drive or gear type. Limited to liquids with sufficient lubricity for bearing lubrication. Temperature limited to 150–175°C.	Same temperature limit as screw type.	2	1
External Timing Gears and Bearings	Most versatile construction for low-viscosity high pressure (low lubricity, solids, etc.). Screws can be replaceable or integral with shaft. Timing gear may be at outboard or coupling end. Temperature limited to 370–400°C.	Differential pressure limit for standard construction, 1400 kPa; can be extended to 2800 kPa with special construction. Temperature limit same as screw type.	2	4

Table 22. Differential Pressure Limitations for Rotary Pumps Due to Viscosity Levels

Kinematic Viscosity (mm^2/sec)	Differential Pressure Limitations
1.4–7.4	Standard designs with external bearings limited to 500–700 kPa.
7.4–32⁺	Novel designs extend differential pressure to 2800 kPa.
32–44	Standard designs limited to 1030 kPa.
44–75	Standard designs limited to 2400 kPa.
75–110	Standard designs limited to 3450 kPa.
>110	Novel designs extend range to 24,000 kPa.
130–640	Screw pumps reach maximum efficiency in this viscosity range.

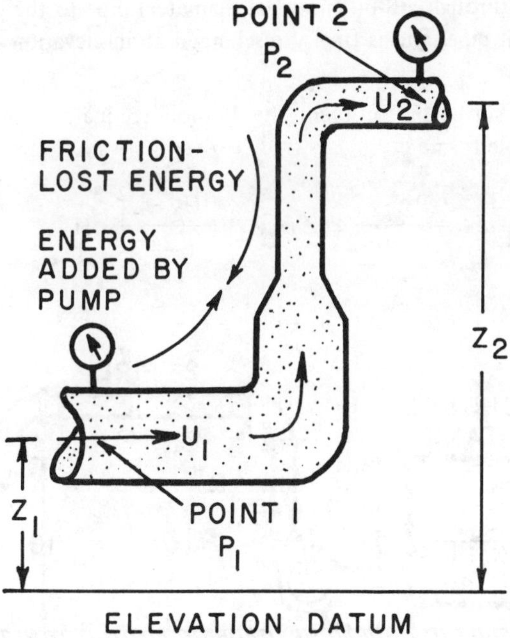

Figure 38. System defining the law of conservation of energy.

The final term on the RHS of Equation 87, H_f, is the friction loss between Points 1 and 2.

The continuity equation is based on the principle of conservation of mass, which states that the total mass of fluid flowing into any closed conduit is equal to the total mass of fluid flowing out of the system. More simply, this means that the flow capacity at Point 1 is the same as at Point 2.

$$Q = U_1A_1 = U_2A_2 \tag{88}$$

where A is the cross-sectional area of the pipe.

The application of these two equations represents a generalized first step to specifying pumping requirements. Their use is best illustrated by a typical pumping problem.

Example: Figure 39 gives conditions for a simple pumping situation in which the total head required by the pump must be determined. The pump must be sized to pump 100 gpm and overcome the friction of the piping (which includes entrance losses, elbows, etc.) of 16 ft at that capacity. The pump takes suction from a closed vessel, which is under a vacuum of 28 in. Hg. The height of the water in the tank is 3 ft above the tank bottom. The fluid (water) flows through a 3-in. (nominal diameter) pipe to the pump and then through a 2-in. pipe, and is finally discharged at an elevation of 10 ft above the floor.

Solution: To solve this problem the Bernoulli equation must be restated in a more usable form:

$$E_p = \frac{P_2 - P_1}{\rho} + (Z_2 - Z_1) + \frac{U_2^2 - U_1^2}{2g} + H_f \tag{89}$$

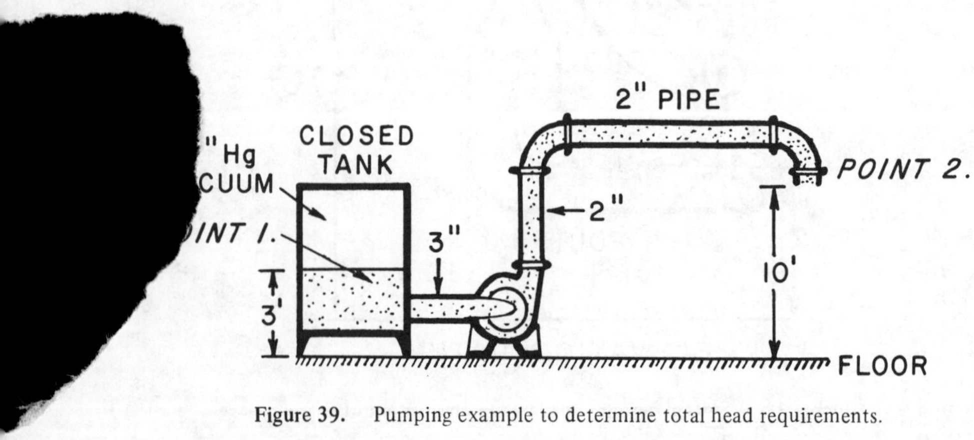

Figure 39. Pumping example to determine total head requirements.

In the equation, P_2/ρ is the atmospheric pressure (i.e., zero in gauge pressure). The static elevation $Z_2 = 10$ ft and $Z_1 = 3$ ft; hence $\Delta Z = 7$ ft.

The velocity head is determined by first evaluating the fluid velocity in the 2-in. pipe. This can be obtained from the continuity equation, in which the velocity of 100 gpm in a 2-in. pipe is 9.5 fps (the details are left to the reader). The term $U_1^2/2g$ is very small since the fluid in the suction chamber is moving with such a low velocity and, hence, for practical purposes may be ignored.

The friction loss in the system between the suction chamber and the discharge side is 16 ft. Adding these values, an answer of 56 ft is obtained. Hence, a pump must be selected on the basis of a total head developed of 56 ft.

In the above example, problem information on the friction loss of the system was given. However, in many situations the determination of the friction loss in a pipe system is an important part of the problem. The calculation procedures outlined in Chapter 2.2 are applicable. A second problem will help illustrate friction loss calculations for pumping problems.

Example: Figure 40 defines the system for the problem. Water is being pumped at a rate of 200 gpm through a 4-in. steel pipe. The water is introduced to the pipe system through a foot valve and discharged upward by the

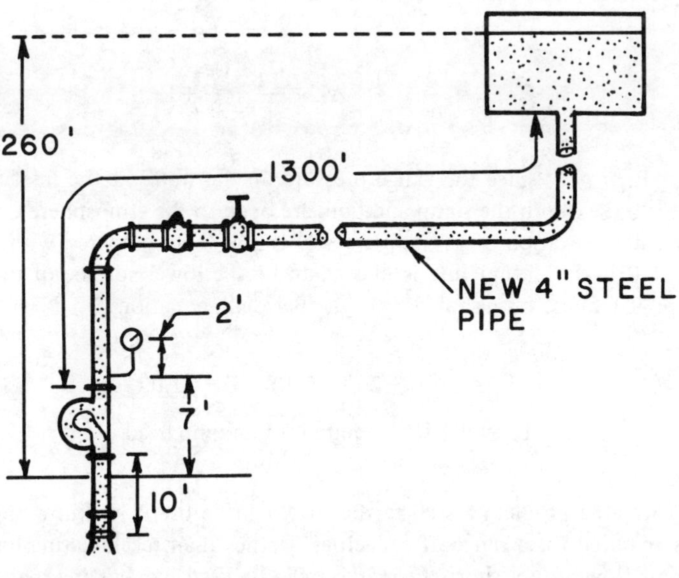

Figure 40. Pumping problem example to determine friction losses.

pump through an elbow, a swing check valve, a gate valve, another elbow and, finally, a sudden enlargement to an open vessel. The total length of pipe above the pump is 1300 ft. Determine the total friction loss for the system.

Solution: First determine the friction loss in the total length of pipe:

Total Pipe Length = 1300 ft pipe above the pump
+ 10 ft pipe below the pump = 1310 ft

Using the methods outlined in Chapter 2.2 for a flowrate of 200 gpm through 4-in. steel pipe, the pipe friction loss is 29.74 ft of flowing fluid.

Other losses in the problem can be determined by the friction loss formula, $H_f = KU^2/2g$:

$$U^2/2g = 0.395 \text{ ft for 200 gpm and 4-in. steel pipe.}$$

Values of coefficient K for various fittings are found in Chapters 2.2 and 2.3. Substituting into the friction loss formula and adding up all losses, the total friction loss through all pipe and fittings is 31.02 ft.

Next the pump energy needed to pump 200 gpm through the pipe system must be determined through the Bernoulli equation:

$$E_p = \frac{P_2}{\rho} - \frac{P_1}{\rho} + \Delta Z + \frac{U_2^2}{2g} - \frac{U_1^2}{2g} + H_f$$

$P_2/\rho = P_1/\rho = 0$, since the gauge pressure on the water at the inlet and the outlet is 0 (i.e., both the sump and tank are open to the atmosphere).

The static elevation of the upper vessel is 260 feet compared with a static level of 0 if the datum line is chosen to be the lower surface of the water (Figure 40). Substituting values into the Bernoulli equation,

$$E_p = 0 - 0 + 260 - 0 + 0 - 0 + 31.02$$

$$E_p = 291 \text{ ft} = \text{required total pump head}$$

To carry the problem a step farther it would be useful to know the pump heads required for a range of capacities. Rather than recalculating losses for different flowrates, a short-cut method can be used by realizing that friction loss varies as the square of the velocity in the pipe, or as the square of the volumetric flowrate ($H_f = fU^2/2g$). For example, for half the flowrate, or 100 gpm, the friction loss is the square of ½, or ¼ X 31.09 = 7.8 ft.

Similarly, 300 gpm is 1.5 times the original 200 gpm. Hence, the new friction loss is $(1.5)^2 \times 31.02$, or 70 ft.

We now have three points and can prepare a curve of friction loss versus flow, as shown in Figure 41. Note the change in static elevation of 260 feet is a constant regardless of flow, and is thus just added to the three values. Figure 41 is referred to as the system head curve. It depicts the pump energy or total head required for any capacity.

Not all head curves are alike. Various types of systems present different system head curves. Also note that while a system head curve can be prepared for most systems, its accuracy may not always be precise. Distortions affect the real system head curve in many ways. Undetermined obstructions in the system, such as a clogged valve, clogged strainers in the line or unexpected pipe roughness, can contribute to higher pipe losses. Also, the real system head curve may be lower than calculated due to more conservative estimates. For example, losses might be computed based on old pipe with scale buildup, when in fact the system uses new pipe.

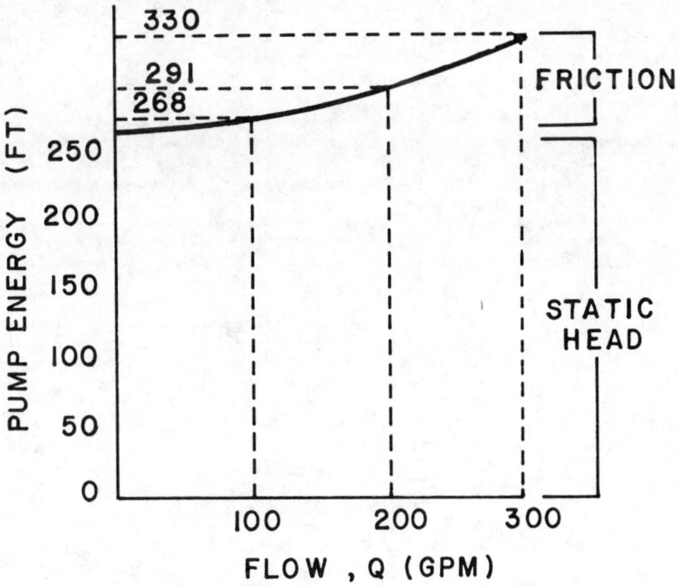

Figure 41. Friction loss curve for example problem.

CHAPTER 2.5

SYSTEM ANALYSIS FOR PUMPING EQUIPMENT

Technical Staff

Eastern Operations
Peerless Pump, an Indian Head Co.
Indianapolis, IN

CHARACTERISTICS OF CENTRIFUGAL PUMPS

The selection of a centrifugal pump for an energy-efficient pumping system requires an understanding of the principles of mechanics and physics that can affect the pumping system and the pumped liquid. The efficiency of a centrifugal pump is also dependent on the behavior of the liquids being pumped. The principles of centrifugal pump operation that govern head and flow must be understood clearly before pump performance can be evaluated accurately.

The behavior of a fluid depends on its state–liquid or gas. Liquids and gases offer little resistance to changes in form. Typically, fluids such as water and air have no permanent shape and readily flow to take the shape of the containing enclosure when even a slight shear loading is imposed. Factors affecting behavior of fluids include:

- viscosity
- specific gravity
- vapor pressure

In Section 1 it was noted that viscosity is the resistance of a fluid to shear motion–its internal friction. The molecules of a liquid have an attraction for each other. They resist movement and repositioning relative to each other. This resistance to flow is expressed as the viscosity of the liquid. Dynamic viscosity also can be defined as the ratio of shearing stress to the rate of deformation. The viscosity of a liquid varies directly with temperature; therefore, viscosity is always stated at a specific temperature.

Liquid *viscosity* is very important in analyzing the movement of liquids through pumps, piping and valves. A change in viscosity alters liquid handling characteristics in a system; more or less energy then may be required to perform the same amount of work. In a centrifugal pump, an increase in viscosity reduces the pressure energy (head) produced while increasing the rate of energy input. In piping systems, a liquid with a high viscosity has a high energy gradient against which a pump must work, and more power is required than for pumping low-viscosity liquids.

Specific gravity is the ratio of the density of one substance to that of a reference substance at a specified temperature. Water at 4°C is used as the reference for solids and liquids. Air generally is used as the reference for gases. The specific gravity of a liquid affects the input energy requirements, or brake horsepower (bhp), of centrifugal pumps. Brake horsepower varies directly with the specific gravity of the liquid pumped. For example, water at 4°C has a specific gravity of 1.0. Table 23 includes some specific gravity values for water at selected temperatures.

Specific gravity affects the liquid mass but not the head developed by a centrifugal pump, as shown in Table 24. The specific gravity also affects the energy required to move the liquid and, therefore, must be used in determining the pump's horsepower requirement.

Vapor pressure is the pressure at which a pure liquid can exist in equilibrium with its vapor at a specified temperature. Fluids at temperatures greater than their specified (critical) temperatures will exist as single-phase liquids (vapors), with no distinction between gas and liquid phases. At less than the critical temperature, two fluid phases can coexist; the denser fluid phase exists as a liquid and the less dense phase as a vapor. At a specific

Table 23. Specific Gravity Values for Water at Selected Temperatures

Water Temperature		Water Specific Gravity	Effect of Specific Gravity on Energy Input Constant Flow	
(°C)	(°F)		(kW)	(bhp)
4	39.2	1.0	74.6	100
60	140	0.983	73.3	98.3
100	212	0.958	71.5	95.8
125	257	0.939	70.0	93.9
150	302	0.917	68.4	91.7

Table 24. Effect of Specific Gravity on Pump Head

Specific Gravity	Effect of Specific Gravity on . . . Linear Head (ft)	(m)	Gauge Pressure (psi)	(bar)
0.75	246	75	79.9	5.51
1.00	246	75	106.5	7.34
1.20	246	75	127.8	8.81

temperature, the liquid phase is stable at pressures exceeding the vapor pressure and the gas phase is stable at pressures less than the vapor pressure.

For a fluid to exist in a liquid state its surface pressure must be equal to, or greater than, the vapor pressure at the prevailing temperature. For example, water has a vapor pressure of 0.1781 psia at 10°C and 14.69 psia at 100°C. The vapor pressure of a volatile liquid (such as ether, alcohol or propane) is considerably higher than that of water at the same temperature; consequently, much higher pressures must be applied to maintain volatile materials in their liquid states. The surface pressure of a liquid must be greater than its vapor pressure for satisfactory operation of a centrifugal pump.

Centrifugal pump characteristics remain constant unless an outside influence causes a change in operating conditions. Three conditions can alter pump performance:

1. changes in impeller or casing geometry;
2. increased internal pumping losses caused by wear; and
3. variation of liquid properties.

For example, if the impeller passages become impacted with debris, the head-flow relationship will be reduced. Similarly, performance will decline if mechanical wear increases the clearance between the rotating and stationary parts of the pump.

Except for specially designed pumps, most centrifugal pumps can handle liquids containing approximately 3–4% of gas (by volume) without an adverse effect on performance. An excess of gas will reduce the flow of liquid through the pump and, under certain conditions, flow will cease, setting up a condition that may damage the pump.

The function of a pump is to move liquids by imparting pressure energy (head) to the liquid. The ability of the pump to perform its function is based on the Bernoulli theorem, which states that energy cannot be created or destroyed, but can only be converted in form. A pump converts mechanical energy into pressure energy. Part of the converted energy is required to overcome inertia and move the liquid; most of the remaining energy is stored in the liquid as elevated pressure, which can be used to perform useful work outside the pump. A centrifugal pump is basically a velocity machine designed around its impeller. The interaction between the impeller and its casing produces the characteristics of head or pressure energy.

To review, the *developed head* is a function of the difference in velocity between the impeller vane diameter at entrance and the impeller vane diameter at exit. The expression of theoretical head can be related to the law of a falling body:

$$H = \frac{u^2}{2g}$$

where H = height or head (ft)
u = velocity of moving body (fps)
g = acceleration of gravity (32.2 ft/s^2)

When the height of fall is known (for example, H = 100 ft), the terminal velocity can be determined (in this case u = 80.3 fps). Conversely, if the direction of motion is reversed, a liquid exiting through an impeller vane tip at a velocity of 80.3 fps reaches a velocity of 0 fps at 100 ft above the impeller tip (Figure 42).

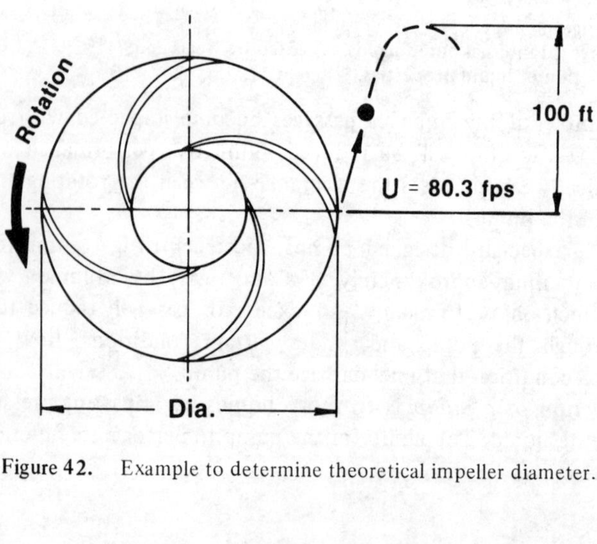

Figure 42. Example to determine theoretical impeller diameter.

When the developed head required is known, the theoretical impeller diameter for any pump at any rotational speed can be determined from the equation for peripheral velocity of a round rotating body:

$$D = \frac{(229.2)u}{N} \tag{90}$$

where D = unknown impeller diameter (in.)
u = velocity (derived from $u = \sqrt{2gH}$) (fps)
N = rotational speed of the pump (rpm)

If pump head is 100 ft and rotational speed is 1750 rpm, the formula indicates that a 10.52-in. theoretical diameter impeller is required.

The function of a centrifugal pump is to transport liquids by transferring and converting mechanical energy (foot-pounds of torque) from a rotating impeller into pressure energy (head). By the transfer of this energy to a liquid the liquid can perform work, move through pipes and fittings, rise to a higher elevation or increase the pressure level. Because a centrifugal pump is a velocity machine, the amount of mechanical energy per unit of fluid weight transferred to the liquid depends on the peripheral velocity of the impeller, regardless of fluid density. This energy per unit of weight is defined as pump head and is expressed in feet of liquid. If the effect of liquid viscosity is ignored, the head developed by a given pump impeller at a given speed and flowrate remains constant for all liquids.

Head is the vertical height of a column of liquid. The pressure this liquid column exerts on the base surface depends on the specific gravity of the liquid. A 10-ft column of liquid with a specific gravity of 0.5 exerts only 2.16 psi; a 10-ft column of liquid with a specific gravity of 1.0 exerts 4.33 psi.

The formula for converting feet of head to pressure is

$$H = \frac{P \times 2.31}{Sg} \tag{91}$$

where P = pressure (psi)
Sg = specific gravity

The proper selection of a centrifugal pump requires that pressure be converted to feet of head. If not, a pump that is incapable of imparting the required energy to a liquid may be installed before the inadequacy is discovered. The heads required to produce a pressure of 100 psi for three liquids of different specific gravities are given in Table 25.

Table 25. Heads Required to Produce 100 psi Pressure for Different Liquids

Head equivalent of 100 psi for liquids of different specific gravity	
Specific Gravity	Head (ft of liquid)
0.75	308
1.0	231
1.2	193

From this analysis a single-stage pump selected for pumping a liquid with a specific gravity of 1.2 will have the lowest peripheral velocity at the impeller tip, while the pump for a liquid with 0.75 specific gravity will have the highest peripheral velocity.

The volumetric flowrate is a function of the peripheral velocity of the impeller and the cross-sectional areas in the impeller and its casing; the larger the passage area, the greater the flowrate. Consequently, the physical size of the pump increases with higher flow requirements for a given operating speed. The liquid flowrate is directly proportional to the area of the pump passages and can be expressed as

$$Q = \frac{u \times A}{0.321} \tag{92}$$

where Q = flow (gpm)
u = velocity (fps)
A = area (in.^{2})

For example, a centrifugal pump having a 15-fps liquid velocity at the discharge flange can pump approximately 330 gpm through a 3-in.-diameter opening. If the flow requirements are increased to approximately 1300 gpm and the velocity remains at 15 fps, a 6-in.-diameter opening will be required. Liquid flow is also directly proportional to the rotational speed that produces the velocity. The four preceding equations establish two relationships:

1. Head is directly proportional to the square of the liquid velocity.
2. Flow is directly proportional to the peripheral velocity of the impeller.

Cavitation in a centrifugal pump can be a serious problem. Liquid pressure is reduced as the liquid flows from the inlet of the pump to the entrance to the impeller vanes. If this pressure drop reduces the absolute pressure on the liquid to a value equal to or less than its vapor pressure, the liquid will change to a gas and form vapor bubbles. The vapor bubbles will collapse when the fluid enters the high-pressure zones of the impeller passages.

This collapse is called cavitation and results in a concentrated transfer of energy, which creates local forces. These high-energy forces can destroy metal surfaces; very brittle materials are subject to the greatest damage. In addition to causing severe mechanical damage, cavitation also causes a loss of head and reduces pump efficiency. Cavitation will also produce noise.

If cavitation is to be prevented, a centrifugal pump must be provided with liquid under an absolute pressure that exceeds the combined vapor pressure and friction loss of the liquid between the inlet of the pump and the entrance to the impeller vanes. Chapter 2.8 discusses cavitation more completely.

NET POSITIVE SUCTION HEAD (NPSH)

NPSHA (net positive suction head available) is the absolute pressure of the liquid at the inlet of the pump. NPSHA, a function of the elevation, temperature and pressure of the liquid, is expressed in units of absolute pressure (psia). Any variation of these three liquid characteristics will change the NPSHA. An accurate determination of NPSHA is critical for any centrifugal pump application.

The net positive suction head required (NPSHR) by a specific centrifugal pump remains unchanged for a given head, flow, rotational speed and impeller diameter, but changes with wear and liquids.

Specific speed is a correlation of pump flow, head and speed at optimum efficiency. It classifies pump impellers with respect to their geometric similarity. Specific speed is usually expressed as

$$Ns = \frac{N\sqrt{Q}}{H^{3/4}} \tag{93}$$

where Ns = pump specific speed
Q = flow at optimum efficiency (gpm)

The specific speed of an impeller is defined as the revolutions per minute at which a geometrically similar impeller would run if it were of a size that would discharge 1 gpm against a head of 1 foot. Specific speed is indicative of the impeller's shape and characteristics.

Centrifugal pumps are divided into three classes:

1. Radial flow
2. Mixed flow
3. Axial flow

There is a continuous change from the radial flow impeller (which develops head principally by the action of centrifugal force) to the axial flow impeller (which develops most of its head by the propelling or lifting action of the vanes on the liquid.) Typically, centrifugal pumps also can be categorized by physical characteristics relating to the specific speed range of the design (see Table 26). Once the values for head and capacity become established for a given application, the pump's specific speed range can be determined and specified to ensure the selection of a pump with optimal operating efficiency.

GENERAL PUMP CHARACTERISTICS

Differing pump hydraulic characteristics will result in one pump being better suited for a given application than another. This section examines head-flow curves, pump performance curves, variations in curve shape as a function of specific speed and pump affinity laws.

Head-Flow Curve

The head-flow curve of an ideal pump with ideal (frictionless) fluid is a straight line whose slope from zero flow to maximum flow varies with the impeller exit vane angle. For example, a given impeller and casing combination with an impeller vane exit angle of 25° will have a greater maximum flow than a similar impeller with a 15° vane exit angle (Figure 43). However, the actual head-flow characteristic of a centrifugal pump is not an ideal straight line. Its shape is altered by friction, leakage and shock losses that occur in impeller and casing passages.

Friction losses in a centrifugal pump are proportional to the surface roughness and the wetted areas of the impeller and casing. Leakage losses result from the flow of liquid between the clearance of rotating and stationary parts, such as impeller to case wear ring clearances. Shock losses occur as the liquid enters the impeller entrance vanes and as the liquid flows from the impeller into the casing. These internal losses characteristically reduce pump performance from the ideal to the actual total head-flow curve shown in Figure 44. The flow at which the sum of all these losses is the least determines the point of maximum efficiency.

Mechanical losses in bearings, packings and mechanical seals further reduce pump efficiency. Although mechanical losses may be calculated, the results are generally not accurate; actual pump performance can be determined only by testing.

A centrifugal pump is designed around its impeller. The function of the casing is to collect the liquid leaving the impeller and convert its kinetic

Table 26. Centrifugal Pump Configurations and Specific Speed Ranges

Single suction, horizontal or vertical centrifugal pumps with narrow port impellers have low capacities and deliver high heads. Specific speed range is 500 to 1000.

Single or multistage, double (illustrated) and single suction, volute and diffuser design centrifugal pumps deliver medium capacity and medium heads. Specific speed range is 1000 to 2000.

Single or multistage, and single (illustrated) and double suction, Francis-type impellers, operating in volute or diffuser type casings, produce medium to high capacity at medium to low speeds. Specific speed range is 2000 to 4000.

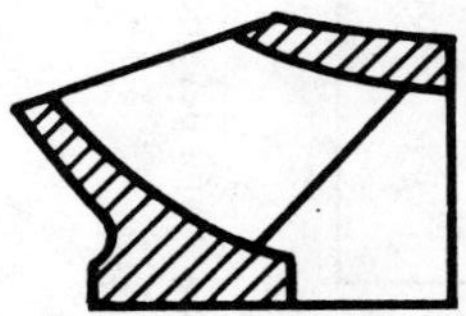

Single or multistage single suction centrifugal pumps, in volute or diffuser-type casings with mixed flow impellers, deliver high capacity at low head. Specific speed range is 4000 to 6000.

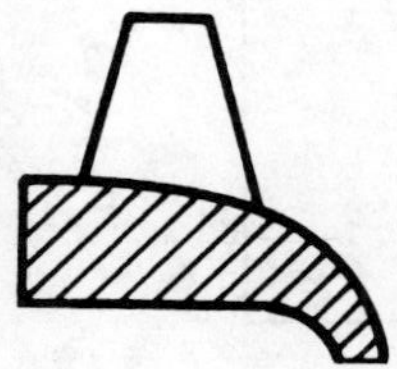

Single or multistage pumps with mixed flow and propeller-type impellers have very high capacities and deliver very low heads. Specific speed range is 6000 to 10,000.

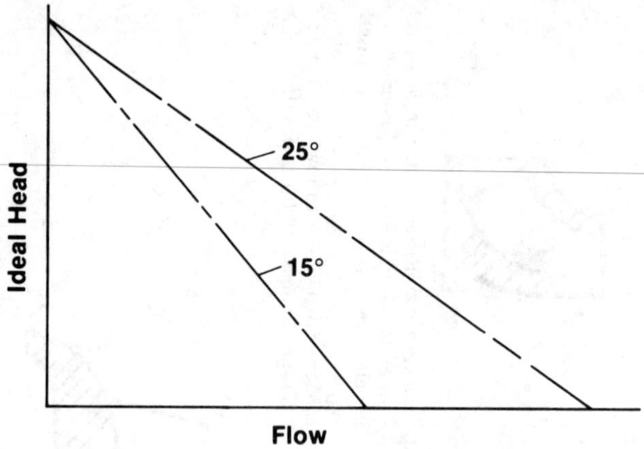

Figure 43. Illustrates the ideal head curve.

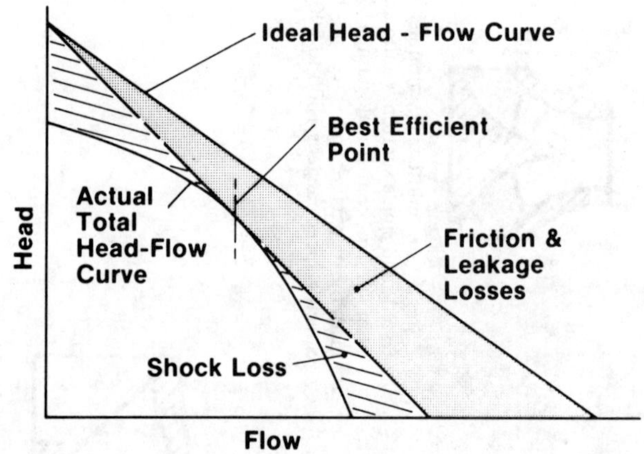

Figure 44. Illustrates deviation from the ideal head-flow curve due to internal losses.

energy into pressure energy. A case is designed in conjunction with an impeller to achieve the most efficient match for conversion of liquid velocity to pressure energy. The interaction between the impeller and its casing for a given pump determines the pump's unique performance characteristics.

Performance Curves

Pump performance curves depict the total head developed by the pump, the brake horsepower required to drive it, the derived efficiency and the net positive suction head required over a range of flows at a constant speed. Pump performance can be shown as a single line curve depicting one impeller diameter (Figure 45) or as multiple curves for the performance of several impeller diameters in one casing (Figure 46).

The performance characteristics of pumps are classified by discharge specific speed and have the approximate curve shapes shown in Figure 47. According to shape, centrifugal pump head-capacity and horsepower curves are classified as follows:

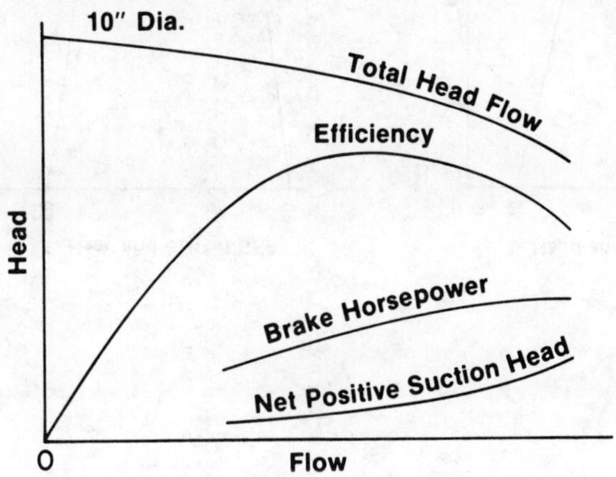

Figure 45. Pump performance curve for a single impeller diameter.

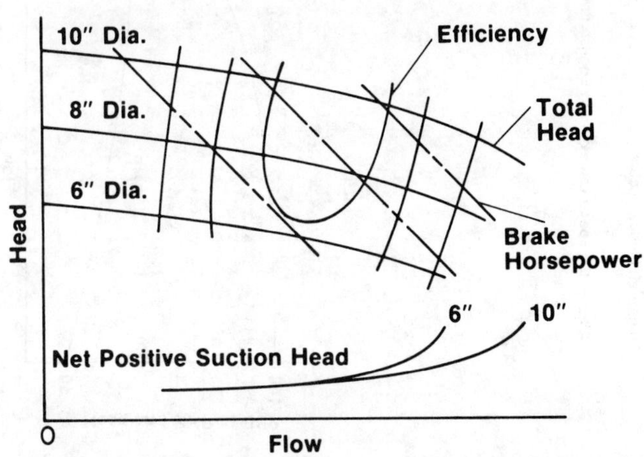

Figure 46. Pump performance curves for several impeller diameters.

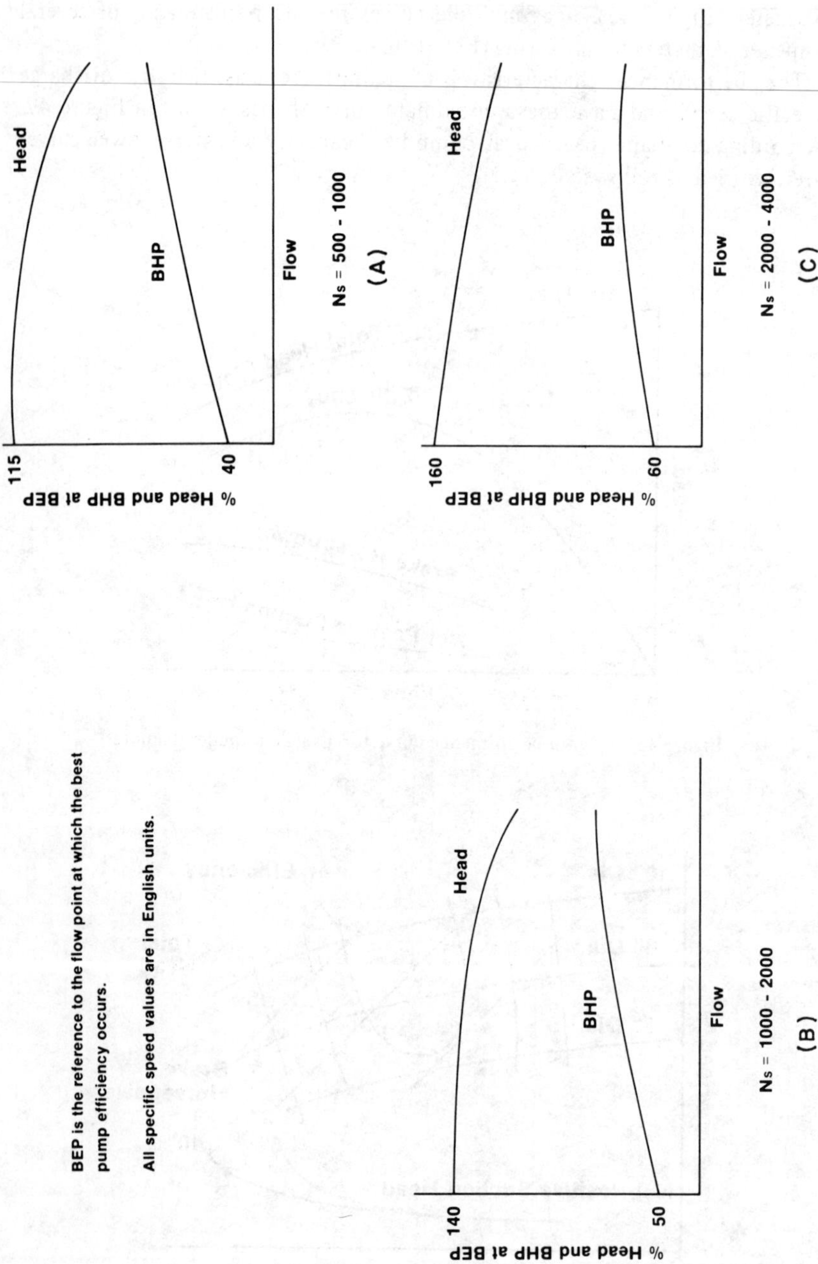
% Head and BHP at BEP
115
40
Head
BHP
Flow
Ns = 500 - 1000
(A)
% Head and BHP at BEP
140
50
Head
BHP
Flow
Ns = 1000 - 2000
(B)
% Head and BHP at BEP
160
60
Head
BHP
Flow
Ns = 2000 - 4000
(C)
BEP is the reference to the flow point at which the best pump efficiency occurs.
All specific speed values are in English units.

Figure 47. Performance curve shapes for centrifugal pumps having different specific speeds.

Drooping head characteristics are those in which the head at zero flow is less than the head developed at some greater flow (Figure 48).

Continually rising head characteristics are those in which the head rises continuously as the flow is reduced to zero (Figure 49).

Steep head characteristics are those in which the head rises steeply and continuously as the flow is reduced. Curve steepness is only a relative term because there is no defined value of curve slope for comparison (Figure 50).

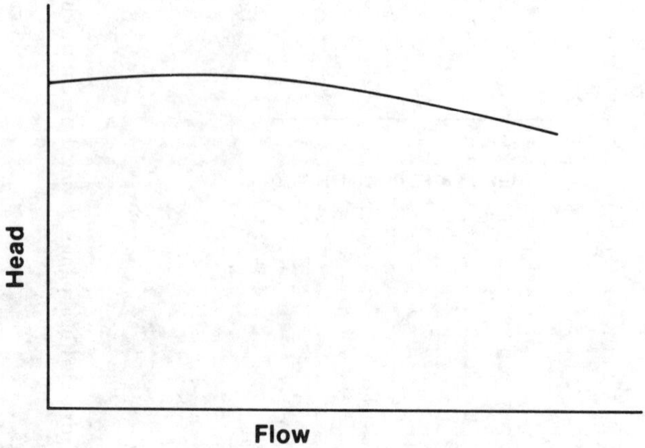

Figure 48. Drooping head curve.

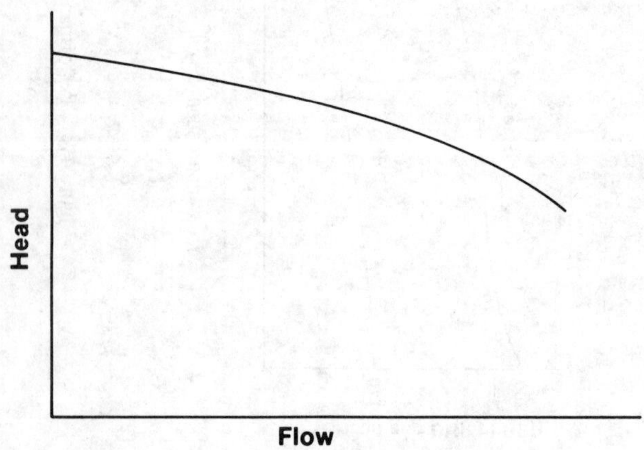

Figure 49. Continuously rising head curve.

Flat head characteristics are those in which the head rises only slightly as the flow is reduced. As with steepness, the magnitude of flatness is a relative term (Figure 51).

Discontinuous head characteristics are those in which a given head is developed by the pump at more than one flowrate. Many pumps in the high specific speed ranges have this characteristic (Figure 52).

Continually increasing horsepower characteristics are those of low specific speed pumps where the horsepower increases at flows greater than the best efficiency point (BEP) and decreases at flows to the left of BEP (Figure 53).

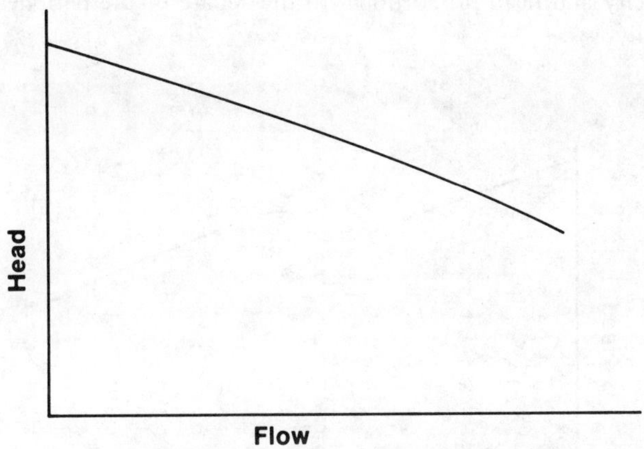

Figure 50. Steep head curve.

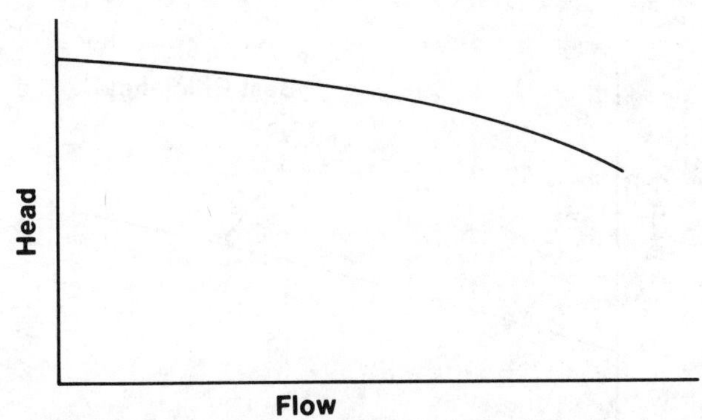

Figure 51. Flat head curve.

Peaking horsepower characteristics are those of medium specific speed pumps where maximum horsepower occurs in the BEP range and decreases at all other values of flow (Figure 54).

Discontinuous horsepower characteristics are those of high specific speed pumps where horsepower increases at flows less than BEP and decreases with flow to the right of BEP. The discontinuity usually occurs at flows less than BEP (Figure 55).

PUMP AFFINITY LAWS

The pump affinity laws state that flow is proportional to impeller peripheral velocity and head proportional to the square of the peripheral velocity. Specifically,

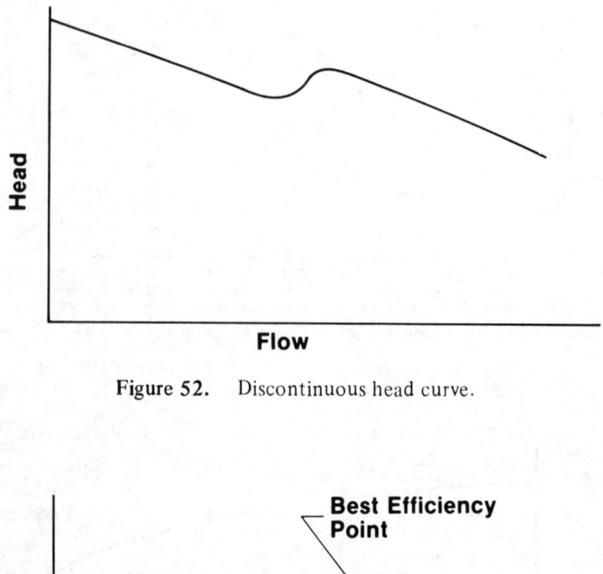

Figure 52. Discontinuous head curve.

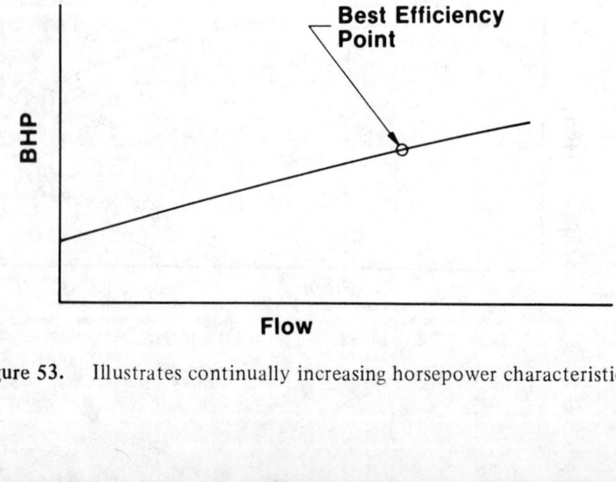

Figure 53. Illustrates continually increasing horsepower characteristics.

Flow, Q, will vary directly with the ratio of change of speed, N, or impeller diameter, D:

$$\frac{Q_2}{Q_1} = \frac{N_2}{N_1} \text{ or } \frac{Q_2}{Q_1} = \frac{D_2}{D_1} \tag{94}$$

Head, H, will vary as the square of the ratio of speeds or impeller diameters:

$$\frac{H_2}{H_1} = \left(\frac{N_2}{N_1}\right)^2 \text{ or } \frac{H_2}{H_1} = \left(\frac{D_2}{D_1}\right)^2 \tag{95}$$

The pump required horsepower (BHP) will vary as the cube of the ratio of speeds or impeller diameters:

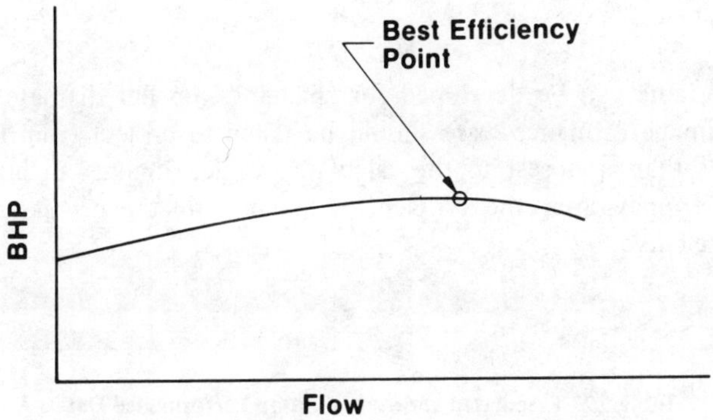

Figure 54. Illustrates peaking horsepower requirements.

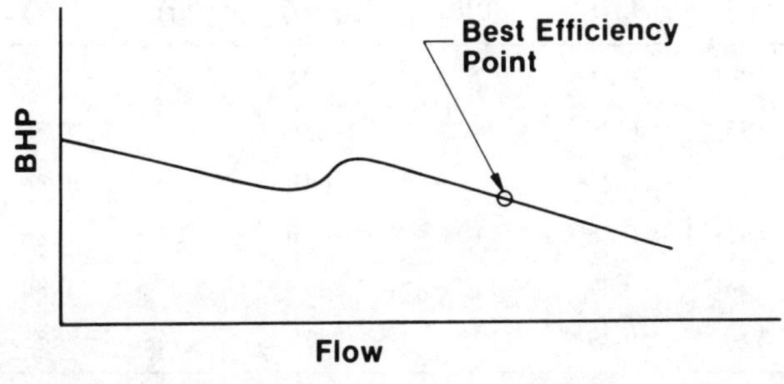

Figure 55. Illustrates discontinuous horsepower characteristics.

$$\frac{BHP_2}{BHP_1} = \left(\frac{N_2}{N_1}\right)^3 \quad \text{or} \quad \frac{BHP_2}{BHP_1} = \left(\frac{D_2}{D_1}\right)^3 \tag{96}$$

A basic premise throughout these analyses is that pump efficiency remains constant for speed or impeller diameter changes. For example, the data in Table 27 show a projection of pump performance from 1760 rpm to 1450 rpm. The subscript 2 is used for the calculated performance and subscript 1 for the values selected from the 1760-rpm curve:

$$Q_2 = \left(\frac{N_2}{N_1}\right) Q_1 \tag{97}$$

$$H_2 = \left(\frac{N_2}{N_1}\right)^2 H_1 \tag{98}$$

$$BHP_2 = \left(\frac{N_2}{N_1}\right)^3 BHP_1 \tag{99}$$

A similar table can be developed for changing impeller diameter. In projecting pump performance, care should be taken to project from the speed or diameter that is closest to the calculated value. Changes in both speed and diameter may alter the efficiencies taken from the originating pump performance curve.

Table 27. Calculated and Actual Pump Performance Data

Tabulated Performance Data from 1760–rpm Curve				Calculated Performance Data for 1450–rpm Operation			
Q_1 (gpm)	H_1 (ft)	Eff_1	BHP_1	Q_2 (gpm)	H_2 (ft)	Eff_2	BHP_2
1000	184	0.61	76.2	824	125	0.61	42.6
1500	175	0.76	87.2	1236	119	0.76	48.9
2000	166	0.84	99.8	1648	113	0.84	56.0
2500	151	0.86	110.8	2060	103	0.86	62.3
3000	128	0.82	118.3	2478	87	0.82	66.4
3250	110	0.73	123.7	2678	75	0.73	69.5

Pump performance curves are developed as a means of conveniently viewing the complete characteristics of a given pump and selecting a pump that will successfully meet system requirements. The pump is a source of pressure energy and, within its limits of design and operating speed, has the capability of providing the liquid in a system with the potential to do work.

Pressure energy capability is dependent on the velocity the impeller imparts to a liquid. For a given pump, this velocity can be changed by varying the impeller diameter (Figure 56) or by changing the operating speed. Therefore, by controlling impeller angular velocity a centrifugal pump can be used as a variable energy source to satisfy pumping systems having variable pressure energy requirements.

When system requirements dictate that the centrifugal pump *must* have particular curve shape characteristics, the approximate discharge specific speed index range may be used to define the general requirement for the specified pump.

PUMP SYSTEM CHARACTERISTICS

A pumping system is the arrangement of pipe, fittings and equipment through which liquids flow. Only a few pump applications are such that a pumping unit can operate alone, one example being a pump lifting water

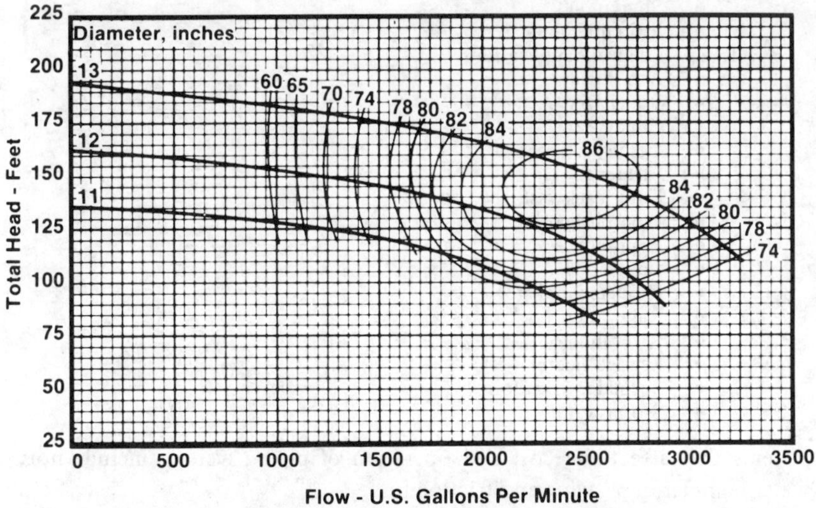

Figure 56. Pump head curves as a function of impeller diameter.

from a well into an irrigation ditch. The ditch acts as a conduit distributing the water by gravity flow. Pumps are required when liquids need to be transported from lower to higher elevations, moved over long distances, or distributed in grids, circulated in loop or pressurized systems to perform work.

Mechanically, systems can be classified by path of liquid flow as follows:

1. *Nonreturn systems*, in which all the liquid is discharged from the system.
2. *Return systems* in which (a) none of the liquid is discharged from the system; and (b) some of the liquid is discharged from the system.

Functions of systems can be classified by the characteristics of work performed as follows:

1. Thermal exchange, in which some form of thermal exchange is conducted for the purpose of satisfying a design condition.
2. Removal/delivery, in which the system is designed to remove the liquid from, or deliver it to, some point to satisfy a specific design service.

The following lists illustrate typical pump application names classified by system function:

- Thermal exchange applications
  - Chilled water
  - Cold well
  - Condenser water
  - Cooling tower
  - Heat recovery
  - Hot well
  - Mill roll cooling
  - Plant circulating water
  - Plant cooling water
  - Spray pond
  - Strip mill quench
- Removal/delivery applications
  - Ash sluice
  - Boiler feed
  - Condensate
  - Domestic water
  - Effluent
  - Filter backwash
  - Flood irrigation
  - High service
  - Low service
  - Municipal booster
  - Raw water
  - River intake
  - Sewage ejector
  - Sprinkler irrigation
  - Storm water

Items that affect the cost and operation of piping systems include noise, water hammer, pipe aging and friction loss.

Flow noise, or "hiss," contains all sound frequencies at an evenly distributed sound level, called "white noise." This noise is generated by turbulent flow through system pipes and increases with velocity; or it is generated by cavitation in the system. Flow noise occurs in pumps, valves, elbows, tees or wherever flow changes direction or velocity.

Conduction of noise proceeds through the liquid, as well as through the pipe material, and is usually controlled by flexible fittings, sound insulating pipe supports and pipe insulation.

Pipe resonance can result in objectional sound in a piping system. Resonance is a single tone sound magnified and emitted by a length of piping. It occurs when the natural frequency of a length of piping is matched by the frequency of some regular energy source such as the vanes in an impeller passing the volute tongue (cutwater) as shown in Figure 57. An illustration of this type of resonance is shown in Figure 58. Other regular energy sources, such as the rotating elements in a bearing or the power supply frequency,

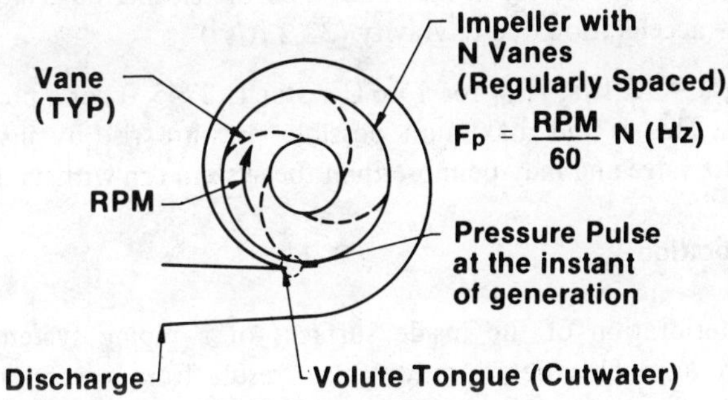

Figure 57. Illustrates noise problem in pumps generated by vanes in an impeller passing the volute tongue.

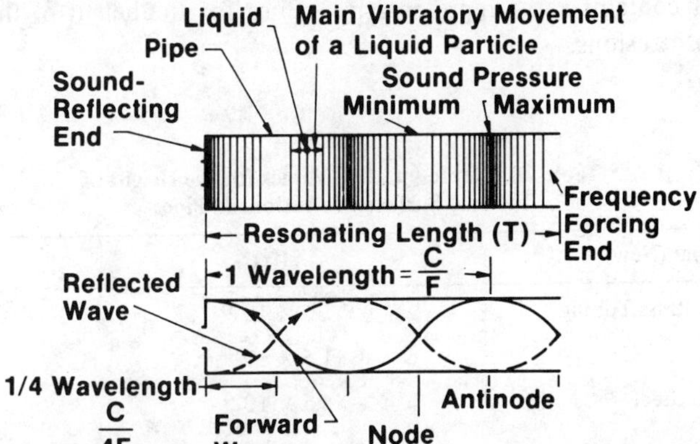

Figure 58. Illustrates noise generated by pipe resonance.

also can cause pipe resonance. Pipe resonance can be corrected by increasing or decreasing the resonating length or weight of pipe under the direction of an acoustical consultant.

Water hammer is the result of a strong pressure wave in a liquid caused by an abrupt change in flowrate. As an illustration, for the maximum possible instantaneous head increase above the normal head due to a water hammer pressure wave the following expression is applicable:

$$H_{wh} = \frac{Cu}{g} \text{ (Joukowsky's law)} \qquad (100)$$

where C = velocity of sound in the liquid (ft/s), e.g., C for 15.6°C water is 4820 ft/s.
u = normal velocity in the conduit before closing valve (ft/s)
g = acceleration due to gravity (32.2 ft/s^2)

Assuming u = 15 ft/s, H_{wh} for 15.6°C water is 2245 ft, or a pressure surge of 972 psi. This is the maximum possible pressure rise by instantaneous closing of the valve and may be more than the system can withstand.

Pipe Deterioration

Some deterioration of the inside surfaces of a piping system normally occurs with age. This deterioration may result from corrosion, deposits or a combination of both, which impede flow by increasing the relative roughness ratio, ϵ/D

where ϵ = height of protrusion on a side in feet or inches
D = inside diameter of pipe in feet or inches

Table 28 contains some representative values for the height of the internal surface protrusions.

Table 28. Representative Values for the Height of Internal Surface Protrusions in Pipes

Material (New)	ϵ(ft)	ϵ(in.)
Copper or Brass Tubing	5×10^{-6}	6.0×10^{-5}
Steel Pipe	1.5×10^{-4}	1.8×10^{-3}
Galvanized Steel	5×10^{-4}	6.0×10^{-3}
Concrete	4×10^{-3}	5.0×10^{-2}

For old material, the values listed may be increased by 100 or more times due to corrosion or deposits. Past experience is generally the most accurate guide to determine the extent of pipe deterioration caused by aging.

Pipe Friction

Pipe friction is resistance to flow and results in a loss of head, which is expressed as friction head H_f, in feet of liquid. Methods of calculating friction head have been discussed in previous chapters and only supplementary notes are given. First, proper viscosity conversion is sometimes confusing in evaluating friction losses. If absolute viscosity is expressed in centipoise (cp), the following conversions should be used:

$$2.089 \times 10^{-5} \times \text{cp} = \text{lb-s/ft}^2$$

which, in turn, can be converted into kinematic viscosity (ft^2/s) by dividing by the mass density ($\text{lb-s}^2/\text{ft}^4$).

If kinetic viscosity is expressed in centistokes (cs), use the following conversion:

$$1.076 \times 10^{-5} \times \text{cs} = \text{ft}^2/\text{s}$$

If viscosity is expressed in SSU, there is no distinction between the absolute and the kinetic designations. Convert SSU to kinematic viscosity as follows:

For SSU larger than 100:

$$1.076 \times 10^{-3} \times \left(0.0022\ \text{SSU} - \frac{1.35}{\text{SSU}}\right) = \text{ft}^2/\text{s}$$

For SSU equal to 100 or less:

$$1.076 \times 10^{-3} \times \left(0.00226\ \text{SSU} - \frac{1.95}{\text{SSU}}\right) = \text{ft}^2/\text{s}$$

The following examples illustrate these conversions:

Example 1: Given–40°C water with kinematic viscosity of 1.58 cs. Convert viscosity into ft^2/s units.

Solution:

$$1.076 \times 10^{-5} \times 1.58\ \text{cs} = 1.70 \times 10^{-5}\ \text{ft}^2/\text{s}$$

Example 2: Given–Ethylene glycol at 21.1°C with a viscosity of 88.4 SSU. Convert viscosity into ft²/s units.

Solution:

$$1.076 \times 10^{-3} \times \left(0.00226 \times 88.4 - \frac{1.95}{88.4}\right) = 1.912 \times 10^{-4}\ \text{ft}^2/\text{s}$$

Example 3: Given–Ethyl alcohol at 20°C with specific weight of 49.4 lb/ft³ and absolute viscosity of 1.2 cp. Convert viscosity into ft²/s units.

Solution:

$$2.089 \times 10^{-5} \times 1.2 = 2.507 \times 10^{-5}\ \text{lb-s/ft}^2$$

$$\text{Mass Density} = \frac{49.4\ \text{lb/ft}^3}{32.2\ \text{ft/s}^2} = 1.5342\ \frac{\text{lb-s}^2}{\text{ft}^4}$$

$$\frac{2.507 \times 10\ \text{lb-s/ft}^2}{1.5342\ \text{lb-s}^2/\text{ft}^4} = 1.63 \times 10^{-5}\ \text{ft}^2/\text{s}$$

Once proper viscosities have been determined, friction factors can be obtained.

Through many years, simplified variations of the Darcy-Weisbach formula have been devised for limited areas of application. The accuracy of such variations is supported by successful use and allows the user to bypass the diagram interpretation and calculation necessary with the basic formula. The use of specific formula variations outside their known areas of successful application may lead to inaccurate projections of friction head.

This simplified derivation of the Hazen-Williams formula can be used to calculate friction head for liquids having a kinematic viscosity of 1.1 cs. (Water at 60°F has a viscosity of 1.13 cs.)

$$\text{Friction Head, } H_f, = 10.45 \left(\frac{\text{gpm}}{C}\right)^{1.85} \left(\frac{L}{D^{4.87}}\right) \tag{101}$$

where gpm = gallons per minute
C = pipe roughness factor ranging from 60 to 160
D = inside diameter of pipe (in.)
L = length of pipe (ft)

The Hazen-Williams formula is generally used for cast iron pipes of 3-inch and larger diameter. As pipes deteriorate, the roughness factor, "C," decreases. This decrease in C depends on the pipe material, pipe linings, pipe age and the characteristics of the liquid.

For liquids with viscosities other than 1.1 cs, methods more accurate than the Hazen-Williams formula should be used in determining friction head. For example, friction head computed with the Hazen-Williams formula will increase by 20% at a viscosity of 1.8 cs and decrease by 20% at a viscosity of 0.29 cs. In addition, friction loss will occur in fittings such as elbows, tees and valves. Many handbooks and manuals provide means for determining these losses. Manufacturers' data must be used to determine friction loss for mechanical equipment installed in the system.

SYSTEM HEAD CURVES AND COMPONENTS

A system head curve is the graphic representation of the head required at all flows to satisfy the system function. Regardless of the mechanical configuration, function, or means of control, the system head curve or system head band is used to define total head versus flow for any piping system. System head curve development is a combination of calculations and the designer's "best feel" for the variable conditions.

The three components that make up total system head are static head, design working head and friction head. These components are defined as follows:

1. *Static head* is the vertical difference in height between the point of entry to the system and the highest point of discharge.
2. *Design Working Head* is the head that must be available in the system at a specified location to satisfy design requirements.
3. *Friction head* is the head required by the system to overcome the resistance to flow in pipes, valves, fittings and mechanical equipment.

Total system head at a specified flowrate is the sum of static head, design working head and friction head. The values of static head and design working head may be zero. Figure 59 illustrates a total system head curve made up of the friction head and the design working head and static head.

For most systems, the total system head requirements are best illustrated by a band formed by two total system head curves. This band of system requirements is a result of variable factors that affect total system head calculations. The following list of variables should be considered in projecting system requirements.

Static head will vary as a result of change in elevation of highest point of discharge of the system.

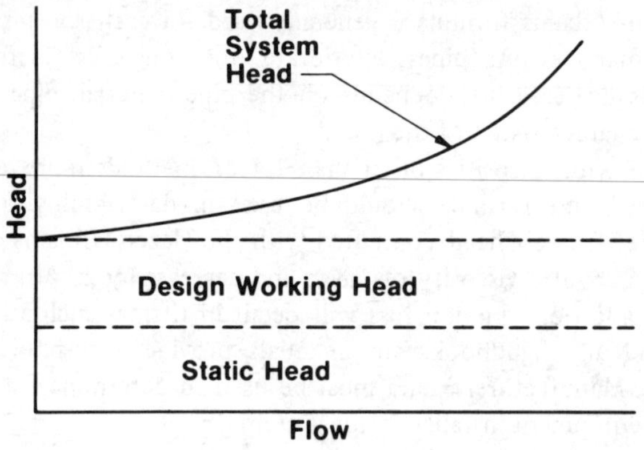

Figure 59. Illustrates total system head curve components.

Design working head is usually treated as a constant component of total system head.

Friction head at any specified flow will vary as a result of:

- method of calculation or source of tabulated data,
- change in viscosity resulting from a change in liquid temperature,
- deterioration of the piping system,
- load distribution,
- systems differences between design and "as-built,"
- manufacturing tolerances of mechanical equipment, and
- accumulation of solids in the system.

Most systems are not completely defined by a single line system head curve. Those variables that are applicable determine the band of maximum and minimum system head requirements. The band between maximum and minimum head requirements at any given flow will be significant in some systems and not in others. For all systems, interaction of total head available and total head required must be evaluated to achieve satisfactory operation.

SYSTEM OPERATION AND CONTROL

For a system to operate at design flow, total head available to the system must be equal to the total system head required. For proper evaluation of system operation at flowrates other than design, a comparison of total available head and system required head must be made based on the means selected to control system flow. The reference point used to determine total available head must be the same as that used to determine total system head required. The point of entry to the system, which is also the pump discharge connection, is used in this text as the reference point.

Total available head at the point of entry to the system is the sum of static head and pump head.

Static head available is the vertical difference in height between the point of entry to the system and the liquid supply, minus any friction head in the supply pipe and fittings. Static head may be relatively constant. The decrease in static head with increasing flow is the result of friction head in the supply pipe. In some applications static head will vary, and is normally illustrated by a band of maximum and minimum values.

When the liquid supply is below the point of entry to the system, as shown in Figure 60, static head available is a negative value.

The difference in elevation between the fluid supply and the point of entry to the system must be developed by the pump. Therefore, pump head must include static head and friction head in the supply pipe and fittings.

In systems in which the liquid supply is from a pressurized main, the pressure must be expressed in feet of liquid. The equivalent height of the liquid above the system entry point is static head available. In Figure 61, with a 100 psig pressure at point A and a liquid with a specific gravity of 1.0, the static head available will be 231 feet minus elevation height, X. In return piping systems, in which none of the working fluid is discharged from the system, static head available is zero.

Pump head is the total head developed by a pump. Its value at any given flowrate must be obtained from the pump manufacturer's performance curve.

A curve of total available head versus flow is developed by adding static head and pump head at several values of flow. The total available head curve

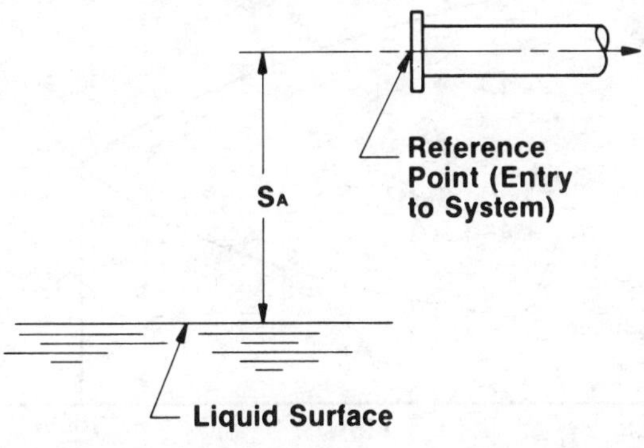

Figure 60. Illustrates a case in which the static head available is a negative value.

is slightly steeper than the pump head curve due to the slope of the static head curve. To evaluate the operation of a system, the total system head curve or curves and the total available head curve are plotted on a common graph. The two curves will intersect at maximum flow for that system. As shown in Figure 62, system flow greater than 100% cannot occur since total system head required exceeds the total available head.

Consider a system where the total available head must be represented by two curves forming a band as shown in Figure 63(A). This is a common situation that is a result of maximum and minimum values of static head available. It is now possible for system flow to exceed design flow of 100%

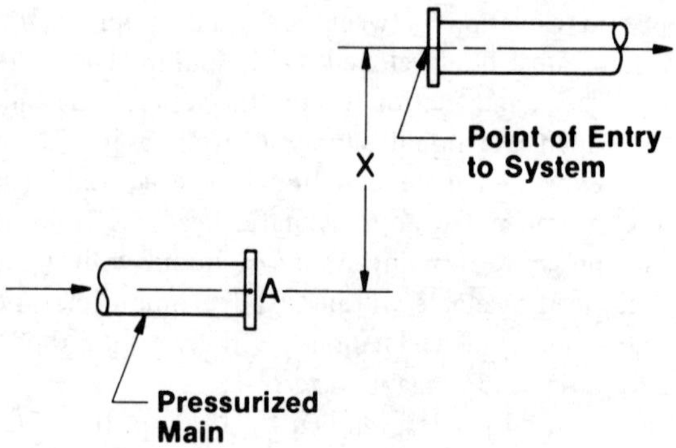

Figure 61. Example problem to determine static head available.

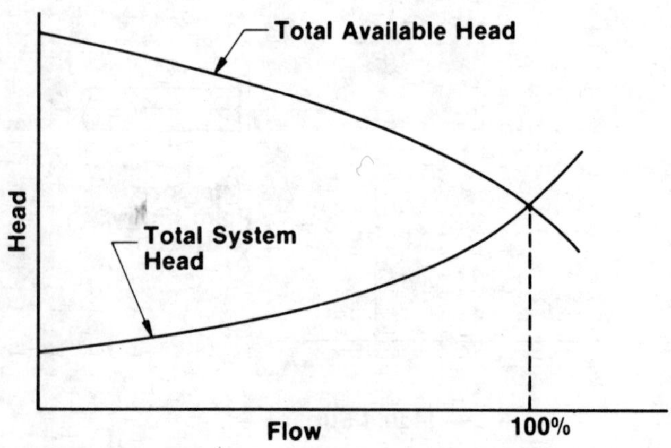

Figure 62. Shows that the total system head required exceeds the total available head above 100% flow.

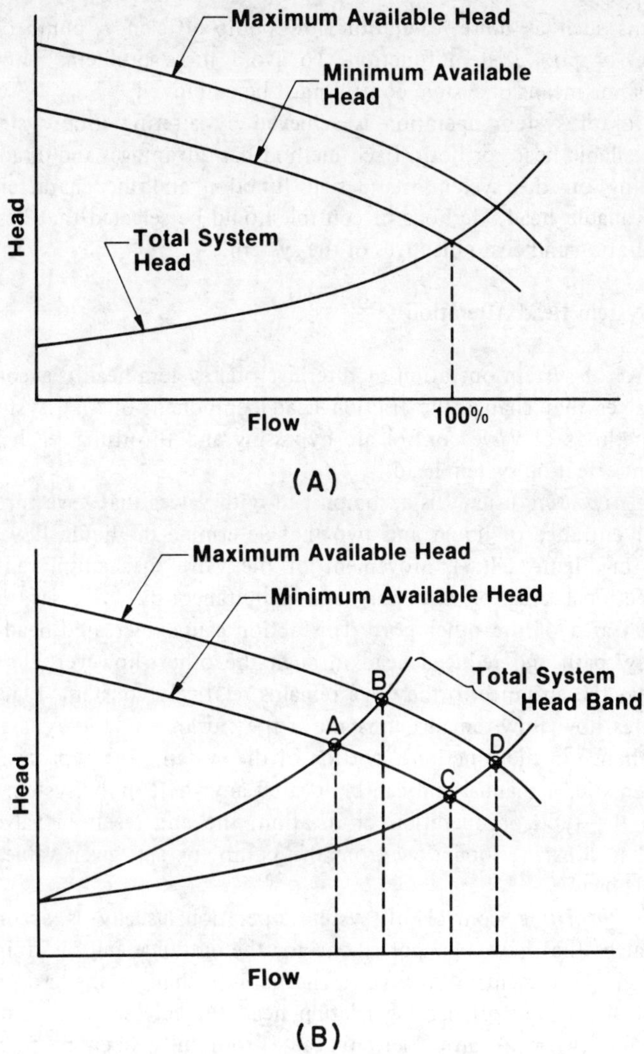

Figure 63. Illustrates interpretation of flow head curves.

since the pump selection was based on the minimum value of static head available at 100% flow.

Figure 63(B) illustrates a system with a band of total system head and a band of total available head. At any given time, system operation can occur anywhere within the area A-B-D-C-A. While the pump is selected for proper operation at design flow, operation at flows other than design may result in problems such as noise, cavitation, low pump efficiency, pump or system damage, or poor system function. To avoid these problems, should they occur, some means of system control must be employed.

Control of system operation is achieved by altering total system head, total available head, or both. Each method has advantages and disadvantages depending on the system, its design function and the characteristics of total available head. Methods of control should be selected that best satisfy the operation and cost objectives of the system.

Total System Head Alteration

Control of system operation by altering total system head is accomplished with valves that change the friction head component of total system head. Two methods of valve control are bypassing and throttling, each having a different effect on system head.

Valve bypassing usually is accomplished with valves that have three ports: one for entrance of liquid and two that determine the liquid flow path, as shown in Figure 64(A). Movement of the valve mechanism reduces the cross-sectional area of one port and simultaneously increases the cross-sectional area of the other port. This action causes increased head through one flow path and reduced head through the other; however, the head required at the entrance to the valve remains relatively constant. Figure 64(B) illustrates how flow remains constant at the entrance of the bypass valve as flows change in the functional portion of the system. The bypass valves may be automatic or manual and can be located anywhere in the system. Regardless of the type of operation or location, the end result of valve bypass control is constant total flow from the pump for any given value of total available head.

Valve throttling control of system operation usually is accomplished with valves that have two ports: one for the entrance and one for the exit of liquid. Movement of the valve mechanism changes the cross-sectional area of the valve port causing friction head to increase to a greater value than the full open valve friction head. From fully open to fully closed position, valve throttling produces a series of system head curves as shown in Figure 65.

System friction head will vary depending on relative valve opening, and resulting system flow will occur at the point of intersection between the system head curve and the total available head curve.

Total Available Head Alteration

Control of system operation by altering the pump head portion of total available head can be accomplished by changing pump impeller diameter, selecting pumps for series operation or parallel operation, or by changing

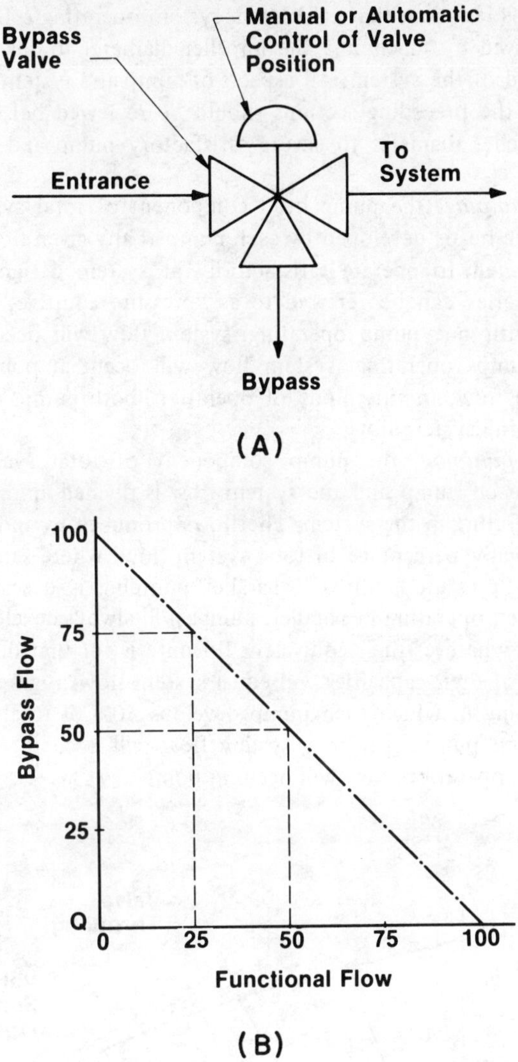

Figure 64. Illustrates valve bypassing. Flow remains constant at the entrance of the bypass valve as flows change in the functional portion of the system (see (B)).

the pump operating speed. Each method of control will have a different effect on system operation.

Change of impeller diameter will alter the pump head component of total available head. In systems in which constant speed pump impeller diameters have been selected for the *calculated* system design head, and the *actual* system head is less than that calculated, system operating efficiency usually can be improved by machining the impeller diameter to develop the actual operating head of the system. All aspects of pump and system characteristics explained in the preceding sections should be reviewed before considering changing impeller diameter to ensure satisfactory pump and system operation.

In *series pumping*, the pump head component of total available head is the sum of the heads developed by each pump at any given flow. Each pump must be selected to operate satisfactorily at system design flow. Pumps operated in series can be referred to as "pressure additive," illustrated in Figure 66. With one pump operating, system flow will occur at point A. With both pumps operating, system flow will occur at point B, which is system design flow. In this mode of operation, both pumps are developing equal head at full system flow.

In *parallel pumping*, the pump component of total available head is identical for each pump and the system flow is divided among the number of pumps operating in the system. The flows produced by individual pumps can represent any percentage of total system flow. Where series pumping is described as "pressure additive," parallel pumping is described as "flow additive." When operating in parallel, pumps will always develop an identical head value at whatever their equivalent flowrate is for that developed head, and the sum of their capacities will equal system flow. Figure 67 illustrates parallel pumping in which each pump develops 50% of total flow at 100% head. With one pump operating, system flow will occur at point A; with both pumps in operation, flow will occur at point B.

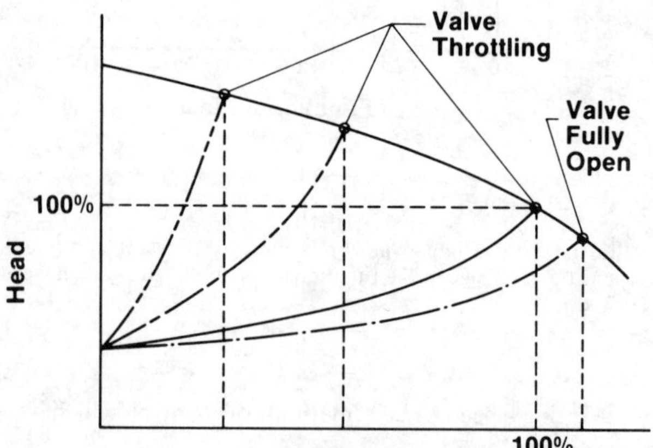

Figure 65. Illustrates valve throttling effect on head curve.

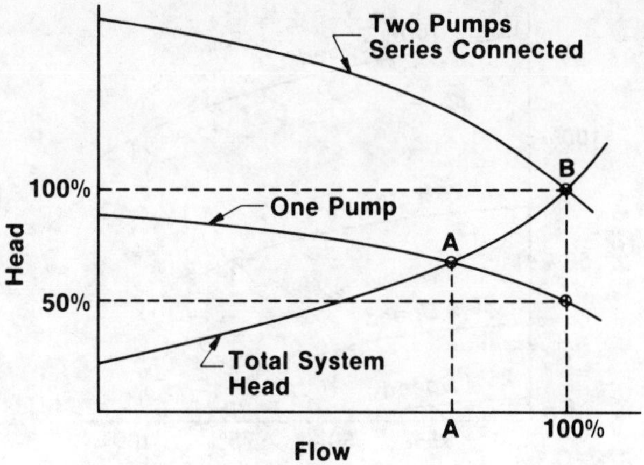

Figure 66. Effect of series pumping on the head curve.

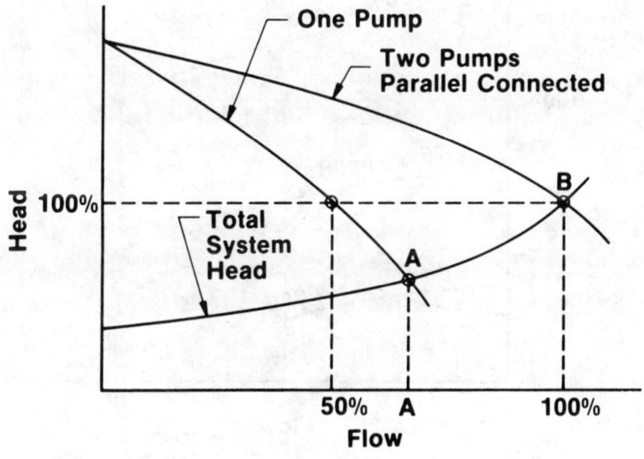

Figure 67. Effect of parallel pumping on the head curve.

Changing pump speed will alter the pump head component of the total available head. The performance characteristics of a pump at various speeds can be calculated by use of the affinity laws and will result in a pump head-flow curve for each operating speed, as shown in Figure 68(A). The application of variable-speed pumps in systems can result in system operation at the intersecting points shown in Figure 68(B). The number of intersection points will be dependent on the number of pump speed changes.

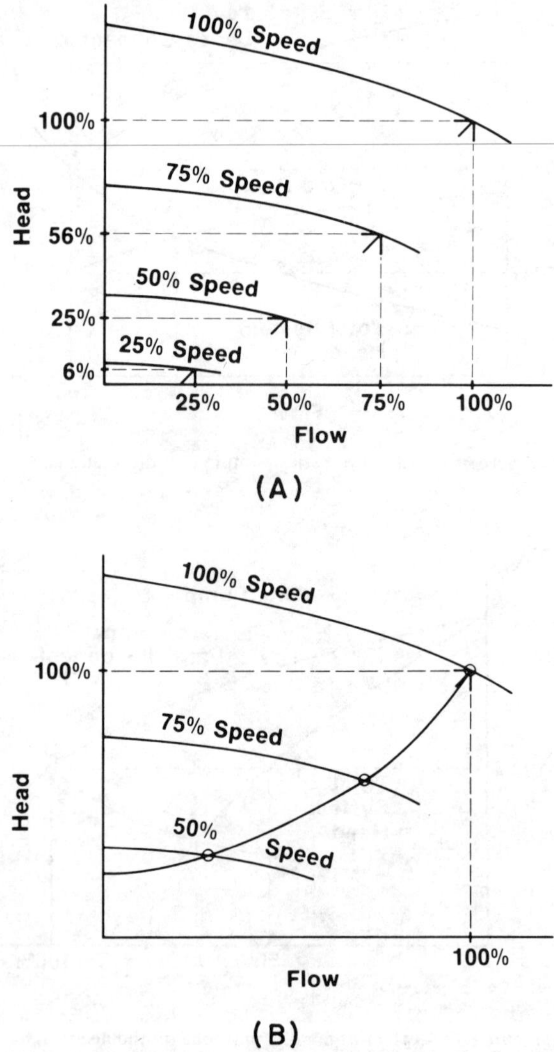

Figure 68. Effect of changing pump speed on head curve.

There are several devices available to vary pump operating speeds. Some cause fixed increments of speed change, such as multiple-speed motors, while others operate at an infinite number of speeds within the design limits of the device. With drives that have fixed increments of speed change, a fixed number of intersection points occur between the pump head curves and the system head curve. Pump drives that have an infinite number of operating speeds can produce an infinite number of intersecting points on the system head curve.

VARIABLE-SPEED PUMPS

When variable speed pumps are employed as a means of system control, the system sensing method that will control the pump speed must be determined. This is necessary because the total available head at the point of entry to the system is a function of the type and location of the monitor.

Functions

The sensing equipment must perform three functions: (1) monitor the system variable; (2) compare the value of the system variable to the required value (set point); and (3) cause the variable-speed pump to restore the variable to the required value. System values typically monitored include level, flow, pressure, differential pressure, temperature and differential temperature.

Types

Sensing equipment may be as simple as a piece of tubing connecting the piping system to the variable-speed drive or as complex as electronic telemetry equipment that monitors variables many miles from the pump.

Location

The total operation of the system must be evaluated at the location where the system variable is being monitored.

When equipment in the system causes a change in flow, the path of pump operating points will not follow the unaltered system head curve. An infinite number of new system head curves will occur when system components cause flow change. For example, in a system that requires a fixed design working head, the path of pump and system operating points can be predicted once the location of the system monitoring device is established. This path of operation will be controlled by the location of the monitor and will affect the operating speeds and power requirements of the pump.

Figure 69(A) illustrates a simple system in which flow change is caused by a valve near the end of the system. The system head curve for this example is shown in Figure 69(B) and includes static head (S_1) and friction head (H_f) between points A and C, with the valve in a fixed position.

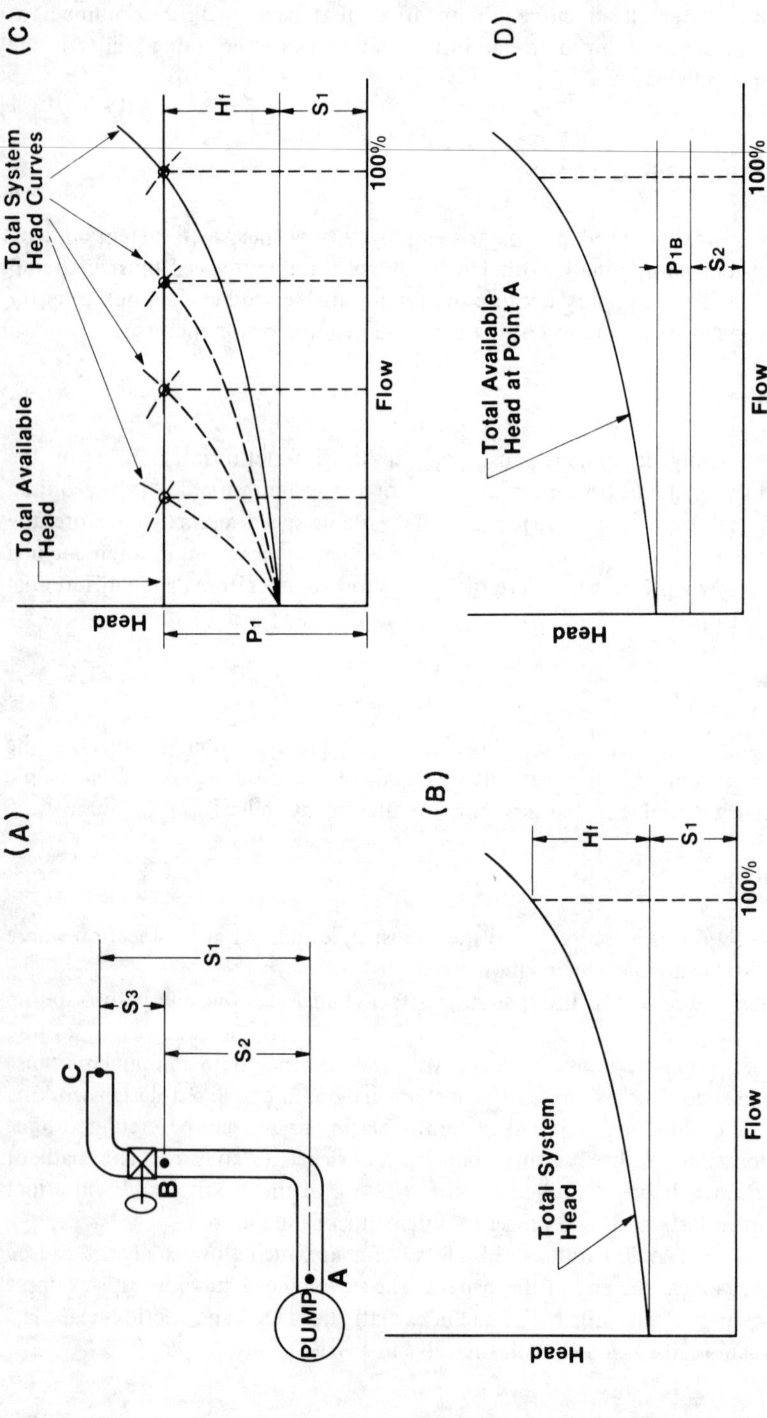

Figure 69. Pumping problem example.

A variable-speed pump with a pressure monitor at the point of entry to the system, A, will produce a constant total available head at A for any system flow. The setting for the pressure monitor is determined from the following formula:

$$P_1 = P_2 + H_f + S \tag{102}$$

where P_1 = total available head, in feet, to be maintained at the pressure monitor

P_2 = design working head, in feet

H_f = full flow friction head, in feet, between the monitor and the location in the system where P_2 is to be maintained, or from the monitor to the end of the system (A to C) when P_2 is zero

S = static head, in feet, between the monitor and the location in the system where P_2 is to be maintained, or from the monitor to the highest point of discharge; static head, S, is applicable only in systems employing pressure monitors

In Figure 69(A), P_2 is zero and $P_{1A} = H_f + S_1$.

As system flow is throttled, a series of new system head curves will be produced, as shown in Figure 69(C). System flow will occur at those points where the system head curves intersect the available head curve. At any flow of less than 100% there is more head available than required, and the excess is consumed as friction head across the throttling valve. Now consider the same system with the pressure monitor located in Figure 69(A) at point B. The system has not changed, and the total system head required is still as shown in Figure 69(B). The pressure to be maintained at the monitor is determined from the same formula:

$$P_1 = P_2 + H_f + S$$

Design working head, P_2, is zero, and static head is equal to S_3. H_f is the full flow friction from B to C. The pressure setting is $P_{1B} = H_f + S_3$. The total head that must be available at point A to satisfy the monitor at point B will be equal to the pressure monitor setting, P_{1B}, plus the static head, S_2, plus the friction head from A to B. This is illustrated in Figure 69(D).

As system flow is throttled, a series of system head curves will be produced, as shown in Figure 70. System flow will occur at those points where the system head curves intersect the total available head curve. By comparing Figures 69(C) and 70, the throttling losses are less with the pressure monitor at B than they are with the pressure monitor at A for specific values of flow. This can result in lower pump operating speed, lower horsepower required and lower operating cost.

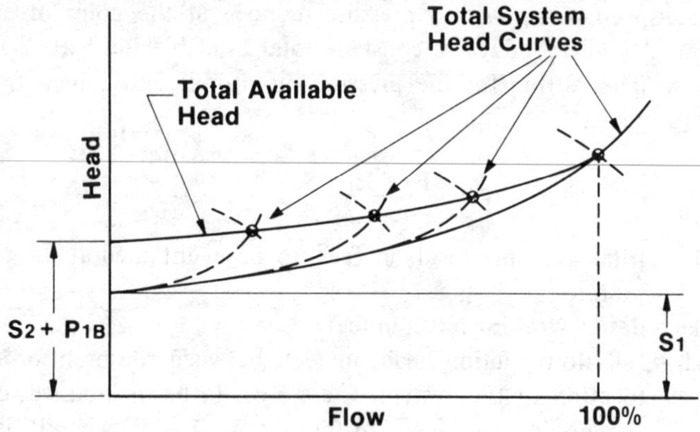

Figure 70. Series of head curves generated for example problem.

Sensing Multiple Branch Systems

In parallel piping systems with two or more branches, a single pressure monitor must be set to satisfy the branch with the maximum calculated friction head. In some instances, load variation in the system can cause the pressure in nonmonitored branches to drop below acceptable limits, even though the monitored branch is satisfied. This type of problem can be resolved by analyzing the various branches based on the location and setting of a single pressure monitor. Additional pressure monitors can be added to those branches that experience the low pressure condition to restore system operation to acceptable levels.

Evaluation of the total available head band and the total system head band over the expected range of system flow is necessary to make decisions that will affect the cost and operation of a pumping system. The decisions are based on problem statements and evaluation of the possible solutions with respect to the cost and functional objectives of the system design. The following questions outline some of the considerations that should be reviewed to assure satisfactory operation of a pumping system:

1. If the system is throttled for reduced system flow, does the maximum total available head exceed the system working pressure? If so, the alternatives include:

a) Increase system working pressure.
b) Select pump with a "flatter" head capacity characteristic.
c) Consider variable-speed pump drive.
d) Add bypass valve or pressure-reducing valve to system.

2. Does pump horsepower increase with decreasing flow? If it does, the alternatives include:

a) Add bypass valve to system.
b) Consider variable-speed pump drive.
c) Select several lower specific speed pumps to operate in parallel.

3. Will operation at reduced flows occur for significant periods of time? If so, pump efficiency will be reduced and the service life of pump bearings, packing, mechanical seals and close clearance wearing rings usually will be shortened, and the alternatives include:

a) Select smaller pump for the reduced flow operation.
b) Add bypass valve to system.
c) Consider variable-speed pump drive.

4. Will actual system head at design flow be considerably less than the calculated value? If so, the system flow will exceed design flow, and the alternatives include:

a) Reduce the pump impeller diameter.
b) Install valve to throttle system flow.
c) Consider variable-speed pump drive.

5. In systems planned for two or more pumps operating in parallel, have all the operating possibilities been considered? For example, one pump alone can operate at higher flow than its rating, as selected for parallel operation. For the higher flow operation the pump will require more net positive suction head and may be subject to undesirable hydraulic loads, motor overload, increased noise level and shortened life.

These are only a few of the many problems that can be encountered in the course of designing a pumping system. In many instances, several solutions can provide equally satisfactory results if operating power requirements are not considered. If design objectives include minimum power consumption, an energy evaluation of the pumping system must be made.

TOTAL SYSTEM EVALUATION

Total system evaluation involves developing information about the system and its pumping equipment necessary to satisfy a system's design requirements, concluding with a projection of total power requirements. To perform the evaluation, system head and total available head must be established. These two components are combined with pump power requirements and system load profile to project total system operating costs.

Load profile expresses the measure of work executed in a system compared to a unit of time. Work performed has a direct relationship to flow; therefore, load expressed as flow provides a common base for any system energy requirement projection. System load profile can be illustrated with a curve as shown in Figure 71; however, it is generally easier to use the flow/time relationship in tabular form, as illustrated in Table 29. The tabular format organizes total operational time at specific flow values and simplifies identifying high flow/time concentration areas, which should be the focal point for preliminary pumping equipment selection.

Figure 72 (top) illustrates how the flow/time relationship can be combined with a system head curve to aid in pump selection procedures. In this example, a significant amount of system operational time will occur in the flow ranges of 20% to 30% and 60% to 75% of the total design flow.

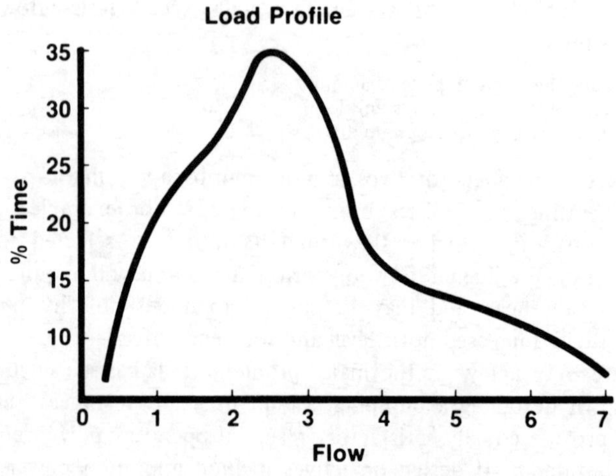

Figure 71. System load profile plot.

Table 29. Tabulated Flow/Time Data

Flow (gpm)	Hours	Flow (gpm)	Hours
280	87	4480	788
840	1138	5600	613
1680	1841	6440	350
2520	2629	7000	87
3640	1227	---	8760

Pumping equipment should be selected to provide maximum operating efficiencies in these flow ranges. One possible pump equipment selection for this system would include a single pump selected for the system head and flow requirements in the 20% to 30% range, as well as two identically sized pumps to share the total system capacity equally at full system flow, but selected so that each will be in its best efficiency range when operated independently in the 60% to 75% system flow range (Figure 72, bottom).

The time base for computing load requirements should be of sufficient length to incorporate all variables affecting system operation. The minimum practical load profile time length is one complete cycle, from minimum to

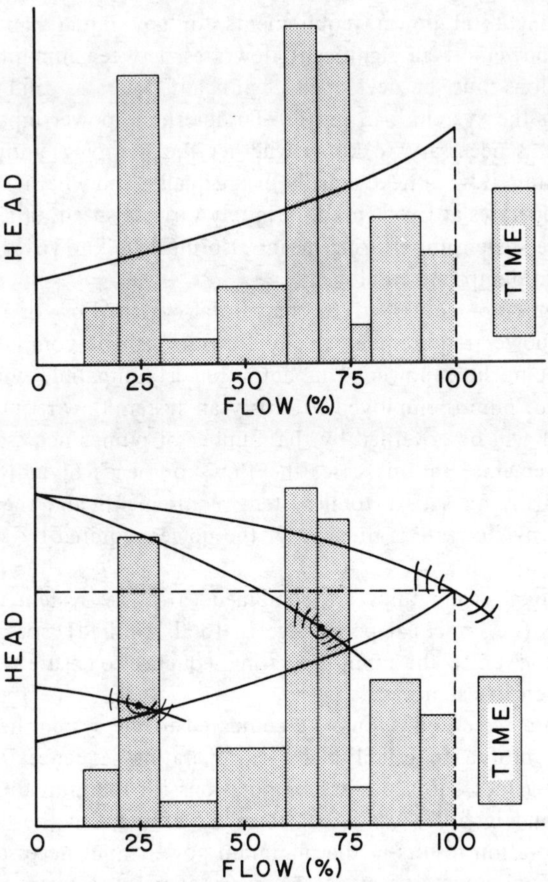

Figure 72. Illustrates how the flow/time relationship can be combined with a system head curve.

maximum flow requirements. A twelve-month time base for total system load evaluation generally encompasses all production, load and climatic variables for a system and can be used to project system operating requirements and costs. System load profile is the central component of any energy evaluation process.

In projecting total system energy requirements, total available head must be reviewed. If static head available is variable with respect to system flow, time, or both, a profile of these relationships should be made. This projection should represent the magnitude of change compared to the length of time the change is expected to occur, thus developing a correlation with system load profile. If the accuracy of these data is questionable, a band of minimum to maximum values should be established from which a total system power requirement band can be developed.

In projecting total power requirements for any liquid handling system, pump horsepowers at all significant flowrates between minimum and maximum conditions must be developed. Significant flowrates can be determined by reviewing the system load profile. To determine power inputs at various flowrates, it is necessary to know whether the pumping equipment shares the total system flow or head equally or unequally, and whether the pumping equipment operates at fixed speed, infinitely variable speed, or a combination of both. The system monitoring means also must be known when infinitely variable-speed pumps are employed.

When pumps are combined to share total system flow or head equally, the system power requirement at any flowrate can be computed based on individual pump horsepower. This definition is compatible with any range or quantity of pumps employed to satisfy all system flow requirements. The energy input will be governed by the number of pumps necessary to satisfy those requirements. In this case, the flows or heads of individual pumps combine equally to satisfy total system requirements and the total power required at any flowrate is the sum of the power required by the individual pumps.

When unequal size pumps are combined to satisfy system flow or head requirements (i.e., unequal percentage of total required), special consideration must be given to the pump operating sequence to ensure that maximum system efficiency is achieved.

Initial pump selections should be compared to the system head band and system load profile to establish a basic operating sequence for significant system flow values. Horsepower computations for the individual pumps are then made and recorded on a data form. To ensure that the selected order of pump operation results in the minimum power input necessary to satisfy the system requirements, rearrange the pump operating sequence at individual system flowrates, compute new pump horsepower requirements and record this information.

This procedure will illustrate that for any system flowrate there is a single combination of pumps that requires minimum power input to satisfy the system requirements.

PROJECTING PUMP HORSEPOWER

Constant Speed

To compute constant speed pump horsepower for parallel, series, equal or unequal combinations, all pump characteristics necessary are available on the manufacturers' performance curves. For pump operation with fixed-speed change increments, the characteristics of flow, head and brake horsepower can be calculated from the manufacturers' curves using the affinity laws. Therefore, power requirements for all constant-speed pumps and pumps of fixed-speed increments can be projected from available data.

Variable Speed

To compute variable-speed pump horsepower, it is first necessary to determine system monitor type and location. Type and location will determine head available, and head available will determine power required. Unfortunately, there is no simple mathematical relationship that can be developed for computing the interacting characteristics of systems and variable-speed pumps. The derivation of the affinity laws,

$$Q_2 = Q_1 \sqrt{\frac{H_2}{H_1}}\,, \tag{103}$$

is helpful in computing performance for individual variable-speed pumps. In this formula, Q_1-H_1 represents any single value of flow and head taken from the total system head curve. Q_2-H_2 represents a point on a pump total head curve of known operating speed that will satisfy the formula. The relationship of these values is shown in Figure 73. When the formula has been satisfied, the pump operating speed necessary to produce system conditions Q_1-H_1 can be calculated.

In using the proposed formula, values known include system flow, Q_1, system head, H_1, and full pump operating speed, N_2. Values unknown are the equivalent pump flow, Q_2, equivalent pump head, H_2, and the unknown speed, N_1, at which the pump must operate to produce the required system flow and head. When making variable-speed pump brake horsepower calculations, it is necessary to know the equivalent point on the full-speed

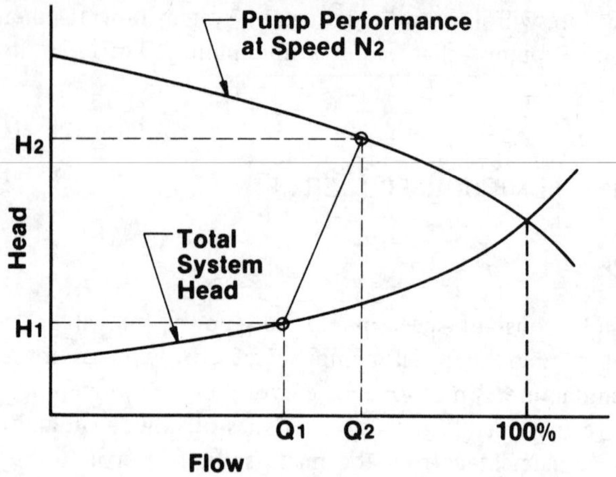

Figure 73. Shows relation of pump affinity law to total system head curve.

characteristic curve that equates to the required system flow and head, since the efficiency at the equivalent point will reoccur at the reduced flow and head condition. Variable-speed pump brake horsepower is computed from the following formula:

$$BHP_1 = \frac{Q_1 \times H_1 \times \text{Specific Gravity}}{3960 \times \text{Eff.}_2} \tag{104}$$

where Eff._2 is the efficiency from the full-speed pump curve at Q_2, H_2 expressed as a decimal value, Q_1 is U.S. gpm, and H_1 is feet of head. The reduced pump operating speed necessary to satisfy the system flow and head can be projected from the affinity law derivation:

$$N_1 = \left(\frac{Q_1}{Q_2}\right) N_2 \tag{105}$$

This value is used when computing brake horsepower requirements for variable-speed equipment where drive efficiency is a function of output to input speed ratio. The expected range of pump operating speed should be determined and compared by the pump manufacturer to the critical speed of the pump. If operating speed occurs at or near pump critical speed, it will be necessary to modify the pump or to select a different pump for the application.

Determining the precise Q_2-H_2 point on the pump curve usually will require reiterative calculation. The triangulation procedure illustrated in Figure 74 locates an intercept point on the pump curve that closely approximates the actual flow and head solution, Q_2-H_2. The accuracy of that approximated point is affected by the relative magnitude of H_1, H_2 and the slopes of the pump and system head curves. To satisfy the formula, and thus determine the point on the pump curve of the known speed that produces the required system values Q_1-H_1 at reduced speed, it is first necessary to calculate the initial Q_2-H_2 condition as a reference point. This is done by entering known system conditions Q_1-H_1 and an initial value of H_2 in the formula. The initial value of H_2 is taken from the known pump performance curve at system flow, Q_1. By solving the formula for the initial value of Q_2, a reference point for initial Q_2-H_2 is established. By connecting the initial (reference) values of Q_2-H_2 with required system condition Q_1-H_1, an intersection point with the known pump curve is established. This intersection point becomes the approximate Q_2-H_2 value that will, at reduced operating speed, produce system conditions Q_1-H_1.

In a parallel pumping system where system flow is shared equally by all pumps, the procedure is identical to that for a single pump system, except that system flow, Q_1, must be divided by the number of pumps, P_n, operating. Figure 74(B) illustrates the triangulation process for a system incorporating two equally sized pumps operating in parallel.

When pumps selected to share unequal portions of total system flow are operating, the individual pump solution formula must be modified to $Q_1 = Q_2\sqrt{H_1/H_2}$. This is necessary since the flow produced by each pump for a selected system flow, Q_1, is unknown. To determine the flow and head developed by unequal size variable-speed pumps when they are both contributing to total system flow, it is first necessary to solve the basic triangulation process for system flow, Q_1, on the combined pump characteristic curve (Figure 74(C)). The intersection point that occurs on the combined pump curve establishes the H_2 value for the individual pumps. Use of the formula $Q_1 = Q_2\sqrt{H_1/H_2}$ determines individual pump flow, $Q_1(P_1)$ and $Q_2(P_2)$, for the system operation at Q_1. System power requirements at flow Q_1 are determined by adding the horsepower requirements of each pump.

When a variable-speed pump is combined with a constant-speed pump, the constant-speed pump will always produce flow at the point where the developed head matches the required system head. When system flow requirements are such that the constant-speed and variable-speed pumps must be operated in parallel, flow produced by the variable-speed pump will be the difference between required system flow, Q_1, and constant-speed pump flow at system head, H_1.

Figure 74(D) illustrates a two-pump system in which both pumps are of equal size—one constant speed and one variable speed. With a required

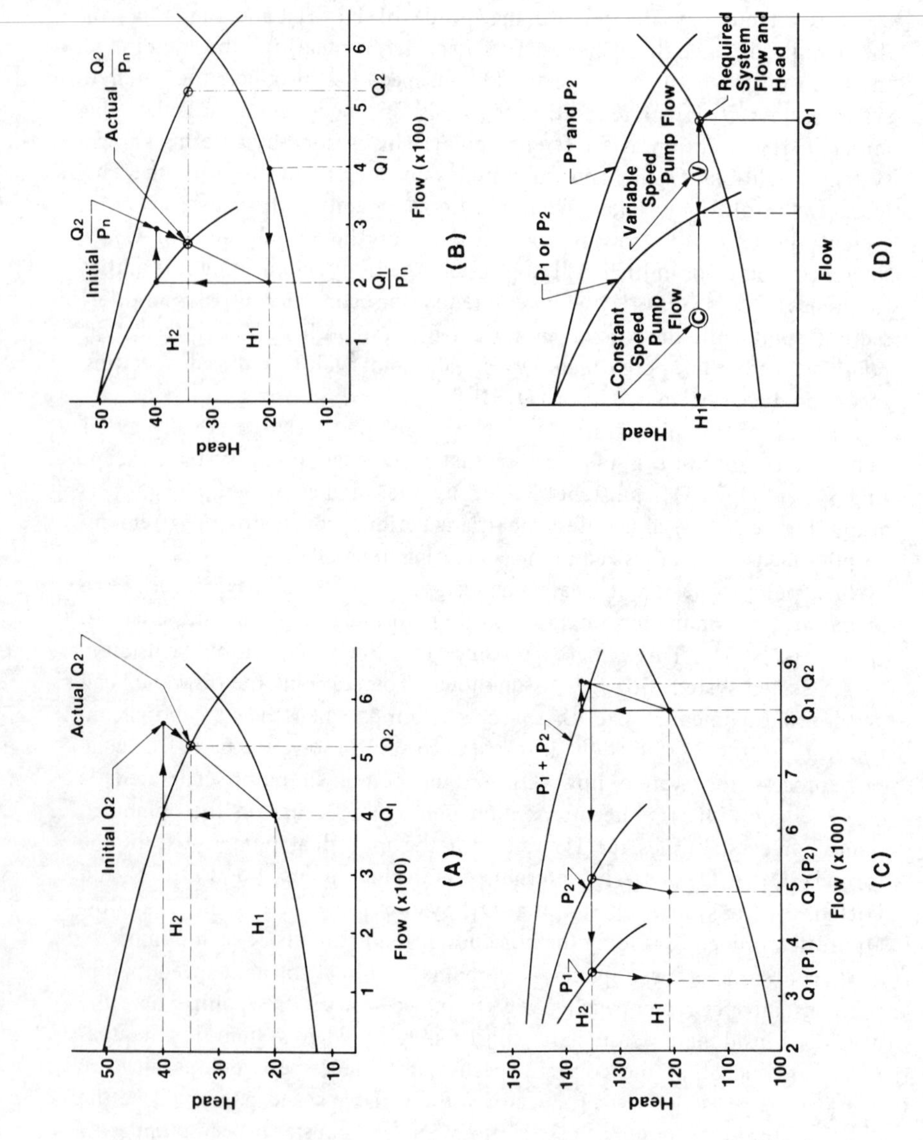
(A)
Head
Flow (x100)
Initial Q2
Actual Q2
H2
H1
Q1
Q2
(B)
Head
Flow (x100)
Initial Q2/Pn
Actual Q2/Pn
H2
H1
Q1/Pn
Q1
Q2
(C)
Head
Flow (x100)
P1 + P2
P2
P1
H2
H1
Q1(P1)
Q1(P2)
Q1 Q2
(D)
Head
Flow
P1 or P2
P1 and P2
Constant Speed Pump Flow
Variable Speed Pump Flow
Required System Flow and Head
H1
Q1

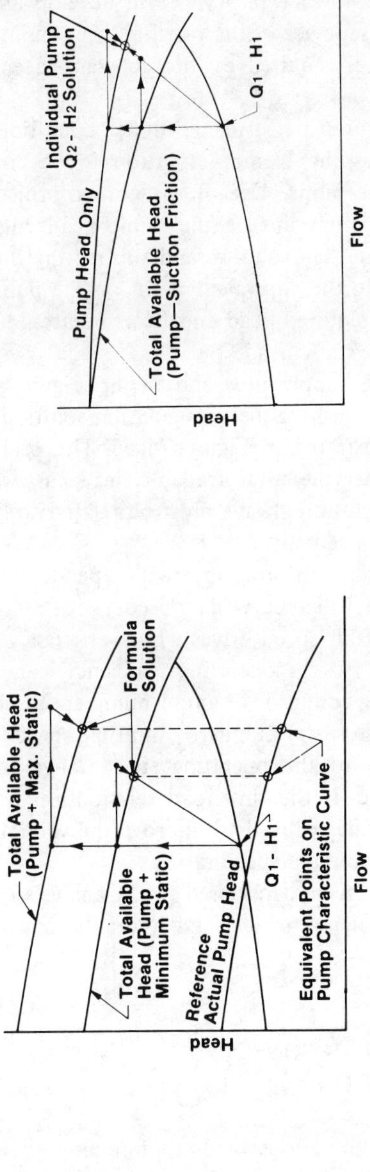

Figure 74. Triangulation procedure to determine actual flow-head solution for pumps.

system flow of Q_1 and head of H_1, the two pumps will develop unequal percentages of the total required flow. As system flow changes, the percentage of flow contribution by each pump will change.

To determine total system power requirements under these circumstances, it is necessary to exercise the basic triangulation procedure for the variable-speed pump based on the known capacity it will develop at the required head of the system. The horsepower of the variable-speed pump is calculated and added to the horsepower required by the constant-speed pump at its developed head and flow.

Projecting variable-speed pump performance under conditions of variable static available head requires the formula solution to be equated to the actual developed head of the pump. Use of the normal projection formula is sufficient, and the equivalent point on the pump head curve will result from projecting the actual Q_2, H_2 solution and subtracting the static head. This will determine accurately the equivalent point on the pump head curve that equates to the system required head and flow. Figure 74(E) illustrates the effect of variable available static head.

When friction head in the supply pipe and fittings significantly reduces pump head, it is necessary to project the triangulation solution to the pump head curve only. This is illustrated in Figure 74(F). The projection values can be computed from either the total available head curve or the pump curve, but the intersection solution always must be related to the equivalent point on the individual pump head curve.

The accuracy of the triangulation procedure will depend to a great extent on the accuracy of pump and system curve data used.

In the employment of variable-speed drives whose efficiency is a function of output to input speed, system operating efficiency can be improved when major system operating conditions require pump speed changes below the nominal speeds available from standard, multiple-speed ac induction motors. Careful comparison of the operating speed range to the system load profile should be made. If the time/load relationship is of sufficient magnitude, motor speed can be reduced to improve the variable-speed drive efficiency by altering output to input speed ratios.

Numerous contingencies must be considered in the analysis of this type of application to ensure that all mechanical and hydraulic considerations are taken into account.

EQUIPMENT POWER COMPARISONS AND TOTAL ENERGY PROJECTION

Accurate comparisons of the various kinds of pumping equipment under consideration for a given application make it necessary to establish a common point of reference. That reference point is the power that the prime mover must supply at its output shaft:

1. *Constant speed*. To project the total system power required for a constant-speed pump, divide the pump horsepower by the prime mover efficiency.
2. *Variable speed*. The horsepower at the output shaft of the prime mover for a pump driven by a variable-speed device is derived by dividing the pump brake horsepower by both the prime mover efficiency and the variable-speed device efficiency. This applies whether the variable-speed device is between the pump and the prime mover or ahead of the prime mover.

Do not be misled by the lower net efficiency of the equipment, which includes the variable-speed device. Via pump operation, which is closely aligned to system needs, significantly valuable operating cost savings can be obtained, particularly in systems that have variable loads. The greater initial cost of the variable-speed equipment often can be amortized in a relatively short time, after which a net reduction in overall system operating costs can be obtained.

To project the total energy requirements for any system with any combination of pumping equipment and to ensure that all answers are projected from a common base, a structured procedure for the analysis should be followed; such a procedure is outlined in Figure 75. Each item in the procedure will have a significant impact on the total energy requirement of a specified pump-equipment combination and system operating procedure.

To make comparisons of dissimilar equipment or to evaluate the impact of system operation change, it is necessary to follow the evaluation procedure for each specified condition. The individual steps of the procedure are defined as follows:

Step 1. At the point of entry to the evaluation procedure, it is necessary to have established and assembled all information relative to the system design requirements. An accurate projection of system flow and head requirements, as illustrated by a system head curve or band, is necessary at this stage of evaluation.

Step 2. The load profile of the system should be reviewed to make preliminary pump equipment selections and to establish the evaluation base for the system under consideration. If future system conditions such as expansion, etc., will significantly alter the load profile, these considerations should be confirmed at this time.

Step 3. Pump selection evaluation. Beyond the normal mechanical and hydraulic criteria associated with pump selection for a specific application, the selection procedure should encompass the primary operating efficiency at the point of selection and the amount of efficiency deviation that will be caused by system flow variation. These values then should be compared to the operating time at specified flowrates to ensure that the selection offers

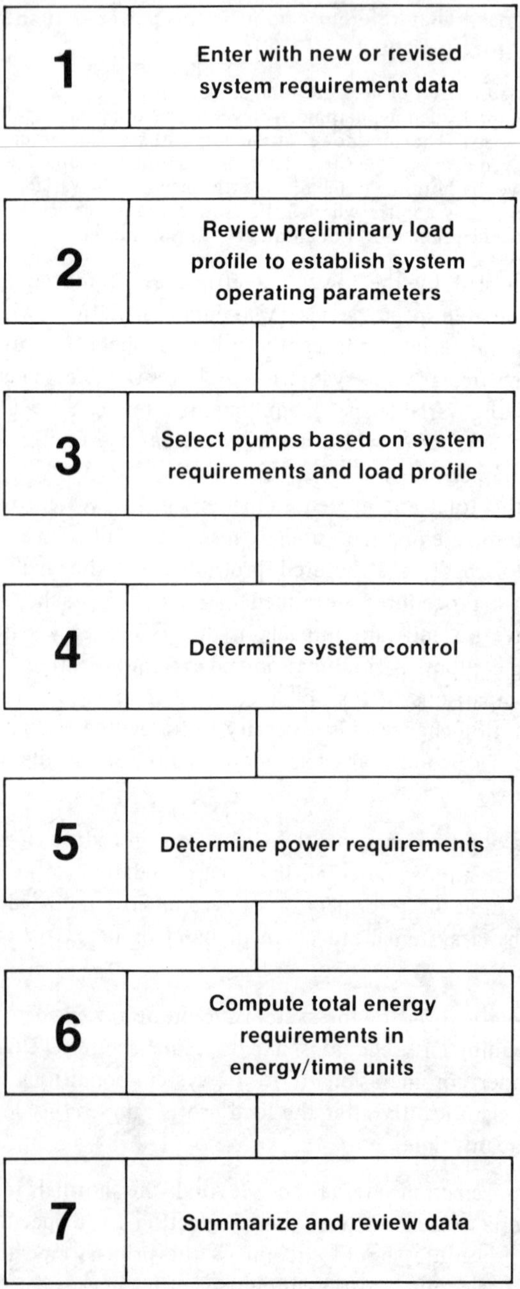

Figure 75. Summary of procedure to project total energy requirements.

the maximum potential operating efficiency through the range of maximum flow/time concentration.

Step 4. Preliminary decisions must be made relative to the control means that will be employed to satisfy system flow and head requirements. The decision to control system required head, total available head, or both, will be a significant factor in the total energy requirements projected for the system being evaluated.

Step 5. This stage of the evaluation requires a projection of the pumping equipment power requirements over the entire system operating flow range. The pump required power can be organized in tabular or graphical form, as best suits the user. In tabular form, the power requirements of the pumping system, compared to required system flow, should be of sufficient quantity to allow the user an analysis of all significant operating areas, as dictated by the flow/time profile.

Step 6. The load profile is used to compute the energy requirements of the selected pumping equipment and system operating procedures. When compared to the power projection in Step 5, the load profile will determine the total energy requirements of the system, thus concluding the evaluation in energy time units (brake horsepower hours, kilowatt hours, etc.).

Step 7. Determination of the total energy required to satisfy all projected system operating conditions is the basis from which all other considerations and comparisons can be made. This is the reviewing and decision-making phase of the total system evaluation procedure. If the energy summary is adequate, the evaluation procedures are concluded; if not, the evaluation can be repeated with new or modified data to determine what effect the change will have.

When the total system energy answer has been determined, normal decision-making processes to select that combination of pumping equipment and system operating procedures that most effectively satisfy the design goals of the system can begin. Total system evaluation ensures that all projections, comparisons and decisions are based on as much factual information as possible.

When comparisons of various combinations of pump equipment and system operating procedures are desired, total system evaluation allows the comparisons to be made on an equal basis—the requirements of the system. Total system evaluation is a structured, fundamental process for determining the pump and system combination that best suits the cost-benefit-functionality requirements of the user.

Figure 76 is a recommended data sheet for recording the information for the energy analysis.

POWER COMPARISON DATA

FROM:
INDIVIDUAL ______________________________

COMPANY ______________________________

ADDRESS ______________________________
(Street) (City) (State) (Zip)

TELEPHONE ( ) ______________________

SYSTEM FLOW

Maximum __________ GPM

Minimum __________ GPM

Liquid __________

Specific Gravity __________

Viscosity __________

SYSTEM HEAD

Constant Pressure

Static __________ (FT)

Delivery __________ (FT)

Friction __________ (FT)

Total __________ (FT)

PUMP SUCTION SUPPLY

☐ Constant

☐ Variable

Minimum __________ (FT)

Maximum __________ (FT)

LOAD PROFILE—ESTIMATED OR MEASURED

GPM									
Hours									

(Fill in as many conditions as applicable.)

APPLICATION, SERVICE OR PROCESS BEING CONSIDERED:

____

____

(Please provide simple piping sketch.)

PRESENT ELECTRICAL COST:

________________ ¢ Per KWH

TYPE OF COMPARISON TO BE MADE
(check one)

☐ Peerless variable speed drives with existing pumps *to* existing constant speed pumps.

☐ Peerless variable speed drives with new Peerless pumps *to* existing constant speed pumps.

☐ Peerless variable speed drives with new Peerless pumps *to* proposed constant speed pumps.

PUMPS PROPOSED FOR/OR EXISTING IN SYSTEM

Quantity ________________

Design Capacity ________________ (GPM EACH)

Design Head ________________ (FT EACH)

Driver ____________ HP ____________ RPM ________

(Please provide performance curve for each pump)

Figure 76. Power comparison data sheet for pump energy comparison evaluation.

CHAPTER 2.6

OSCILLATING DISPLACEMENT PUMPS

Gerhard Vetter

Professor of Engineering
LEWA Herbert Ott GmbH and Co.
Stuttgart, West Germany
American Lewa, Inc.
Natick, Massachusetts

The oscillating displacement pump uses one of the oldest pump principles. Known to have been used in Greece long before the birth of Christ, piston pumps are still among the most extensively applied in a wide variety of technical services.

The present widespread application of the oscillating displacement pump is due to its unique properties. Conspicuous among these is its high efficiency—a characteristic that is particularly valuable in today's era of energy shortages. Although it has been replaced in some services by the less costly centrifugal pump, the oscillating displacement pump is finding many new applications, especially in high-pressure engineering. Here, the basic principle of oscillating displacement is employed in varied designs for such services as processing, power hydraulics and high-pressure purification.

The following is a brief survey of applications and types of oscillating displacement pumps:

1. Drinking water supply
   - Deep-well piston pumps
   - Pressure-booster piston pumps
2. Crude oil production
   - Deep-hole reciprocating piston pumps
   - Hydraulic deep-hole pump systems
   - Drill-hole, piston-type scavenger pumps
3. Power hydraulics
   - Axial piston pumps
   - Radial piston pumps
   - Inline piston pumps

4. Natural gas exploration and chemistry
   Natural gas-driven piston pumps
   Air-driven diaphragm pumps
5. Homogenizing by throttles split friction
   Homogenizing pumps of inline piston-type construction
6. Boiler feed water supply
   Steam-driven piston pumps
   Electric piston pumps
7. Lubrication
   Lubricating pumps for inline piston-type arrangement
8. Heart prostheses
   Oscillating or pulsating blood pumps
9. Fuel injection in combustion engines
   Inline piston pumps
10. Volumetric liquid metering
    Mostly stroke-adjustable diaphragm, piston, bellows and tube-type metering pumps
11. High-pressure pumping of process technology liquids
    Multicylinder piston pumps
    Single- or multicylinder diaphragm pumps

Because of their importance in process technology and to permit detailed treatment of their design and application, only the last two types, i.e., metering pumps and high-pressure discharge pumps, are discussed in this chapter.

Process technology imposes the highest requirements for proper choice of materials because of the different media involved. Toxic, abrasive and corrosive media are to be controlled at high and low temperatures, as well as in the form of suspensions, abrasive slurries and pastes. The pressure ratings encountered range up to several thousand bar.

The differences between metering and discharge (conveying) pumps are dictated by the job to be performed. Metering means conveying, measuring and adjusting the metering flow. That is, compared to discharge pumps, which convey only, two additional tasks have to be completed. As will be shown, this has a strong influence on the technical design of pumps.

OPERATING PRINCIPLES

Kinematics of Displacement

The chronological progression of the stroke motion of frequently used oscillating drive units (Figure 77) more or less corresponds with harmonic motion. The rotary cam-type drive unit exhibits harmonic motion exactly [14,15]. At a not too large rod-ratio with respect to the displacement speed, the direct-thrust, crank-type drive unit hardly deviates from harmonic motion; however, it markedly deviates from it in regard to acceleration values. The pulsation of the single cylinder ($i = 1, i = 2$) can be smoothed out

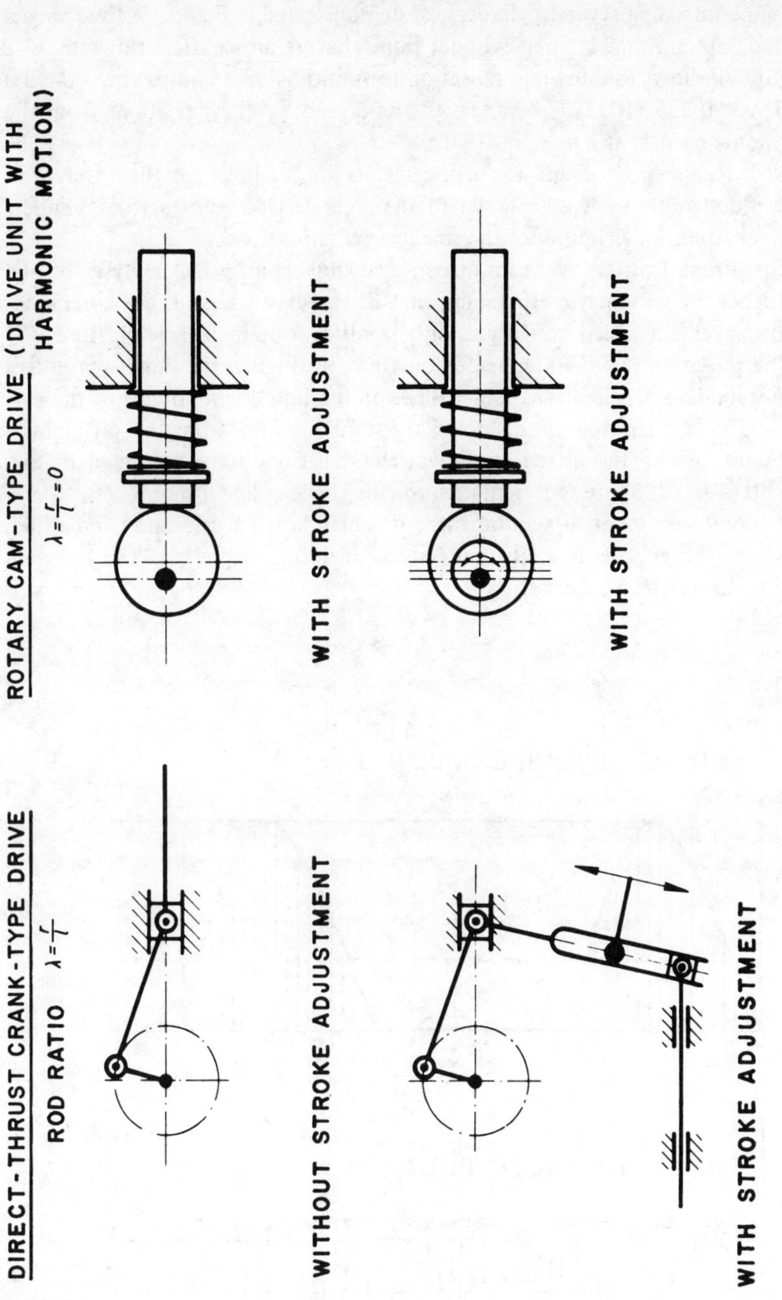

Figure 77. Frequently used oscillating drive units.

by superimposing several cylinders, as demonstrated in Figure 78 for a single-cylinder pump and a three-cylinder pump having an identical flowrate, Q_m. Figure 79 shows the flowrate pulsation for multicylinder pumps at $\lambda = 0$ (LH) and $\lambda = 0.225$ (RH). The variances are caused by the deviations from the harmonic motion at $\lambda = 0$.

With regard to i = 5 and i = 6, the results are identical for the different λ. The fluctuation of Q_m is shown in the figures. Odd numbers of cylinders (triplex, quintuplex) produce the smallest pulsations.

For simplification, we can dispense at this point with analysis of the influence by volumetric efficiency; that is, the elasticities on the kinematics of displacement which, roughly, is only significant up to average pressures.

The pulsation in flowrate, as demonstrated in Figures 78 and 79, produces corresponding accelerations and delays of the liquid compounds in the system. The acceleration or delay of the liquid results from the differential quotient of the instantaneous flowrate, Q(t). Conditions as shown in Figure 80 ($\lambda = 0.225$) are representative for the single-cylinder pump.

At commencement of suction there is a high initial acceleration and a final delay at the end, as is easily understood in terms of physics. The opposite is true during the pressure stroke.

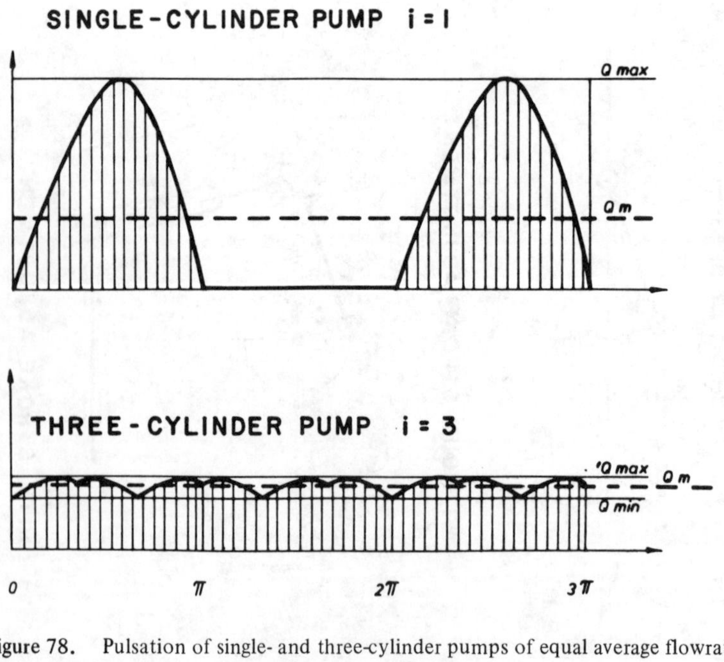

Figure 78. Pulsation of single- and three-cylinder pumps of equal average flowrate.

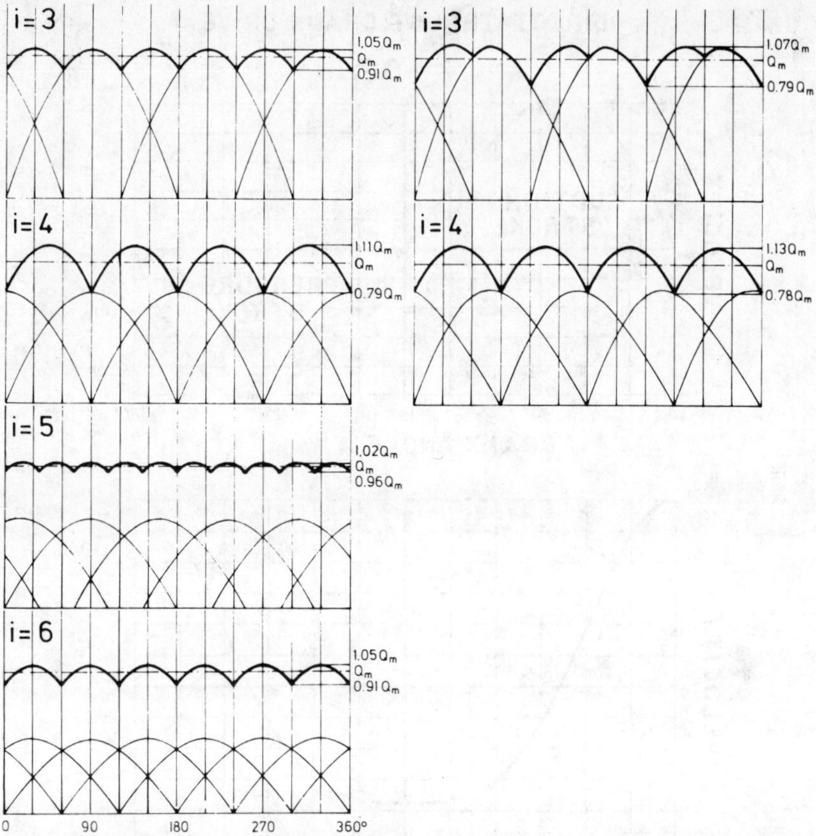

Figure 79. Pulsation of the flowrate in multicylinder pumps.

Flowrate Characteristics

The average rate of flow is approximated as follows [16,17]:

$$Q_n = n \times h \times F_K \times \rho \times \left(1 - C \frac{\rho}{h/h_{100}} \times \eta_G\right) \qquad (106)$$

where n = stroke frequency
h = stroke length for pumps with constant stroke length is $h/h_{100} = 1.0$
F_K = cross section of piston
ρ = density

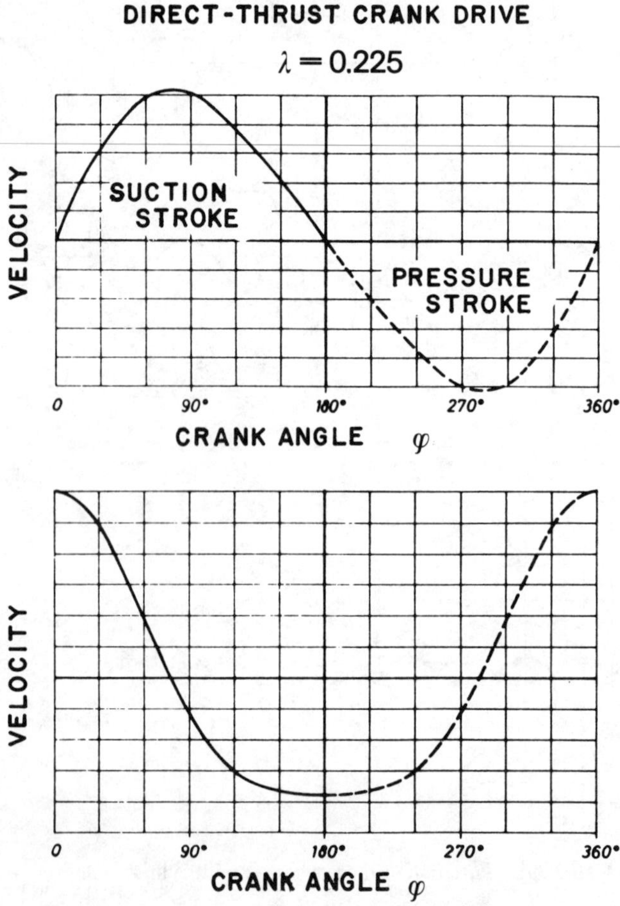

Figure 80. Velocity and acceleration of a single-cylinder pump.

The individual variables can be summarized into three factors:

$$Q_m = Q_{mth} \times \eta_E \times \eta_G \tag{107}$$

1. The theoretical flowrate, Q_{mth}, includes the rating variables of the pump and can be approached only under ideal conditions.
2. The extent of elasticity, η_E, takes into consideration the elasticity influence of the pumped medium and of the pump chamber, which depends on pressure and length of stroke. The constant, C, is dependent on the relative compressibility of the liquid, $\Delta V/\Delta O$, and on the elasticity constant of the pump chamber [17].
3. The internal and external leakage losses, which are also influenced by the viscosity, are included in the quality level, η_G.

Elasticity level, η_E, and quality level, η_G, combine to form volumetric efficiency, $\eta_F = \eta_E \times \eta_G$.

Usual values for water at average performance for η_F are as follows:

Piston pumps	up to 100 bar	95–98%
	up to 500 bar	93–96%
Diaphragm pumps	up to 100 bar	85–95%
	up to 500 bar	80–90%

The effect of elasticity on the displacement process is shown in Figures 81a and 81b. It is easy to find one's way through the indicator diagram by using the explanations (Figures 81b and 81c). Note that condition $Q_m = 0$ occurs when the parenthetical expression of Equation 106 becomes zero, which is the case in the limiting stroke length:

$$h_0 = C \times \rho \times h_{100}$$

This is of extreme importance for stroke-adjustable metering pumps because an adjustment like $h < h_0$ no longer has any influence on the flowrate. The relation $Q_m = f(h)$ is, admittedly, linear as per Equation 106. However, it bisects the axis of abscisses offset by $h = h_0$; thus, the flowrate depends on the stroke length (Figure 81c) in a linear, but not proportional, mode.

Figure 82 provides a collection of measured results for pumps of very different sizes. As shown by this figure, there is a substantial increase of h_0, that is, the parallel shift to the right at small stroke volume and high pressure (for instance, line f). With regard to stroke-adjustable metering pumps, it is particularly necessary that the characteristic form, $h_0 1$, be very acutely observed because of the available range of adjustment. At lower pressures and larger stroke volumes $h_0 = 0$, and the Q(h) line almost runs through the zero point.

The influence of the pumped medium on the Q(h) characteristic is great because of the very different elasticity rate of the liquid (Figure 83). Particular attention must be paid to the liquified gases whose elasticity rate is both large and temperature-dependent. With small stroke volumes, the relative clearance volume is rather large, which increases the elasticity effect.

For different pressures, the $Q(h,\rho)$ performance characteristic of a small oscillating metering pump is represented in Figure 84. The lines of resultant pressure run in a parallel mode in a variant displacement motion, $h_0(\rho)$.

At low pressure the parallel shift (Figure 82) is infinitely small, so the pressure influence can almost be neglected. How pressure-rigid the flowrate pressure characteristic can quantitatively be in oscillating displacement pumps in comparison to other pump types is shown in Figure 85. This is the reason why oscillating displacement pumps are used so successfully as metering pumps.

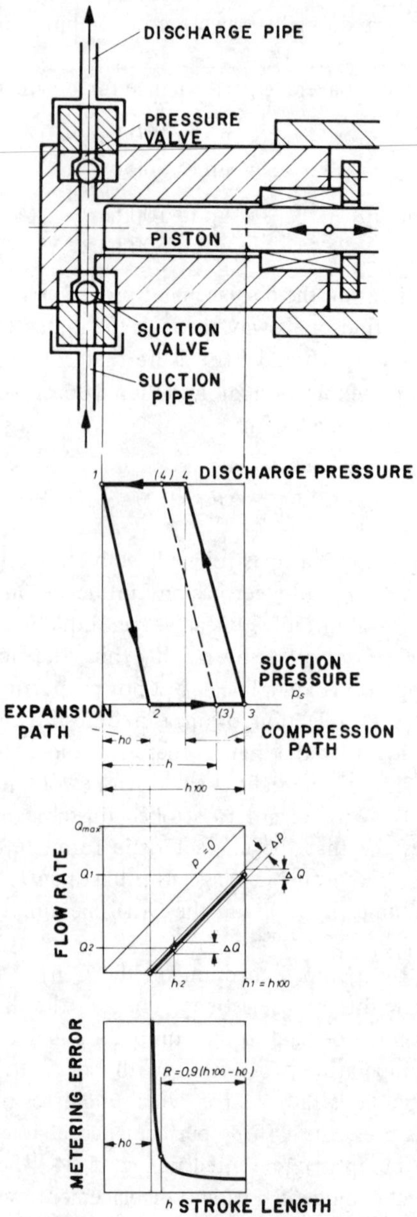

Figure 81. Displacement action and its impact on characteristics. Legend: 1 = pressure valve closes; 1-2 = release from discharge pressure to suction pressure; 2 = suction valve opens; 2-3 = suctioning of pumped liquid; 3 = suction valve closes; 3-4 = compression from suction pressure to discharge pressure; 4 = pressure valve opens; 4-1 = displacement into pressure pipe.

Curve	Symbol	Pump Type	Flowrate (liter/hr) $h/h_{100} = 1.0$	Outlet Pressure (bar)
a	●	MH	2000	5.0
b	○	MH	500	30
c	□	MH	100	300
d	■	MH	100	600
e	▽	MH	15	500
f	▼	MH	2	2000
g	x	FM	500	3.5
h	△	MM	50	5.0
i	▲	K	70	1000

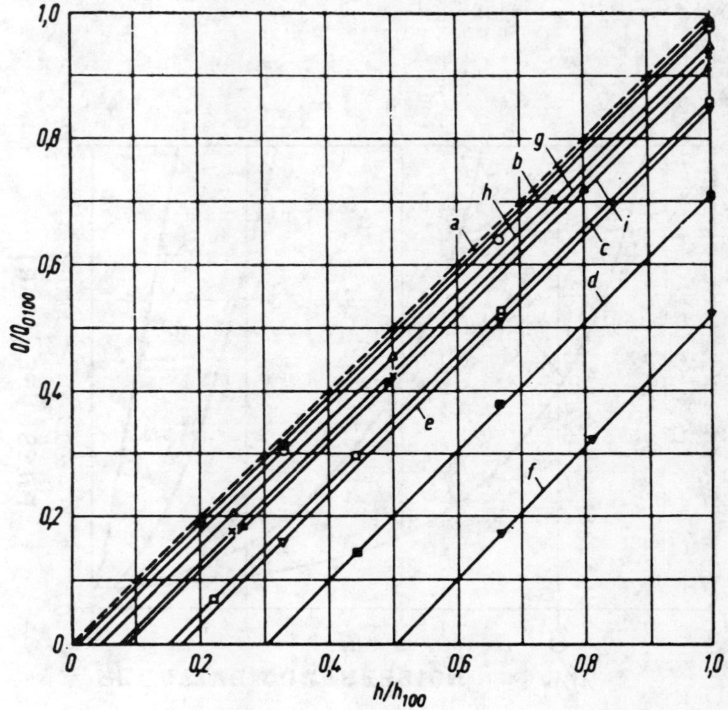

Figure 82. Flowrate dependent on stroke length in nondimensional plotting. Pump types: MH = diaphragm pump head with hydraulic pump drive; FM = bellows-type pump head with mechanical bellows drive; MM = diaphragm pump head with mechanical diaphragm drive; K = piston pump head.

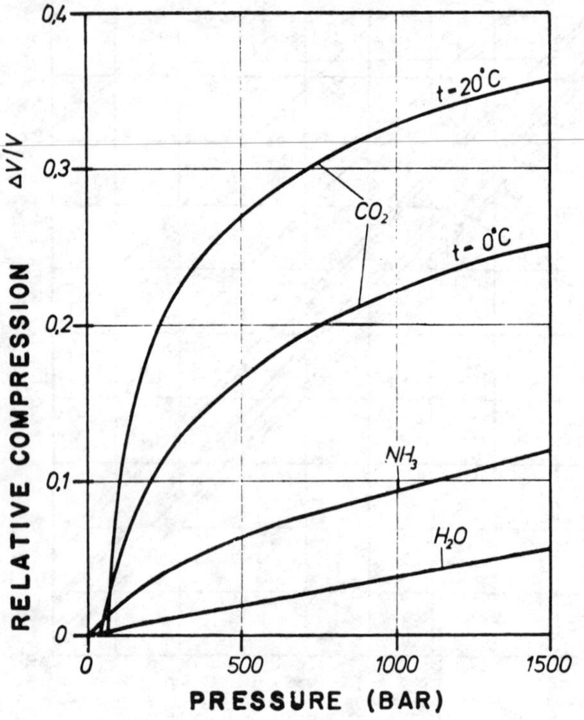

Figure 83. Relative compressibility of liquids dependent on pressure.

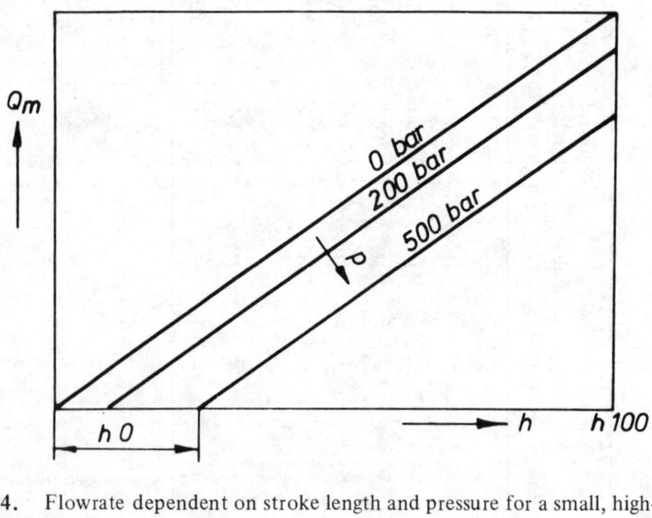

Figure 84. Flowrate dependent on stroke length and pressure for a small, high-pressure metering pump.

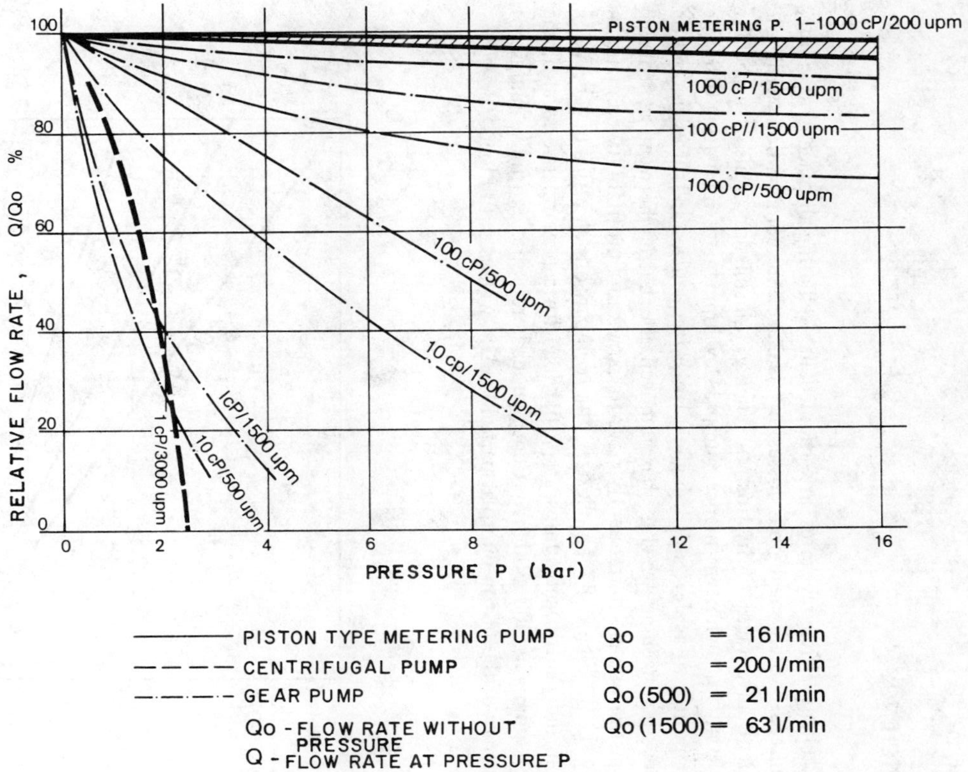

Figure 85. Flowrate versus pressure characteristics for different pump types (gear, centrifugal, oscillating piston).

According to Equation 106, the relation Q = f(n) is linear and proportional, as is demonstrated in Figure 86 at different stroke adjustment, h/h_{100}. The range of adjustment can, therefore, be increased by adjusting stroke *frequency* when it is no longer possible to adjust stroke *length*. The reader should also refer to the literature [18-21] for discussions and data on oscillating displacement pumps and volumetric metering.

EFFICIENCY

Excellent efficiency rates are attained in oscillating displacement pumps because of their negligible losses by leakage, in combination with the inherent advantages of hydrostatic energy transmission. This is especially important for high-performance pumps. Development of oscillating displacement pumps to achieve much higher stroke frequencies and, thus, reduction of specific machine weights, has removed the barrier between centrifugal and piston pumps. It frequently pays to carry out a profitability comparison study [22]. Figure 87 emphasizes this comparison of efficiency between piston and centrifugal pumps at partial load for a performance level of the piston pump of 200–300 kW. Conspicuous here is the enormous margin of efficiency of the piston pump at partial load, as well as at full load.

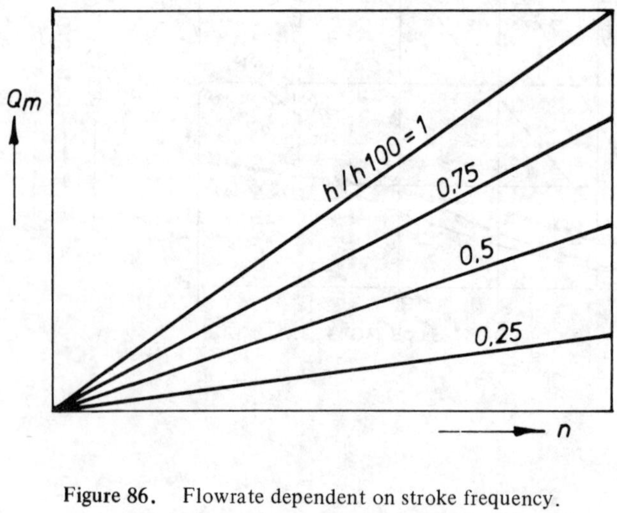

Figure 86. Flowrate dependent on stroke frequency.

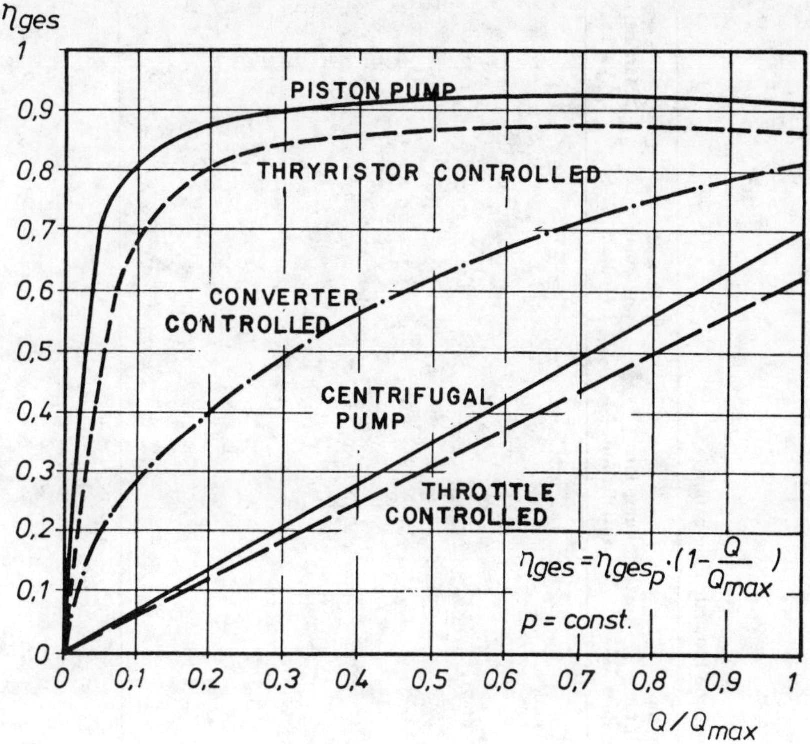

Figure 87. Efficiency comparison between oscillating displacement and centrifugal pumps [22].

Metering Error

Analyzing the variables that influence metering error is of great importance in determining the application of metering pumps. Table 30 shows which influences affect the individual factors of Equation 106 and, thus, can vary the flowrate. Once the average erratic value of all individual variables in Equation 106

$$\pm\Delta A_K, \pm\Delta\rho, \pm\Delta n, \pm\Delta h, \pm\Delta p, \pm\Delta\eta_G$$

is known, the mean mass flow fluctuation rate then can be determined in accordance with the law of error propagation:

Table 30. Influences on Flowrate Fluctuation

$$Q_m = A_K \cdot \rho \cdot n \cdot h \left(1 - C \frac{P}{h/h_{100}}\right) \cdot \eta_G$$

Analytical Connection	Influences on the Fluctuation of the Individual Variables	Pump Plus Installation	Caused by Pump Medium	Marginal Value
η_G	Viscosity, stroke frequency, pressure, temperature	X	X	X
h	Adjustment accuracy, load (delivery pressure)	X		X
n	Type of drive, load (delivery pressure)	X		X
ρ	Pressure, temperature, gas content		X	X
A_K	Production allowances, wear and tear	X	X	

$$\frac{\Delta Q}{Qm} = \sqrt{\left(\frac{\Delta F_K}{F_K}\right)^2 + \left(\frac{\Delta \rho}{\rho}\right)^2 + \left(\frac{\Delta \eta_G}{\eta_G}\right)^2 + \left(\frac{\Delta h}{h}\frac{1}{1 - C\frac{p}{h/h_{100}}}\right)^2}$$

$$+ \sqrt{\left(\frac{\Delta p}{p}\frac{C\frac{p}{h/h_{100}}}{1 - C\frac{p}{h/h_{100}}}\right)^2} \quad (108)$$

If it is further supposed that the erratic value of all individual variables corresponds to the Gaussian distribution theorem (bell curve)–this always prevails whenever chance has a hand in it–99.7% of the measured values, in other words, almost all of them, are within the triple mean error [19].

The metering error is defined as the triple mean error of the mass flow in relation to its mean value:

$$S_T = 3\frac{\Delta Q}{Q_m} \quad (109)$$

Figure 88 demonstrates what the metering error is composed of as per Equations 108 and 109. In comparison, Figure 89 represents the metering error measured for a piston pump; close attention was paid to maintaining nearly as much as possible constant marginal values [18]. The graph shows that a stroke-adjustment ratio of 1:10 still produces good metering accuracy. Some important features of volumetric metering pumps can be read from Equation 108 and from the plotted characteristics:

The smaller the stroke length, h, the larger the metering error becomes. At $h = h_0 = Cph100$ (see Figure 81d), it heads toward infinity. Therefore, it is not possible to operate a metering pump at high metering accuracy over a broad stroke-adjustment range at random (perhaps 1:100 or even 1:1000). The solution to the problem must be adjustment through stroke length and stroke frequency of distribution of the flowrate range among several individual pumps. To maintain a high metering accuracy across as large a flowrate range as possible, it is necessary to aim at the following targets:

- smallest dead spaces independent of stroke length
- working spaces of small elasticity
- constant marginal values

In most metering pumps, and in piston pumps in particular, the elasticity value, C, is small. According to Equation 108, the pressure fluctuations within a broad stroke length range have only a very weak effect on the metering error. This provides the explanation for one of the known properties

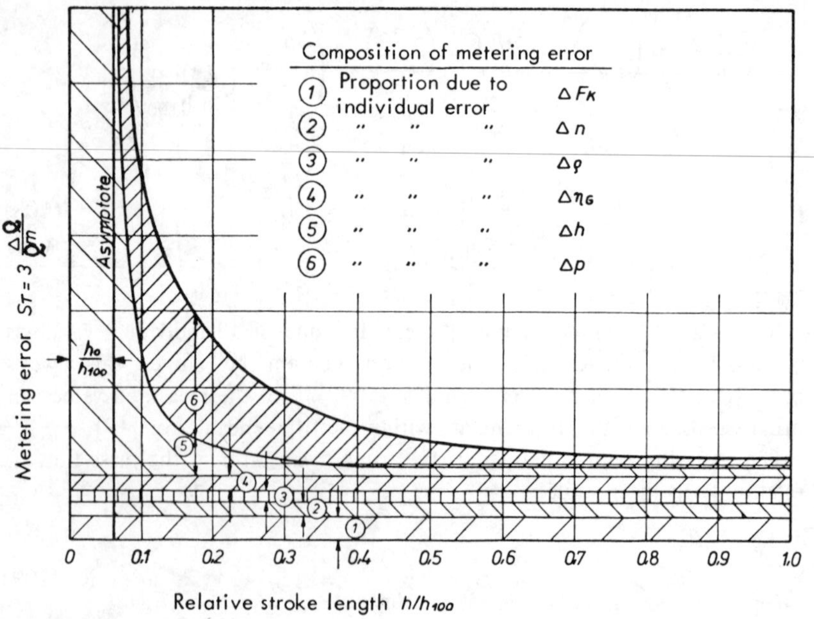

Figure 88. Composition of metering error.

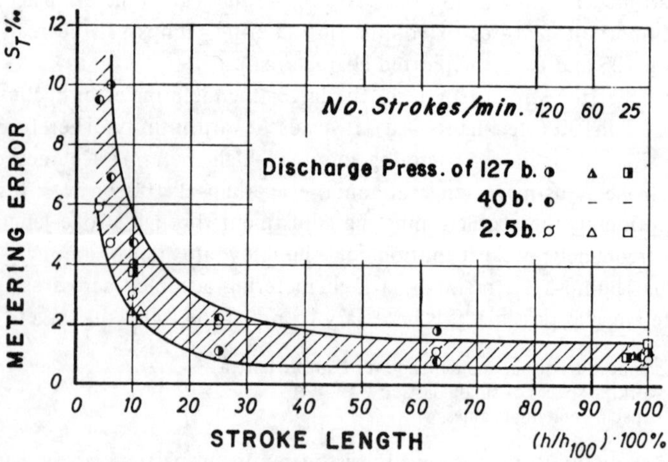

Figure 89. Measured metering error dependent on stroke adjustment, h/h_{100}, counter-pressure and stroke frequency.

of the volumetric metering pump, namely, its metering accuracy nearly independent of pressure fluctuations.

EFFECT OF INFLUENCING VARIABLES ON FLOWRATE AND METERING ERROR

Density

Oscillating pumps operate on a volumetric basis. The flow of the material conveyed is directly influenced by its density, ρ, which is dependent on pressure and temperature (Figure 90). In liquified gases in particular, this

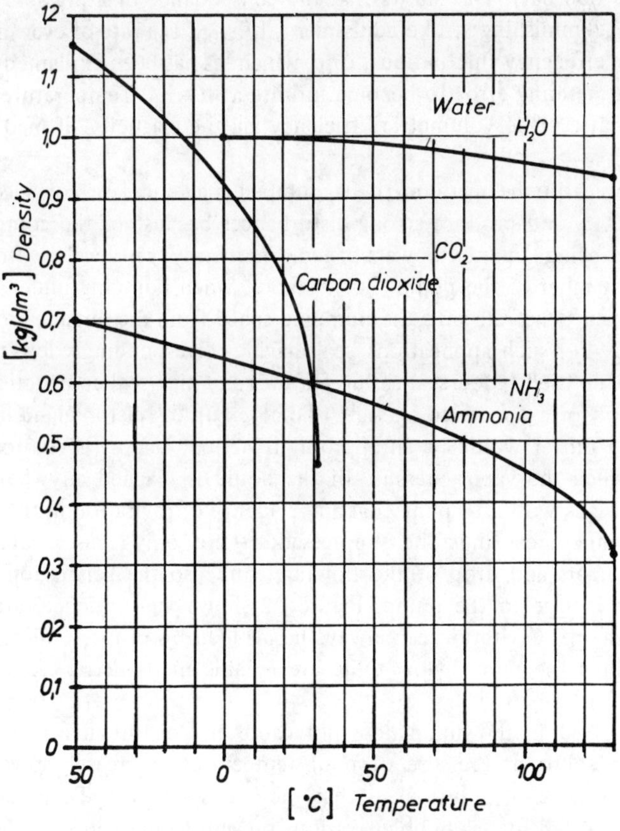

Figure 90. Specific gravity of some liquids.

dependence is considerable so that, for instance, an accurate mass metering operation requires constant levels of pressure and temperature. The metering error due to density variation of the liquid is reproducible at constant marginal values (pressure and temperature). Gas bubbles in the pumped liquid vary the average density to the degree that the gas portion changes. The only measures possible to eliminate the density influence are the constant stabilization of the density or the use of a density-dependent control system.

Pressure and Elasticities

The effect produced in accordance with Equation 106 has already been described. In Figure 91, taking a small high-pressure diaphragm pump as an example, the performance characteristic "volumetric efficiency/discharge pressure" is explained in a nondimensional plotting (maximum discharge-pressure is 600 bar). This shows that the consequence of a pressure fluctuation, ΔP_{at}, diminishing stroke adjustment, h/h_{100}, is a rate of ever increasing volumetric efficiency fluctuations, $\Delta\eta_F$, which, as has been explained already, causes the metering error to become infinite at $h = h_0$. Temperature fluctuations also affect the volumetric efficiency via the elasticity of medium (see Figure 83).

Gas bubbles in the medium to be pumped may increase considerably its total elasticity and produce strong disturbances because of the compressibility of gases being larger by powers of ten. Frequently, gas develops first in the working chamber of the pump. For instance, when pumping under vacuum, gas or air can penetrate into the pumping space from the outside around the sealed area. For such applications, seals having correspondingly lower leakage rates must be used (good seal-casing features and high seal compaction). (See Chapter 2.7 for a discussion on seals.) To block up the piston gland by means of a liquid buffer would be an appropriate action. Vapor (cavitation) starts to form when the vapor pressure of the liquid is reached anywhere in the system. It goes without saying that any evidence of cavitation in the working and hydraulic areas must be suppressed. It not only causes catastrophic metering errors and drop of the flowrate, but also the generation of loud noise and damage to the pump. Figure 92 shows the influence rate on the cylinder charge. Cavitation can always be avoided by properly dimensioning and installing the pump. Now there are reliable methods of calculation in existence [20].

Gas dissolved in the liquid does not cause any disturbance as long as it remains in solution. Pressure drop or temperature increase, however, can shift the balance of solution so that the gas is released. To preclude disturbances created by gas bubbles, various preventive measures can be taken:

1. Keep gas bubbles away from the pump by appropriately sized piping or by superimposing a degassing reservoir.

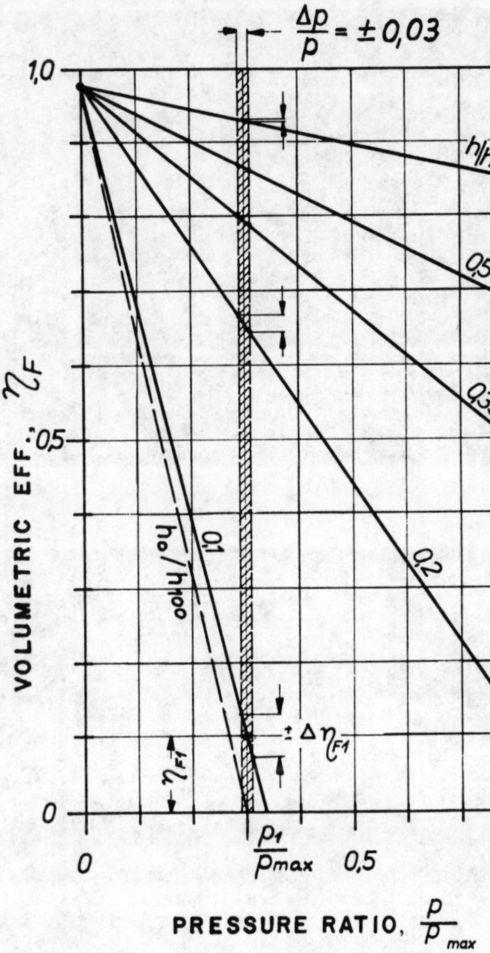

Figure 91. Volumetric efficiency dependent on pressure and stroke length for a small diaphragm pump.

2. Avoid conditions that cause or promote the formation of gas in the pump housing by a correct selection of the variable parameters during layout.
3. Prevent the settling of gas bubbles by optimally designing the working chamber of the pump in a flow-technical respect. This step is particularly important for metering the smallest flowrates because the long dwell time of the liquid in the pump cylinder promotes gas evaporation. Vetter et al. [17] discuss the influence of elasticity on the walls of the working chamber of the pump.

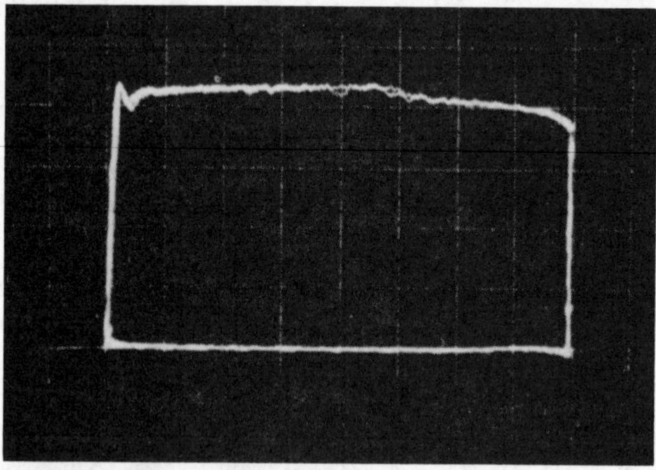

(A)

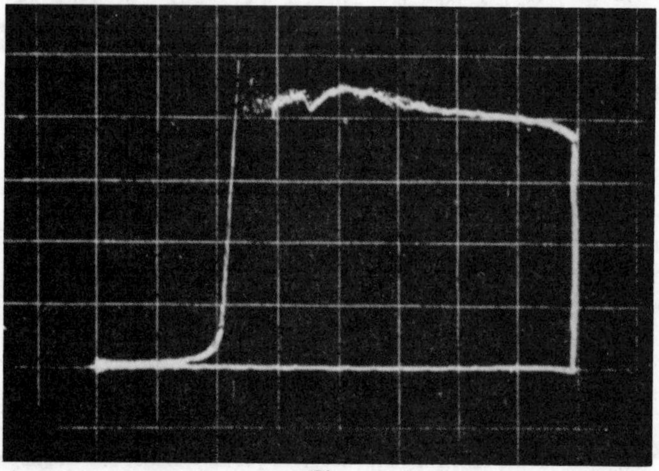

(B)

Figure 92. Cavitation in indicator diagram: (A) no cavitation; (B) insufficient cylinder charge due to strong cavitation.

Volumetric Efficiency η_G

The volumetric efficiency takes leakages and reflux losses into account. In accordance with Table 30, this efficiency is dependent on viscosity (therefore, also on temperature), stroke frequency and pressure.

Leakage Losses in Moving Glands

In extreme cases, leakage flow in piston glands can lead to a total stoppage of pumping (for instance, if there is no maintenance over an extended period of time). If the unit is subjected to competent maintenance and proper rating, the leakage flow can, however, be kept to a minimum so that its influence on the flowrate can be neglected. Where media are involved that are hard to seal, the problem, more often than not, is solved by the use of glandless pumps.

Reflux Losses of Valves

If liquid refluxes into the working chamber of the pump through the pressure valve or into the suction pipe through the suction valve, the conveying performance of the pump is diminished; this may result in considerable metering errors. There are two essential requirements for proper functioning of the valves:

1. The moving valve body (ball cone, disc) must close tightly during the period of time anticipated for this purpose:

Valve leakages are caused by wear and tear, corrosion, or jamming solid particles between guide and seat areas. By means of suitable materials (hardness, corrosion resistance), or by using suspension valves (Figure 93) specially designed for polluted media, these causes of error almost always can be eliminated, or at least reduced to tolerable conditions. Due to the small flow speed in contaminated media or suspensions, there is the increased risk in smaller stroke lengths for solid particles to settle in the valve and cause disturbances. The higher the stroke frequency of the pump, the smaller the

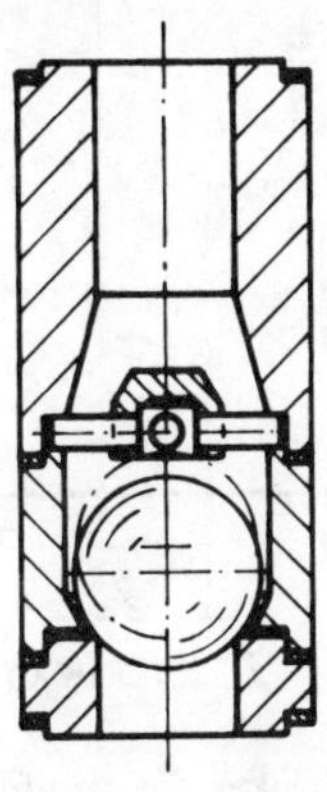

Figure 93. Suspension valve.

reflux losses at the valve-seal areas. In this respect, a smaller high-speed pump is superior to a larger one of lower stroke frequency.

2. The valve must open or close at the proper timing:

The influence of the valve kinematics on flowrate and metering error is hard to understand and has not yet been clarified entirely. Whereas the opening timing plays a subordinate role–it is obtained by force by the displacer via the hydraulic coupling–the closing timing and the timed progression of the valve-stroke curve have a distinct effect in the proximity of the closing point (see Figures 13 and 14 and Chapter 2.3 for different types of valves).

INFLUENCES ON VALVE KINEMATICS

Figure 94 shows some stroke curves for valves that close without delay, that is, they close at the timing scheduled. The lower the stroke frequency and the

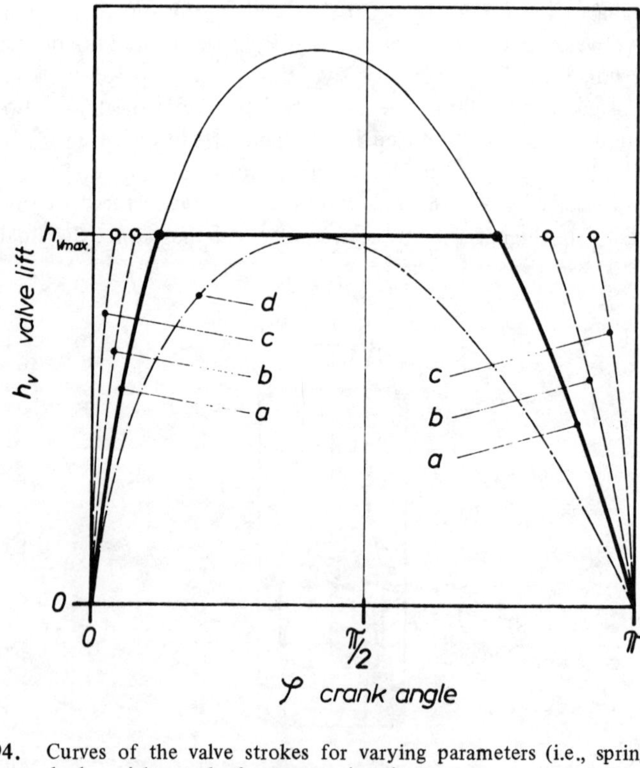

Figure 94. Curves of the valve strokes for varying parameters (i.e., spring loading, obturator body weight, stroke frequency, viscosity).

viscosity of the medium and the heavier the weight of the obturator (closing body) and its spring load, the flatter the curve's progression.

In essence, the closing delay of a valve with fixed geometry depends on the influence rates as shown in Table 31. The same tendency as in Table 31 is also shown in Figure 95, which represents the relationship determined from measurements between closing angle, ζ = wt (without delay means = 180°), viscosity and spring load at average stroke frequency.

A large closing delay leads with certainty to large reflux losses and, thus, to a smaller flowrate and metering errors. It is a question whether the pump has been properly rated to be able to keep the closing delay time so small that it no longer has any appreciable influence. There is no problem with the design rating where low-viscosity media are involved. Highly viscous media require a careful synchronization among the freely selectable variables such

Table 31. Influence Rates on Closing Delay

Closing Delay	High	Low
Viscosity	High	Low
Stroke Frequency	High	Low
Weight of Closing Body	Low	High
Spring Load	Low	High

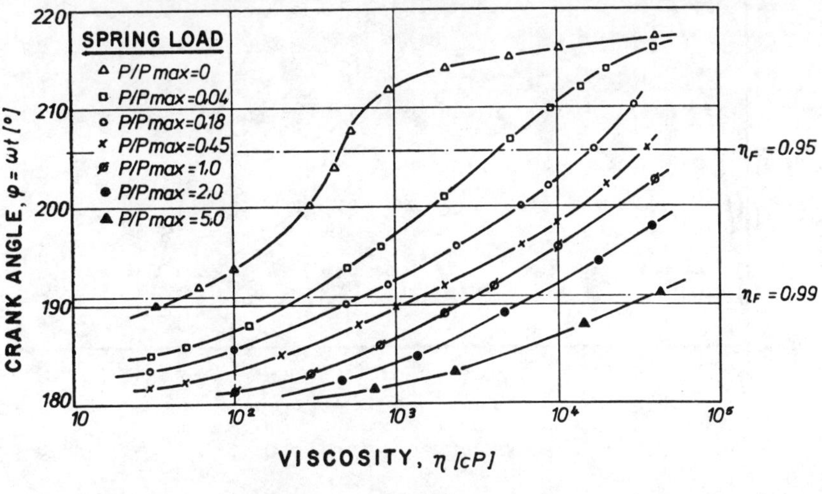

Figure 95. Closing delay of valve dependent on viscosity and spring load.

as stroke frequency and spring loading, whereby the layout latitude is narrowed by the size of the pump (low stroke frequencies result in larger and, thus, more expensive pumps) and by the risk of cavitation during the suction stroke (high spring-load rates generate high entry-pressure losses). Figure 96 represents a stroke curve with closing delay. A heavily delayed closing action also can be seen in the indicator diagram.

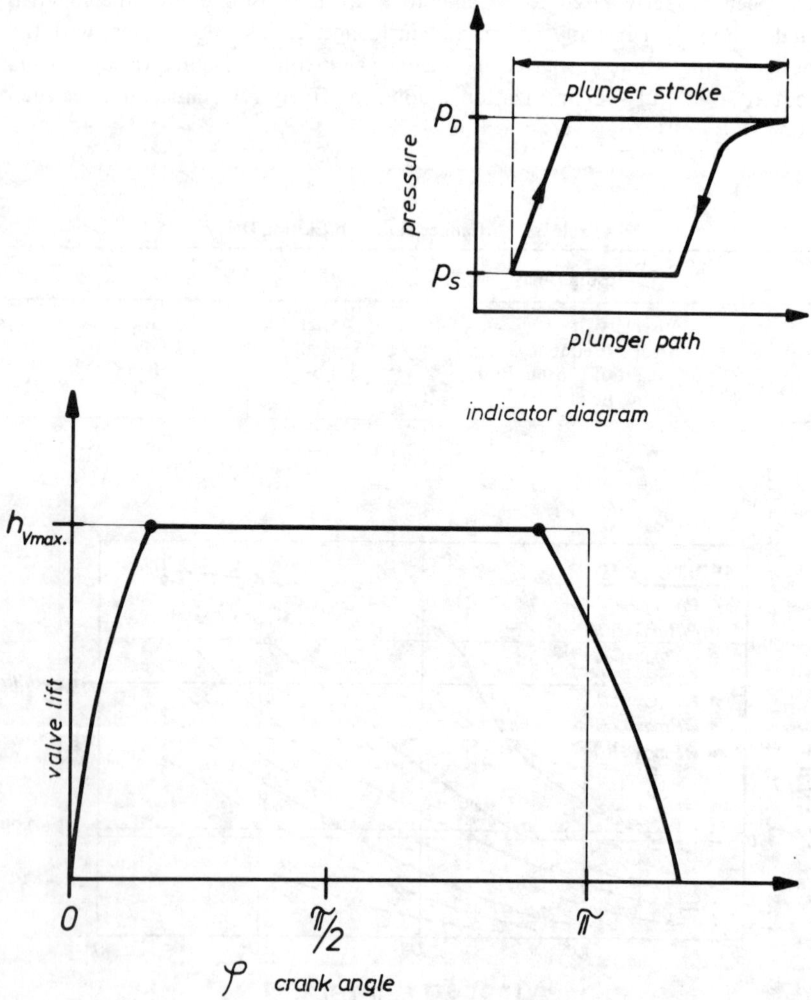

Figure 96. Closing delay of valve (indicator diagram and stroke curve).

SUCTION ACTION, ENTRY PRESSURE LOSS, CAVITATION

The objective in examining the suction action is to determine the pressure progression in the pump's working chamber. Figure 97 is a schematic of a pump head and Figure 98 describes this suction action with a piston pump head having a spring-loaded ball valve as an example, assuming a low operating pressure p_{DF} of the pump. Obviously, conditions for other compression ratios are quantitatively different. However, in principle, the described presentation is applicable to all cases. When considering pressure loss, it is appropriate to examine separately the opening phase, p_{E1}, and the flow phase, p_{E2}. For consideration of the opening phase, the piston is in the upper dead position (O.T. in Figure 97). Let us assume that the working space pressure (see Figure 98, point a) has the value p_{DF}. Due to the elasticities of liquid and working space walls, as well as leakage of the pressure valve, the return expansion curve, b, does not progress in a vertical direction, but does in an inclined mode. In order to open, the suction valve first requires equilibrium of the static forces at the moving part (for instance, the ball).

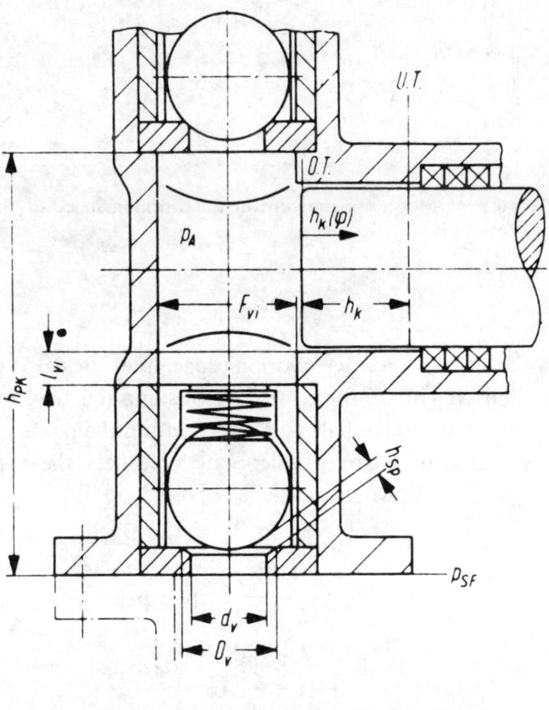

Figure 97. Schematic diagram of a pump head.

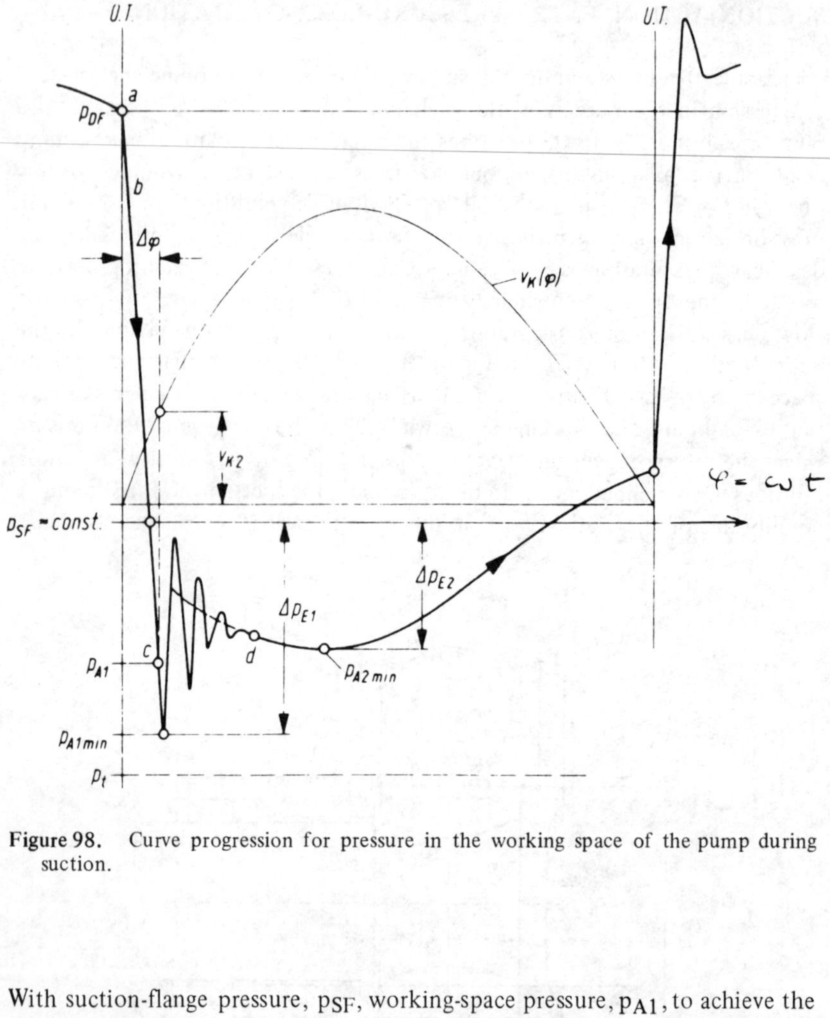

Figure 98. Curve progression for pressure in the working space of the pump during suction.

With suction-flange pressure, p_{SF}, working-space pressure, p_{A1}, to achieve the static equilibrium weight of forces and the assumption that vapor pressure, p_t, is reached when the valve ball pulls away within the phase of sealing (this assumption is easily understood when one considers the rapidity of the opening process), the following equations are obtained:

$$p_{A1}\frac{D_v^2\pi}{4} + G + p_F = \frac{D_v^2 - d_v^2}{4}\pi p_t + p_{SF}\frac{d_v^2\pi}{4} \tag{110}$$

$$p_{A1} = p_{SF}\frac{d_v^2}{D_v^2} + \left(1 - \frac{d_v^2}{D_v^2}\right)p_t - \frac{G + p_F}{D_v^2\pi} \tag{111}$$

Equation 111 can be simplified for the frequent special case, $p_t - p_{SF}$, as follows:

$$p_{A1} = p_{SF} \left(\frac{d_v}{D_v}\right)^2 - \frac{4(G + p_F)}{D_v^2 \pi} \tag{112}$$

The pressure drop, $p_{sf} - p_{A1}$, arises due to gravity, G, and resilience, p_F, less buoyancy and adhesive force by a finite width of the valve seat. To begin with, the pressure drops to p_{A1} in the working space, with the total elasticity of the working space permitting one finite piston motion with the suction valve still closed. Then comes the time of pulling away (see Figure 98, point c). By means of the finite piston speed, v_{K2}, the liquid masses are suddenly hydraulically coupled to the piston, causing movement at the suction side. Depending on conditions such as size of mass, size of delay angle, etc., an additional pressure drop to $p_{A1\,min}$ will develop although the adhesive force collapses and becomes zero. Due to the liquid now quickly flowing into the relaxed space, the suction valve is fully opened:

$$p_{A1\,min} = p_{A1} - \Delta p_{m\infty} \tag{113}$$

The pressure drop, $\Delta p_{m\infty}$, is that portion created by jerky acceleration of the valve ball and by the total liquid beyond the pressure decrease on the suction side due to the adhesive force. After the suction valve has opened and the fadeout process has terminated (see point d in Figure 98), the flow phase will commence during which the pressure in the pump's working space is determined almost solely by the variations of friction and mass pressure of the quasistationary flow in the valve:

$$p_{A2}(\varphi) = p_{SF}(\varphi) - \Delta p_{E2r}(\varphi) - h_{pk} \cdot \varphi \cdot g > p_t \tag{114}$$

where $\Delta p_{E2m}(\varphi)$ = mass pressure during flow phase
$\Delta p_{E2r}(\varphi)$ = friction pressure during flow phase
h_{pk} = geodetic extension of pump's working space (see Figure 97)

$$p_{A2}(\varphi) = p_{SF}(\varphi) - \Delta p_{E2}(\varphi) > p_t \tag{115}$$

Amplitude and duration of the oscillations between opening and flow phases depend on the prevalent damping rate, natural frequency of elasticity and mass of the oscillating system. It follows from the explanations about the pressure drop at the entry to the pump head that to determine the pressure progression during the suction action, both phases will have to be observed. If a test is made for cavitation, the lowest absolute pressure can arise in the pump's working space during one or the other phase, depending on conditions, and thus cause cavitation. It is known from measurements [20] that for determination of the cavitation limit in oscillating displacement pumps only the flow phase will have to be observed. The opening phase and the minimum pressure, $p_{A1\,min}$, arising therein are dominating so briefly that possibly occurring vapor bubbles disappear again by the end of the suction stroke.

Furthermore, we know from the measurements mentioned that the entry pressure loss, $\Delta p_E = \Delta p_{E2}$, can be determined from static pressure drop measurements. To this end, one determines the drag coefficient, ξ (e.g., see Figure 99, f, ball valve) in dependence on the Reynolds number and from it one determines p_E [20], which describes in detail the formation of bubbles under various cavitation conditions. It is only in the so-called "remaining cavitation," where vapor bubbles do still exist at the end of the suction stroke, that one finds indicator diagrams similar to those shown in Figure 92(B). Disturbing to the pumping function is the beginning remaining cavitation for which the disturbance can be determined by means of the

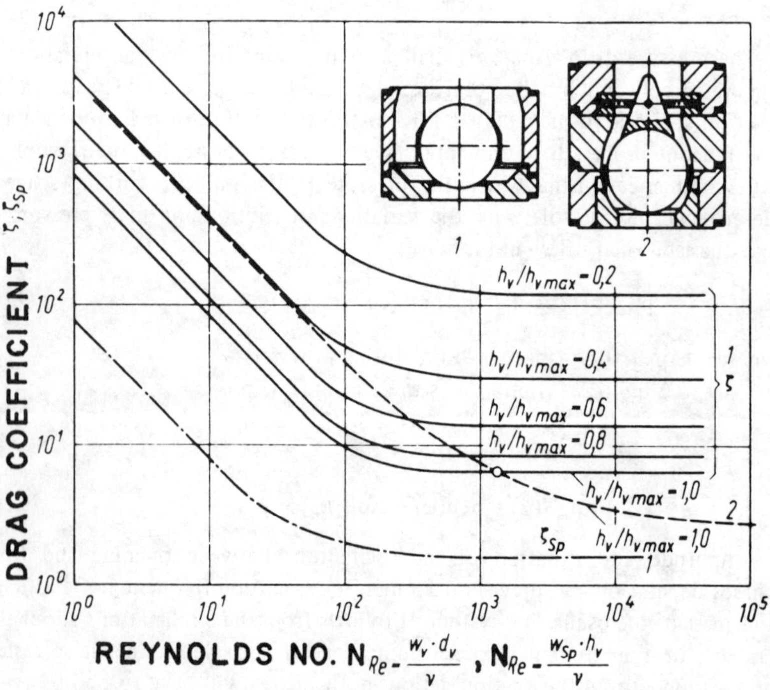

Figure 99. Drag coefficient $\xi = f(N_{Re}, h_v)$ for ball valves of different geometry (1 = standard valve; 2 = valve with nearly constant-flow cross section).

indicator diagram. However, material damage or overload jolts already happen at an earlier point of time when, for instance, vapor bubbles occur during the stroke action, but will have regressed, partially under heavy jolts, by the end of stroke.

DESIGN OF OSCILLATING DISPLACEMENT PUMPS

To highlight the differences of design, metering pumps and conveying pumps will be treated on a separate basis, although there are, of course, flowing transitional intergrades. We can characterize the essential differences as follows:

Metering Pumps

The tasks to be performed by metering include such functions as pumping, measuring and adjusting. Thus, a metering pump is a combination of working machine (pump), measuring device and regulator [21]. Primarily, metering pumps must be rated for their metering function; therefore the stroke volume of the displacer must be as consistently accurate as possible. Exact valve operation, smallest elasticities and minimum leakages are important. This is why stroke frequencies, as a rule, are less than 300 strokes per minute. Due to the required recipe-metering operation, the unit design is applied customarily, mostly the modular block system providing a large selection of different materials and types. Precision adjustability (stroke, stroke frequency) is the characteristic quality feature. The measuring device and the regulator functions of the metering pump make it partly dependent on control circuits and impose demands on its dynamic behavior. To compete with other metering systems, especially with centrifugal pumps having through-flow measuring provision, the capacity of oscillating displacement pumps seldom exceeds 200 kW and is usually in the 20-kW range.

Conveying Pumps

The job required of these pumps is the maximum possible economical conveyance of the pumped medium against the existent pressure. Economy, and efficiency in particular, are the dominant parameters.

The three-plunger pump (triplex pump) is predominant because the rather pulsation-free flowrate and even torque allow maximum stroke frequencies, which exceed 1500 strokes per minute (spm). The variants in materials and types are not so numerous as they are for metering pumps. By contrast, their performance range extends far beyond 100 kW and imposes extreme demands on heavy precision machinery construction.

Detailed Designs of Metering Pumps

The most economical solution to the problems existent for widely different requirements is the use of modular systems, by means of which the metering pump most suitable for the application can be put together from series modular groups on a "tailor-made" basis.

The basic modules are shown in Figure 100. As shown in Figure 101, almost unlimited numbers of metering elements of quite different size and design can be mounted side by side. The horizontal arrangement allows low installation cost, better serviceability, cleaner leakage disposal and simple disassembly. The most important modules are the drive, motor, actuator and pump head.

Drive Unit and Motor

Successful designs for stroke-adjustable drive units include such diverse constructions as eccentric systems, crank-type swivel drive units, rocker-arm systems and spring-action cam systems (rotary cam). Stroke-adjustable drive units are designed for an approximate performance range from 0.1 to 100 kW.

As an example, Figure 102 illustrates the proven rocker-arm system (crank slot). Bearance fulcrum A of rocker arm 3 can be infinitely varied by handwheel during standstill and operation. At the same time, the upper dead position of piston rod 4 remains unchanged; this maintains constant clearance volume and very favorable characteristics. The kinematics remains quasi-harmonic for all stroke lengths. Where higher performance is required, the horizontally opposed arrangement (by removing cover, second pump head can be added to same drive unit) would permit a substantial increase in the machine's output (Figure 103).

Quite popular in the area of application for smaller outputs is the spring-action cam-drive system shown in Figure 104 as a diaphragm pump head with direct diaphragm-link-rod control. Here, the kinematics is discontinuous because of the phase-angle control (Figure 105) at partial stroke, which causes jerking accelerations. This is why the range of application remains restricted to small flowrates. For micrometering, the jerking piston motion at partial stroke presents advantages for the valve control [15].

Besides three-phase asynchronous motors, variable-speed dc and ac motors, as well as mechanical and hydraulic variable-speed drives, are used to drive metering pumps. From the fairly large number of special metering pump drive units–hydraulic [23,24], pneumatic, pulsation-free (by superimposing two cylinders with special cams [15]–we have chosen the explosion-proof solenoid drive (Figure 106) to illustrate a practised and popular type of digital proportioning of small quantities.

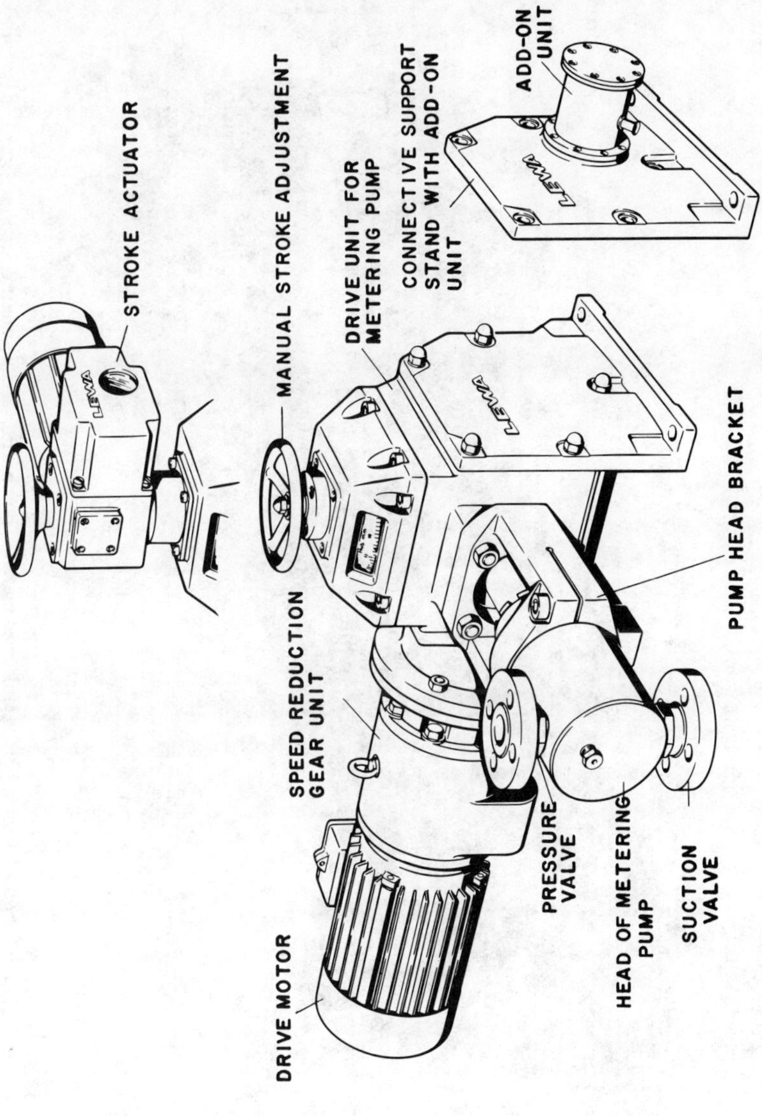

Figure 100. Modules of metering pumps.

Figure 101. Multihead metering pump with variable-speed gear.

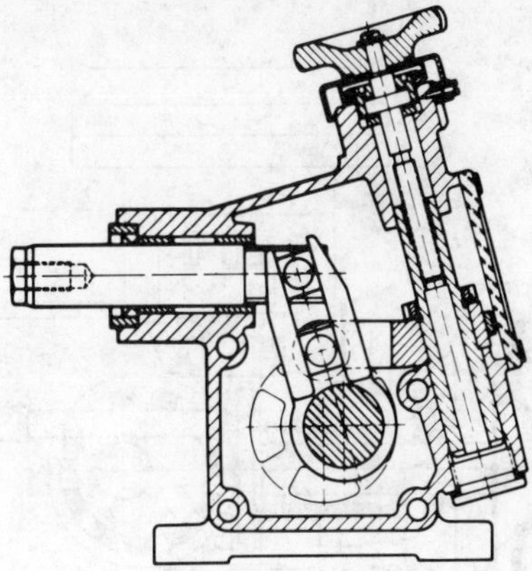

Figure 102. Rocker-arm system (American Lewa Inc. design).

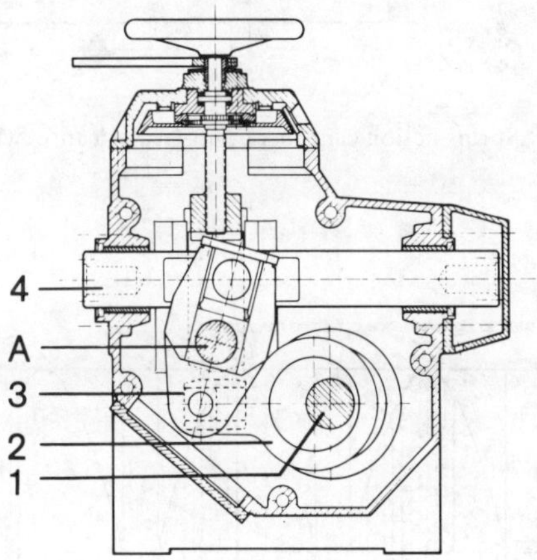

Figure 103. Rocker-arm system in possible horizontally opposed arrangement (American Lewa Inc. design).

The oil-flooded solenoid 3 drives directly via a semiconductor control system, piston 1 of a diaphragm pump head having hydraulic diaphragm actuation. The stroke length adjustment 5 allows control of the flowrate. Solenoid-type metering pumps are the most economical solution to the digital proportioning of small flowrates. The main flow-proportional pulse sequence of the flowmeter actuates the solenoid metering pump via an electrical control unit (Figure 107).

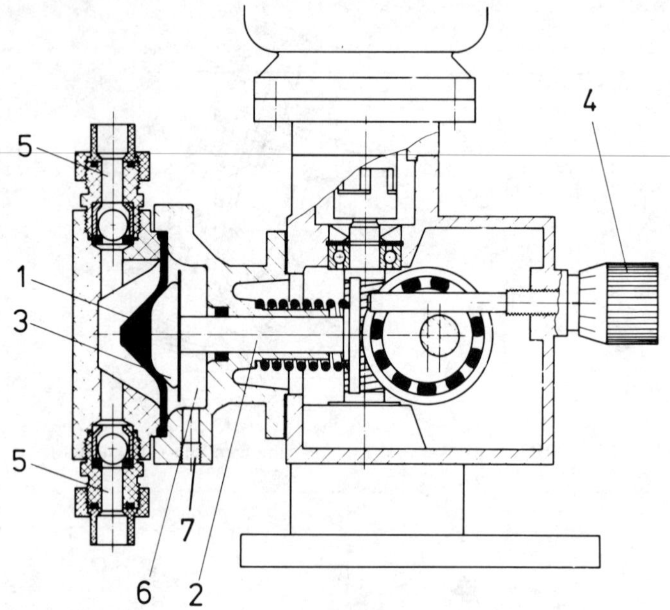

Figure 104. Spring-action cam-drive unit (American Lewa Inc. design).

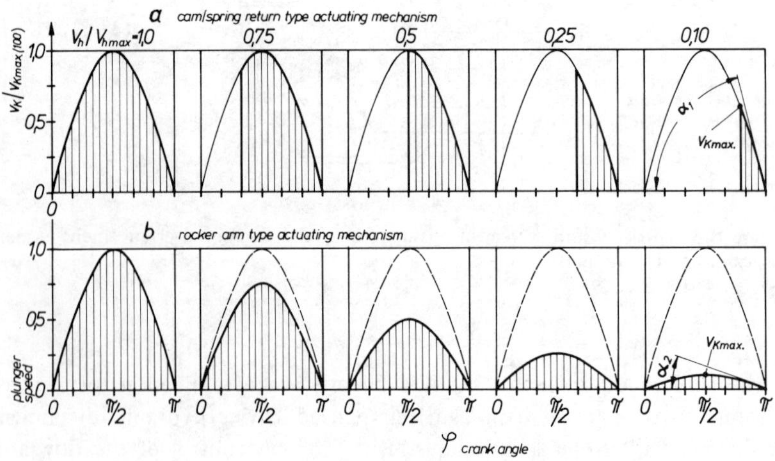

Figure 105. Chronological progression of piston speed at partial stroke for (a) spring-action cam-drive units, and (b) rocker-arm drive units.

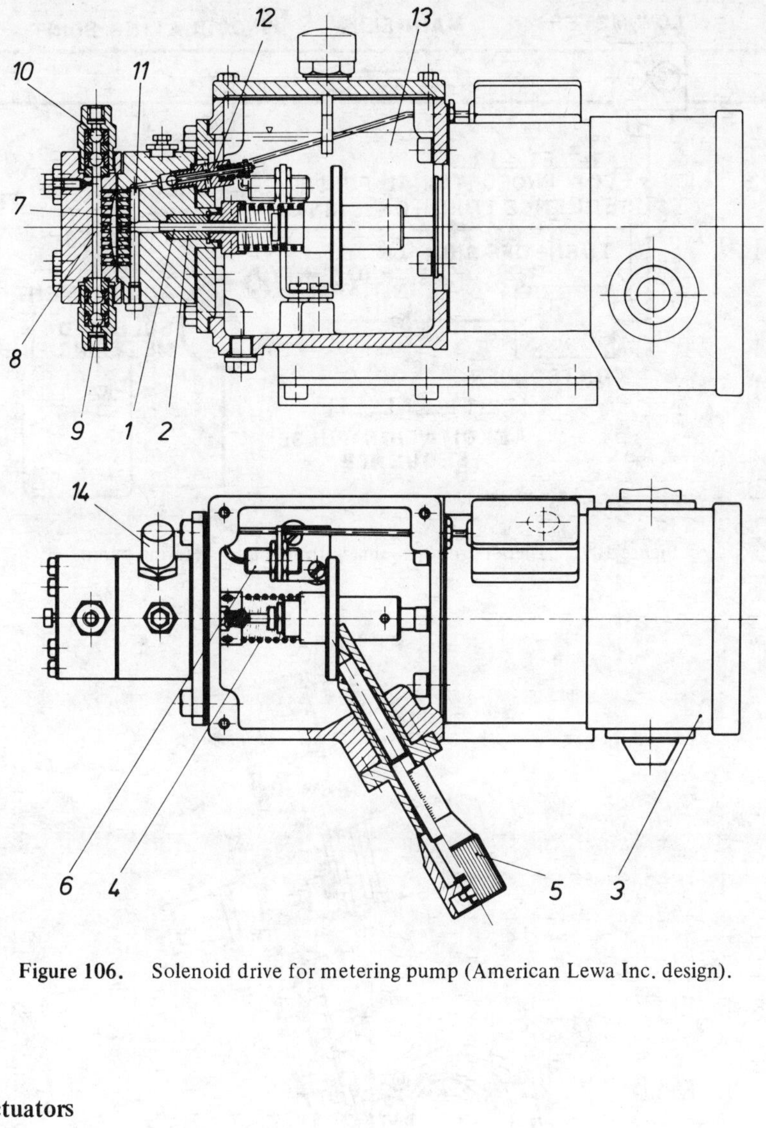

Figure 106. Solenoid drive for metering pump (American Lewa Inc. design).

Actuators

Heavy-duty demands are made on stroke actuators with respect to instrumentation and actuation behavior. Linearity, reproducibility, continuous actuating periods, maximum number of switch operations and smallest actuation steps are advantages provided. Electric actuators (Figure 108) without exception operate on the basis of a positive-negative three-step action (RH operation, LH operation, standstill). In part, they are also

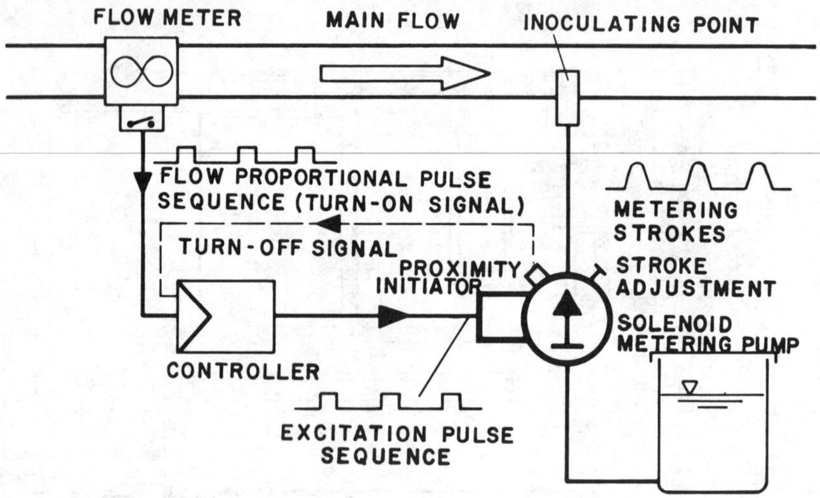

Figure 107. Digital proportioning with solenoid metering pump.

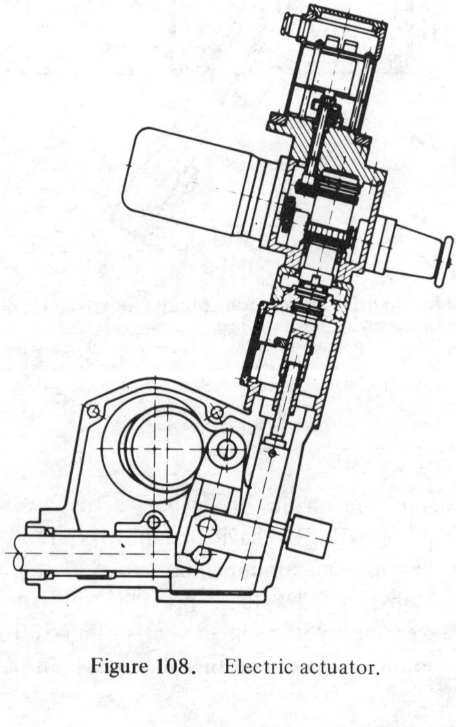

Figure 108. Electric actuator.

directly excitable by means of a 0–20 mA signal. Explosion-proof protection and high impermeability grades are required. However, in the area of explosion-proof protection, pneumatic instrumentation usually is preferred. Preponderantly, for this purpose, double-acting hydropneumatic piston-type actuators (Figure 109), controlled through a position controller, are used in the field of metering pump construction. The operating air acts on the actuating piston across the oil surfaces and via reservoirs.

Pump Heads

Pump heads preferably are made from stainless and acid-resisting steels, nickel alloys, titanium, nickel, ceramics, glass and various types of plastic materials.

Pump Valves

The majority of all metering pumps operate with self-acting valves, which, depending on conditions, are designed as multiball, cone or disc-type valves, with or without spring loading. In particular, ball valves are used for small

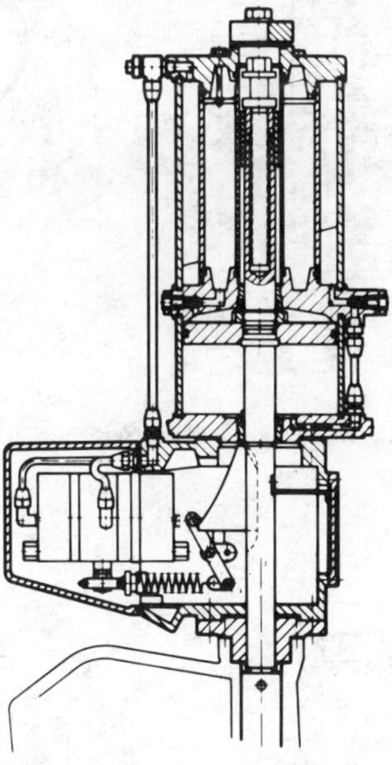

Figure 109. Pneumatic actuator.

nominal widths because they offer optimal properties and features with regard to choice of material, manufacture and resistance to dirt and wear (Figure 110). Pump valves of larger nominal width and, in particular, for higher stroke frequencies, must have light, movable parts for positive operation and, therefore, are mostly designed to be easily spring-loaded ball or disc valves (Figure 111). Spring loading is used to offset unfavorable geodetic conditions and to obtain by force undelayed valve movement where high stroke frequencies or viscous media are involved.

For the metering of suspensions and contaminated media, especially flow-favorable designs of valves are used in which clogging or jamming is avoided (see Figure 93). The following theorem applies: metering accuracy decreases with deteriorating guiding of the ball. On the other hand, however, the operating safety increases. For very unfavorable conditions, such as high viscosity, nonhomogenous media etc., nonself-acting, but separately energy-controlled, valves (mechanical, electrical, pneumatic) [25] are sometimes used. Piston pumps can be used for the great majority of applications because small leakages are permissible.

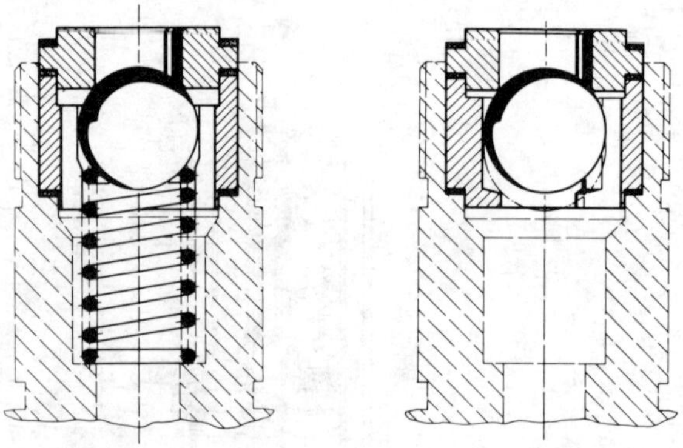

Figure 110. Ball valves with or without spring loading.

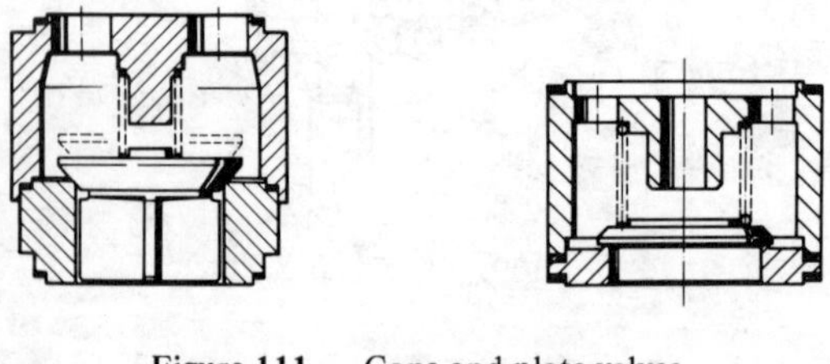

Figure 111. Cone and plate valves.

Pistons and Piston Glands

Pump pistons are sealed by means of differently formed packings or rough-pressed seal rings (Figure 112). Compared to rings, packings have an advantage in that their lips are pressed on the piston by the pressure during the pressure stroke. In other words, they seal off without having to be tensioned from the outside. Seal rings must be pretensioned from the outside and retensioned from time to time. They require the very best casing quality features, especially at the extreme end of the stuffing box. Maintenance of plunger glands having seal rings is costlier than with packings, but they are, by far, not so susceptible to damage. As far as the life of the piston glands is concerned, the decisive factors are sufficient hardness and surface quality of the piston. Pistons of hardened steel with spray-on wear protection, and of aluminum oxide, ceramics and carbide metal, are in use. Because of the more

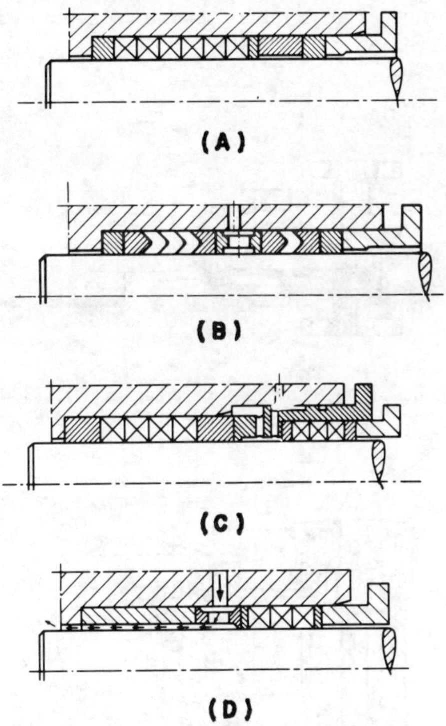

Figure 112. Piston glands: (a) piston gland with seal rings (with flushing possibility); (b) piston gland with profiled seal rings and flushing lantern; (c) packing with separately tensionable medium and flushing liquid seal; and (d) packing with volumetric flushing possibility.

favorable compression distribution, a two-step design (Figure 112(C)) is more appropriate for higher pressures. Here, the seal for the flushing area is separately put under tension. The type as shown in Figure 112(D) is used to serve the injection-type flushing of the stuffing box. A suitable liquid (usually by means of a volumetric pump) is pressed into the pump's working chamber during the suction stroke so that the flushing medium, but not the pumped medium so hard to seal, makes contact with the plunger gland.

DESIGN EXAMPLES

After suitable pump heads have been chosen, it is relatively easy to mount them on, and connect the pistons to, the drive unit (or units). This type of assembly has the advantage of maintaining total separation of the pumped medium from the drive unit—a feature that is absolutely essential when corrosive materials are being handled.

Figures 113 through 116 represent piston-type pump heads of various

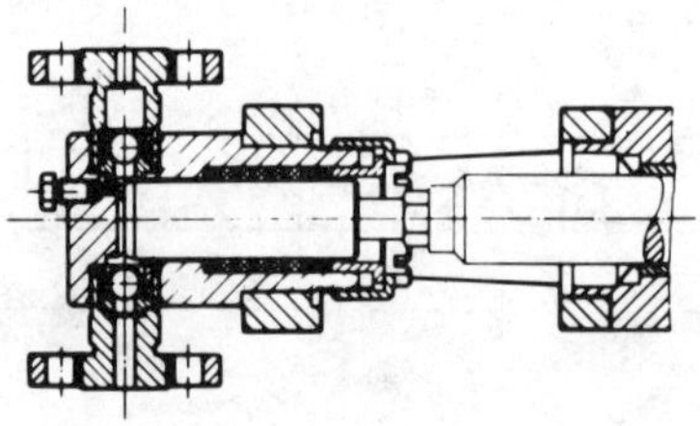

Figure 113. Piston head in metal design for average pressures (300 bar maximum).

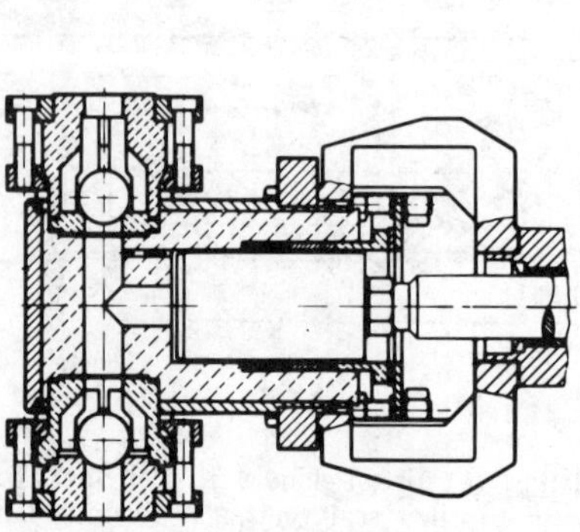

Figure 114. Piston head in plastic design (20 bar maximum).

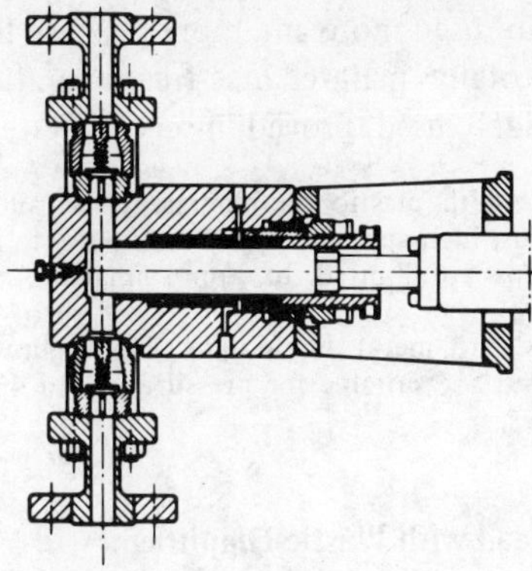

Figure 115. High-pressure piston head (700 bar maximum) with double tensionable piston seal.

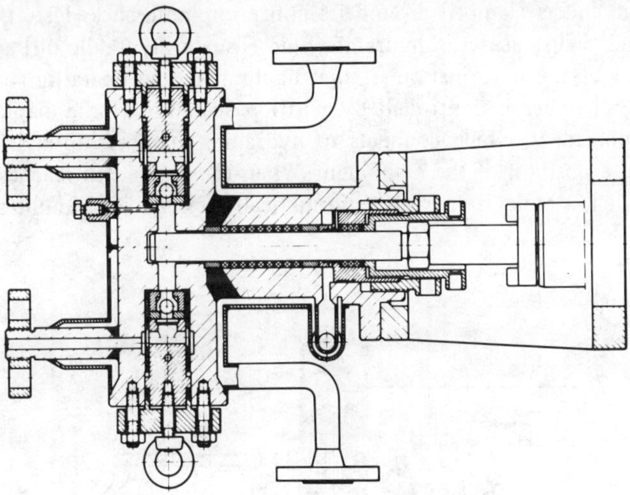

Figure 116. Piston head in cross-head design, completely heat-jacketed for metering melts.

designs. Pump heads for high-pressure application (for instance, 3500 bar) require that special strength problems [26] be overcome.

DIAPHRAGM AND BELLOWS PUMP HEADS

Where critical media (toxic and dangerous ones plus those that are subject to wear) are involved, glandless metering pumps having a diaphragm or bellows as a displacer are used more and more because they offer the distinct advantages of nearly totally maintenance-free operation and of excellent sealing features. Two widely used, proved-in-service systems are:

1. Diaphragm pumps with plastic diaphragms. These allow larger diaphragm deflection and, thus, high specific displacement. As to pressure and temperature, these pumps are limited to approximately 350 bar and 120°C, respectively.
2. Diaphragm pumps with metal diaphragms. These allow smaller diaphragm deflection rates, but are suitable for pressures up to 4000 bar and also for higher temperatures.

Hydraulic Diaphragm Head with Plastic Diaphragm

The principle is illustrated in Figure 117. Pump working chamber 1 is separated from the hydraulic system 3a by a specially shaped plastic diaphragm 2 (usually PTFE). The diaphragm permits a considerable rate of deflection and is clamped around its entire circumference. Piston 4 sealed with gland 5 displaces the hydraulic fluid (usually hydraulic oil) and, thus, effects an elastic deformation of the diaphragm. The operation to fill the hydraulic chamber is controlled by position scanning of the diaphragm via the snifter gate 9, which connects to hydraulic chambers 3b and 3a in the appropriate position of the diaphragm. When the diaphragm reaches the rear contact area, a replenishing valve automatically refills the hydraulic chamber.

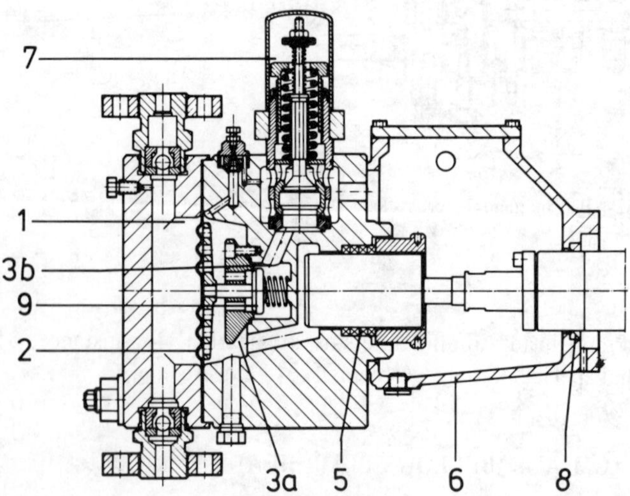

Figure 117. Diaphragm head with PTFE diagram.

As soon as the diaphragm leaves the rear contact, the gate closes and keeps any negative pressure effect away from the replenishing valve.

This is why the system is also safe from excessive pressures without having a front diaphragm contact face. Safety from excessive pressure is provided by pressure relief valve 7. Automatic bleeder valves carry out a permanent venting of the hydraulic system. When the diaphragm is oscillating freely, the pump's working chamber is totally insensitive to dirt. Therefore, suspensions, slurries and other viscous liquids can, without hesitation, be pumped. The system is maintenance-free, automatically secured, and also sealed against outside influences. Changing the diaphragm is a simple operation. Because of their thickness and the small notch-sensitivity of plastic, the diaphragms are insensitive to contamination. The excess pressure safeguard of the system on the hydraulic side automatically prevents damage in cases of accident. The advantages provided by the safety devices not in contact with the medium are self-evident.

Diaphragm Head with Metallic Diaphragm

As per Figure 118, the hydraulic system operates on the basis of leakage replenishment from reservoir o through replenishing valve n. Characteristic is

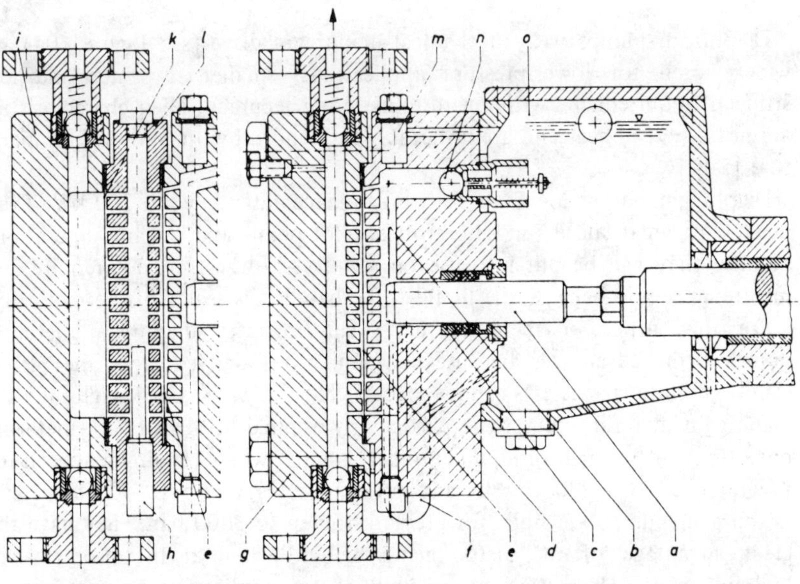

Figure 118. Diaphragm head with hydraulic displacer drive and metallic diaphragm.

the small deflection rate of metallic diaphragms and their amplitude limitation due to the concave or sometimes flat walls of the pump chamber. Safety valve f in the hydraulic system provides protection against excess pressure as well as compensation when the hydraulic space is overloaded. This permits the system to operate reliably and to be independent of excess-pressure or low-pressure conditions. The diaphragm, a thin, cold-rolled sheet (for instance, made from austenitic steel or titanium), is loaded, almost up to the yield point, between lapped surfaces by the use of expansion bolts. The hydraulic drive principle as per Figure 118, and the existing fact that the diaphragm at its extreme positions will contact the contour (or backup plates) requires clean liquids free from dirt and contamination on both process and hydraulic sides. The diaphragm is sensitive to notching to a degree that indentations made by particles will reduce its life. If particles cannot be avoided by the use of dirt traps or if they are, as in suspensions, even scheduled to be part of the medium, the twin-diaphragm model, as per Figure 118 LH, assures a freely oscillating unit. This reduces the system's susceptibility to dirt. Diaphragm pumps with metallic diaphragms have no limit as to pressure. The hydraulic medium sets the temperature limits. However, the application of diaphragm pumps having metallic diaphragms should remain restricted to that area for which plastic diaphragms cannot be used. Metallic diaphragms are much more sensitive than are plastic units. They require more care during installation, in particular.

Diaphragm Head with Mechanically Driven Plastic Diaphragm

Diaphragm pumps with mechanical diaphragm drive (see Figure 104) are widely used for lower-pressure applications, smaller flowrates and less stringent requirements with regard to metering accuracy. Their characteristics are not strictly linear and are noticeably influenced by pressure and temperature [27].

Diaphragm pumps are built for performance rates beyond 200 kW. They meet with great success in laboratories, pilot plants and research operations because they can be put into use on a universal basis, do not require any maintenance and because the liquid to be handled is sealed off hermetically.

For the purpose of metering very small flowrates, diaphragm metering pumps offer advantages because contaminations derived from the piston gland (minute particles) cannot occur. This is why small high-pressure diaphragm metering pumps are particularly applied in "ELUENZ" metering operations in the field of liquid chromatography, where requirements call for flowrates of 100–200 cm^3/hr and pressure rates of up to 400 bar [28]. Diaphragm pumps are built for pressures of up to 3000 bar. They have the great advantage of the piston gland being located at the problem-free hydraulic side. Therefore, the possibilities for application become universal [27].

Bellows Head with Mechanically Driven Bellows

In their application, bellows pumps are almost exclusively of glass/PTFE design. Due to the parts made of glass (Figure 119), the pressure is limited to only a maximum of 5 bar. The characteristics of the metered flow are sharply linear and somewhat pressure-dependent [27].

Life of Diaphragms and Rupture Control

The life of the diaphragms is an essential criterion for the application of diaphragm pumps. The life of the various diaphragm designs differs greatly as described below:

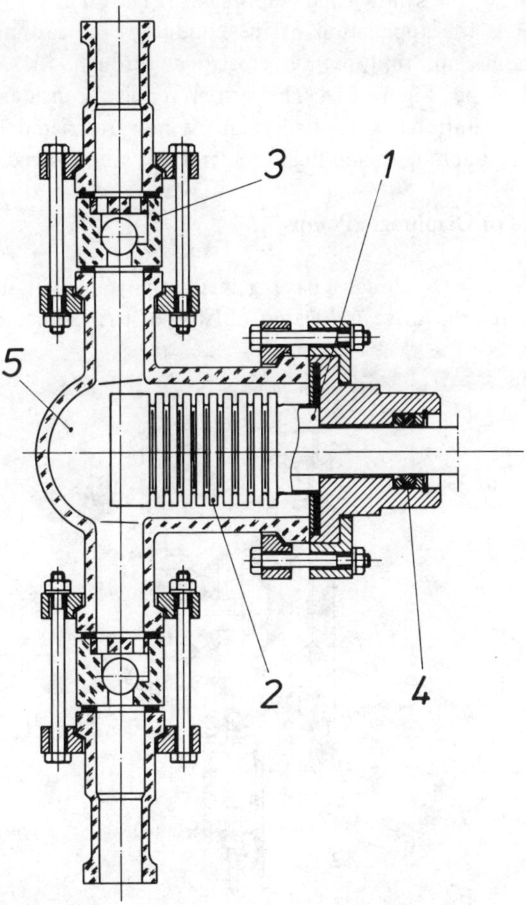

Figure 119. Glass-type bellows head (Lewa/Schott).

Hydraulically driven diaphragm	
PTFE	up to 15,000 hours
Metallic	up to 5,000 hours
Mechanically driven diaphragm or bellows	
Elastomer diaphragm	up to 3,000 hours
PTFE bellows	up to 10,000 hours

Several methods are used to signal rupture of a bellows or diaphragm. In mechanical bellows or diaphragm drives, level switches (Figure 120) are used to signal any ruptures.

Where hydraulic diaphragm pumps are involved, the use of the sandwich-type diaphragm is predominant to signal ruptures [29]. In this method, two or three hydraulically coupled diaphragms, with a connection to sensor 4 and check valve 2, are used to determine selectively the pressure rise during rupture of one of the sandwiched diaphragms (Figure 121).

Much rarer is the application of the conductive or capacitive method of signaling diaphragm rupture by electrode through the twin-diaphragm arrangement, as per Figure 118. This system requires considerable experience and care in operational surveillance and is now restricted to special cases where metallic diaphragms and high temperatures are involved.

Special Types of Diaphragm Pumps

"REMOTE HEAD" designs have gained importance in nuclear technology [30]. Here, the drive is located outside of the radioactively dangerous

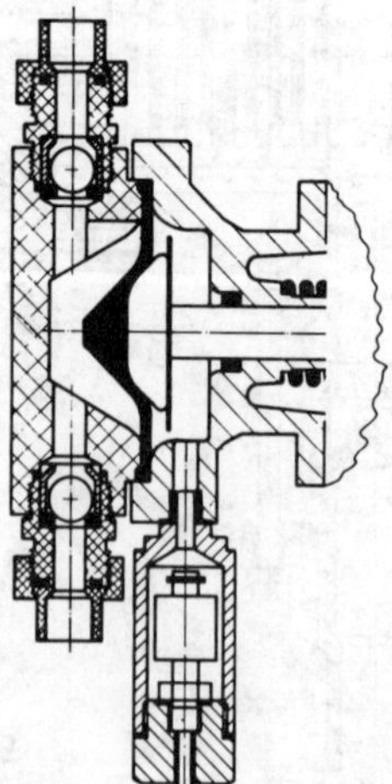

Figure 120. Rupture indication by switch.

area. It is only the liquid part that is exposed to those dangerous conditions and connected to the drive via a hydraulic actuation line. Diaphragm pumps with openable cover for cleaning purposes in hygienic operations [31] are also worthy of notice.

CONVEYING PUMPS

In this class of pumps, the horizontal triplex pump (three-plunger pump) is predominant, but the standup, multicylinder design (for instance, quintuplex), as well as the horizontally opposed arrangement and the double-acting pumps, etc., are in use.

Drive Unit and Motor

Figure 122 shows a typical drive unit in triple-crank construction for external gear mounting. For larger outputs, the forced feed lubrication method is used throughout. The standard drive arrangement provides for

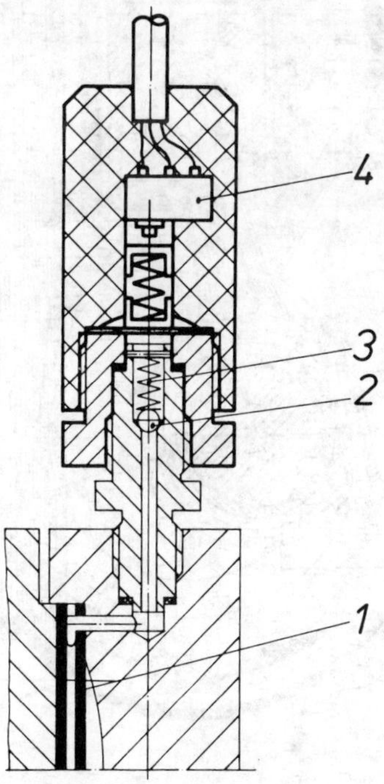

Figure 121. Sandwich-type diaphragm with pressure switch system.

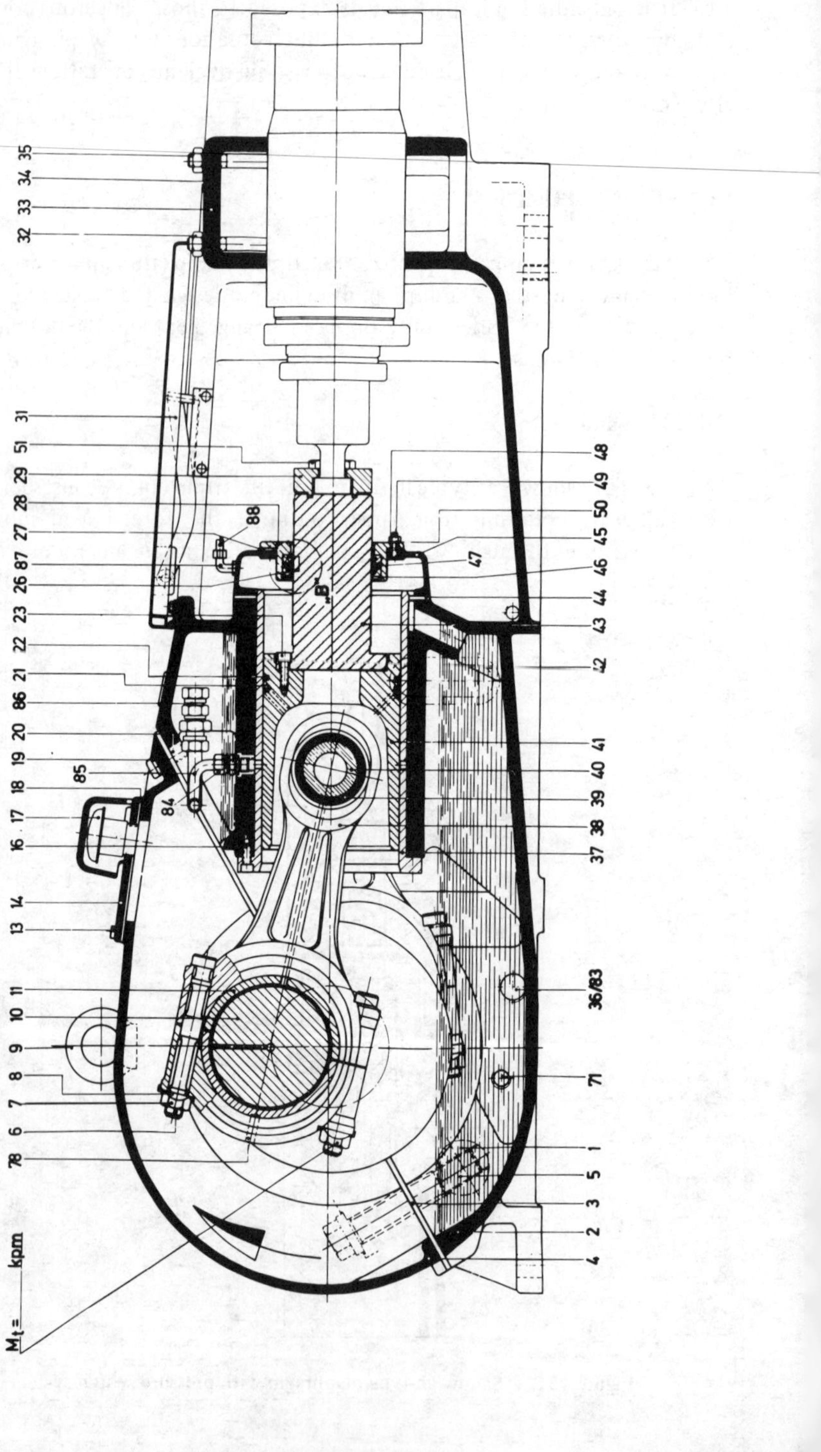

35
34
33
32
31
51
29
28
27
87
26
23
22
21
86
20
19
18
17
16
14
13
11
10
9
8
7
6
78
M_t = kpm
88
85
84
48
49
50
45
47
46
44
43
42
41
40
39
38
37
36/83
71
1
5
3
2
4

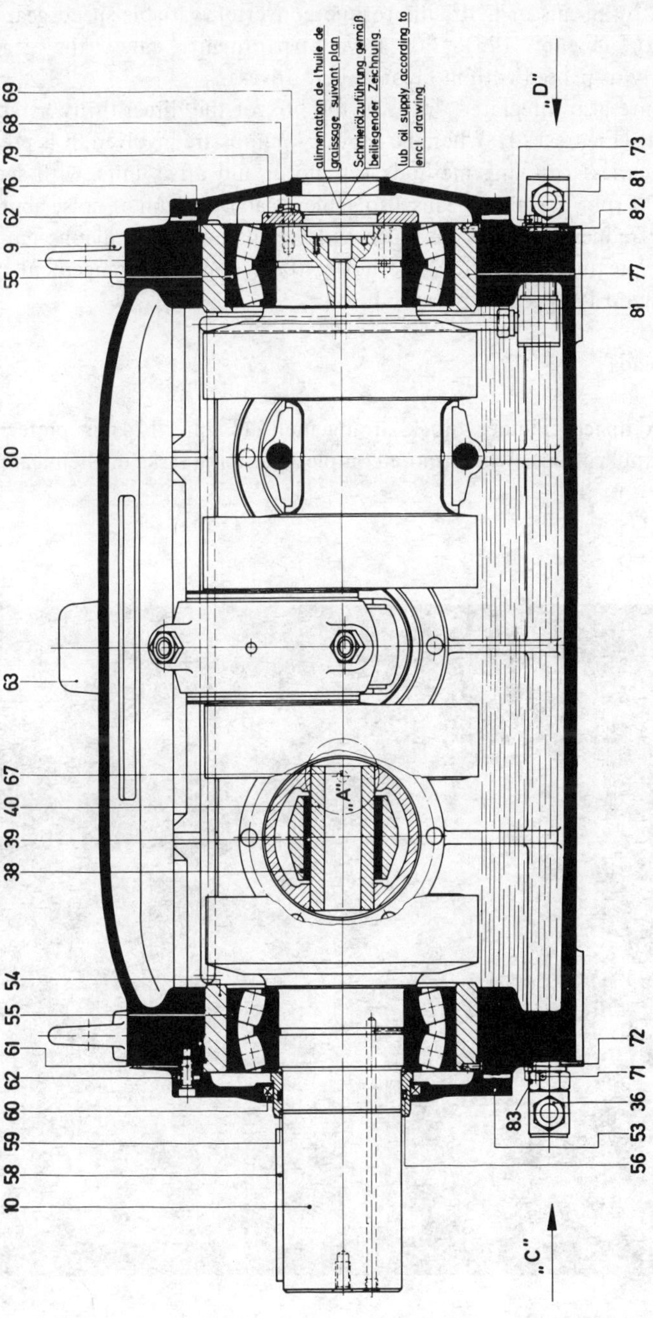

Figure 122. Triplex crank drive unit for heavy-duty application.

external gearing with motor on base plate. Stroke-frequency control is achieved by means of hydraulic torque converter, variable-speed gear unit or dc motor (Figure 123). For lower performance rates, the gearing is frequently dispensed with in favor of belt drives.

There are also integrated drives in the box of the direct-thrust, crank-type gear units (Figure 124). Where heavy-duty pumps are involved, it is preferable to use gearless coupling methods for motor and drive units, with speeds of 750–1500 rpm prevailing. This also achieves an optimum of noise abatement. For less frequent applications, such as large, high-pressure diaphragm pumps and high-pressure circulation, the horizontally opposed arrangement of drive units (Figure 125) generally is used.

Pump Heads

The compact cylinder block arrangement (Figure 124) is preferred for simpler applications. Where more complex pumping tasks in chemical process

Figure 123. Triplex plunger pump with variable-speed drive.

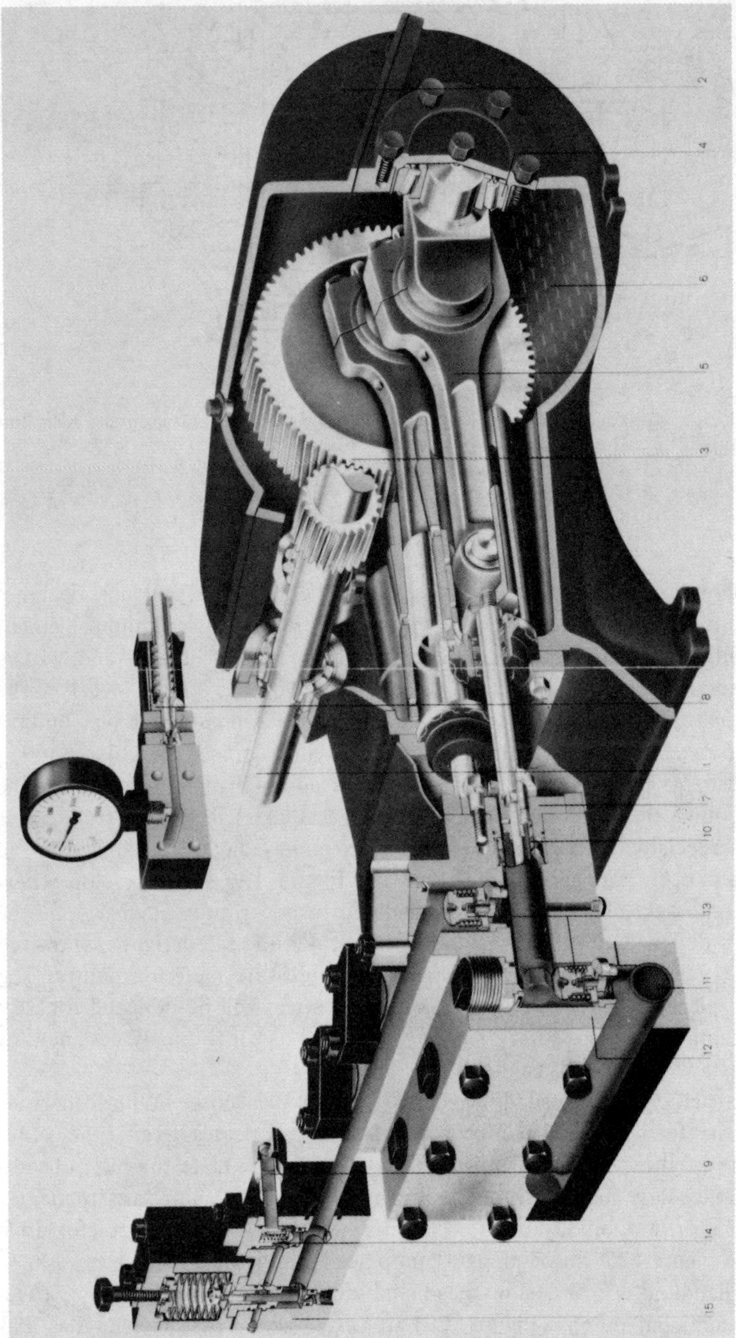

Figure 124. Triplex plunger pump with integrated gear drive.

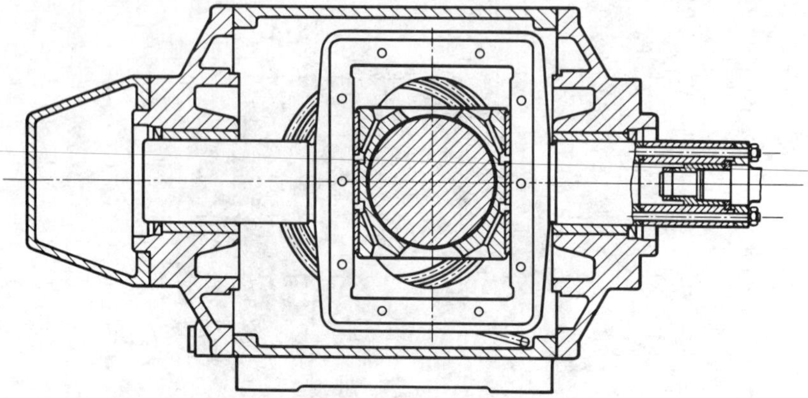

Figure 125. Horizontally opposed drive unit arrangement. When two pump heads are mounted, the RH cover is removed (Lewa).

technology are to be handled, separation of the individual cylinders becomes a necessity because this is the only way to be sure that each pumping element is wholly suited to the characteristics of the various media and to the processing conditions for each of them. As shown in Figure 123, each of the individual cylinders (separated into valve crosshead and stuffing box unit) is mounted by means of the drive-unit bracket and a transverse yoke. Suction and discharge piping are manifolded by the headers (Figure 126). Because the area around the bore holes (valve and piston axis) is subject to extreme stress—especially under high, or even extremely high, pressure over an extended operating period—the item 12, Figure 126, may develop stress-corrosion cracks. Therefore, it is considered to be a throwaway part, and the pump is designed to permit easy replacement. Pump head designs as shown in Figure 127 are customary for extreme high-pressure metering pumps. The part (T piece) being subjected to the highest stress will be designed for easy replaceability. For pressures of more than 2000 bar, it is usually designed for autofrettage (radial expansion).

For specific pump head designs, the three-plunger construction otherwise customary for conveying pumps is not used in most cases because the center distance of the plungers is not sufficient to accommodate the pump heads. This is why large high-pressure, diaphragm-type process pumps are frequently designed for horizontally opposed arrangement, as per Figure 128 (for drive unit see Figure 125; for diaphragm pump head see Figure 118).

For handling slurries, slow-speed and double-acting, low-pressure, tube-type diaphragm pumps (Figure 129) are often used. Their advantage is that

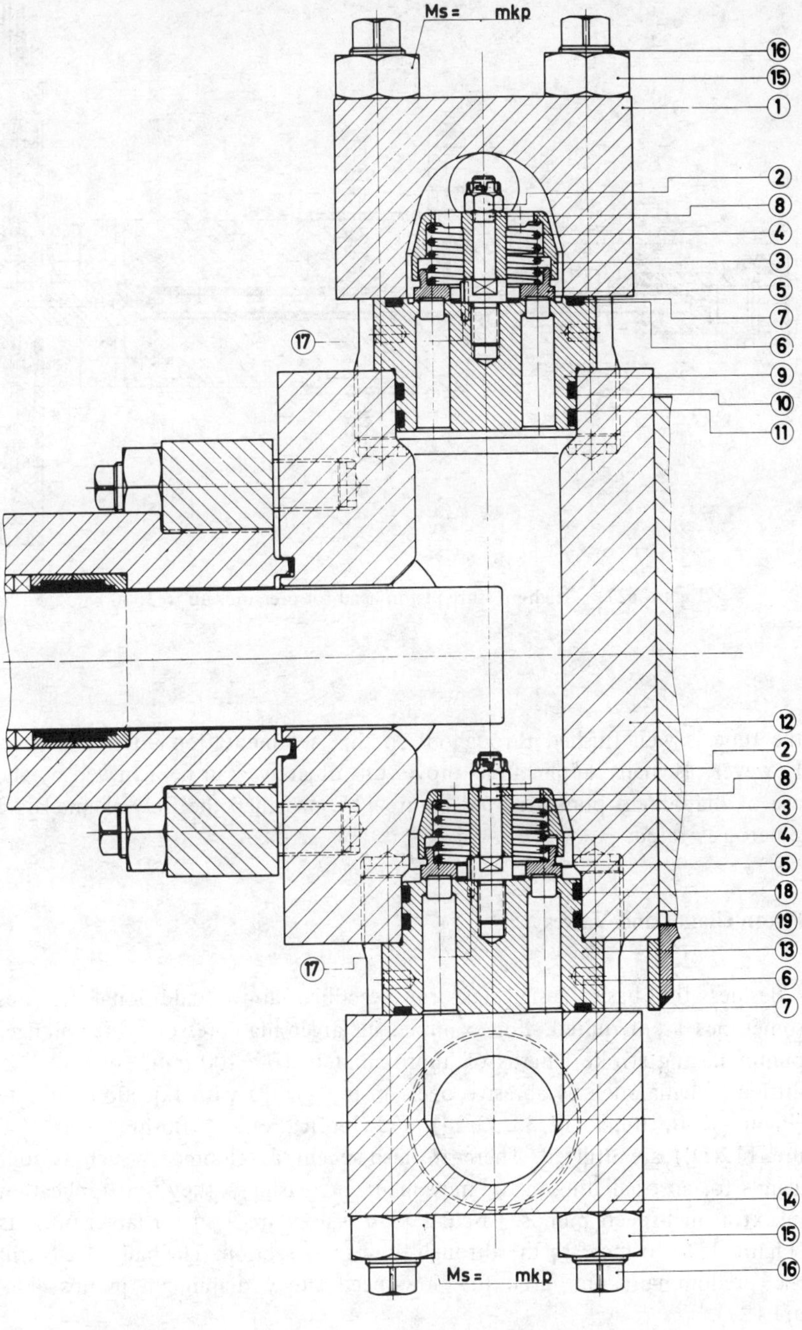

Figure 126. Piston head with separate housing for gland and valve cross head.

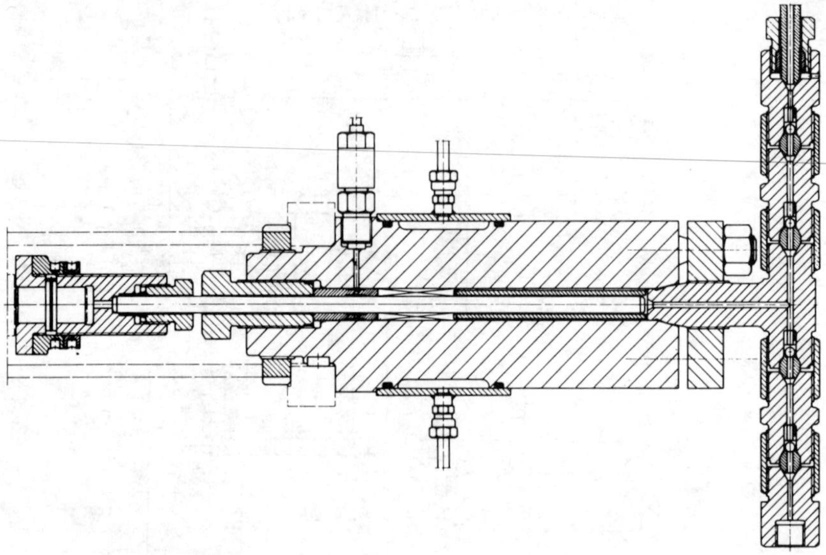

Figure 127. High-pressure piston head for pressures up to 3000 bar.

the tube is well flushed throughout so that sedimentation does not occur. However, as many applications prove, this objective can be achieved by the use of diaphragm pumps as per Figure 117, provided the flow channels are generously rated.

Piston Glands and Valves

Besides the basic design features described above, additional features sometimes are required. For example, in larger high-pressure three-plunger pumps having stroke lengths of approximately 100–200 mm, especially for difficult media such as abrasive ones, piston glands with injection flushing (Figure 130, items 15/33/23/24) and lubrication or flushing (see Figure 112(D)) are utilized. There are also recent developments such as split glands for an oscillating piston movement. Increasingly, they find application in extra high-speed pumps. Circular plate valves are used for larger outputs (Figure 126) to increase the through-flow cross section. The ball valve is still the predominant unit used for slow-speed slurry diaphragm pumps (Figure 129).

Figure 128. Diaphragm pump in horizontally opposed arrangement for high pressure of 350 bar.

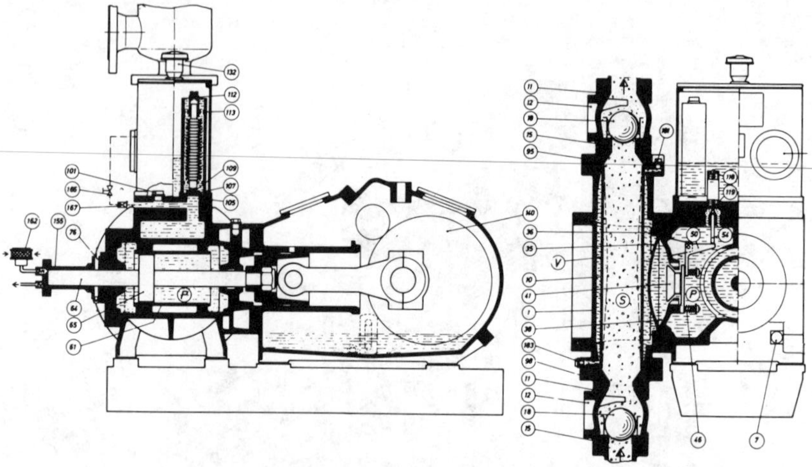

Figure 129. Double-acting (twin-cylinder), low-pressure diaphragm tube pump.

INSTALLATION LAYOUT FOR OSCILLATING DISPLACEMENT PUMPS

Because of the pulsating flow, the layout of the piping for oscillating displacement pumps is much more difficult than for centrifugal pumps. In addition to time-variable flow speeds and because of the oscillating displacement motion, accelerations and delays of the liquid masses emerge, which must be taken into account.

The objective of all calculations for installation is the avoidance of detrimental conditions that disturb or prevent proper operation of the facility. Above all, harmful cavitation must be eliminated. The consequence of cavitation is a drop of the flowrate, erratic operation of the pump, knocking of the pump valves, pressure thrusts and considerable wear and tear.

Other types of disturbances of the pump function concern overloading and such conditions as are known under the term "over-pumping," where the pressure variance in the pump does not suffice for the planned control of the self-acting valves. Under simplifying conditions, all these problems now can be controlled on an analytical basis and solved either manually or by the use of digital computers [14,16].

The kinematics of the oscillating drive units already has been demonstrated. We assume that piston speed, $v_K(\varphi)$, and piston acceleration or delay, $b_K(\varphi)$, in dependence on the crank angle are known factors. Further deliberations are restricted to the case of a total lack of elasticities. This simplifies the

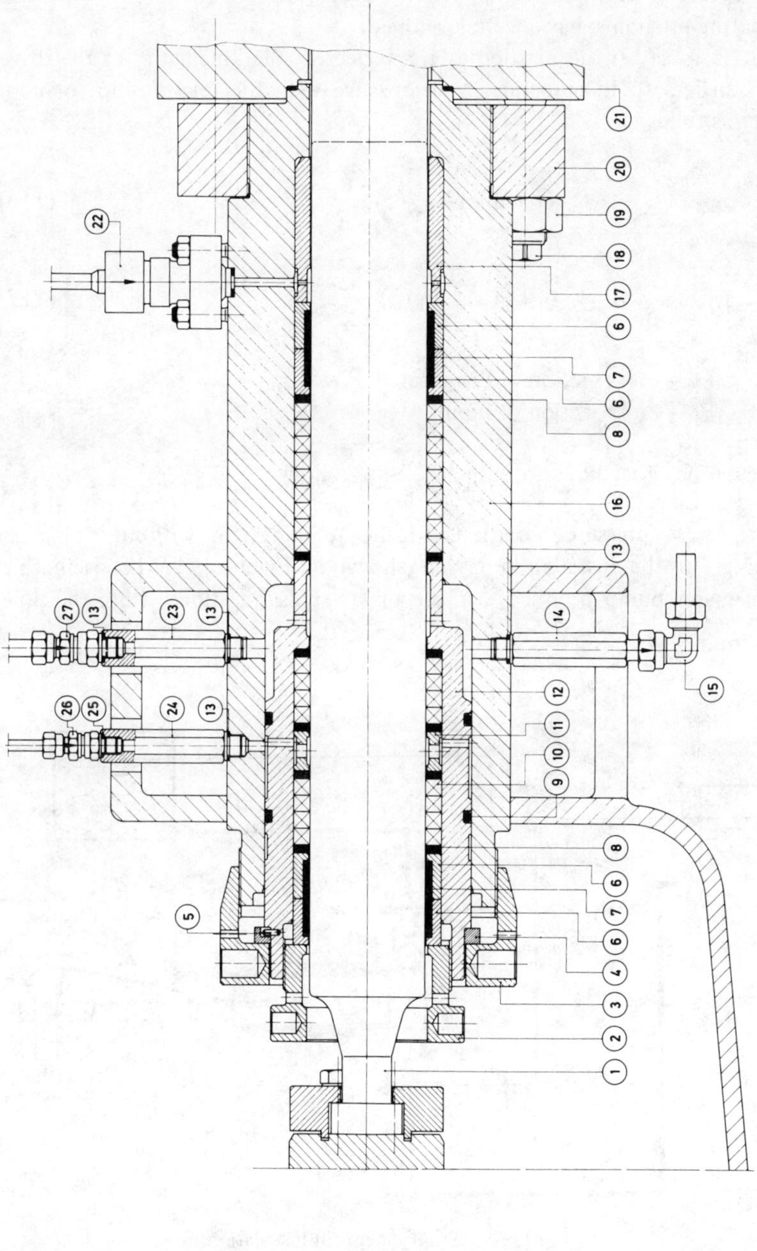

Figure 130. Stuffing box with injection-type flushing.

explanation considerably and provides sufficient data for the objective to be achieved here. In practice, this simplification is frequently not admissible and, therefore, complicated arithmetic processes, together with respective computing programs, have been developed.

That is to say, if all elasticities are neglected, the kinematics of the drive unit transfers to the pipeline in accordance with the relationship for continuity as follows:

$$W_R(\varphi) = v_K(\varphi)\,\frac{F_K}{F_R} \tag{116}$$

$$b_R(\varphi) = b_K(\varphi)\,\frac{F_K}{F_R} \tag{117}$$

where F_K = cross section of displacer
F_R = cross section of piping

Calculation procedures are summarized as follows:

Step 1. Assess pressure variations in the total system without pulsation damping. In the installation layout shown in Figure 131, the oscillating displacement pump produces a more or less pulsating flow within the pipe

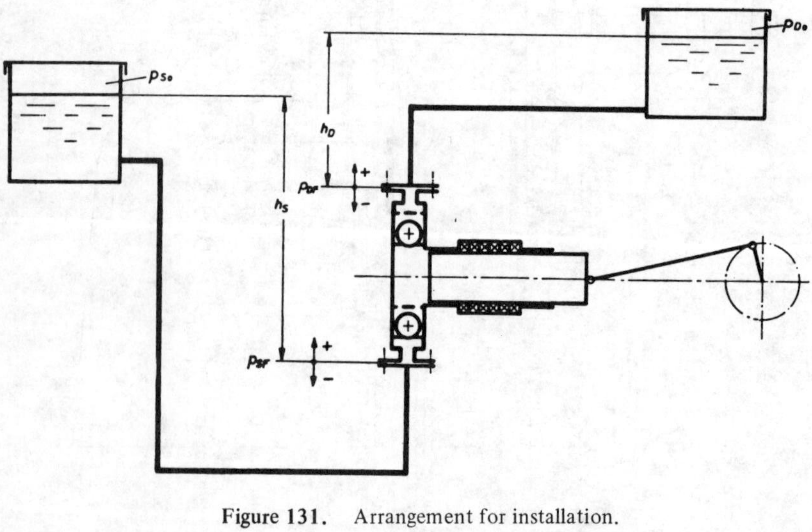

Figure 131. Arrangement for installation.

system. The pressure variations thus caused due to pipe friction and acceleration (or delay) of the mass are functions of the crank angle or time "t," respectively, and must be superimposed. The objective of the calculations is to determine, at any point in time, the pressure in the pump chamber, $p_A(\varphi)$, and the flange pressure ratings, $PSF(\varphi)$ and $PDF(\varphi)$.

Step 2. Ample information is available for calculating pressure fluctuations and their superimpositions. Figure 132 explains, by taking the suction process as an example, how, with a suction-reservoir pressure, p_0, the pump chamber pressure, p_A, is progressing as a function of time. The basic rules for installation can be summarized in a few criteria:

1. At no time must gas or vapor be allowed to form in the pump chamber. Cavitation, in particular, must be avoided.
2. The pressure at the suction flange must at no time exceed the pressure at the pressure flange plus the spring-loading rating of the pressure valve. Otherwise, this will lead to overpumping and, thus, to large metering errors.
3. The pump chamber pressure must at all times remain below the permissible value.

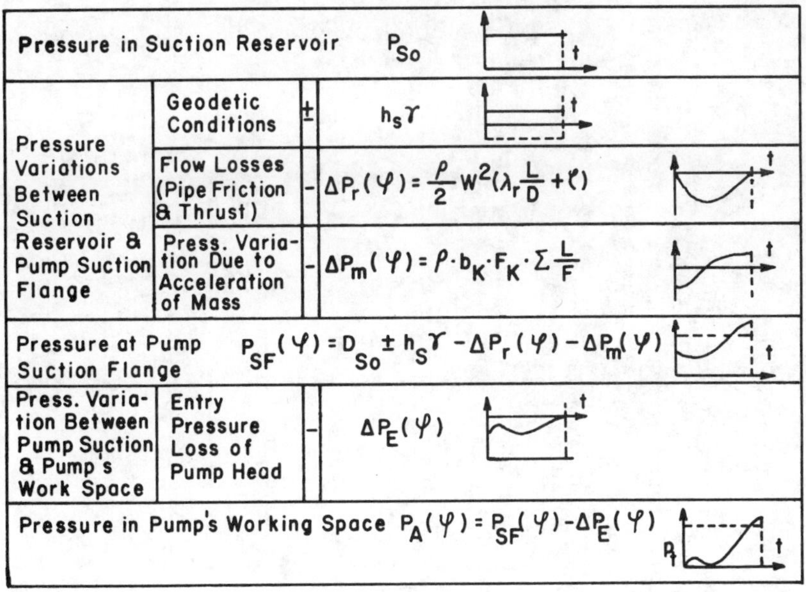

Figure 132. Pressure condition existent between suction reservoir and pump cylinder during suction process.

4. A scheduled, undelayed valve motion (self-acting control) must be assured. Based on the rating of the pump, this means adaptation of the valves to the viscosity and particle charge of the liquids to be metered. The installation is required to make it possible that the medium can flow to the pump in a homogeneous mode.

Step 3. In practice, these calculations are now carried out by computers. With the installation data and other data available, it is easy to determine whether any of the above referenced criteria have not been met. Exact computing instructions can be obtained from some manufacturers [31]. They will also provide hints on installation with respect to special conditions and layout methods for pulsation dampers [12]. The work of Harvey et al. [32] and Plesset [33] also should be consulted.

CHAPTER 2.7

MECHANICAL SEALS FOR PUMPS

Ray Battilana
Manager of Engineering
Durametallic Corporation
Kalamazoo, Michigan

FUNDAMENTALS OF MECHANICAL SEAL OPERATION

The most frequently encountered requirement for sealing a rotating shaft is found on a centrifugal pump. Figure 133 illustrates the liquid end of a typical centrifugal pump. The liquid to be pumped enters the suction inlet at the eye of the impeller. As the impeller rotates at relatively high speed, liquid occupying the space between the vanes is held captive by the close clearance that exists between the front of the impeller and the pump housing or volute. Having no other path of escape, the liquid is centrifugally forced to the outside diameter of the impeller, where it flows out through the discharge nozzle. At the point of discharge, the pressure of the liquid is many times higher than it is at the suction inlet, which accounts for the centrifugal pump's ability to push liquid to heights far above the pump. However, this same discharge pressure flows down behind the impeller to the drive shaft, which is connected to a driver outside the pump. This is the shaft that must be efficiently sealed if the pump is to be of any practical use.

To make the job of sealing the drive shaft somewhat easier, most pump manufacturers relieve some of the pressure behind the impeller. For example, on an open impeller pump this is usually accomplished in either one of two ways, or a combination of both. Considering that the clearance between the back of the impeller and the pump back head is actually much closer than shown in Figure 133, small pumping vanes are often placed on the back side of the impeller. Although not as large or as efficient as the vanes on the front of the impeller, these smaller vanes will significantly reduce the pressure of the liquid trying to escape to the atmosphere from behind the impeller.

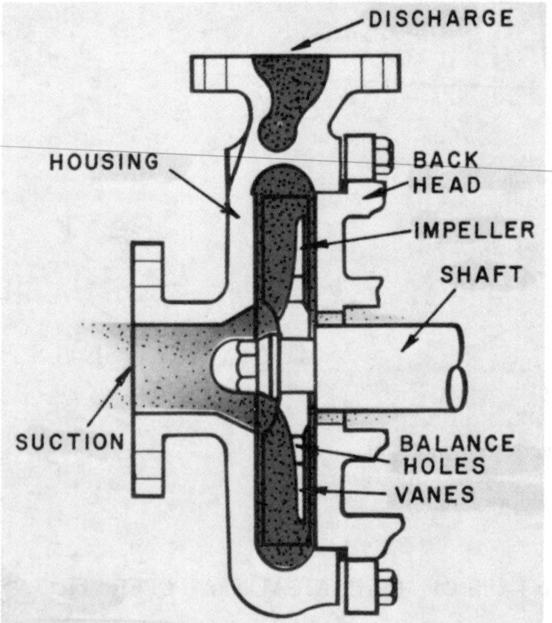

Figure 133. Centrifugal pump liquid end.

In addition to pumping vanes, balance holes can also be drilled through the impeller to the suction eye, thus relieving the high pressure behind the impeller to the low-pressure suction eye of the impeller. By decreasing the pressure differential between the front and rear of the impeller, these methods also greatly reduce the axial thrust on the shaft, thereby prolonging thrust bearing life. Even with the existence of pumping vanes and balance holes, suction pressure, plus usually a small percentage of discharge pressure, always surrounds the drive shaft. Therefore, shafts must be fitted with reliable seals.

Figure 134 shows a typical stuffing box sealed with square rings of compression packing. If this packing were to rub against the shaft, without lubrication present to prevent a buildup of frictional heat it would soon destroy itself. Therefore, a certain amount of the liquid being pumped—or some other liquid injected at high pressure from an outside source—must be allowed to flow between the packing and the shaft. Because of the surface irregularities of packing, eccentricity between the stuffing box bore and the shaft, as well as normal shaft runout, a significant amount of packing must be used to compensate for irregularities. The packing requires a generous amount of lubrication. When this flow of lubrication exits from the packing, it becomes identified as leakage.

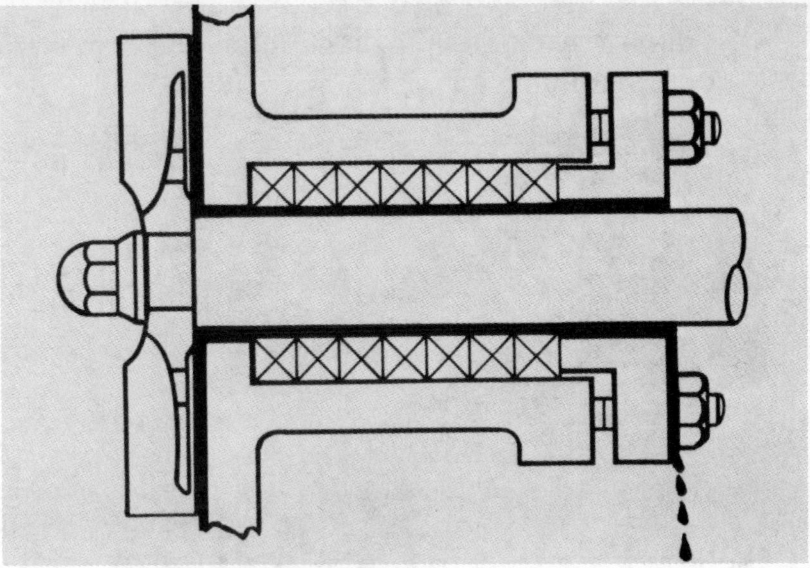

Figure 134. Typical stuffing box with compression packing.

Other inconveniences associated with packing are adjustment and sleeve or shaft wear. It has been suggested already that because of certain irregularities, a particular ring of packing is periodically in contact and out of contact with the shaft. When the packing wears or loses its resiliency, leakage increases. The packing then must be adjusted to bring the leakage under control.

These same irregularities also exert uneven pressure on the shaft and restrict lubrication flow at certain locations, producing wear on the shaft or shaft sleeve. When this wear has been allowed to become acute, stuffing box leakage increases significantly. The only solution to reducing the leakage is to incur the expense of a new or repaired shaft or shaft sleeve.

Figure 135 shows the parts of a simple mechanical seal. The coil spring, "O" ring shaft packing and seal ring fit over the shaft and rotate with it. The spring can be made from any one of a number of materials. The selection of the material depends on the corrosion resistance required to withstand attack from the product being pumped. Similar considerations are evaluated when selecting the "O" ring shaft packing material and seal ring material. The insert and insert "O" ring mounting are fitted into a bore provided in the gland ring. This entire assembly is then attached to the pump stuffing box, remaining stationary as the shaft rotates. The insert is made from carbon graphite, which offers a good bearing surface for a variety of seal ring materials to rotate against. It is inert to corrosive attack by most chemicals at a wide range of temperatures.

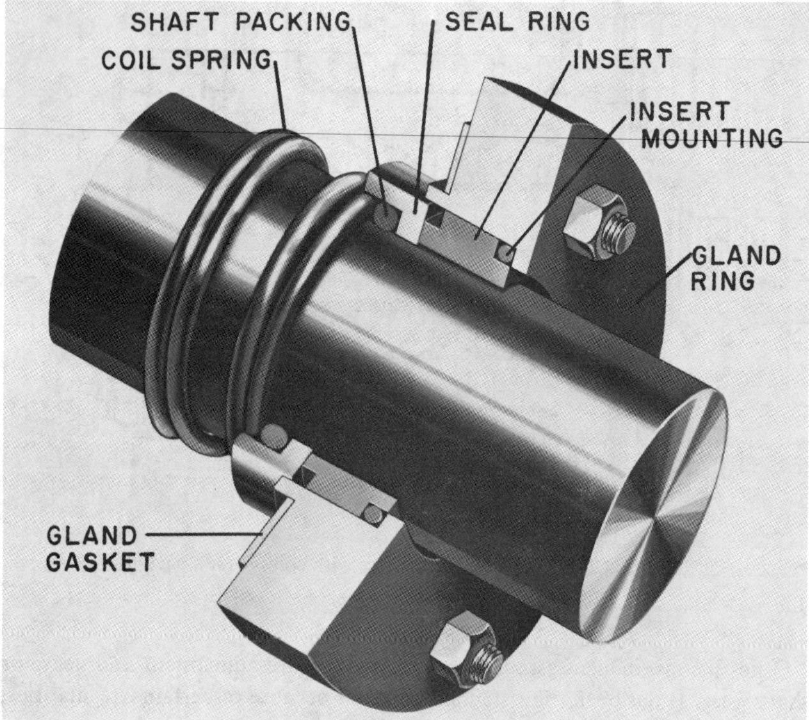

Figure 135. A simple mechanical seal.

Figure 136 shows this simple seal completely assembled in the pump stuffing box. The coil spring rests against the back of the pump impeller, pushing the seal ring forward so that it remains in intimate contact with the stationary insert. Friction between the rotating pump shaft and shaft packing causes the seal ring to rotate with the shaft. Friction between the spring and the impeller, as well as the spring and the compressed "O" ring, is responsible for the spring rotating with the seal ring and shaft packing. The insert is positioned within the gland bore. The gland itself is bolted to the face of the stuffing box.

Friction between the "O" ring insert mounting and the inside diameter of the gland bore holds this part of the seal stationary as the shaft rotates within the bore of the insert.

Sealing Points

The path of the liquid down the shaft is blocked by the "O" ring shaft packing at Point A in Figure 136. Liquid attempting to pass under the seal

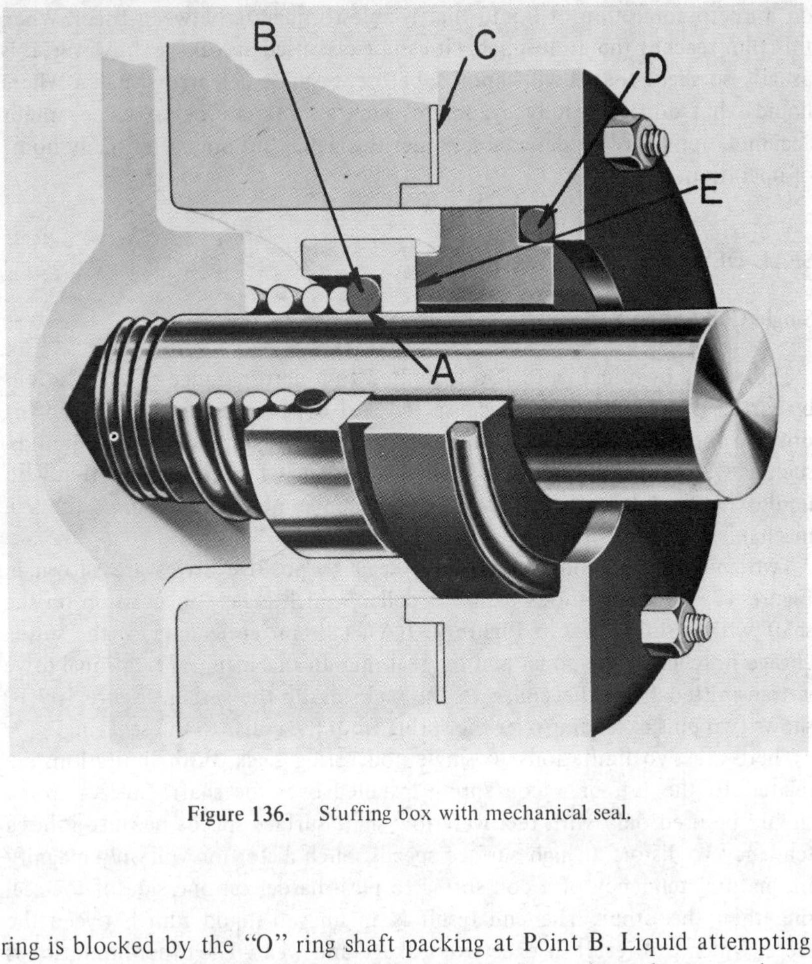

Figure 136. Stuffing box with mechanical seal.

ring is blocked by the "O" ring shaft packing at Point B. Liquid attempting to pass over the seal ring is finally blocked from reaching the atmospheric side of the pump by the gland gasket, Point C, and the "O" ring insert mounting, Point D. The path of escape remaining is the seam between the rotating seal ring and the stationary insert, Point E.

The face of the seal ring and the face of the stationary insert are lapped to a flatness that is measured in millionths of an inch. Therefore, these faces remain in contact throughout their entire contact surface area, thereby providing a nearly positive seal. Just as in the case of packing, the faces must be lubricated. Because the area of the bearing surface is only a fraction of that encountered with packing and the contact pressure is equally distributed throughout the interface, much less lubrication is required. Consequently, a correspondingly smaller amount of lubrication passes between the seal faces to exit as leakage. Whereas most packing depends on a measurable flow of liquid between it and the shaft for lubrication, the mechanical seal faces ride on a microscopic film of liquid that is able to migrate between them. When this film reaches the atmosphere, it can be classified as leakage; however, it is usually so slight that it will vaporize before it can be detected visually. Where liquids that do not readily evaporate, such as wax, are being sealed, minor accumulations may be detectable under the gland, but only after many hours of operation.

SEAL DESIGNS

Single Coil Spring Seals

The seal shown in Figure 136 is limited to a narrow range of applications by virtue of its simplicity. Because the seal depends on friction to impart drive to the rotary unit, it is limited to use with nonlubricating-type liquids, such as water. Therefore, if it is desired to use such a seal in conjunction with liquids having lubricating properties, it becomes necessary to drive the seal mechanically.

Two methods of converting simple seals to positive drives are shown in Figure 137. Both methods utilize a collar, which is held in position on the shaft with a set screw. In Figure 137(A), tabs on either end of the spring engage notches in the collar and the seal ring. In this manner, rotational drive is transmitted from the collar to the seal ring by the spring. Figure 137(B) shows two pins extending over the spring from the collar to the seal ring.

There are two limitations to single coil spring seals. Both limitations are related to the use of a coil spring installed over the shaft. Such a spring should be used only with relatively low shaft surface speeds because it has a tendency to distort at high surface speeds. Such distortion will only magnify the natural tendency of a coil spring to push harder on one side of the seal ring than the other. The end result is an uneven liquid film between the faces, which can result in excessive leakage and wear. The other limitation is economics. Pumps can be purchased in a variety of shaft sizes. Consequently, a user who stores seal parts may find it necessary to carry large inventories of springs to cover the range of sizes required.

Multiple-Spring Seals

Figure 138 illustrates a more sophisticated seal design utilizing a number of small coil springs neatly spaced around the rotary unit. These small springs are not as susceptible to distortion at high speeds and exert an even closing pressure on the seal ring at all times. In addition, the same size springs can be used in conjunction with a wide variety of shaft sizes. This general design is often referred to as a conventional mechanical seal.

The seal in Figure 138 is a positive drive seal. The collar is normally secured to the shaft by set screws. Pins extend from the collar to the compression ring, which presses against, or energizes, the shaft packing. Additional pins extend from the compression ring to the seal ring, completing the drive train from the collar. The collar, compression ring, pins, springs and set screws are usually confined in an assembly that is securely put together at the factory. This assembly is called the compression unit.

The seal points of this multiple spring seal are identical to those outlined for the single-coil spring seal shown in Figure 136.

Bellows Seals

A third basic seal design offered by some manufacturers is the bellows seal, shown in Figure 139. Two forms of the bellows seal are the elastomer bellows (Figure 139(A)) and the metal bellows (Figure 139(B)).

The main advantage of a bellows-type seal is that the seal ring is free to move forward as the insert wears, without being retarded by the friction between the shaft packing "O" ring and the shaft. In most applications this is not an important consideration; however, in some situations the pump shaft or shaft sleeve metallurgy is not compatible with the product and corrosion takes place, freezing the conventional "O" ring shaft packing to the shaft. This leads to leakage. Although a bellows seal is not nearly so sensitive to shaft or shaft sleeve corrosion, such a problem lies not with the conventional seal, but with the improper selection of shaft or shaft sleeve material.

The elastomer bellows seal is energized by a single coil spring, which fits over the shaft and bellows. Consequently, this seal is subject to the same speed and spare parts inventory considerations. Additional limitations are imposed by the bellows itself, which replaces the "O" ring shaft packing found in the conventional designs. While a satisfactory bellows material usually can be selected to meet the environmental requirements of the stuffing box, the bellows itself is not as closely confined in a conventional seal as the shaft packing. Therefore, the bellows is much more susceptible to fatigue failure.

This same lack of confinement adds to the seriousness of a bellows failure, as opposed to a conventional shaft packing failure. If conventional shaft packing begins to deteriorate, its confinement generally dictates that the resulting leakage is restricted and, therefore, growing progressively worse as the shaft packing continues to fail. This often provides the user with sufficient warning to plan for a shutdown to repair the seal. However, in the event of a bellows rupture, the resulting leakage is often great enough to necessitate immediate repairs.

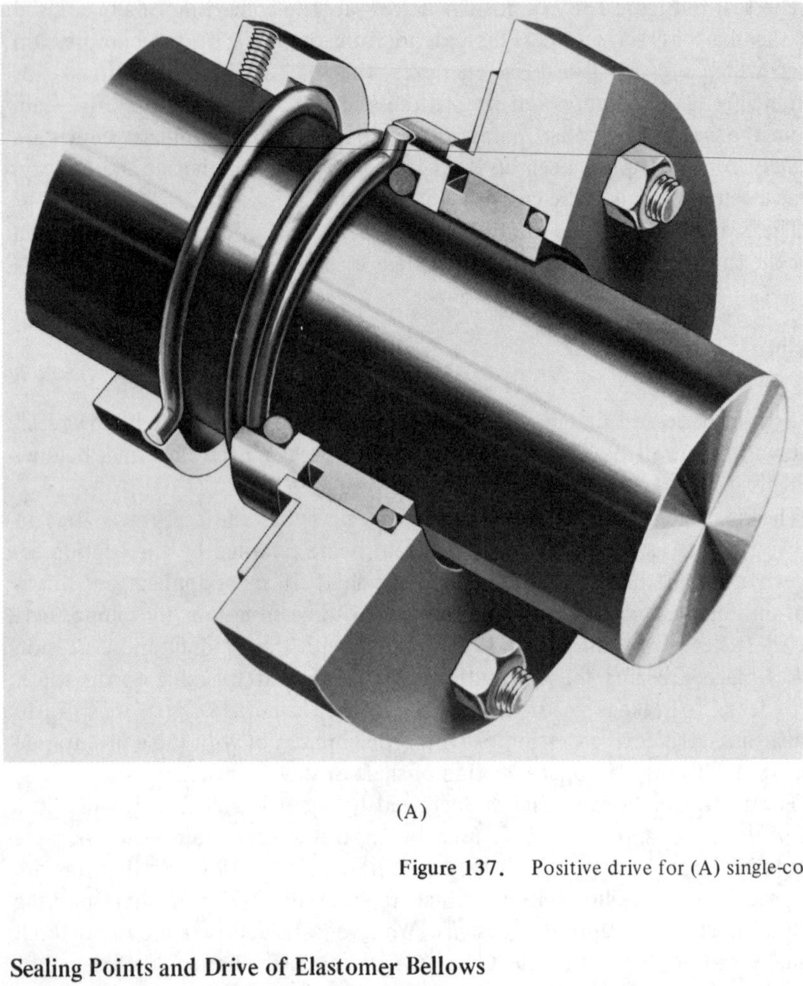

(A)

Figure 137. Positive drive for (A) single-coil

Sealing Points and Drive of Elastomer Bellows

The sealing points of an elastomer bellows seal are primarily the same as encountered in conventional seals. The bellows sleeve grips the shaft at the rear of the rotary unit assembly, transmitting drive to the seal ring. Although the seal illustrated is a friction drive seal, the increased friction offered by the bellows sleeve exempts this seal from the nonlubricating liquid limitation imposed on the conventional friction drive seal. Figure 139(A) illustrates the sealing points and defines the nomenclature of an elastomer bellows seal.

Metal Bellows

The metal bellows seal shown in Figure 139(B) permits the bellows seal design to be used at higher speeds and temperatures than the elastomer

(B)

spring seals and (B) multicoil spring seals.

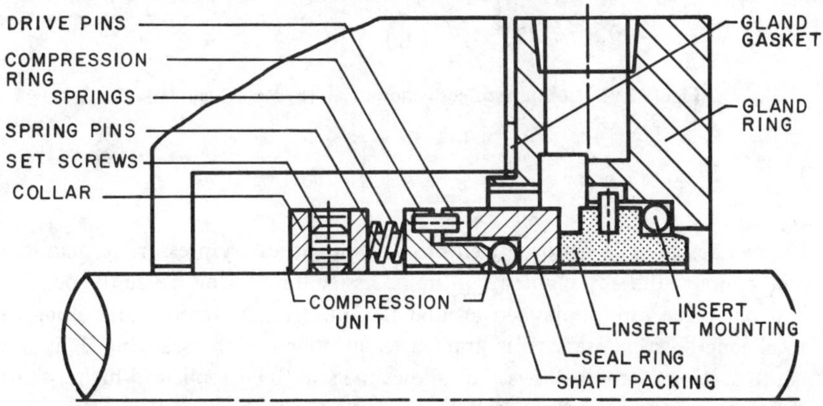

Figure 138. Mechanical seal with multiple springs.

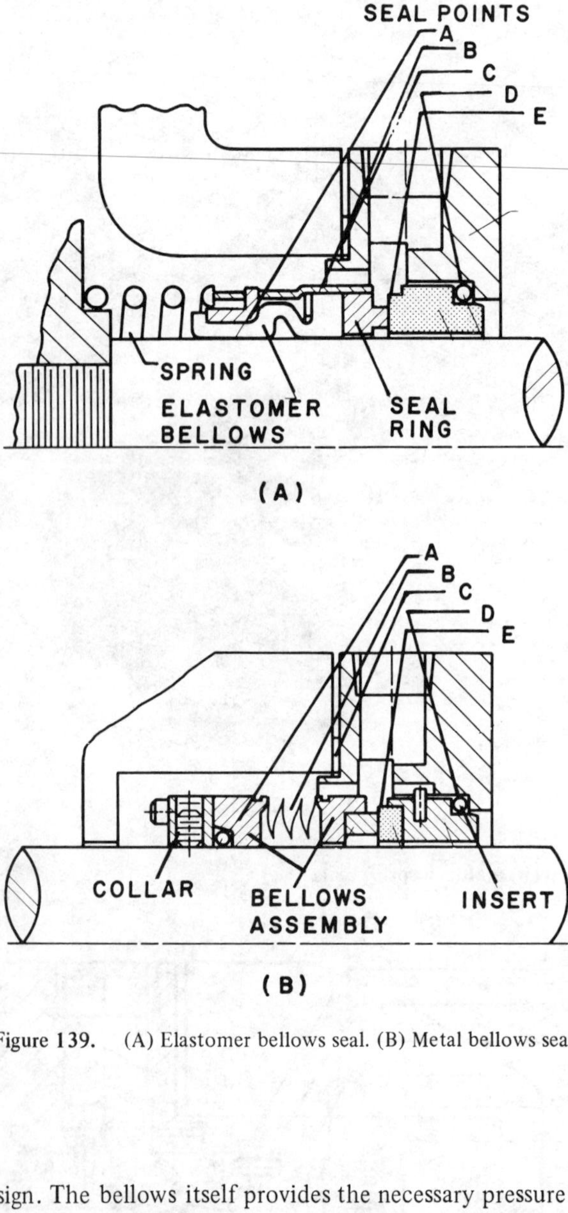

Figure 139. (A) Elastomer bellows seal. (B) Metal bellows seal.

bellows design. The bellows itself provides the necessary pressure to maintain face contact, thereby eliminating the necessity for a spring. Since the bellows is a one-piece unit supported around its entire circumference and along its total length, even pressure is applied to all points of the seal ring. It is this configuration of the bellows that allows the seal to be applied to higher shaft speeds than can be accommodated by a single coil spring over the shaft.

Aside from being able to accommodate higher shaft speeds, the metal bellows also can be applied to high and cryogenic temperatures. Until recently, conventional seals had to rely on elastomers or TFE resins for secondary sealing members (shaft packing and insert mounting). These materials have a rather narrow temperature range, depending on their composition. Even though a particular material might be suitable for use in an extreme temperature, this did not necessarily mean that it has the necessary resistance to chemical attack by the liquid that has to be sealed. Although the combined problem of temperature and chemical resistance could be generally overcome by creating an artificial, compatible environment in the stuffing box, such a solution requires the use of costly environmental controls. Therefore, the metal bellows seal promises an economic solution to many high- and low-temperature sealing problems.

Figure 139(B) shows that the metal bellows is actually a series of plates called convolutions. Each one of these convolutions is a separate ring that has been welded to similar rings on either side. Unfortunately, from a single piece of material it is not possible to make a bellows with the necessary mechanical properties. It is the necessity for welding that currently presents the possibility of fatigue failure, resulting in ruptured bellows. Such a rupture can result in severe leakage. Although there has been considerable effort to improve its reliability, the need for a bellows of this type for high-temperature seal applications has been reduced significantly by the recent introduction of all-graphite secondary packings for conventional seals that are capable of functioning in high- or low-temperature extremes. They are also inert to the deteriorating effects of most chemicals. When considering the bellows seal, note that when the rotary face requires replacement, the entire bellows assembly also must be replaced.

Balanced Seals

One of the most frequently encountered environmental conditions requiring a slight modification in seal design is pressure. If an inside mechanical seal is called on to seal high pressures, provisions must be made to ensure that all of the pressure on the seal is not trying to push the seal ring to the atmospheric side of the stuffing box.

Figure 140 is a cross section of a conventional inside unbalanced seal. Almost all of the stuffing box pressure is exerting a closing force on the seal ring. Only a very small portion of the seal ring face is exposed over the top of the insert, allowing a proportionately small amount of pressure to work against the seal ring in the opposite direction, in addition to the opening force exerted by the liquid film between the faces. If the closing force becomes great enough, the liquid film between the faces is literally squeezed out. Deprived of lubrication and highly loaded, the faces soon destroy themselves. The solution to this problem is to balance the seal.

Figure 141 illustrates a conventional inside seal that has been balanced.

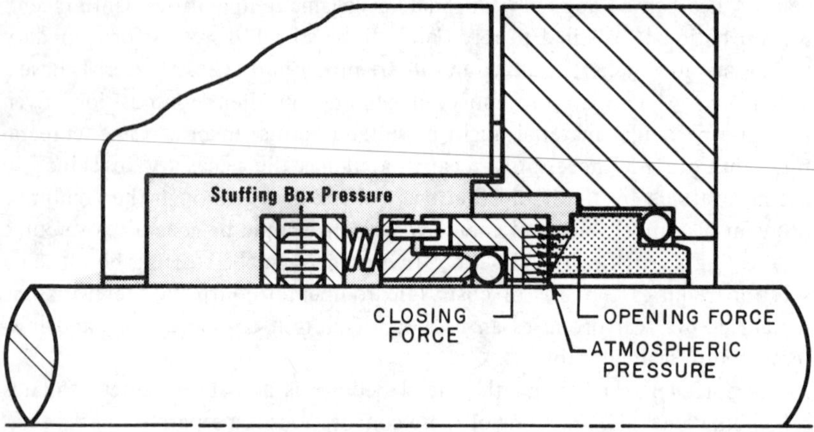

Figure 140. Inside unbalanced seal.

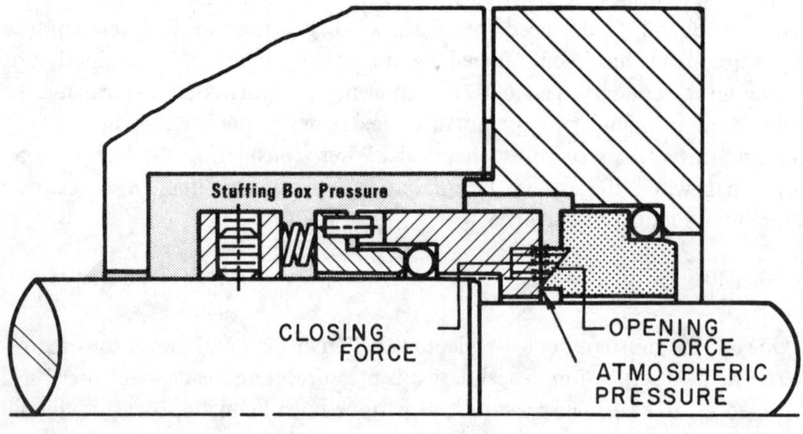

Figure 141. Inside balanced seal.

Notice that a step in the shaft sleeve moves the insert face radially inward without decreasing the width of the face itself. The seal ring, on the other hand, remains mounted on the original shaft diameter, which means that the closing force remains unchanged. Because more of the seal ring face is exposed to the hydraulic pressure working to open the seal, the design is balanced. When comparing the surface areas of the seal ring available for the hydraulic pressure to work against, it becomes apparent that the opening force is slightly less than the closing force. This is intentional to ensure that the faces are kept in contact at all times.

The method of achieving balance on an outside seal is the same as with an inside seal, except that the action is the reverse. Instead of counterbalancing a portion of the closing force imposed by stuffing box pressure, a portion of the stuffing box pressure is added to the closing force, counteracting the opening force exerted by the liquid film between the faces.

Figure 142 is a cross section of an outside balance seal. The shaft packing is forced against the collar, leaving an area under the seal ring exposed to stuffing box pressure. The closing force exerted by the stuffing box pressure, acting against the shoulder of the seal ring, is slightly greater than the opening force exerted by the liquid film between the faces, thereby keeping the faces in contact at all times.

High-Pressure Seals

Seals for extremely high pressures demand more design sophistication than standard balanced seals. Not only is the ratio between the opening and closing force more critical, but the entire seal must withstand elevated pressure and ensure that the faces do not become distorted.

Specifically, the entire rotary unit cross section is thicker. Special springs are provided to ensure proper face loading on this very closely balanced design. Drive pins are also heavier to accommodate more torque. The insert also has a larger cross section to withstand the pressure. The gland thickness

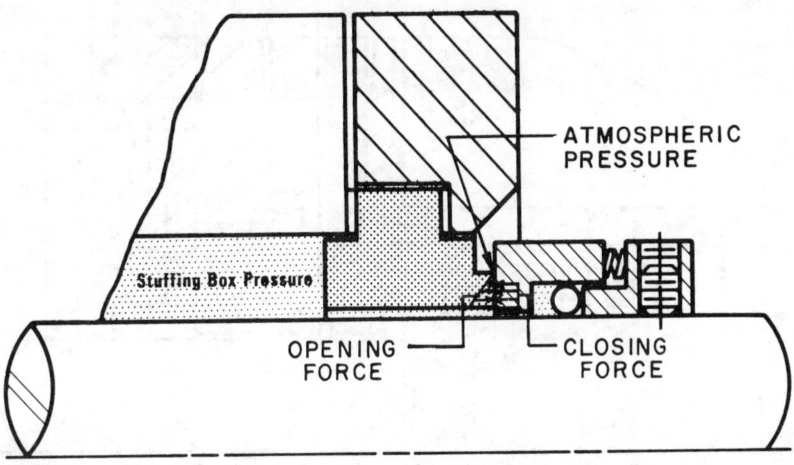

Figure 142. Outside balanced seal.

is greatly increased to prevent distortion. Grinding of the bottom of the gland counterbore, against which the insert is held (insert backup), is important to ensure flatness. At high pressures, any high point on this surface would most certainly be transmitted to the insert face, resulting in leakage. Figure 143 illustrates a high-pressure seal design.

High-Speed Seals

To conserve space and achieve a more efficient weight to horsepower ratio, more attention is being given to high-speed pumps with shaft speeds in excess of 6500 fpm. Because dynamic forces begin to exceed the limitations of a conventional rotary unit at this speed, the roles of the insert and rotary unit are reversed. The springs become stationary, loading a stationary rather than rotating seal ring. The insert is flexibly mounted inside a collar, which is set screwed to the shaft. Rotational drive is transmitted to the insert via lugs in the collar, rather than pins. Similar lugs hold the seal ring stationary. Figure 144 illustrates the configuration of a typical high speed seal.

MECHANICAL SEAL ARRANGEMENTS

Single inside mechanical seals (Figure 140) are the most common type of mechanical seal used. The materials of construction are selected to withstand the corrosive nature of the liquid in the stuffing box. They are easily modified to include environmental controls and can be balanced to withstand extremely high stuffing box pressures.

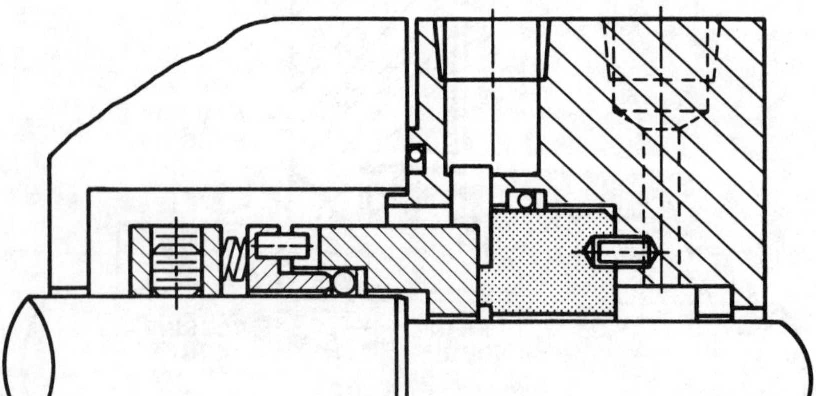

Figure 143. High-pressure seal.

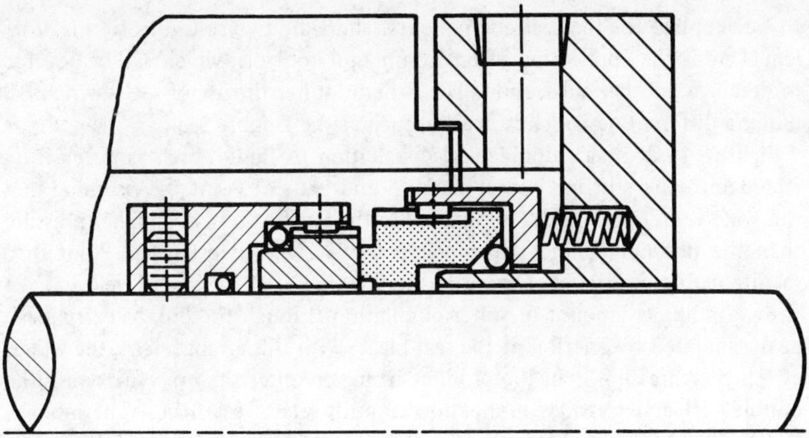

Figure 144. Typical high-speed seal.

Inside seals require a suitable stuffing box housing for installation and cannot be adjusted without dismantling the equipment.

If an extremely corrosive liquid is encountered that has satisfactory face lubricating properties, an outside seal offers another alternative to expensive metallurgies necessary for an inside seal to resist corrosion. Outside seal arrangements have only the insert, seal ring and secondary seals exposed to the product. All of these components can be nonmetallic. The metallic rotary unit parts are exposed only to the atmosphere.

Outside seals also can be an advantage when a piece of equipment is found to have a stuffing box that will not accommodate an inside seal. Although outside seals are easier to adjust and troubleshoot because they are exposed, they have the disadvantage of being vulnerable to damage from impact. Of greater importance, however, is the pressure limitation of an outside unbalanced seal. By contrast to an inside seal, the hydraulic pressure works to open, rather than close, the seal faces. Therefore, the seal design depends entirely on the springs to maintain face contact. Even though a degree of balance can be designed into an outside seal, all outside seals are limited to use in conjunction with moderate stuffing box pressures.

Double Seals

When selecting a seal for a pump, liquids are sometimes encountered that are not compatible with a single inside mechanical seal. Often these liquids carry abrasive materials in suspension that would rapidly wear the faces; or the liquid may be so corrosive that extremely expensive materials would have to be supplied for the seal components. There are two solutions to this problem. One is the application of environmental controls, which will be covered in detail in another subsection. The other requires the use of seals, which fall under a different design classification such as the Double Seal.

Figure 145 depicts a double seal installation. A liquid, such as water, is injected into the stuffing box at Point A and exits at Point B. Assuming that the water entering Point A is available to the stuffing box at a higher pressure than the product behind the impeller, a valve is located beyond Point B to maintain this pressure between the double seal at the desired level, and yet allow a minimal amount of water circulation through the box to carry away heat generated by friction at the seal faces. With this arrangement, the water, at a higher pressure than the product trying to enter the box, surrounds the double seal and provides lubrication to both sets of seal faces. The inboard seal prevents the water from entering the pump while the outer seal prevents the water from escaping to the atmosphere.

The differential pressure across the inner seal is the difference in pressure between the sealing liquid pressure and the product pressure acting on the stuffing box. The differential pressure across the outer seal is the difference in pressure between the sealing liquid and the atmosphere. Either, or both, of the seals may be balanced if the differential pressures exceed the limitations of unbalanced seals.

Another vairation in seal arrangement is the double tandem seal, illustrated in Figure 146. The purpose of this seal is not to create an artifical environment, as is the case with the double seal, but to provide a backup seal in the event the inner seal fails. The inner seal functions in a manner identical to a

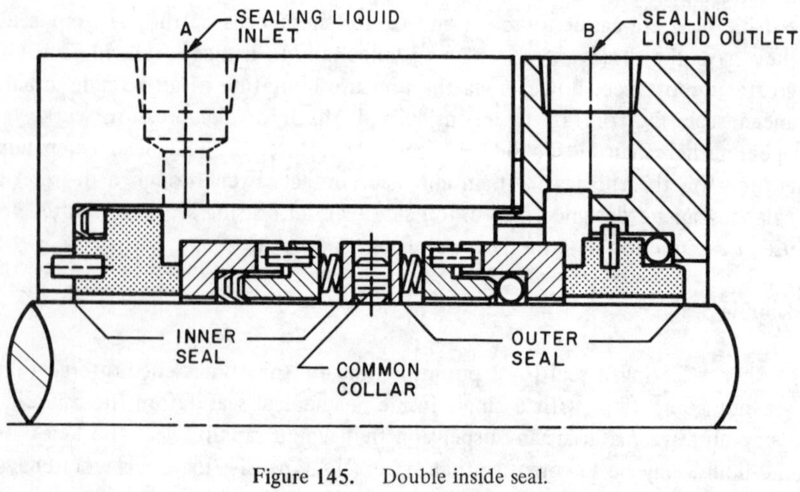

Figure 145. Double inside seal.

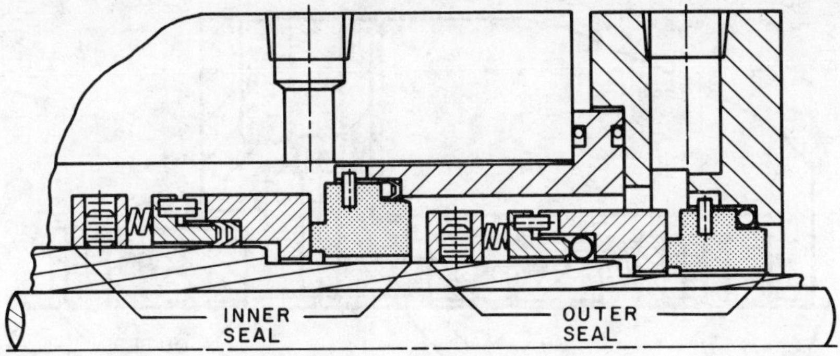

Figure 146. Double tandem seal.

conventional single inside seal. The cavity between the inner and outer seal is flooded from a closed reservoir. The liquid in the reservoir provides lubrication to the outer seal. Because the space between the seals is only flooded and not under pressure, the product, not the liquid in the reservoir, lubricates the faces of the inner seal. If the inner seal fails, the resulting pressure rise in the area between the seals is sensed at the reservoir, where it either can be registered on a gauge or can activate an alarm. In any event, a failure of the inner seal can be detected, while the outer seal assumes the responsibility of sealing the shaft until repairs can be made.

Since the purpose of the tandem seal does not require that the liquid between the seals be at a higher pressure than the product pressure, the inner seal can be balanced for high stuffing box pressures without requiring a higher sealing liquid pressure. If a conventional double seal were used in place of a tandem seal, product pressures would have to be limited to prevent the hydraulic pressure from opening the inner seal faces. However, a tandem seal is capable of being balanced to accept high stuffing box pressures.

Requirements for tandem or double seals are sometimes encountered where the equipment stuffing box is too shallow to accommodate the more conventional varieties of these two fundamental designs. Therefore, it becomes necessary to turn to an alternate seal arrangement. In this case, the alternative is the inside-outside double seal.

The inside-outside double seal is assembled, as the name implies, with one seal inside the stuffing box and one outside the stuffing box. Both seals rotate against opposite ends of the same stationary insert. Figure 147 illustrates an inside-outside double seal assembly, using an unbalanced inner seal and a balanced outer seal.

Whether the inside-outside arrangement is to be considered a tandem or double seal depends on its function. If the liquid used between the seals is at

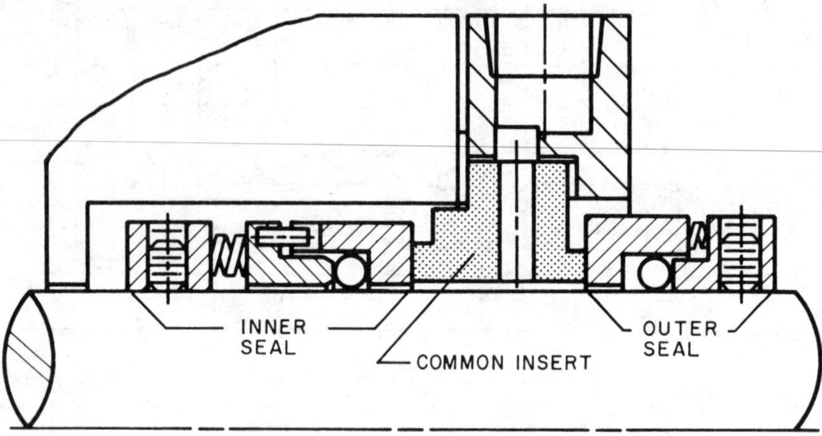

Figure 147. Double inside-outside seal.

a higher pressure than the product in the stuffing box, then the purpose of this design is to lubricate the inner seal with a liquid other than the product. This is the role of a true double seal, creating an artificial environment in which the mechanical seal can operate. However, if the liquid is circulated between the seals at a lower pressure than the equipment stuffing box pressure, the role of the inner seal remains identical to that of any single seal, while the outboard seal simply serves as a backup seal in the event the inner seal fails. A situation such as this would identify the inside-outside assembly as a tandem seal.

MATERIALS OF CONSTRUCTION

The environment in which the seal will operate must be analyzed to determine the proper materials of construction. The corrosive properties of the liquid, temperature and temperature variation must be considered.

Lubricity is another important factor. Liquids with lubricating qualities similar to, or less than, that of water require the use of face materials that are known to be capable of maintaining satisfactory wear rates without the benefit of good lubrication. In addition, whatever face materials are selected must have compatible wearing surfaces. This is not generally a problem since one of the seal face materials is almost always carbon. However, in rare applications in which either the temperature or corrosive properties of the liquid exceed the limitations of carbon, a substitute material must be selected.

Metallic face materials are constructed from Kalamate/Stellite®, Duramate®, Ni-Resist®, bronze or Hastalloy®. Kalamate is a trade name for seal rings made from solid cast stellite. Stellite is a good wearing material with a corrosion resistance greater than that of the 300 series stainless steels. Its composition is primarily chrome and cobalt, with small percentages of nickel, iron and tungsten. Stellite is considered generally satisfactory for use in conjunction with a wide range of liquids with a similarly wide range of lubricating properties. Stellite, however, is not recommended for use in water because the water has a tendency to leach out the cobalt.

Duramate is a hardened stainless steel that is compatible with carbon as a wearing surface in the presence of liquids possessing lubricating properties equal to, or better than, light hydrocarbons. Duramate has a corrosion resistance roughly equivalent to that of 304 stainless steel.

Ni-Resist is an economical material that is sometimes used for mildly corrosive services. Ni-Resist is a 42–67% cast iron material containing 13–32% nickel, 10% copper and small amounts of other metals. When used, Ni-Resist is usually applied to the sealing of petroleum products.

Bronze has proven to be an excellent face material when mated against tungsten carbide in abrasive services. It has greater abrasion resistance than carbon and has been used extensively on water flood, pipeline and crude oil services where abrasion commonly occurs.

Bronze also has been used advantageously when the fragility of carbon limits the use of the latter material. Its higher modulus of elasticity and other improved physical properties dictates its use in high pressure applications. Due to its relatively low corrosion resistance, caution should be used in applying bronze in corrosive liquids.

Hastelloy is almost chemically inert. It is used primarily in compression units requiring a high degree of corrosion resistance. As a seal ring, Hastelloy displays poor wear characteristics and is only used occasionally. Before the introduction of ceramics and cost reductions for tungsten carbide, Hastelloy was used as a seal face where extreme corrosion resistance was required.

Face materials are also constructed from ceramics, which offer excellent wear characteristics due basically to their hardness. They also possess chemically inert properties that often make their use desirable when non-metallic face material combinations are specified for difficult liquids.

Care must be exercised when recommending ceramic seal rings since they are very susceptible to fracture from thermal or mechanical shock. In spite of this disadvantage, ceramic seal rings remain popular because of their economical cost. Ceramic seal rings fall into two major classifications; solid and ceramic faced.

Carbides are another major construction material. A seal face of tungsten carbide possesses qualities of superior hardness, a wide range of chemical inertness and excellent antifrictional properties. Consequently, it is undoubtedly the most versatile face material available. However, cost dictates that the use of tungsten carbide be confined to applications where its performance will prove to be a distinct advantage over that of other face materials. As an example, tungsten carbide is routinely applied to high pressure applications where, in the solid form, its high modulus of elasticity is a valuable asset in preventing face distortion. Tungsten carbide also exhibits outstanding wear characteristics in the presence of liquids with extremely poor lubricating qualities. Table 32 lists different grades of tungsten carbide offered by one manufacturer.

New materials that exhibit superior hardness and wearability are constantly being tested and evaluated for use as seal faces. One of these materials that has shown superior qualities as a seal face and the use of which is increasing rapidly, is silicon carbide.

Silicon carbide's high, uniform hardness and high density make it an extremely abrasion-resistant material that outwears all other materials. Because of its high thermal conductivity combined with its high tensile strength and low thermal expansion, silicon carbide has excellent thermal shock resistance and, unlike tungsten carbide, does not heat check. These properties, along with its unlimited corrosion resistance and low coefficient of friction, make silicon carbide an excellent seal face material.

TFE resin (Teflon®)* is another material with high chemical inertness. It has a tendency to "cold-flow," however, stability is achieved through glass impregnation. Glass-filled TFE resin is a poor wearing material and is, therefore, limited to very low pressure-velocity ratings.

Typical temperature limitations for different types of face materials are given in Table 33.

---

*Registered trademark of E. I. duPont de Nemours and Company, Inc., Wilmington, Delaware.

Table 32. Commonly Used Grades of Tungsten Carbide

Grade	Binder	Remarks
62-6 Tung-Car "M" Tung-Car "A"	Nickel	Standard grade. Wide range of applications. Suited for hot H_2O, petroleum products and most chemicals.
62-1 Tung-Car "M1"	Cobalt	Used in environments where nickel binder of grade 62-6 would be attacked, such as ammonia.
62-3	Tantalum	Called tantalum carbide because of binder. Very expensive, chemically inert. Only used in presence of strong oxidizing agents.

Table 33. Temperature Limitations of Seal Face Materials

Material	Construction	Maximum Temperature (°F)	Maximum Temperature (°C)
Stellite Face	Welded Stellite face on metal ring	350	177
Kalamate	Solid-cast Stellite ring	450	232
Tung-Car	Solid tungsten carbide ring	750	400
Tung-Car "M"	Solid tungsten carbide element mounted in metal body	350	177
Tung-Car "A"	Solid tungsten carbide element mounted in Duramate body	750	400
Duramate	Solid machined stainless steel ring	600	316
Ni-Resist	Solid nickel-cast iron ring	350	177
Bronze	Solid leaded-bronze ring	350	177
Duramic	Solid ceramic ring	350[a]	177[a]
Peramic	Solid pure ceramic ring	350[a]	177[a]
55 Face	Ceramic facing on metal ring	350[a]	177[a]
No. 5 Carbon	Solid carbon-graphite ring	525	275
No. 6 Carbon	Solid carbon-graphite ring	525	275
Silicon Carbide "V"	0.030′ thick conversion of silicon carbide on carbon-graphite substrate	700	371
Silicon Carbide 1	Solid silicon carbide ring	3000	1650
Glass-Filled Duraflon	Solid ring	350	177

[a]Subject to fracture from thermal shock.

SECONDARY SEAL MATERIALS

Secondary seals applied to mechanical seals are defined as the shaft packing and insert mounting. Because they are static and must seal imperfect surfaces, secondary seals must possess a degree of resiliency that will allow them to seat perfectly against adjacent surfaces. The most elastic or resilient of all secondary seals are elastomers, which are applied in the configuration of "O" rings. However, the temperature and chemical resistance of elastomeric materials are often exceeded, thus requiring the use of other less resilient materials. Because these other materials lack the elasticity of elastomers, they assume configurations that tend to compensate for their limited resiliency. Several elastomer "O" ring types are described below:

Buna N (Nitrile)

This material, often referred to as nitrile, is a copolymer of butadiene and acrylonitrile. Buna N has an excellent resistance to petroleum products, also finding wide acceptance in water applications.

Buna N can be exposed to temperatures ranging from -40°F to +225°F (-40°C to +107°C).

Inherently, Buna N does not possess good resistance to ozone, sunlight or weather. Therefore, spare "O" rings should not be stored in direct sunlight or adjacent to electrical equipment that may generate ozone.

Neoprene

Neoprene, also called chloroprene rubber, was among the earliest of the synthetic rubbers available to seal manufacturers. It continues to be used for refrigerants, such as freon and ammonia, and other mild services. Like Buna N, Neoprene can be subjected to temperatures ranging from -40°F to +225°F (-40°C to +107°C).

Butyl

This is an all petroleum product produced by copolymerizing isobutylene and isoprene. Butyl rubber will resist the deteriorating effects of many mild liquids, such as MEK and acetone, but should not be used in petroleum oils. Its temperature range is -40°F to +225°F (-40°C to +107°C).

Silicone Rubber

Silicone elastomers are made from silicone, oxygen, hydrogen and carbon. They usually display poor tensile strength, tear and abrasion resistance. Silicone "O" rings are not recommended for use in most petroleum fluids or ketones. They can be applied successfully to temperatures ranging from -80°F to +400°F (-62°C to +204°C).

Fluorosilicone

A more recent development, fluorosilicone combines the good high- and low-temperature properties of silicone rubber with the resistance to petroleum oils and hydrocarbon fuels. Like silicone rubber, it can be in a temperature range of -80°F to +400°F (-62°C to +204°C).

Hypalon

A chlorsulfonated polyethylene, Hypalon displays good acid resistance. Depending upon the liquid, it can be exposed to temperatures ranging from -40°F to +225°F (-40°C to +107°C). Note, however, that the compression and permanent set characteristics of Hypalon are not as desirable as found in most other elastomeric materials. Therefore, it is used only as a secondary seal when other materials will not stand up when exposed to a particularly difficult fluid.

Viton®

A versatile compound, Viton is a copolymer of vinylidene fluoride and hexafluoro propylene made by Du Pont. Also known as fluorocarbon rubber, it has excellent tolerance for temperatures ranging from 0°F to +400°F (-18°C to +204°C).

Viton is applicable to petroleum oils, diester base lubricants, silicone fluids, halogenated hydrocarbons, water, low-temperature steam, a wide variety of acids and many other fluids too numerous to mention. Fluorocarbon rubber will not tolerate ketones, anhydrous ammonia, amines and hot hydrofluoric, or chlorosulfonic, acids.

Kel-F®

Kel-F is a fluorocarbon rubber made by the 3M Company. It is a material similar to Viton. Therefore, the two materials generally can be used interchangeably.

EPR/EPT

EPR and EPT are made from ethylene and propylene monomers. EPR actually stands for "ethylene propylene rubber." EPT is a very similar compound containing a small amount of a third monomer, making the compound "ethylene propylene terpolymer." Although the ethylene propylene compounds are relatively new, they have found wide acceptance in the sealing industry. EPR's have excellent corrosion resistance to dilute acids and alkalis, ketones, alcohols, water, steam and phosphate ester base hydraulic fluids, but are not recommended for petroleum oils or diester base lubricants. They are applicable for temperatures ranging from -40°F to +300°F (-40°C to 150°C).

Kalrez®

Kalrez is Du Pont's perfluoroelastomer. Kalrez combines many of the elastomeric properties of Viton and the chemical and heat resistance of TFE resin with certain advantages over both. Kalrez can be applied for temperatures ranging from 0°F to +500°F (-18°C to 260°C).

Kalrez has excellent corrosion resistance to solvents, inorganic and organic acid and bases, strong oxidizing agents, metal halogen compounds, hot mercury, chlorine, fuels and heat transfer fluids.

Durafite®

Durafite is a relatively new secondary packing material. It is an all-graphite product containing no resin binders or organic fillers. It is used in environments where the temperature exceeds the limitations of other secondary sealing materials.

Durafite is especially inert. It is vulnerable to attack only by oxidizing liquids. In addition, it is highly resistant to all forms of nuclear radiation. Durafite can be applied in temperatures ranging from cryogenics, -450°F (-268°C), to higher than 750°F (400°C).

Table 34 gives typical temperature limitations of secondary seal materials.

ENVIRONMENTAL CONTROLS

Environmental controls are necessary to control dirt and heat, or to cool the seal area and, under certain conditions, to keep the product away from the seal entirely.

Temperature Control

The temperature surrounding the seal in a stuffing box is an important consideration. Certain parts of the seal, such as the secondary packings and carbon, may deteriorate under exposure to extremely high temperatures. As an example, Kalrez is the most stable elastomer material at high temperatures. Nevertheless, its use is limited to temperatures below 500°F (260°C). The binder in most carbons begins to break down at temperatures above 500°F (260°C). Although the temperature limitations of these two materials are generously high, seal faces and secondary packings cannot be selected entirely on their ability to withstand high temperature. These materials also must be able to resist chemical attack by the liquid surrounding them. Consequently, the mechanical seal specialist is not always at liberty to apply the most temperature-resistant materials. Often a compromise between temperature resistance and resistance to the corrosive properties of the liquid must be made. In making this compromise, it often becomes necessary to protect some of the materials by cooling the seal area.

Occasionally, the material being pumped solidifies at ambient temperatures. In its liquid state, however, this same product may act as an excellent clean lubricant. Such is the case with wax which, in order to be pumped, must be heated. The stuffing box area surrounding the mechanical seal also must be heated to ensure that the seal will operate in a liquid environment.

Another important temperature consideration is the boiling point of the liquid in the stuffing box. Although the product must remain below its

Table 34. Temperature Limitations of Secondary Seal Materials

Material	Form	Minimum Temperature Limit[a] (°F)	Minimum Temperature Limit[a] (°C)	Maximum Temperature Limit[a] (°F)	Maximum Temperature Limit[a] (°C)
Buna N (Nitrile)	"O" ring	−40	−40	+225	+107
Neoprene	"O" ring	−40	−40	+225	+107
Butyl	"O" ring	−40	−40	+225	+107
Silicone	"O" ring	−80	−62	+400	+204
Fluorosilicone	"O" ring	−80	−62	+400	+204
Hypalon	"O" ring	−40	−40	+225	+107
Kel-F	"O" ring	0	−18	+350	+177
Viton	"O" ring	0	−18	+400	+204
EPR/EPT (ethylene propylene)	"O" ring	−40	−40	+300	+150
Kalrez	"O" ring	0	−18	+500	+260
Duraflon (TFE resin)	"V" ring	−100	−73	+350	+177
Glass-Filled Duraflon	"V" ring	−175	−115	+450	+232
Durafite (pure graphite)	Square ring	−450	−268	+750	+400

[a]The minimum and maximum temperature limits listed are general limits and will vary with the liquid in contact with the elastomer.

boiling point if it is to be pumped, stuffing boxes can be heat traps. They may not efficiently dissipate the additional heat that can be expected to be generated at the seal faces and through turbulence created by the seal spinning within the close confines of the stuffing box. Although the amount of heat generated in this manner may not be great, it can be cumulative to a certain degree, thereby eventually raising the temperature of the hot product in the stuffing box to its boiling point. If this occurs, the seal faces do not enjoy the benefits of a liquid film between them. As a general rule, the temperature of the liquid in the stuffing box should be maintained at least 25°F (14°C) below the boiling point of the liquid at the stuffing box pressure. Vaporization also may cause precipitation of salts, minerals or other abrasive impurities. Accelerated face wear is then imminent, ultimately culminating in a complete failure of the mechanical seal.

Cooling normally is accomplished by flushing. Flushing can be defined as the introduction of a liquid into a stuffing box at a higher pressure than stuffing box pressure. To be effective, the liquid being injected must flow into the box at a rate that is sufficiently high to prevent the product behind the impeller from entering the stuffing box. Since most stuffing boxes are restricted at the throat, velocity is imparted to the flush liquid leaving the box, thereby preventing the pumped product from entering the stuffing box.

Two methods of flushing can be employed. The most commonly used is the bypass flush system. When it is desired to use a bypass flush for cooling, a recirculation line is run from the discharge nozzle of the pump through a heat exchanger to a connection in the gland ring located over the seal faces. Since discharge pressure is inevitably higher than stuffing box pressure, circulation is ensured. It is important to note that some pump designs subject the stuffing box to pressures equal to, or very near, full discharge pressure. In such cases, bypass is achieved by running a recirculation line from the gland ring back to the pump suction or other low pressure source. Flushing in the gland ring for an inside seal, shown in Figure 148(A), consists of a drilled and tapped hole in the gland ring, which channels the liquid directly over the seal faces.

Flushing an outside seal is shown in Figure 148(B). It is important to note that in addition to the tapped hole in the gland ring, a hole is drilled through the stationary insert to channel the flush into the space between the insert inside diameter (i.d.) and the shaft. To seal the insert outside diameter (o.d.) to the gland bore, the insert is mounted in two "O" rings.

In the arrangement described, the flush flow is directed through a heat exchanger before entering the stuffing box. However, a simple bypass flush without a heat exchanger is an effective temperature stabilizing system that is also used regularly in conjunction with single mechanical seal installations. Although this type of flush does not cool the liquid being bypassed to or from the stuffing box, it is extremely effective in maintaining the temperature within the stuffing box at the same level as the pumped product. This is

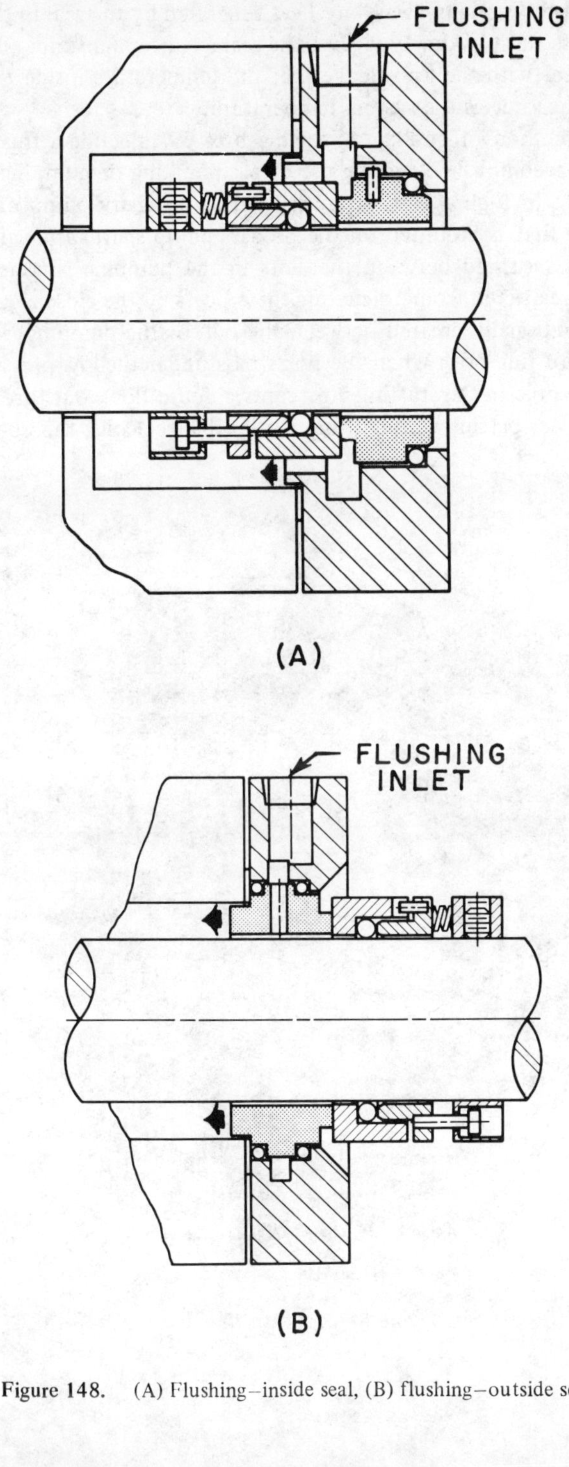

Figure 148. (A) Flushing–inside seal, (B) flushing–outside seal.

accomplished by carrying away any heat generated by the seal in the stuffing box. A bypass flush of this type gives the seal engineer a distinct advantage in selecting a seal with the knowledge that the temperature inside the stuffing box will be the same as the system temperature.

A second method of cooling a stuffing box by injecting a flush over the seal faces is accomplished by the use of a circulating or pumping ring. One type is shown in Figure 149. It consists of an auxiliary pump of low head and capacity that is mounted on the process pump shaft inside the stuffing box. Liquid is carried between the slots in the pumping ring as the shaft rotates. It is restricted from evacuating these ports by the close clearance that has been intentionally created between the o.d. of the pumping ring and the bore of the stuffing box. When the ports pass under the low-pressure area at the discharge hole in the stuffing box, captive liquid flows out through a heat exchanger. After passing through the heat exchanger coils, the cooled liquid

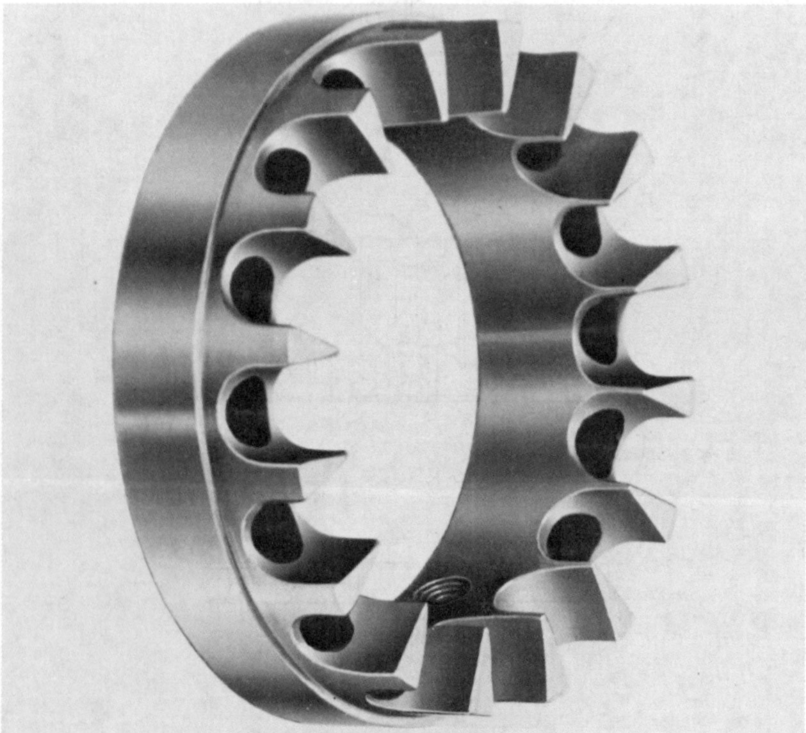

Figure 149. Circulating ring.

returns to the stuffing box via a flushing connection in the gland ring located directly over the seal faces.

The circulating ring is often preferred over a bypass flush system when a heat exchanger must be used. Depending on pump pressures and temperatures, a circulating ring feature may allow the use of a smaller heat exchanger because of its low flowrate. Note that most bypass flush systems originate at the pump discharge; they depend on relatively high flowrates to create enough velocity at the throat of the stuffing box to purge the area of hot process liquid. This increase in flush line pressure, coupled with a shorter residence time in the heat exchanger, often requires the use of a heavier heat transfer device with greater cooling coil area.

When installing a circulating ring cooling system, caution must be exercised in locating the heat exchanger. Unduly long flush lines with unnecessary turns can easily set up frictional losses that exceed the discharge head of the circulating ring. Therefore, the heat exchanger should be mounted as close to the pump stuffing box as possible, preferably directly over the stuffing box.

The capacity of the circulating ring to develop head is limited by the peripheral speed of the shaft. A minimum peripheral speed of 800 fpm is required to deliver 0.5 gpm. Table 35 can be used for computing circumferential speeds in fpm from shaft size and rpm.

Cooling often is used when the liquid is dead-ended in the stuffing box and provisions for flushing are uneconomical. However, the word "cooling" can be misleading since the same design approach can be used to transmit heat to the seal, thereby preventing solidification or crystallization of the product.

Whether this provision is used for cooling or heating, the design and method of heat transfer remain the same. The gland ring is machined to include two drilled and tapped holes leading to an annular groove in the counterbore of the gland ring. The heat transfer fluid, usually water or steam, is circulated through this annular groove picking up from, or imparting heat to, the stationary insert for the purpose of controlling seal face temperatures. Since carbon is a relatively poor conductor of heat, other methods of temperature control prove to be more efficient. However, cooling of this type can be particularly effective when combined with a water-jacketed stuffing box, as shown in Figure 150.

For higher-temperature steam heating, a two-piece welded gland ring is available for steam pressures to 300 psia and temperatures to 417°F (214°C). Most installations that involve any type of steam heating are of such a nature that the heating should be left on at all times, regardless whether the equipment is operating. Careful preheating to dissolve all solids is necessary following a complete shutdown.

Quenching the outside seal is still another approach to temperature control. Quenching an outside mechanical seal involves immersing a large portion of

Table 35. Conversion of Shaft Size and rpm to Peripheral Speed

Diameter in Inches			Revolutions per Minute									
(rpm row 1)			2,400	2,600	2,800	3,000	3,200	3,400	3,600	3,800	4,000	4,400
	(rpm row 2)		1,200	1,300	1,400	1,500	1,600	1,700	1,800	1,900	2,000	2,200
		(rpm row 3)	600	650	700	750	800	850	900	950	1,000	1,100
1/4	1/2	1	157	170	183	196	209	223	236	249	262	288
1/2	1	2	314	340	367	393	419	445	471	497	524	576
3/4	1 1/2	3	471	510	550	589	628	668	707	746	785	863
1	2	4	628	681	733	785	838	890	942	995	1,047	1,152
1 1/4	2 1/2	5	785	851	916	982	1,047	1,113	1,178	1,244	1,309	1,440
1 1/2	3	6	942	1,021	1,100	1,178	1,257	1,335	1,414	1,492	1,571	1,728
1 3/4	3 1/2	7	1,100	1,191	1,283	1,375	1,466	1,558	1,649	1,741	1,832	2,016
2	4	8	1,257	1,361	1,466	1,571	1,675	1,780	1,885	1,990	2,094	2,304
2 1/4	4 1/2	9	1,414	1,531	1,649	1,767	1,885	2,003	2,121	2,238	2,356	2,592
2 1/2	5	10	1,571	1,702	1,833	1,964	2,094	2,225	2,356	2,487	2,618	2,880
2 3/4	5 1/2	11	1,728	1,872	2,016	2,160	2,304	2,448	2,592	2,736	2,880	3,168

3	6	12	1,885	2,042	2,199	2,356	2,513	2,670	2,827	2,984	3,143	3,456
$3^1/_4$	$6^1/_2$	13	2,042	2,212	2,382	2,552	2,723	2,893	3,063	3,233	3,403	3,744
$3^1/_2$	7	14	2,199	2,382	2,566	2,749	2,932	3,115	3,299	3,482	3,665	4,032
$3^3/_4$	$7^1/_2$	15	2,356	2,552	2,749	2,945	3,142	3,338	3,534	3,731	3,927	4,320
4	8	16	2,513	2,723	2,932	3,142	3,351	3,560	3,770	3,979	4,189	4,608
$4^1/_4$	$8^1/_2$	17	2,670	2,893	3,115	3,338	3,560	3,783	4,006	4,228	4,451	4,896
$4^1/_2$	9	18	2,827	3,063	3,299	3,534	3,770	4,006	4,241	4,477	4,712	5,184
$4^3/_4$	$9^1/_2$	19	2,985	3,233	3,482	3,731	3,979	4,228	4,477	4,725	4,974	5,472
5	10	20	3,142	3,403	3,665	3,927	4,189	4,451	4,712	4,974	5,236	5,760
$5^1/_4$	$10^1/_2$	21	3,299	3,573	3,848	4,123	4,398	4,673	4,948	5,223	5,498	6,048
$5^1/_2$	11	22	3,456	3,744	4,032	4,320	4,608	4,896	5,184	5,472	5,760	6,336
$5^3/_4$	$11^1/_2$	23	3,613	3,914	4,215	4,516	4,817	5,118	5,419	5,720	6,021	6,623
6	12	24	3,770	4,084	4,398	4,712	5,027	5,341	5,655	5,969	6,283	6,912
$6^1/_2$	13	26	4,084	4,424	4,764	5,105	5,445	5,786	6,126	6,466	6,807	7,487
7	14	28	4,398	4,764	5,131	5,498	5,864	6,231	6,597	6,963	7,330	8,063
$7^1/_2$	15	30	4,712	5,105	5,498	5,890	6,283	6,676	7,069	7,461	7,854	8,639
8	16	32	5,027	5,445	5,864	6,283	6,702	7,121	7,540	7,959	8,378	9,215
$8^1/_2$	17	34	5,341	5,785	6,231	6,676	7,121	7,566	8,011	8,456	8,901	9,791
9	18	36	5,655	6,126	6,597	7,069	7,540	8,011	8,482	8,954	9,425	10,367
$9^1/_2$	19	38	5,969	6,466	6,964	7,461	7,959	8,456	8,954	9,451	9,948	10,943
10	20	40	6,283	6,807	7,330	7,854	8,378	8,901	9,425	9,948	10,472	

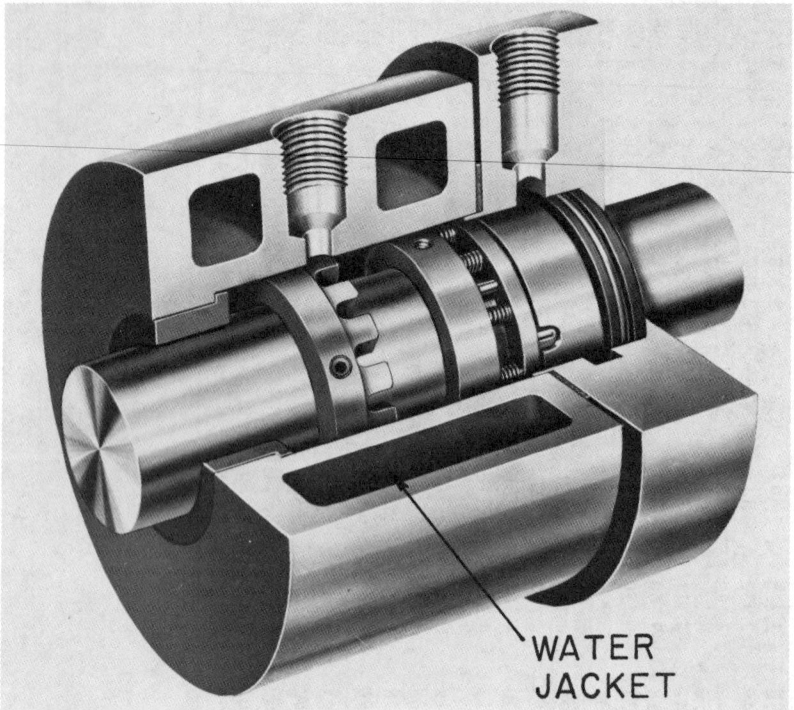

Figure 150. Water-jacketed stuffing box.

the assembly in water, thereby approaching the combined heat transfer efficiency of flushing and gland cooling.

Figure 151 shows an outside, quenched seal. Two pipe taps in the gland, 180° apart, allow water to be run over the seal ring and insert nose. The outlet connection should always be lower than the inlet connection, since the water must drain through the outlet connection. In addition to cooling, this arrangement also has been used as an effective means for washing away any leakage that may occur at the seal faces. The method can be used as a safeguard when handling dangerous liquids or to prevent the accumulation of abrasive solids when handling liquids that crystallize on contacting the atmosphere.

Some designs are available with a quench collar. The unit shown in Figure 152 can accommodate up to 3 gpm of flow at heads of 5–10 feet on the outlet. This means that the inlet and outlet pipe connections can be placed on a horizontal centerline if so desired. The quench collar arrangement is

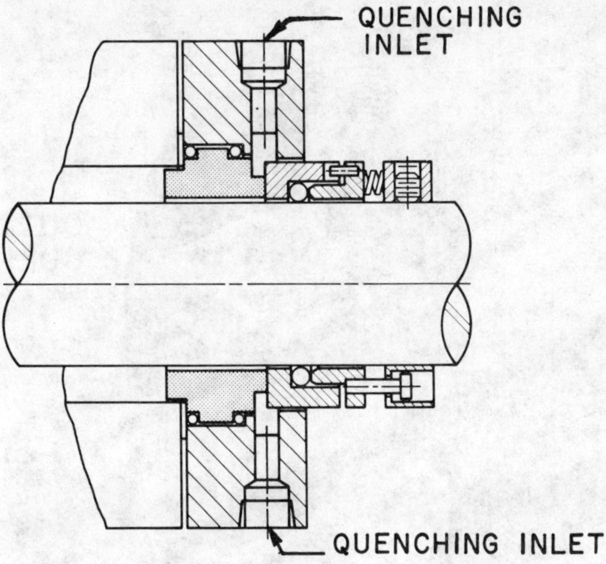

Figure 151. Quenching-outside seal.

also ideal when quenching is used to carry away toxic or noxious vapors emitted from the mechanical seal.

Inside seals also may be quenched. Once more, unpressurized water is allowed to flow through the gland and drain out through a connection at the bottom. The quenching water is retained in the gland by means of a throttle bushing or auxiliary compression packing. However, when an inside seal is quenched, the water naturally tends to follow the path of least resistance, which is directly behind the insert. By contrast to an outside quenched seal, where water is constantly running over the seal faces, exchange of water is at a much lower rate under the insert of an inside seal. Therefore, other forms of cooling inside seals are usually employed, providing greater efficiency when cooling is required. However, this method is effective for use as a safeguard or to prevent the accumulation of solids from crystallization.

DIRTY FLUIDS HANDLING

Quite often, extremely corrosive or dirty liquids are encountered that are difficult to seal even with the use of the most exotic materials. In such cases, the mechanical seal must be isolated from the destructive product.

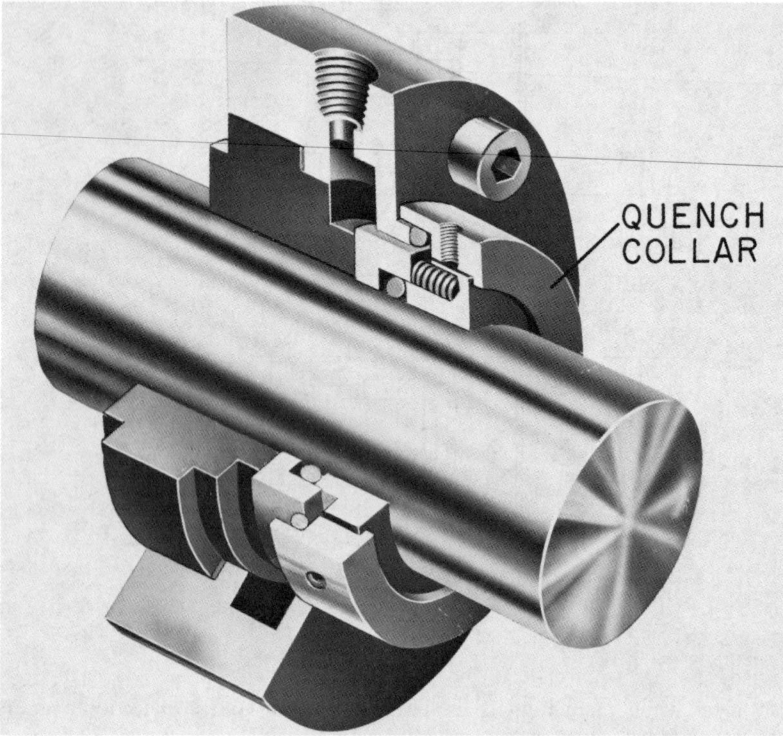

Figure 152. Quench collar.

For dirty fluids, a bypass flush arrangement can be used to flood the stuffing box with clean liquid. However, this can be accomplished only by inserting filters or cyclone separators into the flush line. Caution must be exercised when selecting either of these two options rather than a double seal design. Extremely dirty fluids will rapidly clog the filters. On the other hand, cyclone separators are only 90% efficient in removing solids down to 2.5 μ in size under optimum pressure and flow conditions. Therefore, filters should be called on to clean flush systems that are only lightly contaminated with solid particles. Cyclone separators should be applied only to moderately dirty systems or where the abrasive material is known to be at least 9 μ in size.

External liquid flushing involves the introduction of a clean, clear liquid to the stuffing box from an external source. The first point to be considered with the external flushing arrangement is that to be effective, it must be injected into the box at a pressure greater than the stuffing box pressure. Once more, as is the case with a bypass flush, the stuffing box must be provided with a throat bushing or some effective throat restriction so that it will be completely purged of the pumped product by the external liquid flush.

A definite measurable amount of product dilution will occur when using an external liquid flush. The amount of dilution will depend on the design of the throat restriction and the pressure differential between the flush liquid and the product behind the impeller.

The external flush arrangement has proven satisfactory and is widely accepted as standard procedure by many process industries handling abrasive liquids and slurries where some dilution of the product can be tolerated. The external flush will provide a clean liquid to the seal faces and prevent abrasive solids from entering the stuffing box.

The Stuffing Box Throat

The throat of a pump is that portion of the pump casing that forms a close clearance with the pump rotating shaft. The throat is also the bottom face of the stuffing box and, therefore, is commonly referred to as the stuffing box throat. In addition to forming a close clearance with the shaft to reduce leakage of the pumped liquid to the stuffing box, it also acts as a shoulder to retain the stuffing box packing.

On pumps incorporating packing, the clearance between the throat and the pump shaft must be maintained relatively small to prevent extrusion of the packing.

For most commercially designed centrifugal pumps, the normal throat clearance between the shaft and the stuffing box throat (without throat bushing) is 0.030- to 0.040-inch radial clearance. On pumps with mechanical seals it is often necessary to reduce this clearance to limit flow past the throat, to dead-end the liquid in the stuffing box or to raise the pressure of the liquid in the stuffing box.

When jackets are incorporated into the stuffing box design to facilitate effective cooling or heating of the liquid within, it is normally recommended to dead-end the liquid in the stuffing box. The addition of a restriction device at the throat of the stuffing box dead-ends the liquid in the stuffing box and impedes mix with the pumped liquid.

Bypass lines (recirculation lines) are frequently used to ensure proper lubrication and cooling of mechanical seals when the pressures acting on the stuffing box are near the vaporization pressure of the liquid being pumped. Since most stuffing box pressures are only slightly in excess of suction pressure, it is necessary to pressurize the stuffing box to prevent vaporization of the liquid in the stuffing box due to the small amount of frictional heat being generated at the mechanical seal faces and heat generated by turbulence. Bypass lines under discharge pressure will not raise the pressure in the stuffing box unless there is restriction at the stuffing box throat. In no case, however, can the pressure in the stuffing box be raised to equal full discharge pressure.

The most common use of restriction devices at the stuffing box throat is in conjunction with external flushing to the stuffing box. External flushing involves the introduction of a clean, cool liquid to the stuffing box from an external source. It is used whenever the pumped liquid will not provide a suitable environment for proper mechanical seal operation and life. When the pumped liquid is a slurry containing solids or abrasives that would not be suitable for lubrication of the mechanical seal faces, external flushing at a pressure higher than stuffing box pressure is introduced to prevent the pumped liquid from entering the stuffing box. The mechanical seal then operates in the environment of the external flushing liquid. External flushing also is used to prevent hazardous, toxic, extremely corrosive, hot liquids (or liquids that tend to vaporize or solidify) from entering the stuffing box, and to provide a suitable environment for mechanical seal operation.

When external flushing is necessary to provide a proper mechanical seal environment, and small amounts of dilution of the pumped product can be tolerated, single mechanical seals with a restriction device at the stuffing box throat are used.

To exclude the pumped liquid from entering the stuffing box, it is necessary to create a velocity at the stuffing box throat. Velocities of 10–15 ft/s at the throat have proven satisfactory for this purpose. From Figure 153 it can be seen that with normal throat radial clearances of 0.030–0.040 in., flushing liquid flowrates in the magnitude of 10 gpm are required to obtain the desired velocity at the throat.

The primary purpose of a restriction device at the throat of the stuffing box is to lessen the radial clearance at the throat, thereby reducing the amount of flushing liquid required to create the desired velocity. In no case will the device control the flow.

The simplest type of throat restriction device is the fixed throat bushing. This is a bushing, normally of a nonsparking, nongalling material such as carbon, which is pressed into a bore at the throat of the stuffing box.

The fixed throat bushing shown in Figure 154(A) is the arrangement used in standard chemical-type pumps. The throat of the stuffing box is remachined to accept a 0.003- to 0.005-in. interference fit with the outside diameter of the throat bushing. An alternate arrangement designed for refinery-type pumps is shown in Figure 154(B). Here, the bushing is mounted for opposition to internal pump casing pressure in the event of flushing liquid pressure loss. For horizontally split case pumps, where it is not possible to press the bushing into the throat, the bushing may be mounted in an "O" ring, as shown in Figure 154(C).

The practical radial clearance for fixed throat bushings is 0.006- to 0.008-in. Clearances smaller than this may result in the shaft touching the bore of the bushing due to shaft deflection or shaft whip.

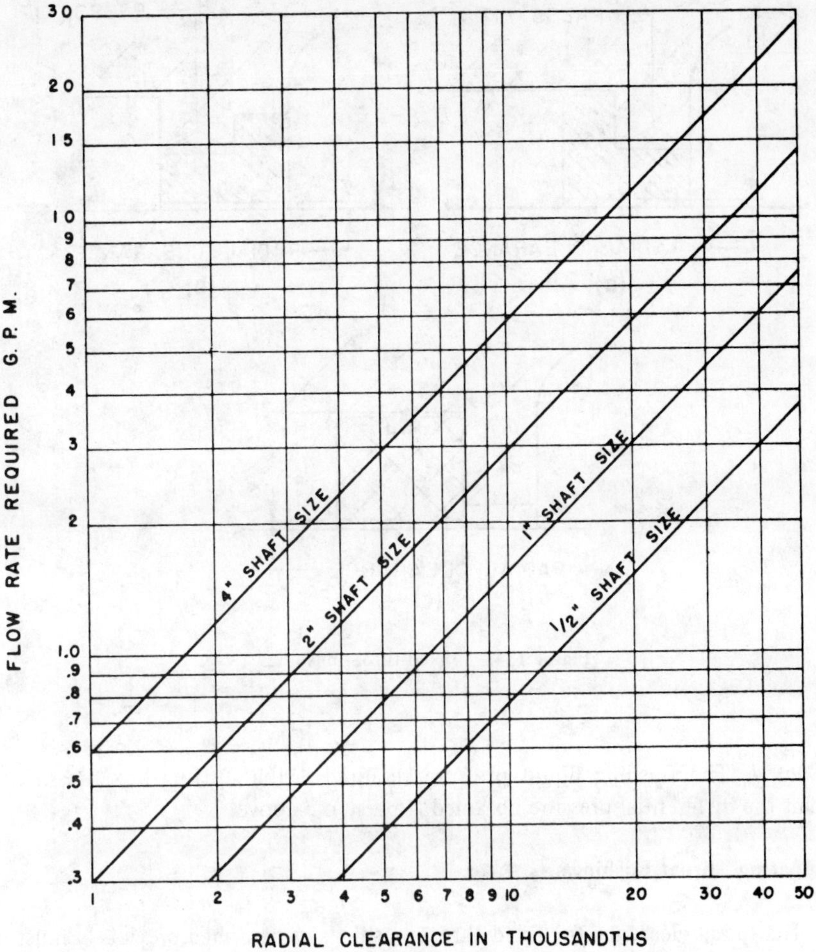

Figure 153. Flowrates required to create 15 ft/s velocity.

The following method is used to calculate the flushing liquid flow rate and flushing liquid pressure required.

Step 1. From Figure 153, determine the flowrate required for the radial clearance and shaft size.

Step 2. From Figure 155, correct the flowrate for shaft size and bushing length.

Step 3. For the corrected flowrate and radial clearance, determine the differential pressure required across the bushing.

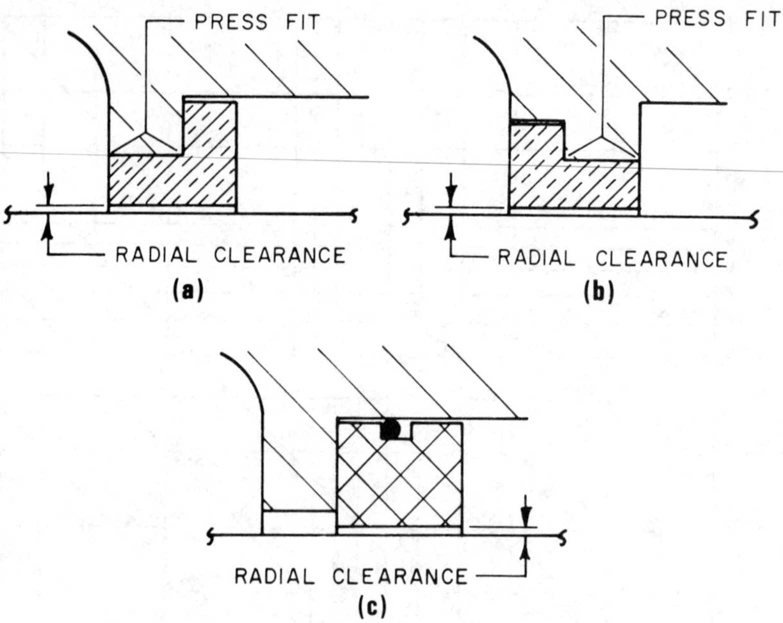

Figure 154. Fixed throat bushings.

Step 4. The flushing liquid pressure required is the stuffing box pressure plus the differential pressure obtained from step 3 above.

Floating Throat Bushings

The radial clearance for fixed throat bushings as described previously must be sufficient to prevent rubbing of the shaft in the inside diameter since the fixed bushing does not compensate for shaft deflection or shaft whip. For smaller radial clearances and resulting lower flushing liquid flow requirements, the bushing must compensate for shaft deflection and/or shaft whip. The floating thrust bushing as shown in Figure 156 is allowed to move radially, thereby centering itself around the shaft, regardless of the location of the shaft with respect to the stuffing box bore.

The carbon bushing is held stationary in the bottom of the stuffing box by spring pressure. There is a liberal clearance between the bushing o.d. and the stuffing box bore so the bushing can move radially, thereby centering itself around the shaft. The springs are held in an expansion collar, which incorporates a tapered set screw tap, which, on tightening of the set screw, expands the collar to grip in the stuffing box bore. Due to the ability of the

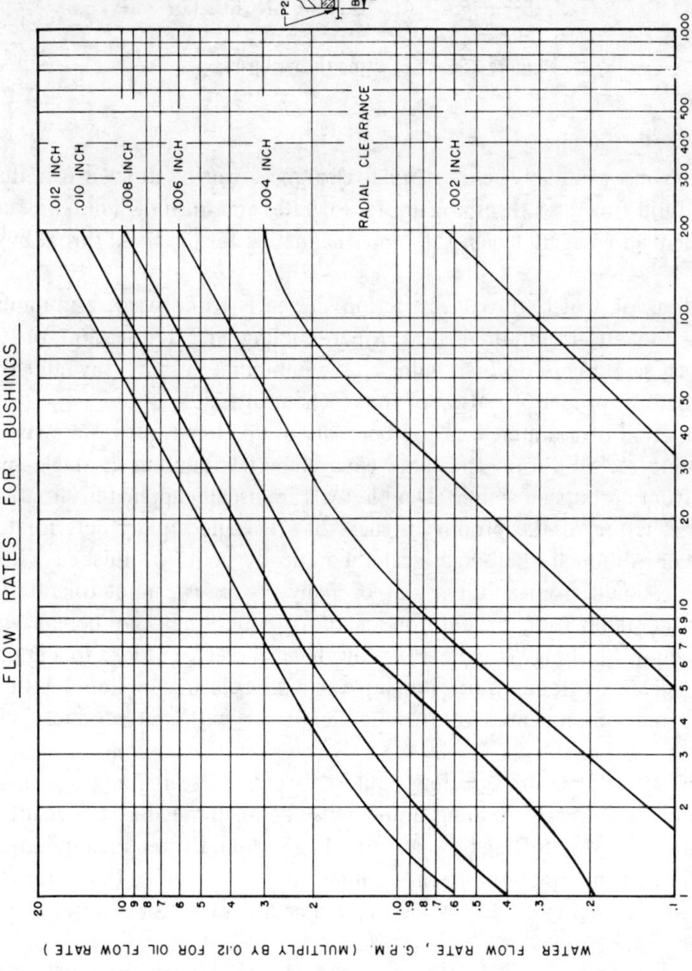

SLEEVE DIAMETER FACTOR	
DIA. (IN)	FACTOR
.50	.50
.75	.75
1.00	1.00
1.25	1.25
1.50	1.50
1.75	1.75
2.00	2.00
2.50	2.50
3.00	3.00
3.50	3.50
4.00	4.00
4.50	4.50
5.00	5.00

BUSHING LENGTH FACTOR	
LENGTH (IN)	FACTOR
.250	2.00
.375	1.60
.500	1.40
.625	1.30
.750	1.20
.875	1.10
1.000	1.00
1.250	0.90
1.500	0.80
1.750	0.75
2.000	0.70
2.500	0.65
3.000	0.60

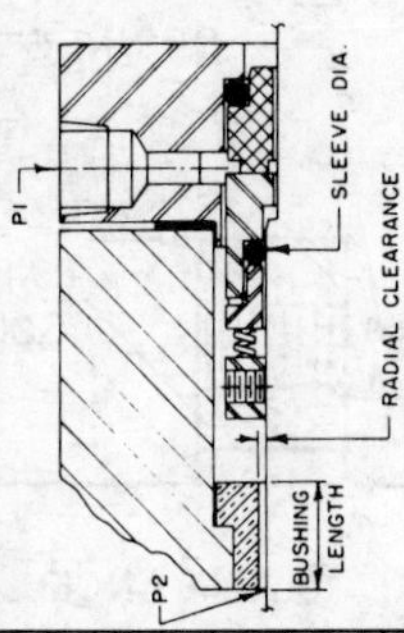

EXAMPLE

BOX PRESSURE = P2 = 50 P.S.I.
FLUSHING FLUID PRESSURE = P1 = 75 P.S.I.
SLEEVE DIA. = 2.50"
RADIAL CLEARANCE = .008"
BUSHING LENGTH = 3.00"
DIFFERENTIAL PRESSURE = P1 – P2 = 25 P.S.I.
FROM CHART FOR 25 P.S.I.G. & .008" RADIAL CLEARANCE, FLOW = 3.2 G.P.M.
SLEEVE DIA. FACTOR = 2.50
BUSHING LENGTH FACTOR = .6
FLOW = 3.2 G.P.M. X 2.5 X .6 = 4.8 G.P.M.

Figure 155. Flowrates for close clearance bushings.

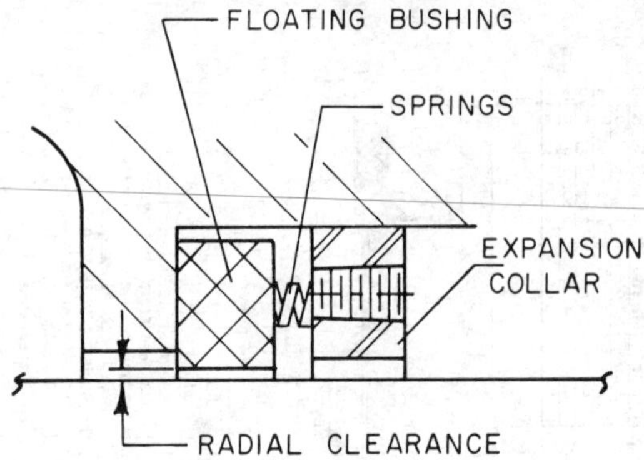

Figure 156. Floating throat bushing.

bushing to move radially, closer radial clearances can be utilized to reduce flushing fluid flowrates. Flushing liquid flowrates and flushing liquid pressure requirements are calculated in the same manner as for the fixed throat bushings.

Regardless of which throat restriction device is to be used, a minimum flushing flowrate is required for proper cooling and lubrication of the mechanical seal. Figure 157 is a guide to the minimum flushing flowrates that are required for proper operation of the mechanical seal.

The function of a double seal has been discussed already. It is noteworthy that double seals perform the same environmental function as single seals flushed from an outside source. Double seals are usually applied in situations where the nature of the product is such that it would rapidly deteriorate a single seal. Although the single seal theoretically could be flushed with a compatible liquid from another source, many processes cannot tolerate the resulting contamination or dilution of the product that is inherent with external liquid flushing. Therefore, the mechancial seal user looks to a double seal as a means of flooding the stuffing box with a clean, cool liquid without incurring the consequences of this liquid mixing with the product being pumped.

All double seals require a sealing liquid between the seals for proper lubrication and cooling. This sealing liquid must be maintained at a pressure of 15–25 psi above the stuffing box pressure. In addition, the sealing liquid must be circulated. The amount of sealing liquid flow required can be estimated from Figure 157. Figure 158 is a suggested arrangement for a water supply system for double seals.

When selecting a sealing liquid system for double tandem seals, one must consider not only the amount of cooling that is required to dispose of the

RPM CONVERSION FACTOR	
RPM	FACTOR
7000	2.0
6650	1.9
6300	1.8
5950	1.7
5600	1.6
5250	1.5
4900	1.4
4550	1.3
4200	1.2
3950	1.1
3500	1.0
3150	0.9
2800	0.8
2450	0.7
2100	0.6
1750	0.5
1400	0.4
1150	0.3
700 & BELOW	0.2

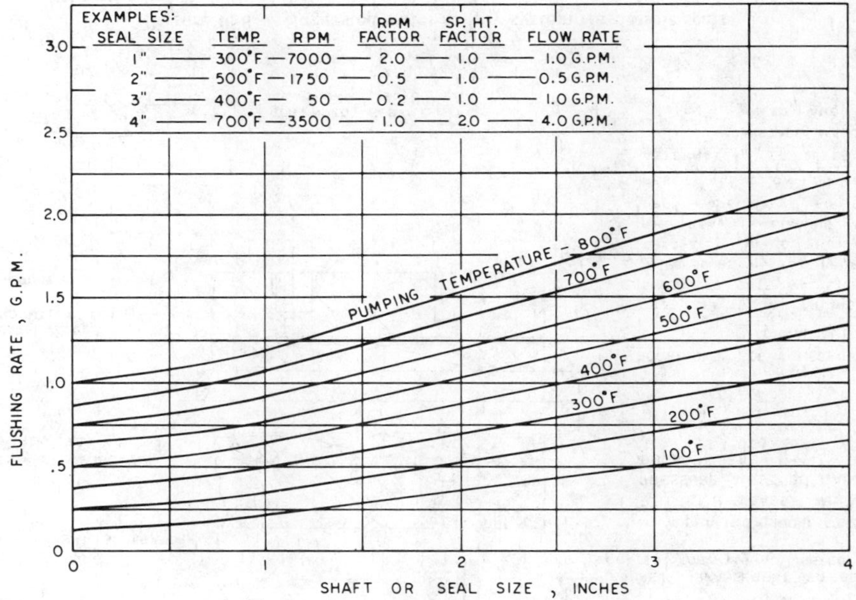

SPECIFIC HEAT CONVERSION FACTOR	
BTU'S/LB·°F	FACTOR
1.0	1.0
0.9	1.2
0.8	1.4
0.7	1.6
0.6	1.8
0.5	2.0
0.4	2.2
0.3	2.4

a) DOUBLE SEAL WITH SEAL FLUID CIRCULATED BETWEEN SEALS AT 15-25 PSIG ABOVE BOX PRESSURE.
b) SINGLE SEAL FLUSHED FROM EXTERNAL SOURCE TO STUFFING BOX.

STUFFING BOX OR SEAL FLUID PRESSURE WITHIN PV LIMITATIONS.
FLUID INLET TEMPERATURE TO STUFFING BOX 125°F MAXIMUM.

Figure 157. Sealing and flushing liquid requirements.

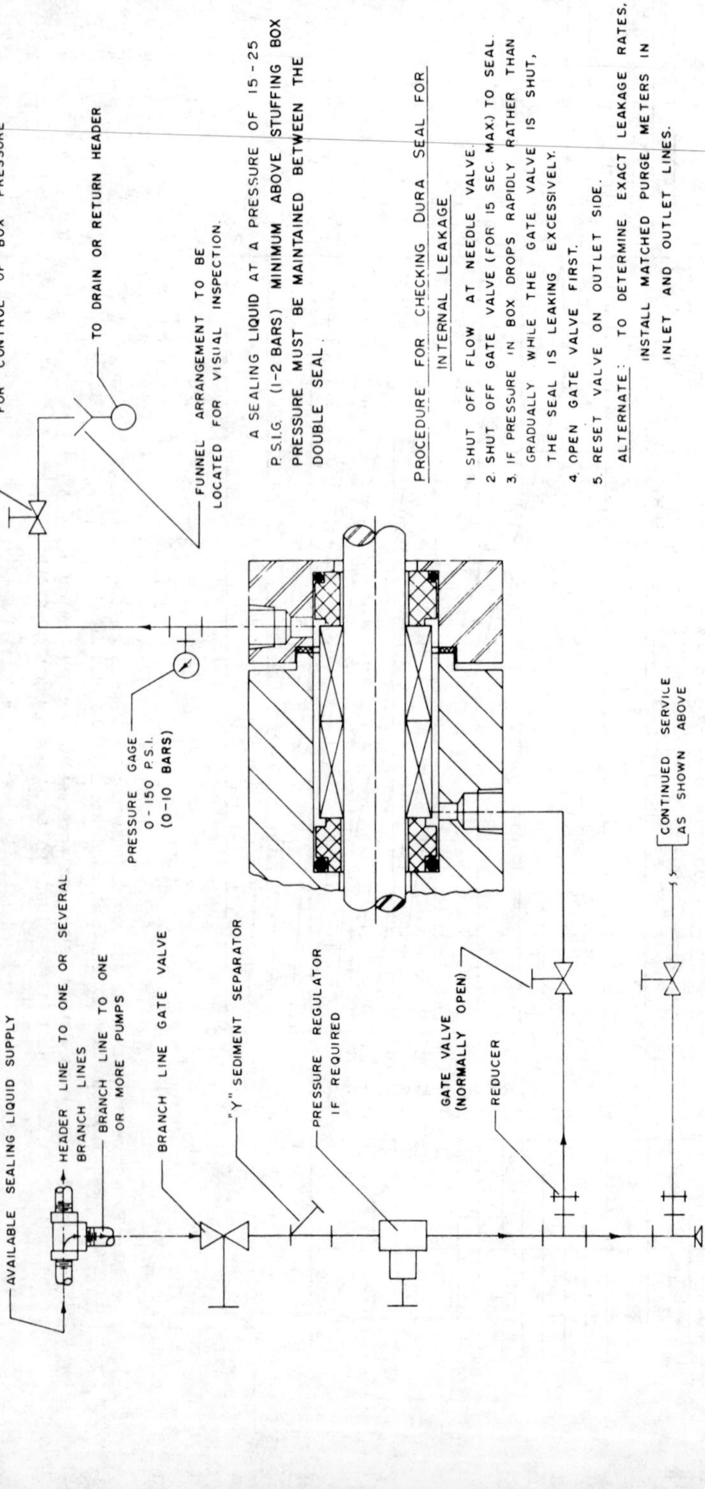

Figure 158. Suggested water supply system for double seals.

heat generated by the outer or secondary seal, but also how to dispose of the primary seal weepage that occurs during normal operation. In a double tandem seal, the primary seal weepage is contained in the sealing liquid system for the secondary seal. Unless this collection of normal primary seal weepage is properly drained or vented, a pressure buildup will occur in the secondary seal system.

SAFETY FEATURES

When a mechanical seal fails, leakage occurs. It is the responsibility of the user and the seal manufacturer to determine whether the leakage resulting from a seal failure poses a threat to the equipment, personnel or environment. If any leakage of the process fluid is hazardous, then provisions must be made for controlling the leakage.

Two options are available to the user for controlling leakage. One is to use double or tandem seals, both of which will prevent leakage of the process fluid to the atmosphere in the event the inner or primary seal fails. The other alternative is to use a single mechanical seal and make provisions for collecting the leakage should a seal failure occur (i.e., a vent and drain).

The vent and drain can be applied to the gland rings of inside seals in addition to other features already discussed. The purpose of a vent and drain is as the term implies. As an example, when handling liquified petroleum gases, any leakage past the seal will be a highly explosive gas that can be "vented" to safe external disposal. The drain connection at the bottom of the gland, 180° from the vent, permits hazardous liquid leakage to be carried away to a point where it can be collected safely.

The vent and drain is provided in the gland ring by machining two drilled and tapped openings behind the stationary insert, as illustrated in Figure 159. Leakage is discouraged from traveling along the shaft to the atmosphere by a restriction, called a throttle bushing, incorporated into the rear of the gland ring. With the vent and drain connections backed up by a throttle bushing, leakage takes the path of least resistance to the point of collection or disposal.

Most throttle bushings used in conjunction with vent and drain connections are of the fixed type, being pressed into a bore at the rear of the gland ring and adequately backed up by the back of the gland ring itself. Throttle bushings are made from either carbon or bronze, both of which are non-sparking materials.

Figure 159 shows a typical fixed throttle bushing application. Since the throttle bushing is not free to float and center itself around the shaft, 0.025-in. diametral clearance is provided between the i.d. of the bushing and o.d. of the shaft. It has been found that this is the optimum clearance to

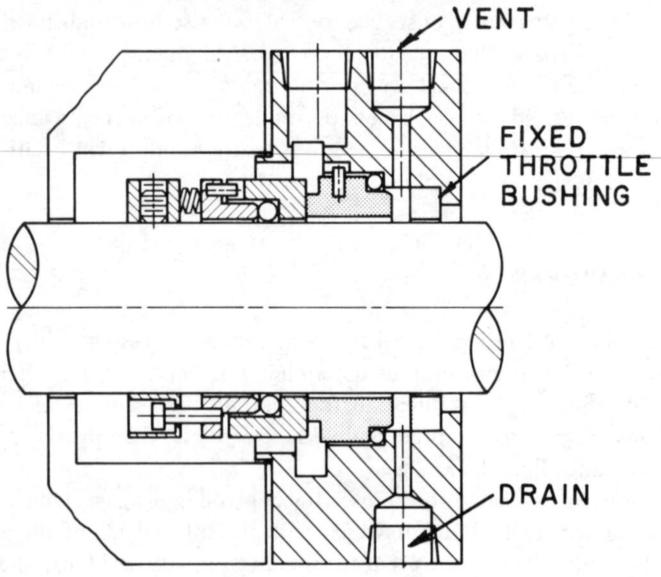

Figure 159. Vent and drain with fixed throttle bushing.

adequately accommodate slight vertical misalignment of the gland ring and/or shaft runout.

An alternate gland throttle bushing arrangement is the floating variety. This type of bushing is free to center itself around the shaft, thereby allowing the application of a closer radial clearance of 0.002-in. between the bushing and the shaft. The additional restricting efficiency provided by the floating throttle bushing is especially popular for applications in which it is desirable to achieve maximum restriction by the use of a throttle bushing.

The application of an auxiliary stuffing box behind a mechanical seal is illustrated in Figure 160. In this case, an auxiliary gland follower flange compresses two rings of suitable packing against the shaft. This arrangement will provide a nearly positive auxiliary shaft seal in the event the primary mechanical seal fails. However, it is necessary that the packing be properly lubricated and cooled to avoid excessive wear on the shaft and packing. This entails circulating a liquid, usually water, through the vent and drain connections.

Although an auxiliary stuffing box provides the user with a tight backup seal, the necessity of furnishing lubrication to the packing may prove objectionable. Therefore, one may prefer to use a throttle bushing design that does not require the circulation of any liquid through the vent and drain connections.

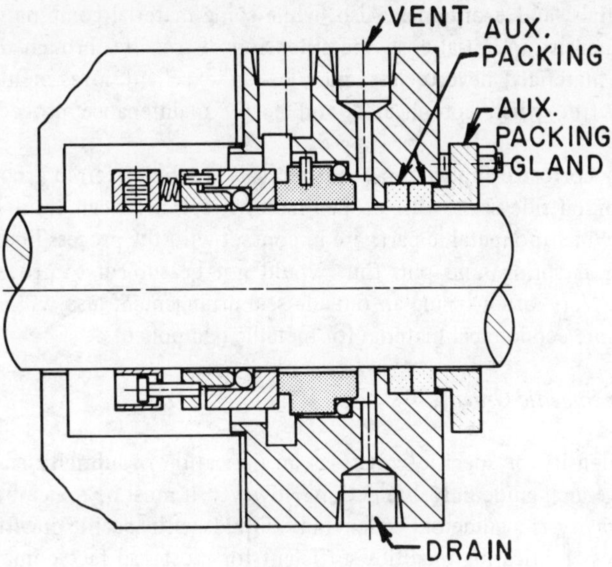

Figure 160. Vent and drain with auxiliary stuffing box.

SEAL SELECTION AND APPLICATION

A number of factors must be considered in selecting the proper mechanical seal for an application. Seal size, rpm, pressure, temperature, chemical and physical characteristics of the process liquid, and the equipment to which the seal is to be applied all come into play in seal selection. The actual selection is often a compromise among several of these factors along with space availability, customer specifications, economics and the availability of environmental controls.

Process Liquid Characteristics

Corrosiveness

The corrosive nature of the process liquid will determine the metallurgy, face materials and type of secondary sealing materials that can be used. Note that while some materials of construction possess excellent resistance to corrosion, they may have poor wearing characteristics, while others that provide excellent wear properties exhibit poor corrosion resistance. Seal manufacturers do not always agree on design and material selection for a given application. This is due to differences in design, standardization of parts and materials, and economics. Also, while some material combinations may be less expensive in initial cost, the extension of seal life through the use of upgraded materials, nevertheless, may be an excellent investment. This is especially true when complicated and costly maintenance procedures are involved.

For very corrosive applications, outside seals are the preferred arrangement. The use of outside seals with proper material selection can result in a seal design in which no metallic parts are in contact with the process liquid. Small parts such as springs and pins thus would not be subject to process liquid corrosion. Very often, using an outside seal arrangement also will allow the use of a more economical material for metallic components.

Density or Specific Gravity

Liquid density or specific gravity is an indication of lubricity and can be used as a general guide in seal selection. However, it must be remembered that specific gravity is a guide only. Generally, liquids with specific gravities above 0.65 possess lubricating qualities sufficient for most seal face combinations.

For liquids with specific gravities below 0.65, balanced seals are required. In addition, tungsten carbide is preferred as one of the sealing faces because of its excellent wearing characteristics in the absence of a good lubricating film between the faces.

Certain face combinations can be used only in good lubricating liquids. For example, ceramic versus ceramic can only be used in liquids possessing lubricating qualities equal to that of oil, concentrated sulfuric acid or concentrated nitric acid.

Vapor Pressure and Boiling Point

Vapor pressure is an important consideration when the process liquid is being handled at a pressure at or near its vapor pressure. Light hydrocarbons, for example, especially liquified petroleum gases (LPG), require a stuffing box pressure-temperature relationship that will ensure a stuffing box pressure of at least 25–50 psi greater than the vapor pressure of the product at the stuffing box temperature.

One method of increasing the stuffing box pressure using the process liquid is through the installation of a bypass recirculation line and a throat restriction device. Note, however, that there must be sufficient pressure differential between the pump discharge pressure and stuffing box pressure to be effective.

An alternative to increasing the stuffing box pressure for proper stuffing box pressure-temperature relationship is cooling the stuffing box. Experience has proven that stuffing box temperatures should be maintained 25–50°F (15–30°C) below the boiling point of the product at the corresponding stuffing box pressure. Stuffing box water jackets, external coolers, etc., can be installed for this purpose.

The volatility aspect of a process stream also must be considered from the safety standpoint. This may involve incorporating seal features such as vents and drains with a gland ring bushing or auxiliary stuffing box.

Viscosity

Viscosity, like density, is an indication of a liquid's lubricating qualities. Liquids with higher viscosities usually possess better lubricating characteristics.

For liquids with viscosities less than 3000 SSU or 700 cp, standard seal designs may be applied without restriction or modification. Special consideration must be given to the sealing of liquids with higher viscosities.

Liquids with viscosities of 3000–7000 SSU (700–1600 cp) require that the seal be of a positive drive design with a "heavy-duty" compression unit. This is to compensate for the increased shear that occurs at the sealing faces under a viscous environment.

Mechanical seals applied to liquids with viscosities higher than 7000 SSU (1600 cp) require both (1) positive drive with "heavy-duty" compression units, and (2) environmental controls for proper lubrication of the sealing faces. With these high viscosities, no liquid film between the sealing faces can possibly exist using the process liquid. Single seals with external flushing, single seals with external lubrication or double seals with an external sealing liquid must be used.

Abrasives

If the liquid being handled is dirty, contains abrasives or is a slurry, all of which would not properly lubricate the seal faces, some arrangement to provide a clean liquid environment for the seal must be applied.

There are numerous approaches to providing an artificial environment for the mechanical seal if the process liquid is unsuitable. Considerable thought should be given to the selection of the most economical environmental control system before the seal selection is made. The selection of the mechanical seal design and materials of construction should be based on the actual conditions in the stuffing box. If the seal is to operate under environmental controls, the selection should be based on the conditions imposed by the environmental control, rather than the operating conditions of the equipment to which the seal is applied.

Physical and Chemical Changes

Process liquids that go through physical or chemical change with slight variations from the temperature at which the liquid is being handled may require some temperature controls in the sealing area to ensure that adequate lubrication is maintained at the seal faces. Likewise, liquids with high melting points or that tend to crystallize on exposure to the atmosphere will require temperature control to ensure that the liquid does not solidify at the faces.

STUFFING BOX PRESSURE

The pressure that exists in the stuffing box varies with equipment design. For example, on most end suction centrifugal pumps the stuffing box is subject to the pump suction pressure or suction pressure plus a small percentage of discharge pressure. Vertical turbine pumps, on the other hand, often subject the stuffing box to full pump discharge pressure. Many multistage pumps have one stuffing box subjected to pump suction pressure while the other stuffing box is subjected to an intermediate stage discharge pressure. All this points out that knowledge of more than pump suction and discharge pressures may be necessary to determine the stuffing box pressure for which the seal must be selected.

The best source of stuffing box pressure information is the original equipment manufacturer. Most pump manufacturers have tested their equipment hundreds of hours to determine stuffing box pressures and can readily provide this information for any given set of operating conditions.

For equipment already in the field, the easiest way to determine the stuffing box pressure is to install a pressure gauge into a tapped hole leading directly to the stuffing box.

In discussing pressure and its influence on seal selection or application, it is necessary to include velocity. Both laboratory testing and field experience have shown that velocity is as important a factor as pressure in determining the performance of mechanical seals. It has been demonstrated that unbalanced seals can be used at relatively high pressures, if the velocities are correspondingly low. On the other hand, when velocities are high, balanced seals are sometimes required even with low pressures.

In the past, pressure limitations of unbalanced seals have been established arbitrarily, regardless of size, speed, mating face materials or liquid handled. Some seal manufacturers have used a value of 200 psig, others 150 or 100 psig, for unbalanced seals. Some refineries have adopted standards that require the use of balanced seals whenever the pressure in the stuffing box exceeds 50 psig. Such pressure limitations are established largely from field experience.

Modern sealing technology recognizes the importance of such factors as shaft or seal size, peripheral speed, contact mating face materials and the characteristics of the liquid handled. It is not to be expected, for example, that for a specific pressure and liquid an unbalanced 3-in. shaft seal operating at 3500 rpm would have the same life expectancy as a 1-in. shaft seal operating under the same conditions at 1750 rpm.

Recognition of these factors has led to the establishment of "pressure-velocity" limitations for inside unbalanced seals. The velocity portion of this relationship is the peripheral speed expressed in ft/min (fpm) which can be arrived at by the formula:

$$\text{Velocity (fpm)} = \frac{\text{Shaft Size (in.)} \times \text{rpm}}{3.82} \tag{118}$$

Figure 161 shows the relationship between shaft diameter, rpm and peripheral speed in fpm in the form of a nomogram.

Note that wear differences exist with different liquids handled, even though seal size, speed and pressure may be identical. Therefore, two sets of curves are provided by seal manufacturers, one for water and water solutions, and one for liquids that could be considered at least equal to gasoline in lubricity. This provides conservative estimates for many acids and corrosive liquids; although they present corrosive problems, they still maintain a good liquid film between the faces.

Pressure-velocity ratings based on peripheral speed now have been established for all of the more popular combinations of mating face materials presently used. From these ratings, pressure-velocity curves have been plotted for both water and water solutions, and gasoline and equivalent liquids, at both 1750 and 3500 rpm. The established ratings are based on a seal life of 15,000 hours on either continuous or intermittent service.

Figure 162 is a pressure-velocity curve for inside unbalanced seals for water solutions when the stuffing box temperature does not exceed 160°F (70°C). For water temperatures in excess of 160°F (70°C), special considerations must be given, as will be discussed later.

Figure 163 is a pressure-velocity curve for inside unbalanced seals for acids, hydrocarbons and other liquids with similar lubricity. This curve may be used for light hydrocarbon applications where the specific gravity is about 0.65. It does not apply to light hydrocarbons or LPG applications in which the specific gravity is below 0.65. Where such a condition is encountered, balanced seals are required regardless of speed, seal size or pressure.

To use the pressure-velocity curves, enter the chart at the correct seal size and read up to the appropriate material-speed curve. At this point, read across to obtain the maximum recommended stuffing box pressure for an inside unbalanced seal.

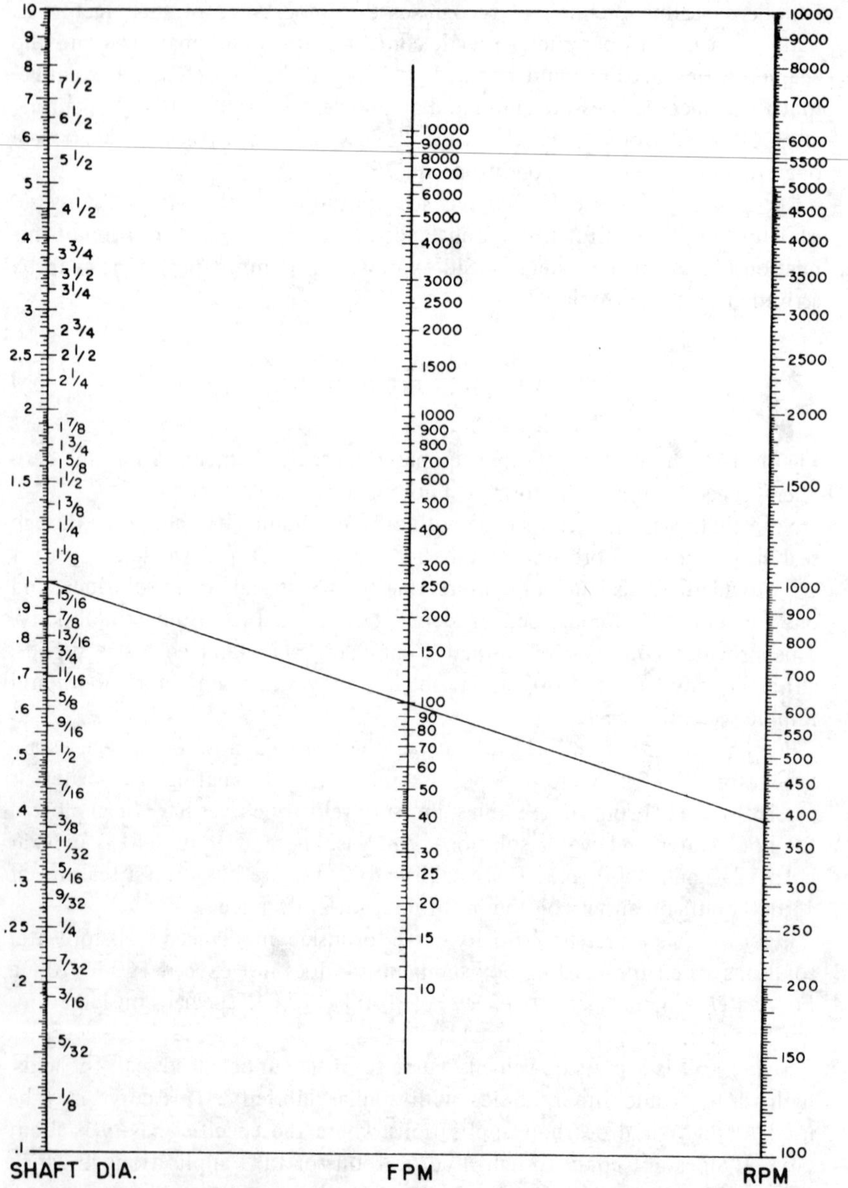

A straight edge line up with the values on any pair of the vertical lines will cross the third line at the corresponding value. For example, a straight edge set at 1" on the shaft diameter line and 382 RPM on the R.P.M. line will cross the FPM line at 100 FPM, the corresponding surface speed.

Figure 161. Nomogram defining the relationship between shaft size, rpm and peripheral speed.

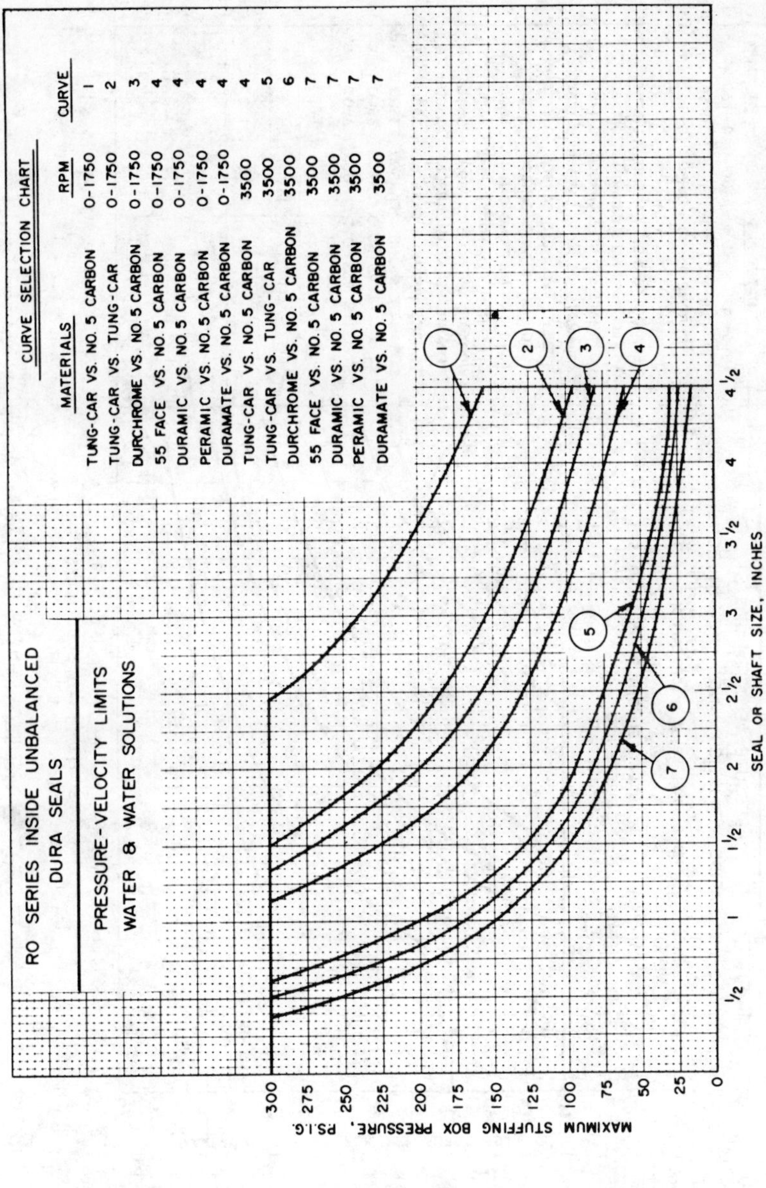

Figure 162. Pressure-velocity curves for water and water solutions.

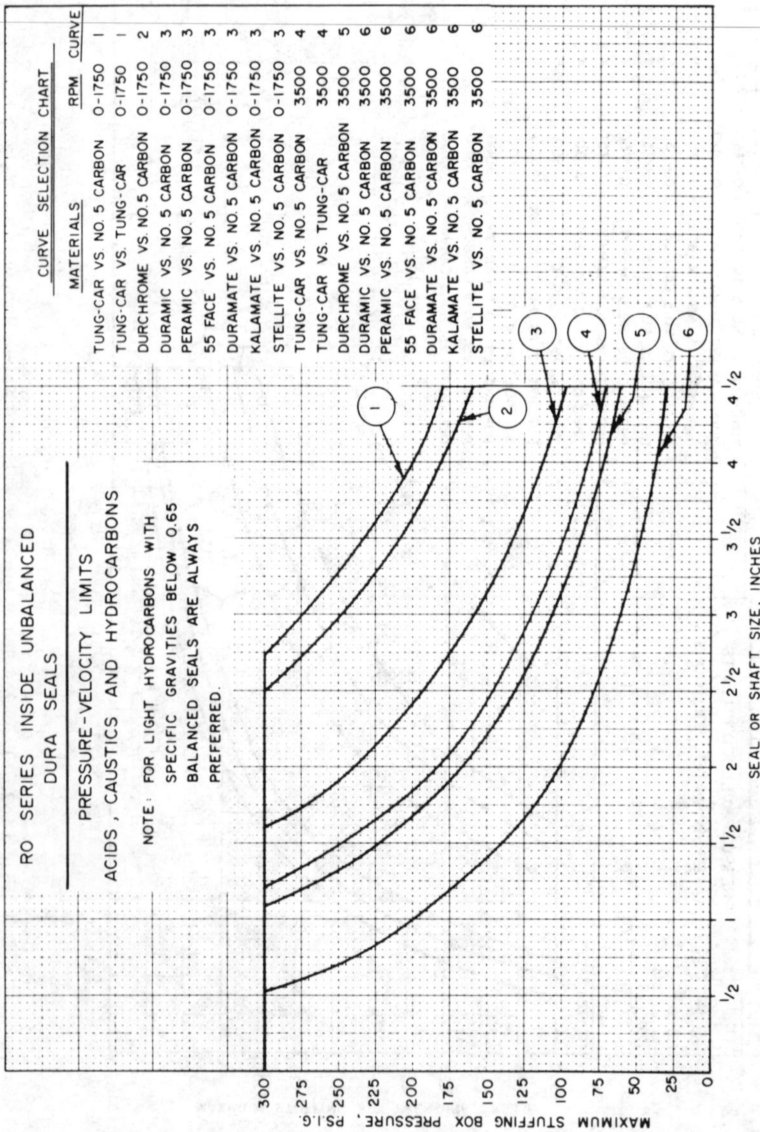

Figure 163. Pressure-velocity curves for acids, caustics and hydrocarbons.

If the actual pressure involved in the application exceeds the maximum recommended stuffing box pressure, then a change in the material selection to one with a higher pressure-velocity rating or a balanced seal is required.

Pressure-velocity ratings and curves also have been established for other seal designs and special material combinations. For example, Figure 164 shows ratings for another series of seals. These ratings are somewhat lower than those in Figure 162 since this series of seals incorporates friction drive.

Vacuum Conditions

Pressure and/or vacuum also determine whether inside- or outside-type unbalanced seals can be used for a given application. Outside unbalanced seals are recommended for use with up to approximately 35 psig of positive stuffing box pressure. They are also recommended for use on negative pressures. In such applications, the vacuum adds to the spring load on the seal.

Vacuum applications require environmental controls such as lubrication or product recirculation to maintain a liquid film between the seal faces. With satisfactory environmental controls, outside mechanical seals are capable of handling very high vacuums.

Outside seals offer ease of assembly, ease of repair and visual performance of the seal since the rotating portion of the seal operates in the atmosphere outside the stuffing box rather than inside the stuffing box.

When space permits, outside unbalanced seals are sometimes preferred in low-pressure, corrosive service. Quench collars provide further protection from corrosive atmospheres.

Balanced seals have also pressure limitations that vary with peripheral speed and seal face materials, although they are substantially higher than the corresponding unbalanced seal limits. In many cases, the pressure limitations of balanced seals are due to inherent design and material limitations, regardless of peripheral speed.

Figure 165 is a pressure-velocity curve for inside balanced seals for water and water solutions.

Outside unbalanced seals are limited to 35 psig stuffing box pressure since one must depend solely on the spring load to keep the seal faces closed. However, outside seals have many advantages, especially in corrosive applications. Further, in many cases the seal must tolerate pressures higher than 35 psig. Balanced outside seals are designed to withstand stuffing box pressures up to 400 psig.

STUFFING BOX TEMPERATURE

Until recently, the use of elastomers for secondary sealing members in a mechanical seal limited the temperature that could be tolerated in the

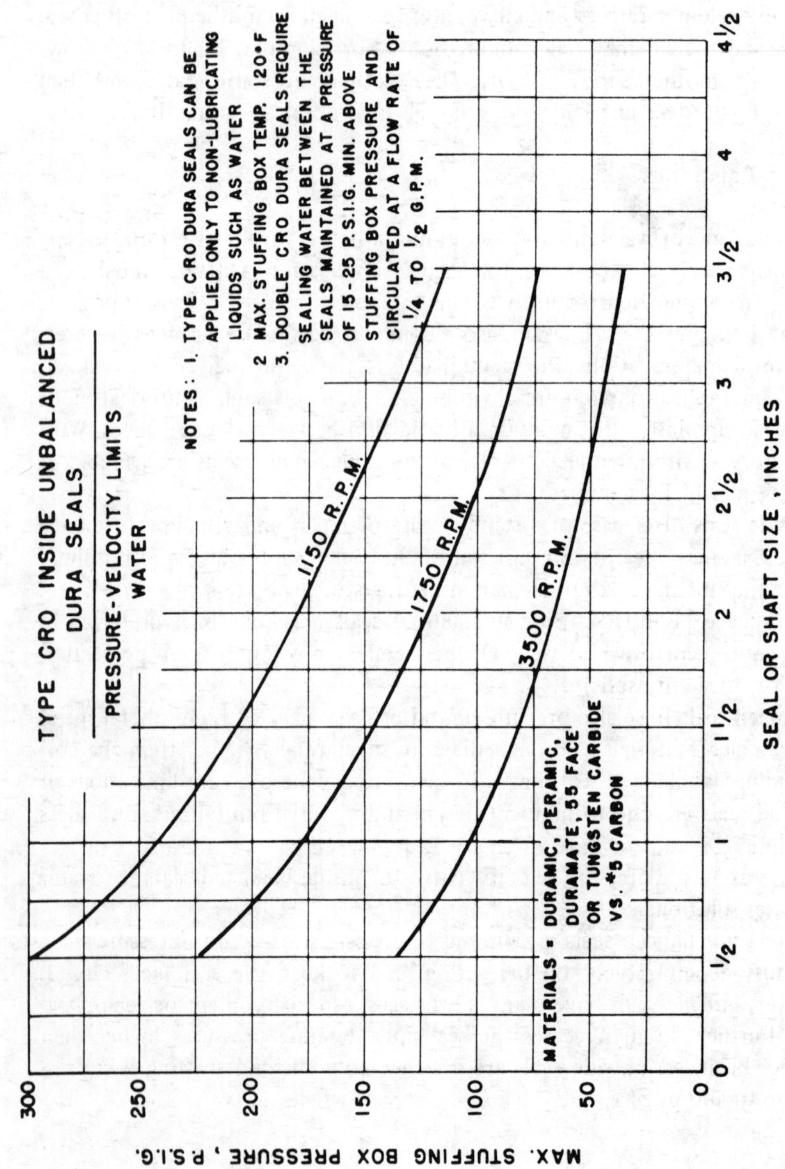

Figure 164. Pressure-velocity curves for another seal series.

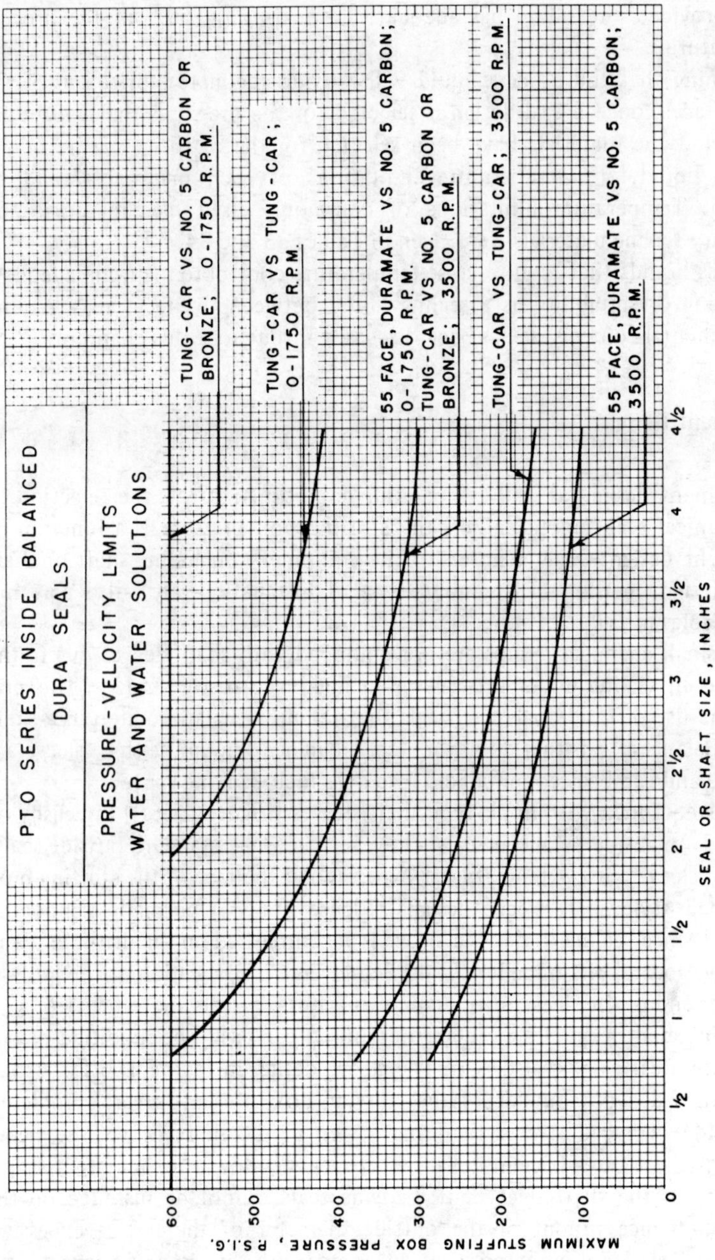

Figure 165. Pressure-velocity curve for inside-balanced seals; water and water solutions.

stuffing box. Advancements in metal bellows technology and flexible non-elastomeric materials have increased the temperature limits of mechanical seals, provided the liquid has adequate lubricating qualities at the process temperature.

Assuming that the process liquid will provide adequate lubrication of the sealing faces, once a seal design is selected for the pressures that will be incurred and the materials have been selected for the corrosive nature of the process liquid, the temperature limitations of each material should be checked. Temperature limitations of commonly used face materials and secondary sealing materials were given in Tables 33 and 34.

Process liquids that require close temperature control to maintain adequate lubricating or suitability as a sealing liquid between the faces may require environmental controls, as has been discussed throughout this chapter.

Equipment Design

Equipment design, as well as operating conditions, affects the selection of a mechanical seal. Noted earlier, the stuffing box pressure is a function of equipment design. Other features of the equipment, including shaft or shaft sleeve design, method of assembly, type of installation, etc., also play important roles in seal selection.

Most small chemical pumps are not equipped with shaft sleeves due to the problems in sealing under the sleeves in highly corrosive services. This prevents the use of inside balanced seals that require a balanced step in a shaft sleeve. Balanced seals that do not require a balance step in the shaft sleeve are available and can be accommodated into a 3/8-inch packing space.

Pumps equipped with hook-type shaft sleeves [Figure 166(a)] can easily be modified for inside balanced seals. However, to apply standard outside seals, the shaft sleeve must extend at least 2 in. beyond the face of the stuffing box.

Clamped style shaft sleeve designs (Figure 166(B)) normally cannot be modified since sealing under the sleeve is accomplished on the driver end and the sleeve locates the impeller on the liquid end. The location of the joint at the outer end of the sleeve must be known if outside seals are to be used.

The threaded sleeve design in Figure 166(C) can easily be modified or cut off to use a cartridge seal design. The sleeve design in Figure 166(D), which uses a threaded nut outside of the stuffing box, presents several problems. For inside balanced sleeves requiring a step in the shaft sleeve, the sleeve thickness must be sufficient to allow proper stepdown but maintain ample thickness in the shaft sleeve nut. Outside seals cannot be installed on the sleeve nut since runout of the outside diameter of the nut is excessive. Cartridge seals are used extensively on pumps with this type of sleeve design.

Vertically Split End-Suction Centrifugal Pumps

Enclosed-type impellers with balance holes and wearing rings generally provide suction pressure to the stuffing box, or suction pressure plus 10–15% of discharge pressure. Enclosed impellers without balance holes or wearing rings impose close to full discharge pressure on the stuffing box.

Fully open impellers or semiopen impellers with balance holes and pump-out vanes on the back of the impeller impose suction pressure on the stuffing box.

Most pumps of this type for chemical services have solid shafts or hook-type shaft sleeves, as shown in Figure 166(A). Pumps for general industrial services have hook-type shaft sleeves or clamped sleeves, as shown in Figure 166(B).

Single-Stage Horizontally Split Case Pumps

These pumps normally have enclosed impellers with balance holes and wearing rings, which provide suction pressure to the stuffing boxes. Threaded sleeves such as those shown in Figures 166(C) and (D) are used normally.

Multistage Horizontally Split Case Pumps

In this design, stuffing box pressures vary with impeller and staging arrangements. Usually the stuffing box on the suction end of the pump is subjected to suction pressure, while the stuffing box on the discharge end of the pump is subjected to discharge pressure of the first or other intermediate stage. Pressure equalizing lines from the high-pressure stuffing box to the low-pressure stuffing box are often furnished to partially equalize the pressure of the stuffing box.

Threaded shaft sleeves like those in Figure 166(C) or (D) are normally used. However, cartridge seal designs are also in common use since they permit seal installation, adjustment and removal without removing the upper half of the casing. Any mechanical seal can be incorporated into a cartridge design, typically shown in Figure 167. The complete seal assembly, including the gland ring, is mounted on a sleeve that is secured to and driven by the shaft with a drive collar. The entire cartridge assembly is installed over the pump shaft and bolted to the stuffing box. The seal setting, adjustment and drive are all accomplished outside of the stuffing box. The total length of the cartridge assembly should be taken into consideration in the design of the spacer on the coupling, so that the assembly on the driver end can be removed without disturbing the driver.

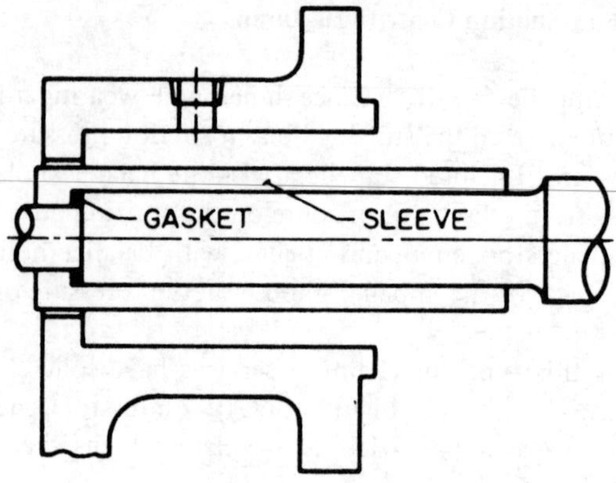

(a) HOOK TYPE SLEEVE

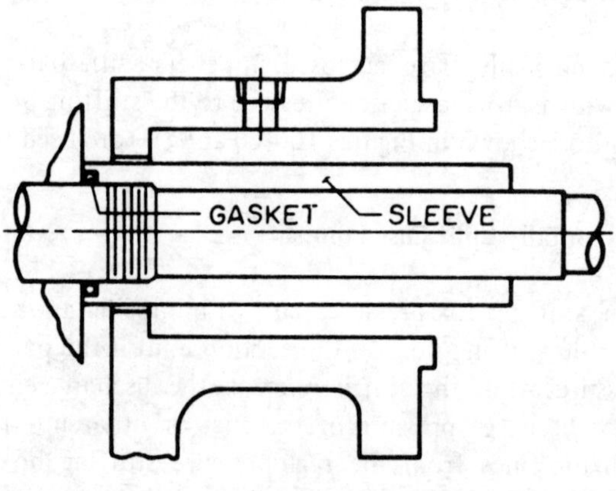

(c) THREADED SLEEVE

Figure 166. Shaft

Vertical Inline Pumps

Vertical inline pumps are vertical pumps with the suction and discharge nozzles inline. The driver is supported exclusively and the pump and the suction and discharge connections have a common centerline, which intersects the shaft axis.

There are three types of vertical inline pumps. The first is a one-piece pump/motor shaft design in which the impeller is mounted directly on an extended motor shaft. The second is a rigid coupling design in which the

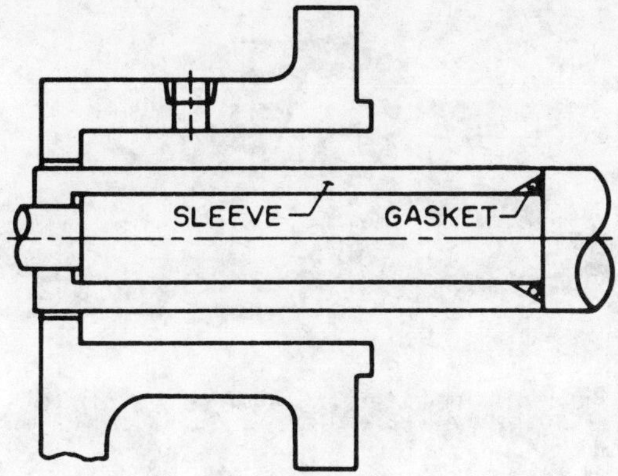

(b) CLAMPED SLEEVE

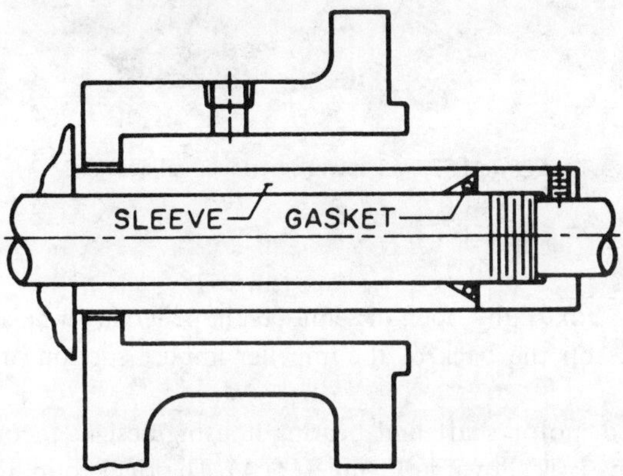

(d) SLEEVE WITH THREADED NUT

sleeve designs.

impeller is mounted on a stub shaft, which is coupled to the motor shaft by means of a rigid coupling. The third design incorporates a separate bearing housing, which makes this design very similar to a horizontal end suction pump.

In most cases, the pressures acting on the stuffing box are similar to horizontal end-suction pump designs. Enclosed-type impellers with balance holes and wearing rings generally provide suction pressure to the stuffing box, or suction pressure plus 10–15% of discharge pressure. Enclosed impellers without balance holes or wearing rings impose close to full discharge pressure

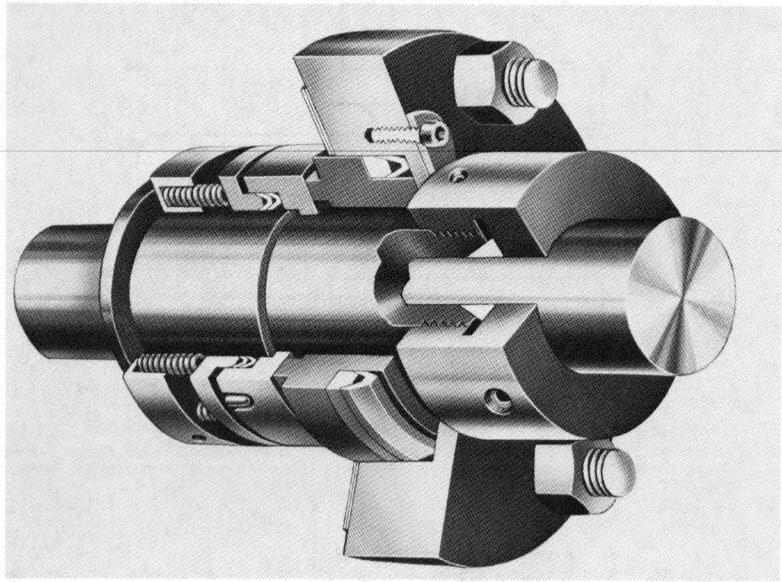

Figure 167. Typical cartridge seal design.

on the stuffing box. Fully open or semiopen impellers with balance holes and pumpout vanes on the back of the impeller impose suction pressure on the stuffing box.

The extended motor shaft and bearing housing designs incorporate hook-type shaft sleeves, as shown in Figure 166(A). The rigid coupling design has a solid shaft. This stub shaft is provided with a balance step to accommodate standard inside-balanced seals.

The stuffing box on all vertical pumps must provide for proper venting. Venting can be accomplished through a flush connection in the gland ring or through a connection in the stuffing box located near the seal faces.

Vertical Multistage Pumps

The stuffing box pressure varies with pump design, specifically with impeller and staging arrangements. Some designs impose full discharge pressure on the stuffing box while others impose only suction pressure. Installations of this type require a thorough knowledge of the actual stuffing box pressure.

This design normally requires adjustment of the impeller clearances after the pump has been completely assembled. This adjustment is accomplished by raising the pump shaft before the coupling is connected. As a result, only cartridge seals or outside seals can be applied since final setting of the seal must be made only after the coupling is connected.

Gland rings for vertical pumps and all other pumps mounted in a vertical position should provide for venting, bleedoff or the equivalent to prevent a vapor trap in the stuffing box area.

Rotary Pumps

Stuffing box pressures vary considerably with this pump design, and thorough knowledge of the actual stuffing box pressure is mandatory.

Many pumps of this type require that the pump be completely assembled before the seal can be installed since the shaft cannot be pulled through the stuffing box. Installation of inside seals requires that an access hole in the stuffing box be provided to tighten the set screws in the seal collar. Double seals are commonly used since rotary pumps are normally used on high-viscosity applications in which the pumped liquid would not lubricate the seal faces properly.

INSTALLATION AND OPERATION

A multitude of seal designs and materials have been devised to meet operating demands and conditions in the chemical and petroleum processing industries. Basically, however, all seal designs consist of two basic elements: one part secured to and *rotating* with the shaft, and the other part *stationary* and secured to the equipment.

The fundamental objective of mechanical seal operation is the development and maintenance of a liquid film between the seal faces. This microscopic film of liquid acts as a cushion for the seal faces to ride on, thus minimizing or preventing wear, as well as minimizing or preventing leakage.

Modern research and manufacturing methods have produced in the mechanical seal a dependable, precision-built device that is designed to be installed in a proper and specific manner in mechanically sound equipment and maintained by recommended practices.

Equipment Check Points

Sealing performance and seal life depend to a great extent on the operating condition of the equipment to which the seal is being applied. Time spent in checking the equipment prior to seal installation has always proved to be a sound maintenance procedure to prevent excessive seal failures and to keep overall maintenance expense at a minimum.

The following conditions of the equipment should be checked:

1. *Stuffing box space.* To properly receive the seal, radial space and depth of the stuffing box must be the same as the dimensions shown on the assembly drawing that accompanies the seal.

2. *Lateral or axial movement of shaft (end play).* Install a dial indicator, as shown in Figure 168(A), so that the stem bears against the shoulder on the shaft. Use a soft hammer or mallet to tap the shaft lightly on one end and then on the other. Total indicated end play should be held between 0.001 and 0.004 in. A mechanical seal cannot function satisfactorily with a great amount of end play or lateral movement. If the hydraulic condition changes, which happens frequently, the shaft may "float," resulting in sealing troubles. *Minimum end play* is a very desirable condition for the following reasons:

- Excessive end play, or lateral shaft movement resulting in the shaft floating, can cause pitting, fretting or wear at the point of contact between the shaft packing in the mechanical seal and the shaft or sleeve outer diameter. This condition may be alleviated if rubber elastomer materials can be used for shaft packing.
- As the mechanical seal driving element is locked to the shaft or sleeve, any excessive end play, or lateral movement, will result in either overloading or underloading the springs, causing excessive wear and/or seal leakage.
- A floating shaft can cause chattering, which results in chipping of the seal faces, especially the carbon element.
- Ideal mechanical seal performance requires a uniform wear pattern and the maintenance of a liquid film between the mating contact faces. Excessive end play, or lateral movement of the shaft due to defective thrust bearings, can thus reduce seal life and performance by disturbing both the established wear pattern and lubricating film between the contact faces.

3. *Radial movement of shaft (whip or deflection).* Install the dial indicator as shown in Figure 168(B) and as close to the radial bearing as possible. Lift the shaft or exert light pressure at the impeller end. If more than 0.002–0.003 in. of radial movement occurs, investigate bearings and bearing fittings. An oversize radial bearing bore caused by wear, improper machining or atmospheric corrosion will cause excessive radial shaft movement, resulting in shaft whip and deflection. *Minimum radial shaft movement* is important for the following reasons:

- Excessive radial movement can cause wear, fretting or pitting of the shaft packing or secondary sealing element at the point of contact between the shaft packing and the shaft sleeve outer diameter.
- Excessive wear at the mating contact faces will occur when excessive shaft whip or deflection is present, due to defective radial bearings or bearing fits. The contact area of the mating faces will be increased resulting in increased wear and the elimination, or reduction, of the lubricating film between the faces, further reducing seal life.
- Defective radial bearings or bearing fits result in excessive vibration of the pump, resulting in reduced seal life and performance.

4. *Shaft run out (bent shaft).* Clamp the dial indicator to the pump housing as shown in Figure 168(C) and indicate the shaft runout at two or

more points on the outer diameter of the impeller end of the shaft. Also, indicate the shaft run out at the coupling end of the shaft. If the runout exceeds 0.0015–0.003 in. remove the shaft and straighten or replace.

A bent shaft can lead to many seal failures and poor sealing performance. Bearing life is greatly reduced and the operating conditions of both radial and thrust bearings can be impaired, resulting in reduced seal life and performance for reasons outlined previously. Bent shafts also can cause vibration.

5. *Stuffing box face squareness or perpendicular to the shaft axis.* With the pump stuffing box cover bolted in place, clamp the dial indicator to the shaft as shown in Figure 168(D), with the stem against the face of the stuffing box. The total indicator runout should not exceed 0.003 in.

When the face of the stuffing box is out of square or not perpendicular to the shaft axis, the result can be serious malfunction of a mechanical seal for the following reasons:

- The stationary gland plate that holds in position the stationary insert, or seat, is bolted to the face of the stuffing box. Misalignment at this point will cause the gland to cock, resulting in cocking of the stationary element. This results in seal wobble, that is, operating in an elliptical plane. This condition is a major factor in causing fretting, pitting and wear of the mechanical seal shaft packing at the point of contact with the shaft or sleeve.
- A seal that is wobbling on the shaft also can cause wear of the drive pins. Erratic wear at face contact causes poor seal performance and life.

6. *Concentricity of stuffing box bore.* With the same dial indicator set up as above, place the indicator stem well into the bore of the stuffing box. The stuffing box should be concentric to the shaft axis to within a 0.005 in. total indicator reading.

If the shaft is eccentric to the box bore, check the slope, or looseness, in the pump bracket fits at location A in Figure 168(E).

Rust, atmospheric corrosion or corrosion due to leakage of gaskets can cause damage to these fittings, which makes it impossible to ensure a stuffing box that is concentric with the shaft.

A possible remedy for this condition is the welding of the corroded area and remachining to proper dimensions. Another possible solution is to center the stuffing box and dowel it in place. Eccentricity alters the hydraulic loading of the seal faces, reducing seal life and performance.

7. *Driver alignment and pipe strain.* Driver alignment is an extremely important consideration, and periodic checkups should be a common practice. Steps to ensure proper coupling and driver alignment are well known and too lengthy to discuss in this chapter. The effect of pipe strain is also well known as to the damage it can cause to pumps, bearings and seals.

In most plants it is customary to blind the suction and discharge flanges of inactive pumps. These blinds should be removed before the pump driver alignment is made. Otherwise the alignment job is incomplete.

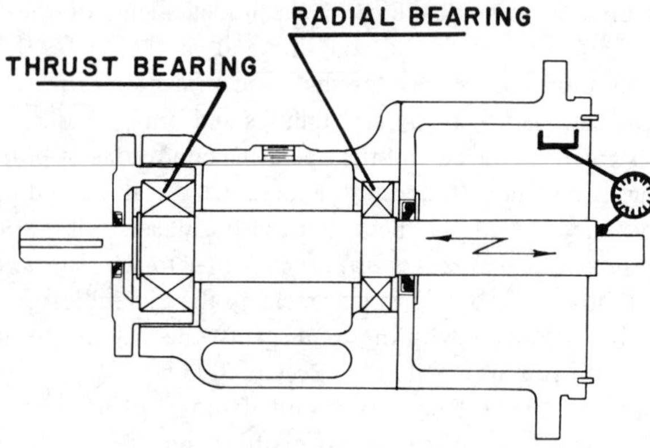
RADIAL BEARING
THRUST BEARING

(A)

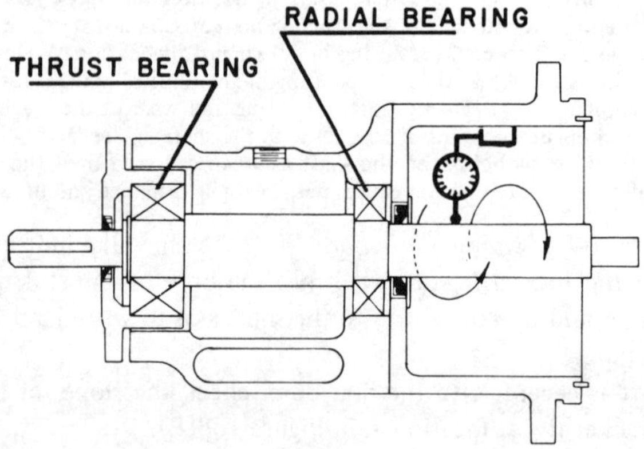
RADIAL BEARING
THRUST BEARING

(B)

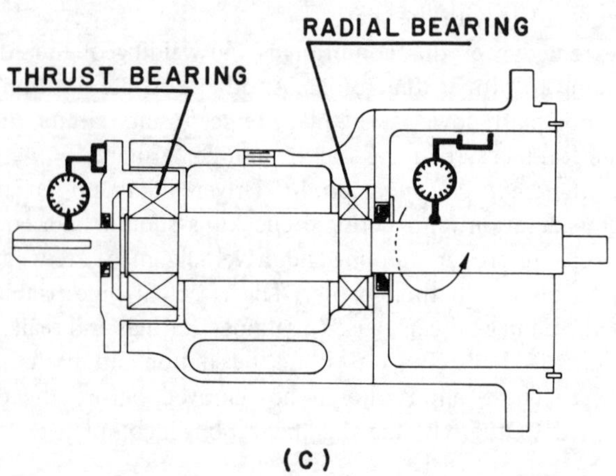
RADIAL BEARING
THRUST BEARING

(C)

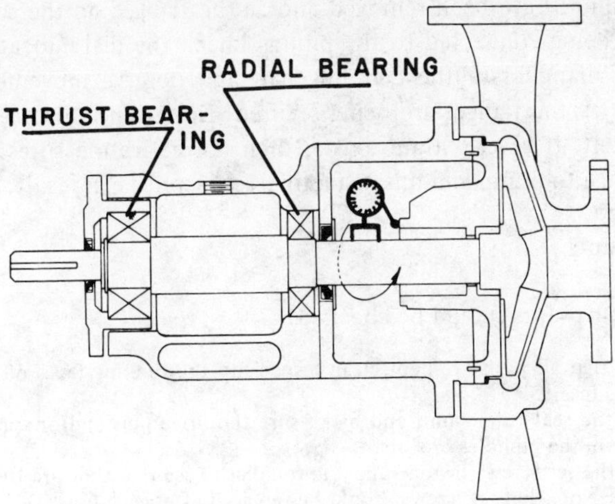

(D)

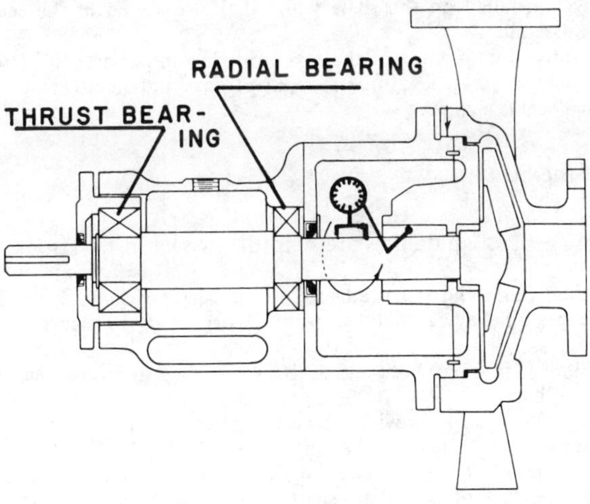

(E)

Figure 168. (A) Illustrates where to check for end play. (B) Checking for whip or deflection. (C) Checking for bent shaft. (D) Checking for stuffing box face squareness. (E) Checking for bore concentricity.

After the blinds have been removed and as the flanges on the suction and discharge are being connected to the piping, check the dial indicator reading on the outer diameter of the coupling half and observe movement of the indicator dial as the flanges are being secured. Deviation indicates that pipe strain is present. If severe strain exists, then corrective measures should be taken or damage to pump and unsatisfactory seal service can result.

Seal Check Points

The following points should be checked:

1. Ensure that all parts are kept clean, especially the running faces of the seal ring and insert.
2. Check the seal rotary unit and make sure the drive pins and/or spring pins are free in the pinholes or slots.
3. Check the set screws in the rotary unit collar to see that they are free in the threads. Note that set screws should be replaced after each use.
4. Check the thickness of all gaskets against the dimensions shown on the assembly drawing. Improper gasket thickness will affect the seal setting and spring-load imposed on the seal.
5. It is always advisable to check the fit of the gland ring to the equipment. Make sure there is no interference or binding on the studs or bolts or other obstructions. Be sure the gland ring pilot, if any, enters the bore with a reasonable guiding fit for proper seal alignment.
6. Make sure all rotary unit parts of the seal fit over the shaft. Particular care should be given to "V" rings. They must be placed on the shaft individually and never installed on the shaft while they are seated in the seal ring or rotating assembly.
7. Check both running faces of the seal (seal ring and insert) to be sure there are no nicks or scratches. Imperfections of any kind on either of these faces will cause seal leakage.

Installation of the Seal

The following general guidelines are helpful for seal installations.

1. Remove all burrs and sharp edges from the shaft or shaft sleeve, including sharp edges of keyways and threads. Worn shaft or sleeves should be replaced.
2. Check the stuffing box bore and stuffing box face to ensure they are clean and free of burrs.
3. The assembly drawing provided by the manufacturer gives the seal setting dimensions from the face of the stuffing box to the back of the rotary unit collar. To physically set the seal to this dimension it is first necessary to scribe a reference mark on the shaft or sleeve that will coincide with the location of the face of the stuffing box. From this reference mark on the shaft it is a simple matter to get the proper dimension to the back of the rotating collar. A flat scribing tool can be used to establish the reference mark on the shaft or sleeve.
4. The shaft or sleeve should be oiled lightly prior to seal assembly to allow the seal parts to move freely over it. This is especially desirable when assembling the seal collar as the bore of the collar usually has only a few thousandths of an inch clearance, and care should be used to avoid getting the collar cocked.

5. Install the rotary unit parts on the shaft or sleeve in proper sequence.
6. Set the back of the collar at the proper distance from the original reference mark on the shaft or sleeve. Tighten all set screws firmly and evenly.
7. Be extra careful when passing the seal gland ring and insert over the shaft. Do not bring the insert against the shaft as it may chip away small pieces from the edge of the running face.
8. Wipe the seal faces *clean* and apply a clean oil film prior to completing the equipment assembly.
9. Complete the equipment assembly, taking care when compressing the seal into the stuffing box.
10. Seat the gland ring and gland ring gasket to the face of the stuffing box by tightening the nuts or bolts evenly and firmly. Be sure the gland ring is not cocked and tighten the nuts or bolts only enough to effect a seal at the gland ring gasket. Excessive tightening of the gland ring nut or bolt will cause distortion that will be transmitted to the running face, resulting in leakage.

If the seal assembly drawing is not available, the proper seal setting dimension for inside seals can be determined as follows:

Establish a reference mark on the shaft or sleeve flush with the face of the stuffing box (point A in Figure 169).

Next, determine how far the face of the insert will extend into the stuffing box bore (B). This dimension is taken from the face of the gasket.

Determine the compressed length of the rotary unit (C) by compressing the rotary unit to the proper spring gap, which is marked on the collar (D).

This dimension (C), added to the distance the insert extends into the stuffing box (B), will give the seal setting dimension (E) from the reference mark on the shaft or sleeve to the back of the seal collar.

When the face of the insert is recessed in the gland ring, it will be necessary to determine the dimension from the face of the stuffing box to the face of the insert (B in Figure 169(B)). This dimension will again be measured from the face of the gasket. Determine the compressed length of the rotating unit as previously explained and subtract the dimension that the insert face is recessed in the gland (B) from the compressed rotary unit length (C) and you will have the seal setting dimension (E) from the reference mark to the back of the seal collar.

Outside seals are set with the spring gap (A in Figure 169(C)) equal to the dimension stamped on the seal collar. Cartridge seals are set at the factory and are installed as complete assemblies. These assemblies contain spacers that must be removed *after* the seal assembly is bolted into position and the sleeve collar is positioned in place.

STARTUP PROCEDURES

The first and foremost consideration in starting up equipment with mechanical seals is to ensure that the seal faces are immersed in a liquid from the very beginning so that they will not be scored or damaged from dry

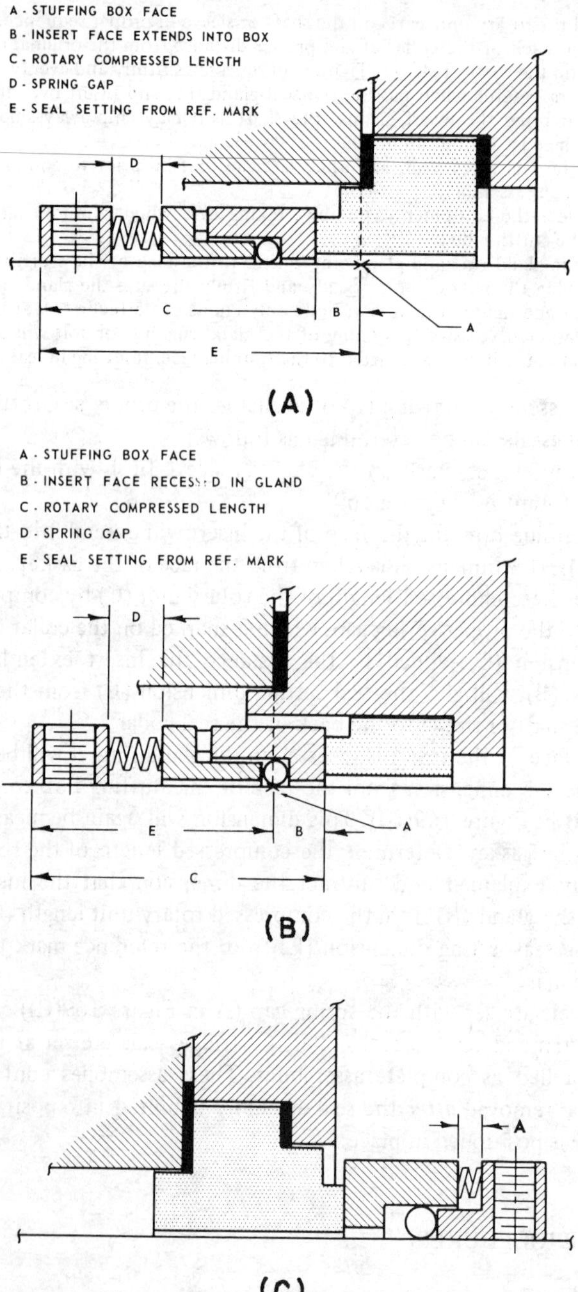

Figure 169. (A) Setting for inside seal. (B) Setting for inside seal with recessed insert. (C) Setting for outside seal.

operation. The following recommendations in reference to seal startup procedure will cover most types of seal installations. If followed where they are applicable, successful seal startup and normal seal life can be expected.

1. Caution the electrician not to run the equipment dry while checking motor rotation. A slight turnover will not harm the seal, but full speed for several minutes under dry conditions will destroy or severely damage the rubber faces.

2. The stuffing box of the equipment, especially centrifugal pumps, should always be vented prior to startup. Even though the pump has a flooded suction, it is still possible that air may be entrapped in the top portion of the stuffing box after the initial liquid purge of the pump.

3. Check installation for need of priming. Priming may be necessary in applications with a low or negative suction head.

4. Where cooling or bypass recirculation taps are incorporated in the seal gland, piping must be connected to or from these taps prior to startup. These specific environmental control features must be used to protect the organic materials in the seal and ensure its proper performance. Cooling lines should be left open at all times or whenever possible. This is especially true when a hot product may be passing through standby equipment while it is not on the line. Many systems provide for product to pass through the standby equipment so the need for additional product volume or an equipment switch is only a matter of pushing a button.

5. On hot operational equipment that is shut down at the end of each day it is highly advisable to leave the cooling water on at least long enough for the seal area to cool below the temperature limits of the organic materials in the seal.

6. Face lubricated-type seals must be connected from the source of lubrication to the tap openings in the seal gland prior to startup. This is another predetermined environmental control feature that is mandatory for proper seal function. Where double seals are to be operated it is necessary that the lube feed lines be connected to the proper ports for both circulatory or dead-end systems prior to equipment startup. This is very important, as all types of double seals are dependent on the controlled pressure and flow of the sealing fluid to function properly. Even before the shaft is rotated, the sealing liquid pressure must exceed the product pressure opposing the seal. Be sure a vapor trap does not prevent the lubricant from reaching the seal faces promptly.

7. Thorough warmup procedures include a check of all steam piping arrangements to be sure that all are connected and functioning, as products that will solidify must be fully melted before startup. It is advisable to leave all heating arrangements on during shutdown to ensure a liquid condition of the product at all times. Leaving the heat on at all times further facilitates quick startups and equipment switchovers, which may be necessary during a production cycle.

Thorough chilling procedures are necessary on some installations, especially LPG applications. LPG always must be kept in a liquid state in the seal area, and startup is usually the most critical time. Even during operation it may develop that the recirculation line piped to the stuffing box may have to be run through a cooler to overcome frictional heat generated at the seal faces. LPG requires a stuffing box pressure that is greater than the vapor pressure of the product at pumping temperature (25–50 psi differential is desired).

8. A squealing noise indicates insufficient liquid at the seal faces (that is, the faces are running dry). To avoid reduced seal life and possible seal damage, the installation should be checked immediately to determine how much liquid can be brought to, or retained in, the seal area. A recirculation line from the discharge to the stuffing box will usually overcome this condition.

9. Dry operation from loss of suction or inadequate suction is sometimes encountered and should be corrected immediately to prevent seal face damage or failure. If suction is lost at startup, it is advisable to again vent the stuffing box chamber before restarting the equipment. Inadequate suction piping should be replaced as soon as possible, as it is not only detrimental to the function of the seal but to the pump itself. Many pump and seal problems are due to improper piping.

10. New plant startup should consider the presence of dirt and foreign materials in the system. Dirt frequently accumulates in the system during construction. This condition cannot be eliminated entirely, but the proper cleansing and flushing of lines and equipment, prior to actual on-the-line running, can eliminate a great many seal failures. The use of strainers, cyclone separators and even filters on critical installations is advisable for new plant initial startup, as a large percentage of seal failures occur at that time.

11. If a seal leaks slightly at startup, allow a reasonable amount of time for it to adjust itself. Liquids that contain good lubricating qualities naturally will take longer to wear in the seal than those with lesser qualities. When a seal starts out with a slight leak and gets progressively less with running, it is indicative of leakage across the seal faces and continued running will cure it. Where leakage occurs immediately and remains constant, unaffected by running, it usually indicates secondary seal (shaft packing) damage or the seal faces that are warped out-of-flat. Out-of-flatness condition will be recognized readily from the wear pattern established on the seal faces where high spots or intermittent contact areas will show up.

12. Unless absolutely necessary, *do not open the seal faces for inspection.* After a seal has been running, a wear pattern is established between the two faces that microscopically mates these two faces. Since it is highly improbable that the two faces can be put back together in their exact original wear pattern, disturbing the seal in any manner probably will necessitate establishing a new pattern.

Concentricity and proper alignment of the equipment are necessary to eliminate undue stresses and strains on bearings and to maintain proper running clearances. Pipe strains also can cause stresses and contribute to misalignment of the equipment to a point where they will cause a malfunction of the seal. Impeller unbalance, caused by improper machining or presence of foreign materials lodged in the vanes, will set up vibrations and a condition of dynamic unbalance that a seal will not be able to tolerate.

Malfunction of seals can further be caused by improper stuffing box machining. The stuffing box bore must be concentric to the shaft, and the face of the stuffing box perpendicular to the axis of the shaft within 0.003 in. total indicator reading. Eccentric boring, or out-of-square gasket surfaces, will cause the seal to be drawn into a position of misalignment. The shaft or sleeve must have a radial surface free from scratches or nicks that allow seal leakage past the secondary seal or shaft packing. Pump throat bushings with proper diametral clearances are also necessary for good seal performance.

Excessive cavitation can cause seal failure, in addition to the damage that may occur to the pump. Cavitation causes shaft vibration, shaft deflection and bearing failures. It can also cause excessive heat buildup in the stuffing box due to lack of product and eventual loss of the liquid film between the seal faces, resulting in ultimate seal failure. (Here again, the shaft packing may be caused to oscillate against the shaft, causing grooving and accelerated corrosion). Pump cavitation is usually indicated by excessive noise in the pump casing. Chapter 2.8 discusses cavitation in pumps.

Ordinarily, the greatest amount of shaft deflection is encountered on single-stage end-suction pumps when they are operated at or near shutoff. The resulting shaft deflection and overheating can, in turn, cause seal failure. The more modern pumps are designed so that this problem is practically eliminated.

The major problem when operating pumps at or near shutoff is the excessive heating of the product in the seal area. The size of the pump, the product being handled and the product pumping temperature all have a definite bearing on when this overheating condition will occur. Some liquids can boil or flash to vapor in the stuffing box, strictly from normal frictional heat at the rubbing faces, plus the heat generated from operating at or near shutoff. When these conditions prevail, ultimate seal failure results. Bypass or recirculation lines frequently can offset this condition.

Standby Equipment

Standby equipment having mechanical seals should not be allowed to stand idle for long periods of time. Rotate them at least once a week, if only for a few moments. Better still, designate a certain time for standby equipment to be placed on the line for a predetermined period of time. This practice will usually require only a minimum of time and effort, as almost all standby equipment has common suction and discharge piping and it is simply a matter of opening and closing the proper valves.

TROUBLESHOOTING

The following notes summarize troubleshooting problems with mechanical seals:

1. *Seal Spits and Sputters in Operation*
   - Product vaporizing and flashing across the seal faces.
   - Take steps to maintain product in a definite liquid condition.
   - Have balance design checked with seal manufacturer.
   - Requires accurate readings of stuffing box pressure, temperature and gravity of product.
2. *Seal Leaks and Ices Gland Ring*
   - Product is vaporizing and flashing across seal faces.
   - Icing may score faces, especially on the carbon insert. Seal faces should be replaced or relapped before starting up again after vaporizing condition has been corrected.
3. *Seal Drips Steadily*
   - First, determine location of leakage.
   - Check gland gasket for proper compression.
   - Faces not flat enough.
   - Gland bolts possibly too tight, causing warpage of gland and insert.
   - Shaft packing nicked or pinched during installation.
   - Carbon insert cracked, or face of insert or seal ring chipped during installation.
   - Seal faces scored from foreign particles between them.
   - Leakage of liquid under equipment shaft sleeve.
4. *Seal Squeals During Operation*
   - Inadequate amount of liquid at sealing faces. A squealing seal is a dry seal, but not all dry seals squeal.
   - Bypass flush line may be needed; if one is already in use, it may need to be enlarged to produce more flow.
5. *Carbon Dust Accumulating on Outside of Gland Ring*
   - There is inadequate liquid at sealing faces.
   - Liquid film is flashing and evaporating between seal faces and leaving residue, which is grinding away the carbon.
   - Pressure in stuffing box is too high for type of seal and product.
   - When this condition is suspected, contact your representative.
6. *Seal Leaks, Nothing Appears to be Wrong*
   - Faces are not flat. This out-of-flat condition will be quite obvious on the wear pattern of seal face, as discussed previously.
   - Carbon insert face distorted from improper or excessive tension on gland ring bolting. This can usually be determined by the wear pattern of seal face.
   - Pipe strain distorting seal faces. When this problem is encountered it is usually on end-suction overhung-type pumps. The design and size of pumps of this type will not tolerate the weight or misalignment of the suction piping on the overhung portion. When this condition exists, seal performance is definitely affected.

- Possible shaft vibration, caused by:
  - Misalignment
  - Impeller unbalance
  - Cavitation
  - Bad bearings

7. *Short Seal Life*
   - Product is abrasive, causing excessive seal face wear. Determine source of abrasives. When abrasives are in suspension in the liquid, bypass flushing, preferably through the gland over the seal faces, will improve the situation by keeping the abrasive particles moving and preventing them from settling out or accumulating in the seal area.
     When abrasives are forming due to the process liquid cooling down and crystallizing or partially solidifying in the seal area, bypass flush will help introduce maximum product to the seal cavity at correct temperature. Additional heat to the seal area will also contribute to overcoming this condition.
   - Equipment mechanically out of line.
   - Check equipment for undue pipe strain, distortion being passed on to seal area.
   - Seal shows signs of running too hot:
     - Recirculation or bypass line may be necessary.
     - Check for possible rubbing of some seal component along the shaft. Throttle bushing in the gland, close clearance of rotary unit outside diameter to stuffing box bore and the bore of the insert are some points to check for this condition.
     - Try to cool seal area more completely:
       - Ensure that all cooling lines are connected.
       - Ensure unobstructed flow in cooling lines (scale formation in cooling jackets and lines).
       - Increase capacity of cooling lines.
   - Possible improper choice of seal type:
     - When there is any indication that the seal being used is an improper type, consult the manufacturer.
   - Possible improper choice of seal materials:
     - Improper seal materials are usually obvious when there is material deterioration of a seal component from corrosion or excessive heat. Consult the manufacturer.

CHAPTER 2.8

CAVITATION

Cavitation is a condition generated by the boiling of a liquid at normal temperatures and low pressures. The condition is accentuated by the presence of dissolved air, which is released when pressures are reduced. It is the combination of air released and vaporization caused by low pressures induced in a fluid.

The physical, chemical and engineering aspects of cavitation have been studied extensively. Studies have been directed at the process of bubble formation and collapse, and at the role of dissolved gases in the liquid. From an engineering standpoint, the formation and collapse of bubbles via cavitation can have a direct impact on the operating efficiency and service life of a variety of equipment and structures such as pumps, turbines and propellers. For example, the formation of bubbles can cause upsets of flow through a process, resulting in a loss of pressure, efficiency and thrust. Also, the collapse of bubbles can produce shock waves, which erode metal surfaces. This chapter gives an overview on the subject of cavitation with particular emphasis of its effects in pumps.

CAVITATION PHENOMENA

The pressure at which cavitation initiates depends on the liquid's physical state. If the liquid contains a large concentration of dissolved gases, then as the pressure is lowered the gas leaves the solution, forming cavities. These cavities produce a pressure that is greater than the liquid's vapor pressure. In many instances, air bubbles are not even visible; however, submicroscopic gas bubbles serve as nuclei causing cavitation at pressures above the vapor pressures. Individual bubbles grow from a nucleation site to a finite size and then collapse.

If nuclei are not present, the liquid may be subject to negative pressures or forces without experiencing cavitation. Many common liquids, such as water,

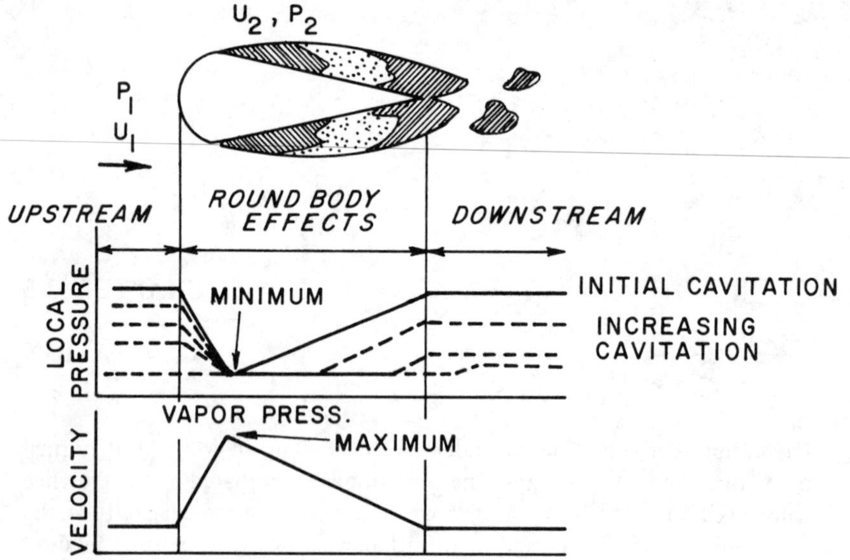

Figure 170. Illustrates the process of cavitation.

are capable of withstanding tensions on the order of thousands of atmospheres [32]. Water without nuclei for example will withstand tensions of 15,000 atmospheres [33].

If bubble sizes are very small, surface tension effects cause them to collapse rapidly. If bubbles are large enough to be visible, they tend to rise to the liquid surface and escape. Often nuclei may consist of irregularities or crevices in the fluid's boundary surface. This explains why cavitation often initiates at or near such boundaries. Cavitation can, however, be induced far from the liquid's containing boundary. For example, suspended particles in the fluid serve as nuclei and replace surface irregularities.

In the final collapse phase of cavitation bubbles, a weak emission of light occurs. The cause of this sonoluminescence is believed due to the recombination of free ions generated by the thermal or mechanical dissociation of molecules on the bubble surface. There is also some evidence that this phenomenon is due to the incandescence of the gas in the bubble resulting from the high pressures and temperatures during collapse [34].

The life of a cavitation bubble involves two changes of phase separated by expansion and collapse, forming a complete thermodynamic cycle. When cavities collapse, high pressures and temperatures are generated in the bubble. Local temperature rises have been estimated in the range of 500 to

800°C [35], and resulting shock waves radiated through the local liquid surroundings may have a pressure differential as great as 4000 atm [36].

The phenomenon of cavitation becomes more complex in flowing systems where liquids are subject to local accelerations and turbulence. In regions of high velocity, low pressures develop and cavities are formed. These cavities may extend until an area of lower velocity and higher pressures occurs where they collapse. This succession of cavities forming and collapsing produces larger, more stable bubbles. Note that these cavities form only when the lower pressure approaches the liquid's vapor pressure.

Local fluid velocity acceleration and resultant pressure fluctuations caused by cavitation are illustrated graphically in Figure 170 for flow around a submerged solid object. The fluid accelerates past the front edge of the object reaching a maximum velocity, while a minimum pressure is obtained at the throat. As shown, the cavity collapse is an unstable process. The ends of the bubbles tend to break off producing a pulsating end to the cavity. In this example, the cavities are symmetrical on two surfaces. This leads to alternate collapse on two surfaces and vortex shedding behind the body. Note also that if the upstream pressure is sufficiently large, the throat pressure will not fall below the vapor pressure; however, as the upstream pressure is lowered, cavitation forms and spreads toward the rear.

The pressure and velocity relation follows from Bernoulli's equation, which can be rearranged to define a *cavitation coefficient*:

$$\sigma = \frac{P_1 - P_c}{\frac{\rho}{2} U_1^2} \tag{119}$$

where P_c is the critical pressure at point P_2 (i.e., where cavitation occurs).

The value of σ at which cavitation initiates is denoted as σ_{crit} (critical cavitation coefficient), also called the *inception point.* The *desinent point* refers to the point where cavitation terminates at increasing pressure. Note that σ_{crit} is a function of a number of parameters, major ones being dissolved air content, scale effects and the nuclei.

Dissolved Gases and Scale Effects

The effect of dissolved air is to increase the cavitation coefficient. Pearsall [37] provides data showing the magnitude of this effect as a function of air concentration in water. At very high air contents a hysteresis effect occurs on cavitation inception, consequently, for comparison purposes, the desinent point, which is less affected by hysteresis, is often used as the criterion.

The time that the fluid spends in the low pressure zone depends on the fluid velocity. Consequently, at low velocities the fluid will be in the low pressure region for a longer time than at higher velocities, and cavitation will occur more readily. Velocity scale effects have been observed experimentally, showing that lower cavitation numbers occur at low fluid velocities [38,39]. Scale effects observed between static bodies and machines such as turbines or pumps generally do not correlate well. Most tests on static bodies show that σ increases with size, whereas other studies on machines show σ to decrease with size [38-40]. A number of correlations have been proposed based on a Reynolds number. One such correlation developed by Holl and Wislicenus [38] is shown in Figure 171.

Temperature Effects

Figure 172 shows the effect of temperature on cavitation in a pump. The cavitation performance of pumps in service with liquids other than water or with water at high temperatures is better than in cold water. The performance of a pump is affected by the size of the cavity. At higher vapor pressures, a higher depression of the total energy at the pump suction (i.e., ΔNPSH) is

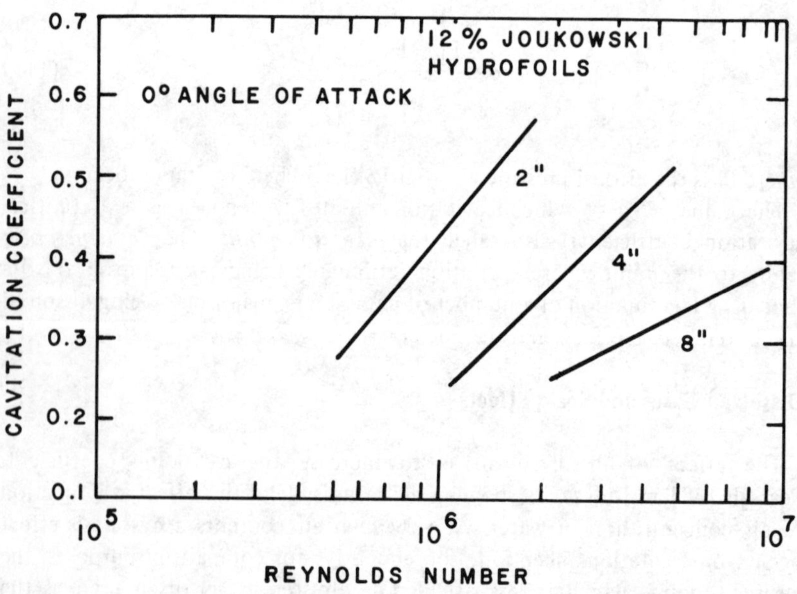

Figure 171. Cavitation coefficient as a function of Reynolds number for a Joukowski hydrofoil [38].

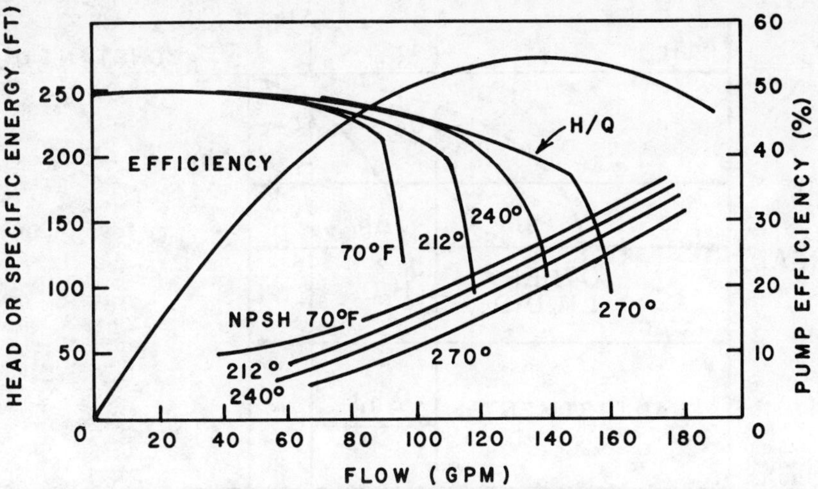

Figure 172. Effect of temperature on cavitation in a pump [41].

needed to produce the same volume of bubbles at the same head loss. Hence, it will require a greater amount of work and a greater temperature increment.

As an example, consider two different liquids handled in a pump. The saturation temperature corresponding to the reduced pressure in the low-pressure zone of the impeller is lower by an amount ΔT (Figure 173). The change in enthalpy for vaporization of the liquid is

$$\Delta\widetilde{H}_f = C_p \Delta T \tag{120}$$

where $\Delta\widetilde{H}_f$ = enthalpy change (Btu/lb)
ΔT = temperature change (°F)
C_p = specific heat of the liquid (Btu/lb-°F)

Equation 120 can be expressed in terms of the latent heat of vaporization:

$$\Delta\widetilde{H}_f = \frac{\overline{V}_v}{\nu_v} \lambda \tag{121}$$

where $\overline{V}$ = volume (ft^3)
ν = specific volume (ft^3/lb_m)
λ = latent heat (Btu/lb_m)

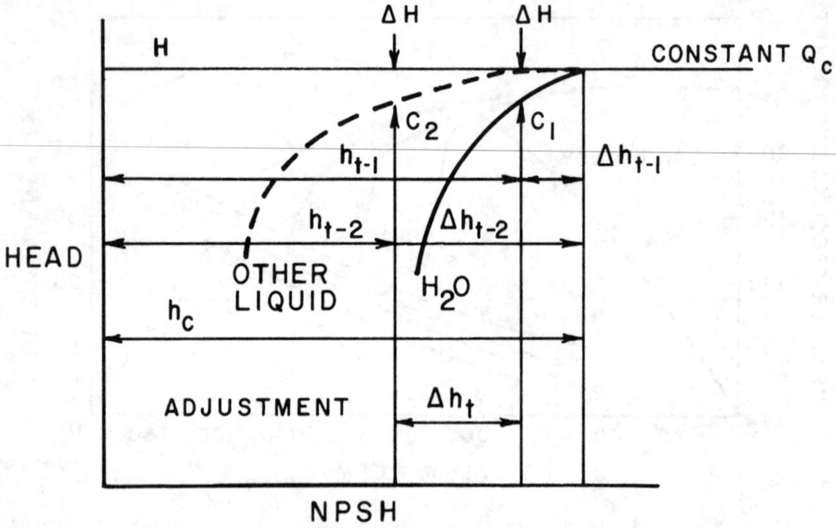

Figure 173. Illustrates cavitation process on two liquids.

Furthermore, parameter ψ can be defined as follows:

$$\frac{\overline{V}_V}{\overline{V}_L} = \frac{\nu_V}{\nu_L} \frac{\Delta\widetilde{H}_f}{\lambda} = \psi \tag{122}$$

where subscripts V, L refer to vapor and liquid, respectively.

From the Clapeyron equation,

$$\Delta\widetilde{H}_f = \frac{(\Delta NPSH)\, T\, \nu_V C_p}{\lambda \nu_L} \tag{123}$$

which can be rearranged to give

$$\psi_1 = \frac{\psi}{\Delta NPSH} = \frac{C_p T}{J' \lambda^2} \left(\frac{\nu_V}{\nu_L}\right)^2 \tag{124}$$

where J′ is the mechanical equivalent of heat.

The parameter ψ_1 has been correlated with vapor pressure, P_v, to give the following empirical expression [42]:

$$\Delta\text{NPSH} = \frac{64}{P_v \psi_1^{4/3}} \tag{125}$$

where ΔNPSH is in feet.

In practice, ΔNPSH is small and most estimations can be based on the correlation of Ward and Sutton [42];

$$\Delta\text{NPSH} \approx \frac{0.8}{\psi_1} \tag{126}$$

The value 0.8 is the approximate ratio of volumes of liquid to vapor.

Noise Effects

As cavitation bubbles collapse, shock waves are created and, thus, noise is produced. This is generally "white noise," which extends over a wide frequency band (as much as 1 MHz). Smaller bubbles are normally associated with a high-frequency noise. The lower end of the frequency spectrum normally is associated with the collapse of larger bubbles.

Noise detection is a common technique used for determining whether cavitation is occurring. Pearsall [43] compares visual methods and noise measurement techniques to detect the onset of cavitation on turbines.

CAVITATION EROSION

High transient pressure fluctuations and thermal shock waves resulting from cavitation cause fatigue to materials in the proximity to collapsing bubbles. Pumps and valves are prime candidates for cavitation erosion in piping systems. Reliable methods for predicting operating conditions that lead to severe erosion are not available; however, there are some general trends that have been observed.

Experimental data on erosion is most easily represented in the form of a plot of rate of material weight loss versus time. Most materials subjected to cavitation attack show a characteristic curve as illustrated in Figure 174(A), in which there is an initial incubation period where no erosion occurs (see portion (I) in small illustration of Figure 174(A), followed by a period of rapid increase in erosion damage (portion (II)), then a relatively stable state (III). Finally, as the material surface becomes pitted, the rate of erosion declines (IV).

From an engineering standpoint, the incubation period and the rate of erosion are of prime interest. By increasing the incubation period and

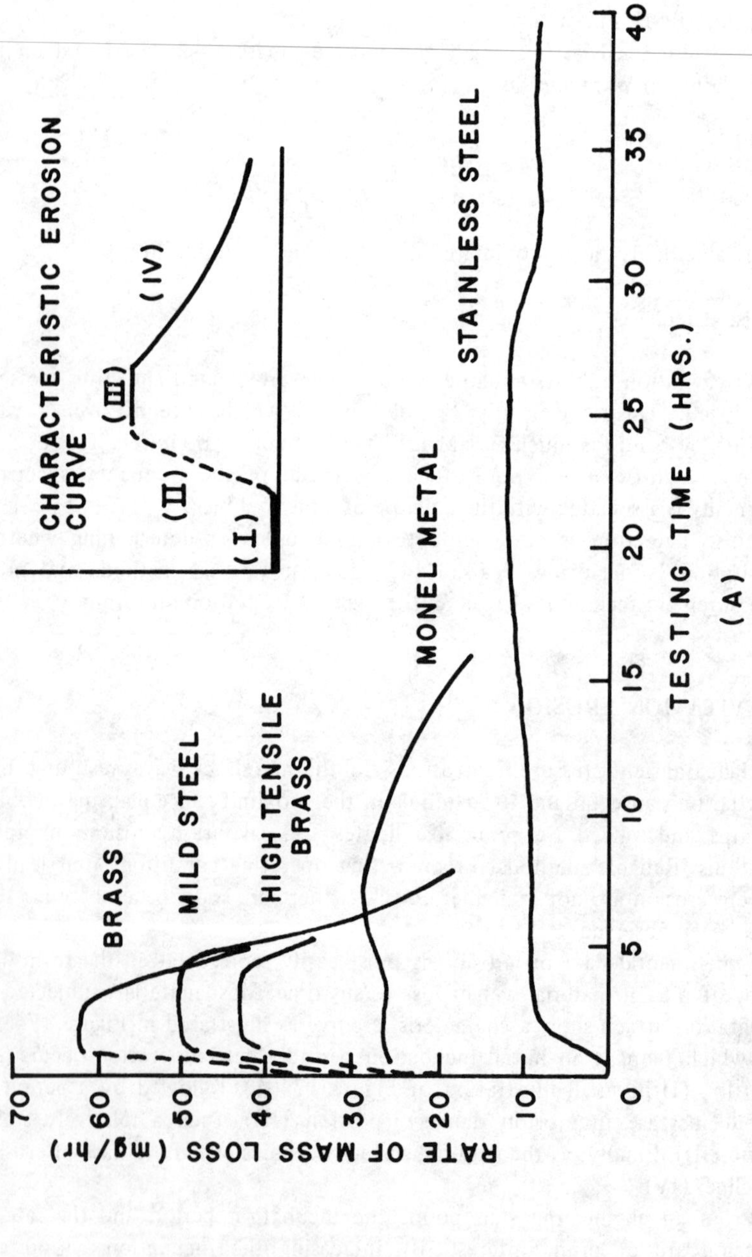
CHARACTERISTIC EROSION CURVE
(I)
(II)
(III)
(IV)
BRASS
MILD STEEL
HIGH TENSILE BRASS
MONEL METAL
STAINLESS STEEL
RATE OF MASS LOSS (mg/hr)
70
60
50
40
30
20
10
0
5
10
15
20
25
30
35
40
TESTING TIME (HRS.)
(A)

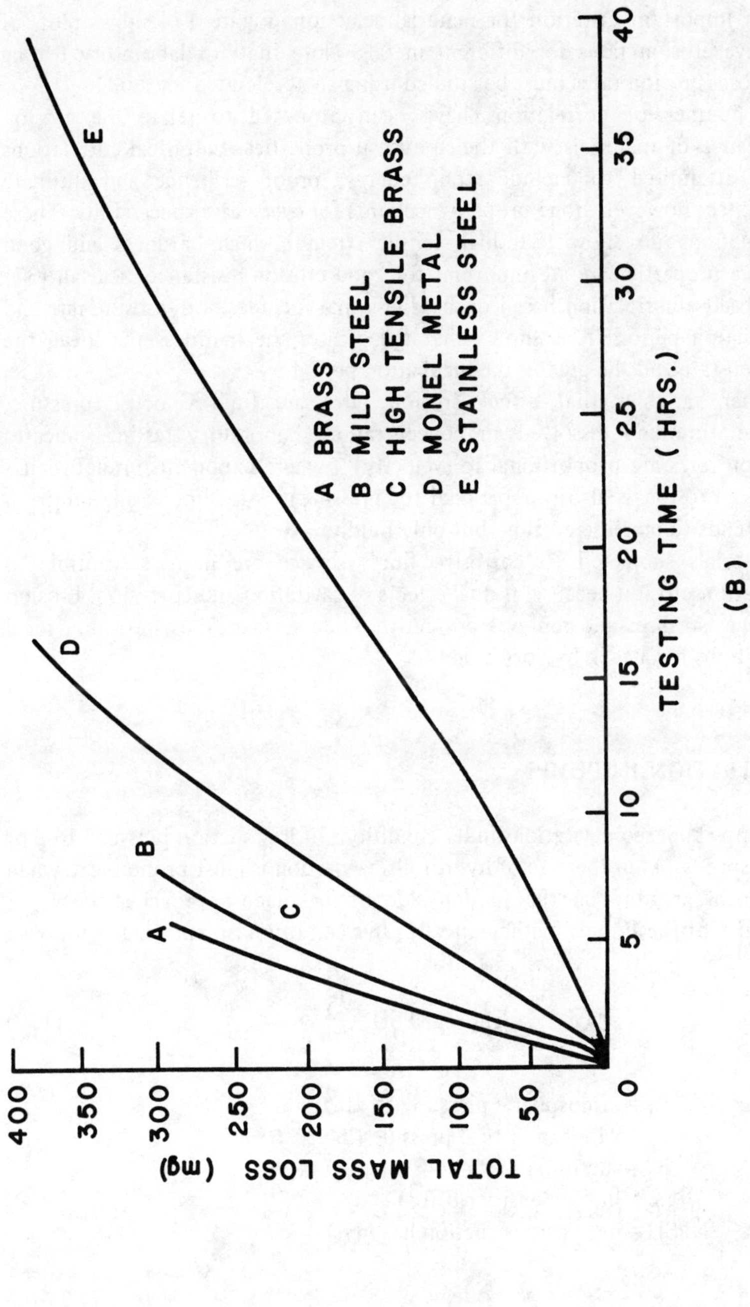

Figure 174. Relative erosion rates of different metals [44].

minimizing erosion rates, the service life of a piece of equipment can be greatly extended. Information on the rates of erosion of different materials helps to establish the relative resistance of these materials. This, then, serves as an important criterion for material selection. Figure 174 shows plots of relative erosion rates for different metals. Note that for laboratory testing purposes erosion rates must be studied using an accelerated method.

A number of correlations have been proposed to relate the erosion resistance of materials with their physical properties. Empirical correlations have attempted to include strain energy, proof resilience and ultimate resilience; however, none properly accounts for every case successfully. These correlations do show that high tensile strength, high hardness and good fatigue properties are all important for good erosion resistance. Pearsall [37] proposed a correlation based on hardness that includes both erosion rate and incubation period. It is shown that the greater the hardness the lower the erosion rate and the greater the incubation period.

Other variables that affect erosion rates are fluid velocity, pressure, temperature and dissolved air concentration. Laboratory testing indicates erosion rates are proportional to $(\text{velocity})^n$, where exponent n is between 5 and 7. Erosion is also proportional to $(\text{pressure})^2$. Also, increasing temperature tends to increase erosion, but only mildly.

Materials subjected to corrosive fluids service are more susceptible to cavitation erosion because of dual effects of cavitation and corrosion. Erosion damages surfaces and removes protective oxide surface films, providing ideal conditions for attack by corrosion.

CAVITATION IN PUMPS

Pumps undergo cavitation under conditions of low suction pressure drop or high speeds. As in the case of hydrofoils, cavitation is most pronounced when the local pressure on the suction side of the blade approaches the vapor pressure of the liquid. Mathematically, this condition is expressed as follows:

$$\frac{P_a}{\rho} - gh_s - gNPSH = \frac{P_v}{\rho} \tag{127}$$

where
P_a = atmospheric pressure (lb_f/ft^2)
P_v = vapor or critical pressure (lb_f/ft^2)
h_s = suction lift (ft)
ρ = liquid density (lb/ft^3)
NPSH = net positive suction head (ft)

The NPSH is proportional to the total head across the pump, h_t, as defined by the Thoma [45] cavitation coefficient:

$$\sigma_{Th} = \frac{NPSH}{h_t} \tag{128}$$

Hence, Equation 127 can be restated as follows:

$$NPSH = \frac{P_a}{\rho_g} - h_s - \frac{P_v}{\rho g} \tag{129}$$

A blade cavitation coefficient can also be defined as follows:

$$\sigma_b = \frac{P_1 - P_v}{\frac{1}{2}\rho U_1^2} \tag{130}$$

where P_1 is the static pressure in front of the blade and U_1 is the fluid's relative inlet velocity. Another important relation is the suction specific speed:

$$S_N = \frac{N\sqrt{Q}}{NPSH^{3/4}} \tag{131}$$

In English units, N (the rotational speed of the impeller) has units of rpm, Q (volumetric flowrate) gpm and NPSH, in feet:

$$S'_N = \frac{\omega\sqrt{Q}}{(gNPSH)^{3/4}} \tag{132}$$

where ω is in units of rad/sec and other units are consistent. Note that in English units,

$$S = 2495\ S' \tag{133}$$

The Thoma cavitation coefficient σ_{Th} is related to S by

$$\sigma_{Th}^{3/4} = \frac{N_S}{S_N} = \frac{K_n}{S'_N} \tag{134}$$

where N_s is the specific speed and K_n is the nondimensional specific speed. For pumps that are geometrically similar at different speeds or of different diameters, the similarity laws defined in Chapter 2.5 can be used. These are restated here for convenience:

$$\frac{NPSH_1}{NPSH_2} = \left(\frac{N_1}{N_2}\right)^2 \left(\frac{D_1}{D_2}\right)^2 \tag{135}$$

Correlations are available for estimating when cavitation inception occurs in pumps. Of greater interest, however, is the point where cavitation begins to have a significant effect on pump performance. During cavitation, as the NPSH is reduced a reduction in performance occurs at diminishing flows. Cavitation affects head and efficiency in a similar manner. The cavitation characteristic of a pump is normally expressed as shown in Figure 175. Figure 175 is a plot of head or efficiency versus NPSH or σ_{Th}. Such a plot is constructed for a pump operating at constant speed and flow (usually for the best efficiency point). Note $NPSH_{crit}$ denotes where the head falls below some designated percentage from the normal noncavitating characteristic. Figure 175 shows that a wide difference in NPSH exists between inception and performance breakdown from cavitation (one may be three to four times the other). In practice, a pump would have to be operated at a speed of one-half to one-third to avoid cavitation altogether, as opposed to operating with cavitation just above the breakpoint. Consequently, a larger (and more costly) pump is needed to satisfy the criterion of no cavitation.

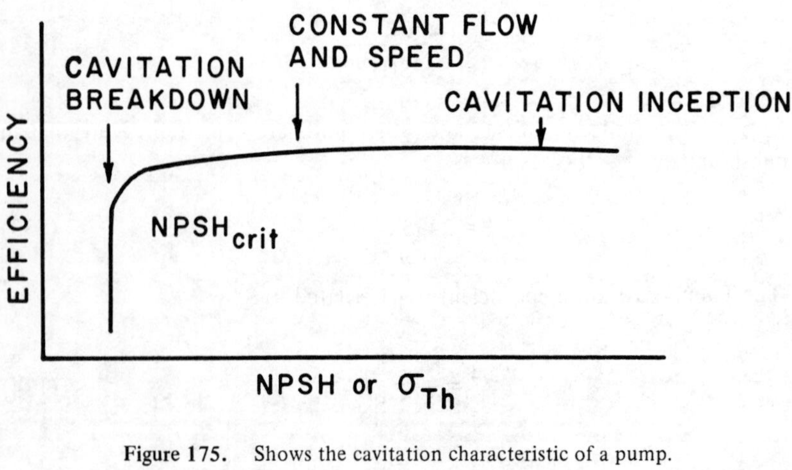

Figure 175. Shows the cavitation characteristic of a pump.

For this reason it is rarely done in commercial practice. Unfortunately, the disadvantage of operating with cavitation is the risk of erosion damage and, thus, short service life.

Centrifugal pumps generally experience a sharp drop in performance because of choking of the impeller passages. By contrast, axial pumps tend to exhibit a much more gradual fall in performance. The American Hydraulic Institute [46] reports cavitation data for different types of pumps.

DESIGN CONSIDERATIONS FOR CAVITATING PUMPS

Applying Bernoulli's equation to a horizontal pump, the following equation is derived:

$$\text{NPSH} = \sigma_b \frac{W_1^2}{2g} + \frac{U_{m1}^2}{2g} + \text{LOSSES} \tag{136}$$

where the velocity terms are defined by the vector diagram for the blade inlet to a pump shown in Figure 176.

For ordinary applications, losses can be approximated by $0.04\ (U_{m1}^2/2g)$; hence, Equation 136 becomes

$$\text{NPSH} = \frac{U_{m1}^2}{2g}\,(1.04 + \sigma_b) + \frac{\sigma_b U_1^2}{2g} \tag{137}$$

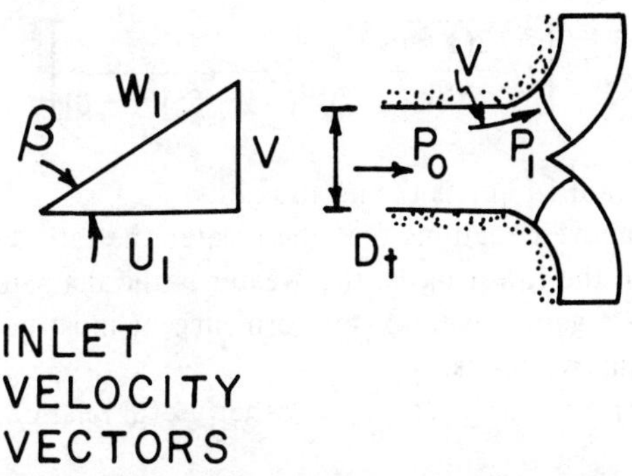

Figure 176. Vector diagram for fluid at the blade inlet of a pump.

or

$$\text{NPSH} = k_1 \frac{U_{m1}^2}{2g} + k_2 \frac{U_1^2}{2g} \tag{138}$$

where k_1 and k_2 are empirical coefficients and

$$K_s = \frac{2.98\sqrt{(1 - \beta^2)}}{\phi((1 + \sigma_b) + \sigma_b/\phi^2)^{3/4}} \tag{139}$$

where β is the hub diameter to tip diameter ratio, D_h/D_t, and ϕ is the flow coefficient defined as the ratio of axial velocity to peripheral velocity, Ua/U.

To determine the condition of minimum NPSH (or maximum suction specific speed), the following relations apply:

$$\phi = \frac{U_{m1}}{U_1} = \left\{\frac{\sigma_b}{2(1 + \sigma_b)}\right\}^{1/2} = \sqrt{\frac{k_2}{2k_1}} \tag{140}$$

and

$$\frac{\text{NPSH}}{U^2/2g} = \frac{3}{2}\sigma_b \tag{141}$$

The optimum tip diameter is

$$D_{t1} = 1.37\left[\sqrt{\frac{2(1 + \sigma_b)}{\sigma_b}}\left(\frac{Q}{\omega}\right)\frac{1}{(1 - \beta^2)}\right]^{1/3} \tag{142}$$

Since σ_b is related to tip values, Equation 142 can also be stated as

$$D_{t1} = 1.72\left[\sqrt{\frac{2(1 + \sigma_b)}{\sigma_b}}\left(\frac{Q}{\omega}\right)\frac{1}{(1 - \beta^2)(1 + \beta)}\right]^{1/3} \tag{143}$$

Note that in Equation 143 σ_b is the mean radius.

Note from the above equations that the greater the ratio Q/ω, the greater the diameter; also, the lower σ_b is, the greater is the diameter. More simply, for a pump to have good cavitation performance, it must have a larger inlet diameter than ordinary pumps.

Equation 143 can be rearranged to give other useful relations:

$$D_{t1} = \frac{4Q}{\pi(1 - \beta^2)}\left\{\frac{3}{2}\frac{(1 + \sigma_b)}{g\text{NPSH}}\right\}^{1/2} \tag{144}$$

and

$$N = 35.2 \sqrt{\frac{(1-\beta)}{(1+\beta)} \frac{(\text{NPSH})^{3/4}}{\sqrt{Q}}} \left\{\frac{1}{(1+\sigma_b)\,\sigma_b^2}\right\}^{1/4} \tag{145}$$

In establishing the optimum design, parameter β is generally kept as low as possible (approximately 0.3). The designer should be provided with information on Q and NPSH; hence, D_{t1} and N can be determined for optimum cavitation performance.

The cavitation coefficient, σ_b, is established based on blade shape and incidence. For performance breakdown, lower limiting σ_b values can be obtained from Figure 177 based on the method of Pearsall and Scobie [47]. A minimum value for σ_b is 0.08, obtainable with low incidence and thin blades. Using empirical methods, approximate equivalent values can be obtained by using $k_1 = 1.41$ and $k_2 = 0.085$. Note that although this method gives optimum performance, it is possible to operate slightly away from the optimum and still achieve an acceptable performance.

As noted earlier, if a pump does not undergo cavitation, no erosion will occur. However, in a practical sense this means that the pump may have to run so slowly and be so large that it becomes uneconomical. Hence, many pumps operate near or in the cavitation range. The present state-of-the-art does not allow accurate predictions as to whether a particular installation will erode. Proper material selection therefore should be based on erosion resistance in addition to corrosion resistance. Pumps in service with fluids other than water are unlikely to experience severe cavitation erosion. Pumps operating at high speeds (typically above 200 fps tip speed) are strong candidates for cavitation erosion.

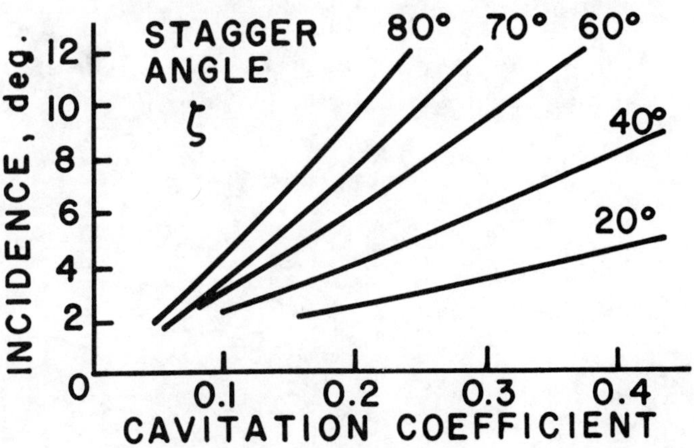

Figure 177. Cavitation breakdown limits as outlined by Pearsall and Scobie [47].

CHAPTER 2.9

FLOW PAST IMMERSED BODIES: NEWTONIAN FLUIDS

This chapter reviews basic principles and equations of motion for characterizing the flow of incompressible fluids past immersed bodies. Incompressible flow around solid objects is of interest to a number of process operations. In tubular heat exchangers, for example, flow occurs parallel and perpendicular to tube bundles. Heat transfer from extended surfaces depends on the nature of the flow past these surfaces. A fluidized bed is another example in which the flow of fluid past solid particles governs both heat and mass transfer mechanisms.

DRAG COEFFICIENTS

Drag is the resistance to movement of a solid in a fluid. It is caused by the shear stresses exerted in the boundary layer of the fluid adjacent to the solid surface. For a completely laminar boundary layer, shear forces are viscous only. If the boundary layer is turbulent in nature, then the resistance results from velocity fluctuations of the fluid in the boundary layer. There are basically two types of drag. If it is caused by stresses in the boundary layer alone, as in the case of flow over a flat boundary, it is surface drag. However, if the flow is around a two-dimensional object such as a cylinder, separation of the boundary layer occurs and the fluid immediately behind the solid body becomes turbulent. This type of drag caused by the turbulent wake is called form drag and depends on the geometry of the body past which the fluid flows.

The drag coefficient is defined by Equation 146 or 147.

$$F = \frac{f \rho U^2 A_w}{2 g_c} \tag{146}$$

$$F = \frac{f_D \rho U^2 A_p}{2g_c} \tag{147}$$

F is the force of resistance exerted on the solid by the flowing fluid. In Equation 146, A_w is the wetted area of the immersed body. In Equation 147, A_p is the projected area of the body on the plane normal to the flow direction. Since the drag coefficient, f, in Equation 146 is based on the wetted area, it is mainly applicable for flat plates or thin streamlined struts. The coefficient f_D (Equation 147) is based on the projected area of the body and, therefore, applies for cylinders and bodies of revolution. Hence, when Equation 147 is used to compute the force, it includes both surface drag and form drag.

FLOW PAST FLAT PLATES

Consider an incompressible fluid flowing past a thin, flat plate (the plate is oriented parallel to the flow direction). The length of the laminar boundary layer will depend on the main stream turbulence and mean velocity, the distance from the leading edge, and the fluid's kinematic viscosity.

The momentum equations of flow in two dimensions were outlined earlier in this section. For convenience, they are restated here for an incompressible fluid with constant viscosity:

$$\frac{\partial u}{\partial t} + u\frac{\partial u}{\partial x} + \nu\frac{\partial u}{\partial y} = -g_c\frac{\partial \Omega}{\partial x} - \frac{g_c}{\rho}\frac{\partial P}{\partial x} + \nu\left(\frac{\partial^2 u}{\partial x^2} + \frac{\partial^2 u}{\partial y^2}\right) \tag{148}$$

$$\frac{\partial v}{\partial t} + u\frac{\partial v}{\partial x} + v\frac{\partial v}{\partial y} = -g_c\frac{\partial \Omega}{\partial y} - \frac{g_c}{\rho}\frac{\partial P}{\partial y} + \nu\left(\frac{\partial^2 u}{\partial x^2} + \frac{\partial^2 v}{\partial y^2}\right) \tag{149}$$

An exact solution to these equations for the laminar boundary layer on a flat plate immersed in a fluid with uniform velocity was first obtained by Blasius [48] by introducing the stream function, ψ:

$$\frac{\partial \psi}{\partial y} = u \tag{150}$$

$$\frac{\partial \psi}{\partial x} = -v \tag{151}$$

Equations 150 and 151 satisfy the two-dimensional Continuity equation, $\partial u/\partial x + \partial v/\partial y = 0$.

Defining U as the velocity at the edge of the boundary layer and noting that $\partial U/\partial x = 0$, the momentum expression over the entire thickness of the boundary layer becomes

$$u \frac{\partial u}{\partial x} + v \frac{\partial u}{\partial y} = \frac{\mu}{\rho} \frac{\partial^2 u}{\partial y^2} \tag{152}$$

Blasius reduced Equation 152 to an ordinary differential equation by introducing the variables η and ϕ:

$$\eta = \frac{y}{2} \left(\frac{U\rho}{\mu x} \right)^{1/2} \tag{153}$$

$$\psi = \left(\frac{\mu U x}{\rho} \right)^{1/2} \phi \tag{154}$$

Using these definitions, the following relations were derived for the terms in Equation 152:

$$u = \frac{U}{2} \phi' \tag{155}$$

$$\frac{\partial u}{\partial x} = -\frac{U\eta}{4x} \phi'' \tag{156}$$

$$\frac{\partial u}{\partial y} = \frac{U}{4} \left(\frac{U\rho}{\mu x} \right)^{1/2} \phi'' \tag{157}$$

$$\frac{\partial^2 u}{\partial y^2} = \frac{U}{8} \frac{U\rho}{\mu x} \phi''' \tag{158}$$

$$v = \frac{1}{2} \left(\frac{\mu U}{\rho x} \right)^{1/2} (\eta \phi' - \phi) \tag{159}$$

ϕ', ϕ'', ϕ''', are the first, second and third total derivatives of ϕ with respect to η.

Using the relations given by Equations 155 through 159 in Equation 152, and applying the following boundary conditions,

$$\left. \begin{array}{l} \text{For } \eta = 0 \text{ (same as y = o)} \\ \phi = 0, \phi' = 0 \end{array} \right\} \tag{160}$$

$$\left. \begin{array}{l} \text{For } \eta = \infty \text{ (same as y} = \infty) \\ \phi' = 2 \end{array} \right\} \tag{161}$$

The solution obtained by Blasius is in the form of a Taylor series solution and is expressed in terms of one of the constants of integration.

$$\phi = \frac{C_2\eta^2}{2!} - \frac{C_2^2\eta^5}{5!} + 11\frac{C_2^3\eta^8}{8!} - \frac{375C_2^4\eta^{11}}{11!} + \ldots \tag{162}$$

where $C_2 = 1.32824$.

Howarth [49] computed values for ϕ, ϕ' and ϕ''' as functions of η. Table 36 gives selected values computed by Howarth along with values of u/U computed from Equation 155. Using these values, flow conditions at any point in the x-y coordinate system may be obtained.

von Karman [50] analyzed the flow in the boundary layer using Newton's second law to develop an integral relationship for the velocity distribution. His analysis considered a two-dimensional region of fluid, which included the boundary layer and had a differential length, dx. It can be shown that in this region the total rate of momentum increase is equal in both magnitude and direction to the forces acting on the boundaries of the region. von Karman's integral relationship also may be derived by integrating the Continuity equation and the simplified two-dimensional equation of motion as written by Prandtl [51]:

$$u\frac{\partial u}{\partial x} + v\frac{\partial u}{\partial y} = -\frac{g_c}{\rho}\frac{\partial P}{\partial x} + \frac{\mu}{\rho}\frac{\partial^2 u}{\partial y^2} \tag{163}$$

The details of the mathematics are given by Knudsen and Katz [4]. von Karman's integral momentum equations for the boundary layer are as follows:

Table 36. Computed Values for Blasius Parameters [49]

η	ϕ	ϕ'	ϕ''	u/U
0	0	0	1.32824	0
0.4	0.1061	0.5294	1.3096	0.2647
0.8	0.4203	1.0336	1.1867	0.5168
1.2	0.9223	1.4580	0.9124	0.7290
1.6	1.5691	1.7522	0.5565	0.8761
2.0	2.3058	1.9110	0.2570	0.9555
2.4	3.0853	1.9756	0.0875	0.9878
2.8	3.8803	1.9950	0.0217	0.9915
3.2	4.6794	1.9992	0.0039	0.9996
3.6	5.4793	2.0000	0.0005	1.0000
3.8	5.8792	2.0000	0.0002	1.0000

$$\frac{\tau_w g_c}{\rho} = \frac{\partial}{\partial x}\int_0^\delta u(U-u)dy + \frac{dU}{dx}\int_0^\delta (U-u)dy \tag{164}$$

or

$$\frac{\tau_w g_c}{\rho} = \frac{\partial}{\partial x}U^2\int_0^\delta \frac{u}{U}\frac{U-u}{U}dy + \frac{dU}{dx}U\int_0^\delta \frac{U-u}{U}dy \tag{165}$$

where fluid thickness, δ, is large enough to include the boundary layer. τ_w is the wall shear stress. To solve Equation 164, the point velocity must be known as a function of y.

For constant U, von Karman's integral momentum equation for the boundary layer on a flat plate becomes

$$\frac{\tau_w g_c}{\rho} = \frac{\partial}{\partial x}\int_0^\delta u(U-u)dy \tag{166}$$

Boundary Layer Thickness

The boundary layer thickness, δ, is the distance normal from the wall where the point velocity is within 1% of the mainstream velocity. That is, δ is the y value where $u/U = 0.99$. Using the values given in Table 36, it can be shown that

$$\frac{\delta}{x} = \frac{4.96}{\sqrt{N_{Re_x}}} \tag{167}$$

where N_{Re_x} is the local Reynolds number.

Pohlhausen [52] modified von Karman's integral equation (Equation 166) to obtain the following expressions for the laminar boundary layer thickness and velocity profile:

$$\delta = 4.64\left(\frac{\nu x}{U}\right)^{1/2} \tag{168}$$

$$\frac{u}{U} = \frac{1.5}{4.64}\frac{y}{\sqrt{\nu x/U}} - \frac{y^3}{199.8(\sqrt{\nu x/U})^3} \tag{169}$$

Drag Coefficients (Laminar Flow)

The Blasius solution predicts the following relation for the total drag coefficient for a plate of length, L:

$$f = \frac{1.328}{\sqrt{N_{Re_L}}} \quad \text{for} \quad 2 \times 10^4 < N_{Re_L} < 5 \times 10^5 \tag{170}$$

where N_{Re_L} is the total Reynolds number for flow over the flat plate $(= UL/\nu)$.

At low Reynolds numbers $(10 < N_{Re_L} < 3000)$, the following empirical equation may be used [53]:

$$f = 2.90(N_{Re_L})^{-0.60} \tag{171}$$

Turbulent Boundary Layer

Using the analogy for flow through circular tubes, von Karman [54] derived the form of the velocity-profile equation describing the turbulent portion of the boundary layer. For any point, x, on the plate, the relation between point velocity, u, and distance normal from the plate, y, is

$$\frac{u}{U} = (y/\delta)^{1/7} \tag{172}$$

Equation 172 is the power law for the velocity distribution over flat plates.

Hansen [55], Burgers [56] and others [57,58] obtained measurements on the turbulent boundary layer and found the exponent on the term y/δ to lie between 0.15 and 0.20. The data were also found to depend on the velocity in the mainstream and the type of plate over which fluid is flowing.

The turbulent boundary layer thickness may be computed from the Blasius friction factor equation for circular tubes, introduced earlier in this section: $f = 0.079\,(N_{Re})^{-1/4}$.

Following von Karman [54], the friction factor equation is used in expressing the shear stress at the plate's boundary:

$$\tau_w = 0.0228\,\frac{\rho U^2}{g_c}\left(\frac{\nu}{U\delta}\right)^{1/4} \tag{173}$$

Combining this expression with von Karman's integral equation (Equation 166),

$$\frac{\partial}{\partial x}\int_0^\delta u(U-u)dy = 0.0228U^2\left(\frac{\nu}{U\delta}\right)^{1/4} \tag{174a}$$

Using the $^1/_7$ power law relation (Equation 172) for obtaining a value of u and substituting into Equation 174a, integration can be performed:

$$\frac{\delta}{x} = 0.376(N_{Re_x})^{-1/5} \quad \text{for} \quad 5 \times 10^5 < N_{Re_x} < 10^7 \tag{174b}$$

For local Reynolds numbers above 6.5×10^5 the following relation may be used [58]:

$$\frac{\delta}{x} = 0.1285(N_{Re_x})^{-1/7} \tag{175}$$

By substituting Equation 174b in Equation 173, it can also be shown that

$$\frac{2\tau_w g_c}{\rho U^2} = 0.0585(N_{Re_x})^{-1/5} \tag{176}$$

$$f' = 0.0585(N_{Re_x})^{-1/5} \tag{177a}$$

The last expression gives the local drag coefficient for the turbulent boundary layer. The total friction factor is obtained by the method outlined earlier:

$$f = 0.074(N_{Re_L})^{-1/5} \quad \text{for} \quad 5 \times 10^5 < N_{Re_L} < 10^7 \tag{177b}$$

Several other expressions for local and total drag coefficients have been derived for turbulent boundary layers on flat plates. These are summarized in Table 37. The last three equations given in Table 37 agree over the entire turbulent Reynolds number range and are recommended for good approximations of the drag coefficient.

Table 37. Local and Total Drag Coefficients for Flow Over a Flat Plate

Local Drag Coefficient	Total Drag Coefficient	Investigator
$f' = 0.0262(N_{Re_x})^{-1/7}$	$f = 0.0306\,(N_{Re_L})^{-1/7}$	Falkner [58]
	$f = \dfrac{0.455}{(\log N_{Re_L})^{2.58}}$	Schlichting [59]
	$f = \left(\dfrac{0.455}{(\log N_{Re_L})^{2.58}}\right) - \left(\dfrac{C}{N_{Re_L}}\right)$ where C is a function of Reynolds numbers of transition (see Table 38)	Prandtl [60]
$f' = \dfrac{0.370}{(\log N_{Re_x})^{2.584}}$	$f = \dfrac{0.427}{(-0.407 + \log N_{Re_L})^{2.64}}$	Schultz-Grunow [61]

Table 38. Values of C to be Used in Prandtl Friction Factor Equation Given in Table 37

Transition Reynolds Number	C-Values
3×10^5	1050
5×10^5	1700
10^6	3300
3×10^6	8700

All drag coefficients given thus far in this chapter describe flow over a smooth plate. Surface roughness for flat plates is described in terms of the *admissible roughness.* This is the roughness above which an increase in the drag coefficient over that for a smooth plate occurs. The following expression can be used to determine admissible roughness [62]:

$$\frac{U\epsilon_{ad}}{\nu} = 10^2 \tag{178}$$

where ϵ_{ad} is the admissible height of roughness projections.

For flat plates having a roughness below the admissible roughness, any of the smooth plate expressions are acceptable for estimating the drag coefficient. The following equation can be used for computing f for $U\epsilon_{ad}/\nu > 2000$.

$$f = (1.89 + 1.62 \log L/\epsilon)^{-2.5} \tag{179}$$

where

$$10^2 < L/\epsilon < 10^6$$

The thickness of the turbulent boundary layer on the surface of a rough plate can be computed from the following expression [63]:

$$\delta = 0.259 \left(\frac{x}{\epsilon}\right)^{2/3} \epsilon(1 - 0.00059x) \quad \text{for} \quad x < 175\text{cm} \tag{180}$$

where the quantities x, ϵ and δ are in cm.

FLOW PAST TWO-DIMENSIONAL OBJECTS

We now consider fluid flowing past a two-dimensional object such as a cylinder. The fluid is accelerated as it passes over the front portion of the object and then decelerated after it passes the thickest portion of the object. This sequence of events causes a separation of the boundary layer.

The boundary layer undergoes separation at a point on the object where the pressure gradient is zero. The reason for this is as follows: The boundary layer thickness increases with distance in the flow direction. The portion of fluid in the mainstream flowing past the object accelerates because of its movement around the object. This acceleration is actually an increase in the fluid's kinetic energy, which must be accompanied by a decrease in the pressure (i.e., $\partial P/\partial x$ is negative). As the mainstream fluid goes past the object, the expanding cross section of flow requires a deceleration of the fluid and, hence, a corresponding increase in the pressure ($\partial P/\partial x$ becomes positive). This means that the boundary layer actually flows against an adverse pressure gradient as it moves around the object, resulting in a distortion in the velocity profile in the boundary layer. To maintain flow in the direction of the adverse pressure gradient, the boundary layer must separate from the solid surface. This point of separation may be thought of as a stagnation point (that is, the velocity gradient $\partial u/\partial y$ at the surface is zero). Hence, at any point on the surface of the object prior to separation, u = 0 and v = 0. Prandtl's [51] momentum equations for the boundary layer are

$$u\frac{\partial u}{\partial x} + v\frac{\partial u}{\partial y} = -\frac{g_c}{\rho}\frac{\partial P}{\partial x} + \frac{\mu}{\rho}\frac{\partial^2 u}{\partial y^2} \tag{181}$$

$$\frac{\partial u}{\partial x} + \frac{\partial v}{\partial y} = 0 \qquad (182)$$

from whence it follows that

$$\frac{\partial P}{\partial x} = \frac{\mu}{g_c} \frac{\partial^2 u}{\partial y^2} \quad \text{at} \quad y = 0 \ . \qquad (183)$$

The significance of Equation 183 is that it states for a change in sign of $\partial P/\partial x$, the term $\partial^2 u/\partial y^2$ changes sign; thus, the velocity-profile curve exhibits a point of inflection.

Flow Past Cylinders

The dashed line in Figure 178 represents the edge of the boundary layer of thickness, δ, for fluid flowing over a circular cylinder. Terms U and P denote velocity and pressure, respectively, for the undisturbed portion of the flowing stream. The outer edge of the boundary layer represents the point at which the velocity gradient, $\partial u/\partial y$, becomes zero. The velocity at the edge of the boundary layer, u_{max}, differs from the mainstream velocity and is a function of positions on the surface of the cylinder in relation to the leading edge:

$$u_{max} = 2U \sin\theta \qquad (184)$$

P_θ is the static pressure in the boundary layer and is constant across thickness, δ.

Bernoulli's steady-flow equation may be applied between the mainstream and the edge of the boundary layer (the implication being that the main fluid stream is irrotational); thus,

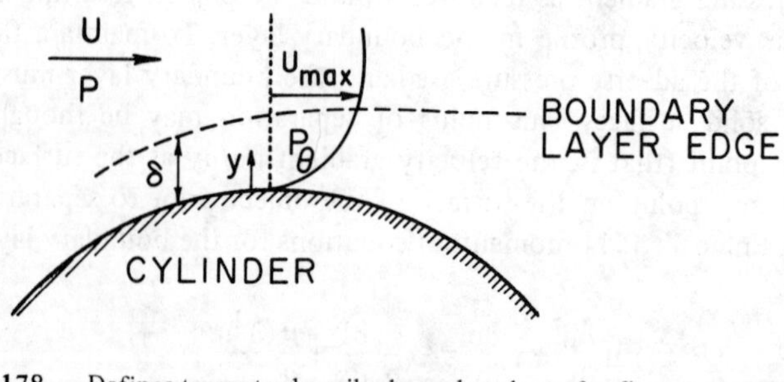

Figure 178. Defines terms to describe boundary layer for flow past cylinders.

$$P_\theta + \frac{\rho u_{max}^2}{2g_c} = P + \frac{\rho U^2}{2g_c} \tag{185}$$

Rearranging to solve for u_{max},

$$u_{max} = U \sqrt{\frac{P - P_\theta}{\rho U^2/2g_c} + 1} \tag{186}$$

Equation 186 can be used to compute the velocity at the edge of the boundary layer from knowledge of the pressure distribution over the cylinder. If the effect of fluid viscosity is negligible, the pressure distribution over the surface of the cylinder can be estimated from Equation 187:

$$\frac{P_\theta - P}{\rho U^2/2g_c} = 1 - 4 \sin^2\theta \tag{187}$$

Drag Coefficients for Cylinders

A two-dimensional cylinder immersed in a flowing fluid has a considerably greater drag than that of a flat plate. This is due to the separation of the boundary layers on two-dimensional objects and the subsequent formation of a turbulent wake behind the body. The drag coefficients were given in Equations 146 and 147. Note that A_p (Equation 147) is the projected area of the cylinder on a plane normal to the flow direction.

A number of investigators [64-67] have measured total drag coefficients past cylinders. Equation 188 is one expression for the drag coefficient for flow past circular cylinders when no boundary layer separation occurs [64]:

$$f_D = \frac{8\pi}{N_{Re_0}(2.002 - \ln N_{Re_0})} \tag{188}$$

Equation 188 applies only for low Reynolds number ($N_{Re_0} < 0.5$). For fluids flowing past immersed cylinders, note that Reynolds numbers are expressed in terms of the cylinder's outside diameter, D_0 ($N_{Re_0} = D_0 U/\nu$).

Certainly one of the most extensive studies on the resistance of cylinders, including noncircular cylinders, is that of Delany and Sorenson [67]. Their investigation covered cylindrical, elliptical, square, rectangular and triangular cylinders. Some of their results are summarized in Table 39. Note that the data summarized in Table 39 also show the effect of rounding sharp edges.

Table 39. Drag Coefficients for Circular and

Flow → Direction	D_0 (in.)	D_1/D_0	$\frac{r}{D_0}$	f_D	Range of Reynolds Number[a]
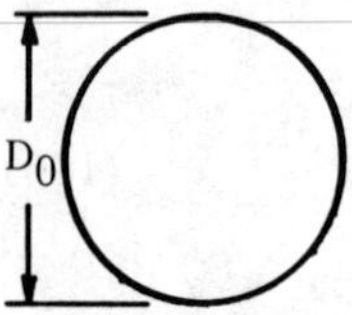	12.00 4.00 1.00	1.00		1.00	$10^4 - 2 \times 10^5$
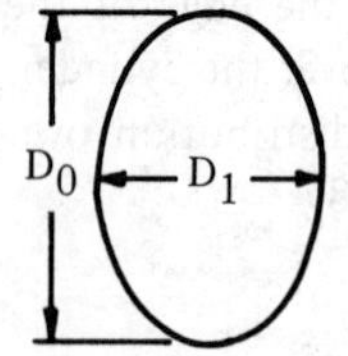	12.00 4.00 1.00	0.50		1.6	$2 \times 10^5 - 8 \times 10^5$
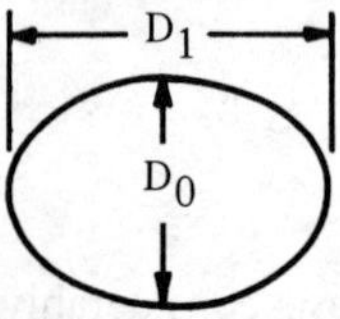	6.00 2.00 0.50	2.00		0.6	$10^4 - 5 \times 10^4$
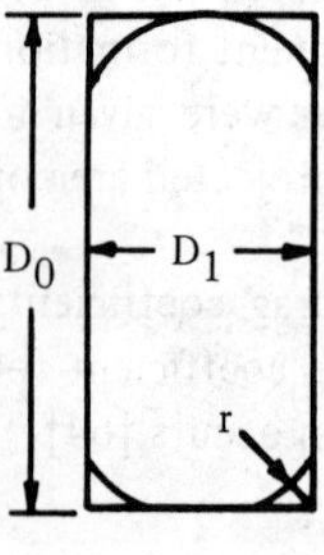	12.00 4.00 1.00	0.50	0.021	2.2	$2 \times 10^4 - 10^6$
	12.00	0.50	0.083	1.9	$1.6 \times 10^5 - 6 \times 10^5$
	12.00 4.00	0.50	0.250	1.6	$3 \times 10^4 - 2 \times 10^5$

[a]Flow is normal to two-dimensional cylinders.

FLOW PAST BODIES OF REVOLUTION

Bodies of revolution of interest are spheres, ellipsoids and disks. Figure 179 shows these various bodies of revolution. The radius in the plane normal to the flow direction is r_0 and the Reynolds number for the flow is D_0U/ν,

Noncircular Two-Dimensional Cylinders [67][a]

Flow Direction →	D_0 (in.)	D_1/D_0	$\frac{r}{D_0}$	f_D	Range of Reynolds Number[a]
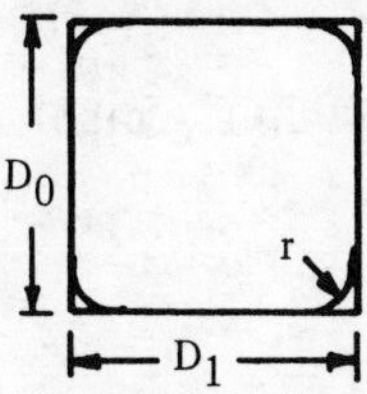	12.00 4.00 1.00	1.00	0.021	2.0	$2 \times 10^4 - 2.8 \times 10^6$
	12.00	1.00	0.167	1.2	$3 \times 10^5 - 10^6$
	12.00 1.00	1.00	0.333	1.0	$5 \times 10^4 - 4 \times 10^5$
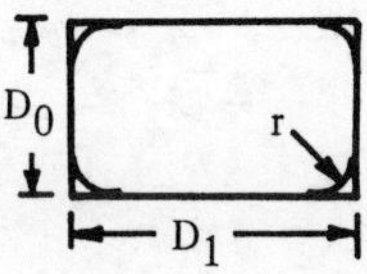	6.00 2.00 0.50	2.00	0.012	1.4	$1.8 \times 10^4 - 7 \times 10^5$
	6.00	2.00	0.167	0.7	$10^5 - 5 \times 10^5$
	6.00 2.00	2.00	0.500	0.4	$10^5 - 10^6$
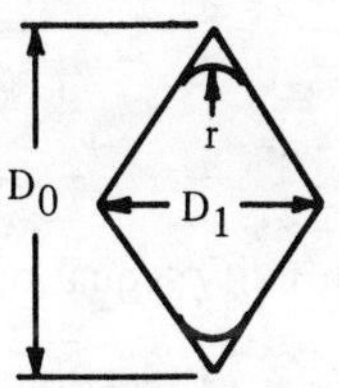	12.00 4.00 1.00	0.50	0.021	1.8	$2 \times 10^4 - 4 \times 10^6$
	12.00	0.50	0.083	1.7	2×10^5
	12.00 4.00	0.50	0.167	1.7	$10^5 - 4 \times 10^5$
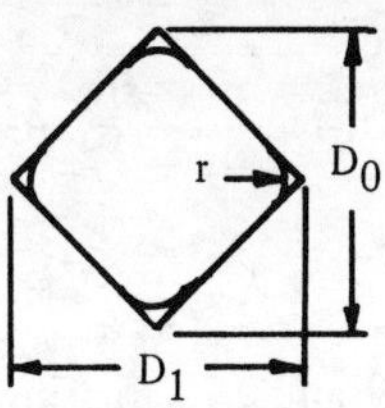	16.97 5.66 1.41	1.00	0.015	1.5	$2 \times 10^4 - 2 \times 10^6$
	16.97	1.00	0.118	1.5	$10^5 - 10^6$
	16.97 5.66	1.00	0.235	1.5	$4 \times 10^4 - 2 \times 10^5$

where $D_0 = 2r_0$. The projected area of the body is πr_0^2, and the overall length of the body in the direction of flow is L.

When flow occurs past any of these objects the fluid motion near the object's surface is three-dimensional with respect to a rectangular coordinate system. The general boundary layer equations for flow over bodies of revolution are

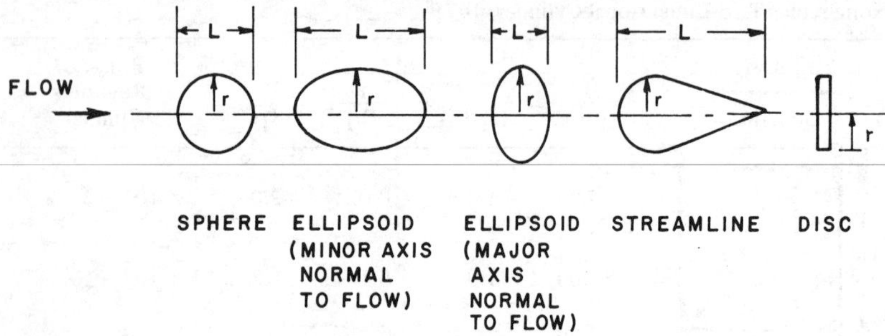

Figure 179. Shows flow past various bodies of revolution.

$$u\frac{\partial u}{\partial x_1} + v\frac{\partial u}{\partial y} = -\frac{g_c}{\rho}\frac{\partial P}{\partial x_1} + \nu\left(\frac{\partial^2 u}{\partial y^2} + \frac{1}{r}\frac{\partial r}{\partial y}\frac{\partial u}{\partial y}\right) \tag{189}$$

$$\frac{\partial (ru)}{\partial x_1} + \frac{\partial (rv)}{\partial y} = 0 \tag{190}$$

Equations 189 and 190 are the momentum and continuity equations, respectively. The terms in these equations are as follows:

x_1 = distance from the leading edge of the body
y = distance normal to the body surface
r = radius of transverse cross section at point (x_1, y).

Flow over Spheres

The definitions of distance terms used in Equations 189 and 190 for flow past spheres are given in Figure 180. The radius of curvature at any point is r_c, and at the sphere's surface (i.e., $y = 0$) the transverse cross section of the body has a radius, r_t. These definitions assume $y \ll r_c$ and that δ is small in comparison to r. These assumptions tend to break down in the region of the leading edge as r goes to 0.

Solutions to Equations 188 and 190 provide expressions for the boundary layer velocity profile, the boundary thickness and local drag coefficients. The following three-dimensional integral momentum equations were developed from Equations 189 and 190 [68]:

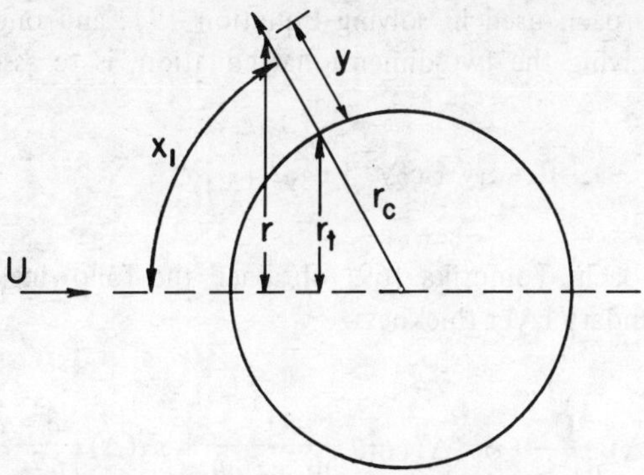

Figure 180. Defines distance terms for three-dimensional equations describing flow over a sphere.

$$\frac{\partial}{\partial x_1}\int_0^\delta u^2 dy - u_{max}\frac{\partial}{\partial x_1}\int_0^\delta u dy + \frac{1}{r_t}\frac{\partial r_t}{\partial x_1}\left(\int_0^\delta u^2 dy - u_{max}\int_0^\delta u dy\right)$$

$$= \delta u_{max}\frac{\partial u_{max}}{\partial x_1} - \frac{\tau_w g_c}{\rho} \tag{191}$$

To solve this expression, information is needed on the pressure distribution over the immersed sphere. The pressure distribution is given by the following expression:

$$\frac{P_\theta - P}{\rho U^2/2g_c} = 1 - \frac{9}{4}\sin^2\theta \tag{192}$$

where θ = angle measured from the forward stagnation point of the sphere
P_θ = static pressure at angle θ
P = static pressure of the undisturbed stream

It can be shown from Bernoulli's equation that

$$\frac{u_{max}}{U} = 1.5\sin\theta \tag{193}$$

A number of empirical correlations are available for estimating u_{max} (the velocity at the edge of the boundary layer).

A simple approach used in solving Equation 191, and one that also has been used in solving the two-dimensional equation, is to assume a velocity profile:

$$u = c_1 y + c_2 y^2 + c_3 y^3 + c_4 y^4 \tag{194}$$

Using this approach, Tomotika [69] obtained the following expression for the laminar-boundary layer thickness:

$$\frac{U}{\nu r_0}\frac{\partial(\delta^2)}{\partial\theta} = \phi_1(\Delta)\frac{U}{u_{max}} - \phi_2(\Delta)\cot\theta\frac{U}{\partial u_{max}/\partial\theta} + \phi_3(\Delta)\frac{\partial^4}{U\nu^2}\frac{\partial^2 u_{max}}{\partial\theta^2} \tag{195}$$

where

$$\Delta = \frac{\partial^2}{r_0\nu}\frac{\partial u_{max}}{\partial\theta} \tag{196}$$

$$\phi_1(\Delta) = \frac{7257.6 - 1336.2\Delta + 37.92\Delta^2 + 0.8\Delta^3}{213.12 - 5.76\Delta - \Delta^2} \tag{197}$$

$$\phi_2(\Delta) = \frac{426.24\Delta - 3.84\Delta^2 - 0.4\Delta^3}{213.12 - 5.76\Delta - \Delta^2} \tag{198}$$

$$\phi_3(\Delta) = \frac{3.84 + 0.8\Delta}{213.12 - 5.76\Delta - \Delta^2} \tag{199}$$

A simpler relation can be obtained by assuming a parabolic velocity profile, as did Millikan [68].

$$\delta^2 = \frac{30\nu}{u_{max}^9 r_t^2}\int_0^{x_1} u_{max}^8 r_t^2 \, dx_1 \tag{200}$$

Drag Coefficients

Consider a sphere moving slowly through a viscous fluid. The drag force exerted on the sphere can be derived from Stokes law:

$$F = \frac{3\pi D_0 \mu U}{g_c} \tag{201}$$

Equation 201 is based on the condition that the boundary layer does not separate from the sphere.

For the flow of a viscous fluid past a sphere, the following relation is applicable in computing the drag force [70]:

$$F = \frac{3\pi D_0 \mu U}{g_c}\left(1 + \frac{3}{16} N_{Re0}\right) \tag{202}$$

For $N_{Re0} < 1.0$, the drag coefficient from Stokes' law is

$$f_D = \frac{24}{N_{Re0}} \tag{203}$$

Combining the above relations, the total drag coefficient describing the motion of spheres (neglecting neighboring wall effects) is

$$f_D = \frac{24}{N_{Re0}}\left(1 + \frac{3}{16} N_{Re0}\right) \tag{204}$$

Equation 204 is applicable in the range $N_{Re0} < 2.0$. For drag coefficients for flow past a sphere contained in a cylinder of diameter D_w, use the following relation:

$$f_D = \frac{24}{N_{Re0}}(1 + 2.4(D_0/D_w)) \tag{205}$$

Equation 205 can be used for rough estimates of the total drag coefficient up to Reynolds numbers of about 100. To estimate drag coefficients at Reynolds numbers outside the range of Equations 204 and 205, Figure 181 may be used. Figure 181 represents best curve fits of Wieselsberger's [65] drag coefficient data for spheres, disks and ellipsoids. It should be noted that there is a distinct effect of aspect ratio on the drag coefficient. Aspect ratio is defined as the diameter to length ratio of the body ($2r_0/L$, refer back to Figure 179). For ellipsoids, for example, the drag coefficient tends to decrease with decreasing values of the aspect ratio. As $2r_0/L$ decreases, the object becomes more streamlined and, hence, the resistance to the flow becomes less.

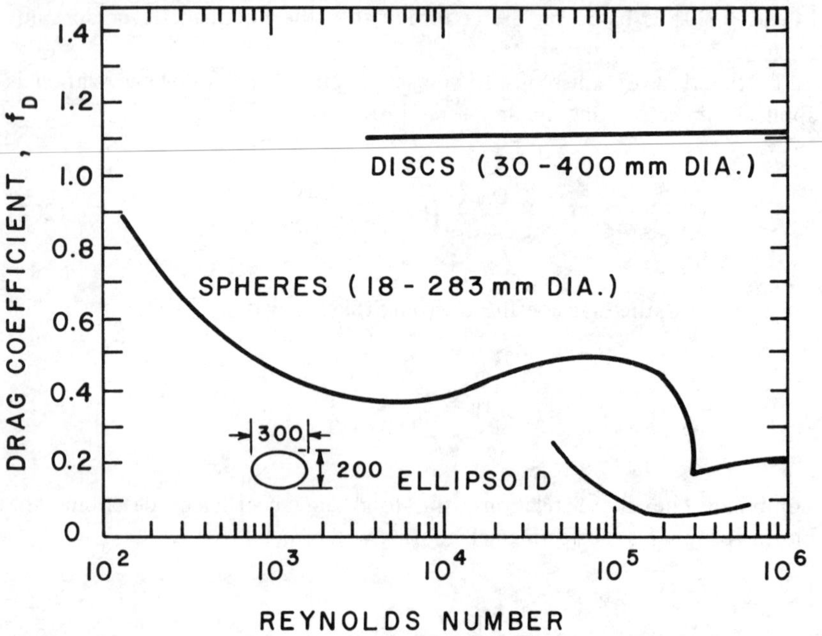

Figure 181. Plot of drag coefficient for flow past bodies of revolution at high Reynolds numbers [65].

CHAPTER 2.10

NON-NEWTONIAN FLOWS

Fluids that are characterized by a nonlinear flow curve plot of shear stress versus shear rate at a given temperature and pressure come under the classification of non-Newtonian. Such fluids are divided into three groups:

1. *Time-independent fluids*–those for which the shear rate at a given condition depends entirely on the instantaneous shear stress.
2. *Time-dependent fluids*–those for which shear rate depends on both the magnitude and duration of the shear.
3. *Viscoelastic fluids*–those that have partial elastic recovery on the removal of a deforming shear stress.

Non-Newtonian fluids are encountered in a multitude of process operations in almost every major industry. Examples of some of these industries include those engaged in the manufacture of rubber, plastics, synthetic fibers, petroleum, paper pulp, soap and detergents, cosmetics, pharmaceuticals, cement, foods, biological fluids, fermentation products, metals refining, printing and solid rocket propellants. Despite the frequency by which chemical engineers encounter these materials, non-Newtonians often pose the most difficult handling problems. Non-Newtonian fluids make the principles of fluid mechanics more complex since the resistance to flow, viscosity, now must be defined through a physical model reflecting process conditions. Inadequate characterization studies of the rheological properties of fluids can defeat the primary objective of the process engineer, which is to design and operate the process at minimum cost. It is essential, then, that a good understanding of the methods of characterizing rheological properties and extrapolating these characteristics to commercial scale be obtained.

This chapter provides information on computational and graphical methods for characterizing non-Newtonian behavior. Correlations are given for describing flow behavior in simple situations such as pipe flow. Section 1 of this volume discusses the properties and characteristics of these type fluids; the reader may wish to review the appropriate subsections in Section 1 before reading this chapter. Those who wish to dig deeper into this subject may refer to the text by Skelland [71].

FLOW PROPERTIES CHARACTERIZATIONS

Rheological properties are characterized with an instrument called a viscometer. There are various methods employed for characterizing non-Newtonian behavior. The major types of viscometers most often used in the laboratory are the capillary tube rheometer, concentric cylinder rotary viscometer, rotating cylinder viscometer and cone-and-plate viscometer.

Capillary Tube Viscometers

These are also called *extrusion rheometers*. A capillary tube viscometer measures the frictional pressure drop produced by the laminar flow of a fluid at a constant volumetric flowrate through a precision-bored cylindrical tube of known length. By obtaining a series of such measurements at different flowrates, a flow curve (a plot of shear stress versus shear rate) can be constructed. The reader should consult the literature [72-74] for details of this instrument.

If the experiment is performed properly, the fluid should approach steady, fully developed flow through the vertical capillary tube. The frictional pressure drop across the tube length, L, can be obtained from a force balance:

$$\frac{\pi D^2}{4} \Delta P_f = \pi D L \tau_w \qquad (205)$$

where D is the tube's inside diameter and τ_w the shear stress at the tube wall. Equation 205 can be rearranged to obtain τ_w:

$$\tau_w = \frac{D \Delta P_f}{4L} \qquad (206)$$

This same expression was developed earlier in Section 1. The rate of shear at the tube wall can be computed from the Rabinowitsch-Mooney equation. Skelland [71] gives the details of the derivation that produces the following expression defining the shear rate at the tube wall for steady laminar flow of a time-independent fluid:

$$\left(-\frac{du}{dr}\right)_w = \frac{8U}{D}\left[\frac{3}{4} + \frac{1}{4n'}\right] = \frac{3n' + 1}{4n'}\left(\frac{8U}{D}\right) \qquad (207)$$

where

$$n' = \frac{d\,\ell n(D\Delta P/4L)}{d\,\ell n(8U/D)} \tag{208}$$

Poiseuille's equation is

$$\Delta P = \frac{8\mu LU}{g_c R^2} = \frac{32\mu LU}{g_c D^2} \tag{209}$$

where viscosity, μ, is constant for a Newtonian fluid. Equation 209 can be rearranged to give

$$\ell n(D\Delta P/4L) = \ell n(8U/D) + \ell n(\mu/g_c) \tag{210}$$

from which it is shown that

$$\frac{d\,\ell n(D\Delta P/4L)}{d\,\ell n(8U/D)} = n' = 1.0 \tag{211}$$

Combining Equations 207 and 211 it can be shown that for a Newtonian fluid

$$(-du/dr)_w = 8U/D \tag{212}$$

where 8U/D is the shear rate for flow in a tube. For a non-Newtonian fluid, 8U/D should not be considered to be the shear rate.

For a non-Newtonian, the variable n' will depend on the position on the logarithmic plot of $D\Delta P/4L$ versus 8U/D. Measurements obtained in a capillary tube viscometer must be represented on a logarithmic plot of $D\Delta P/4L$ versus 8U/D. Parameter n' is evaluated as the slope of the curve at a specified wall shear stress (Equation 206). The corresponding shear rate at the wall, $(-du/dr)_W$, is computed from Equation 207. Repeated calculations are made for various τ_w, and the resultant plot is the flow curve.

Equation 211 can be rearranged to take the form best describing "power law fluids" (see Chapter 1.3, Section 1):

$$\tau_w = \frac{D\Delta P}{4L} = K'\left(\frac{8U}{D}\right)^{n'} \tag{213}$$

On a logarithmic plot of $D\Delta P/4L$ versus $8U/D$, Equation 213 represents the tangent at a point. The parameter K' is the intercept at an $8U/D$ value of unity for any given point on the curve.

The equations outlined above are based on laminar flow conditions only. To check whether measurements are taken under these conditions, Reynolds numbers should be computed for each flowrate and compared to the laminar flow criteria, i.e., $N_{Re} < 2100$. The Reynolds number definition for non-Newtonians as recommended by Skelland [71] is

$$N_{Re} = \frac{D^{n'} U^{2-n'} \rho}{K' \, 8^{n'-1}} \tag{214}$$

The capillary tube technique has several inherent sources of error, which must be accounted for when analyzing measurements. There are four basic effects that must be corrected for:

1. head of the fluid above the tube exit,
2. kinetic energy effects,
3. tube entrance effects, and
4. slip at the tube walls.

The first three effects may be estimated through application of the total mechanical energy balance to the flow in both the feed reservoir and capillary tube. Metzner [75] recommends the following equation for estimating the frictional pressure drop across the capillary tube:

$$\Delta P = P_{gas} - P_a (L + L') \rho \frac{g}{g_c} - 1.12 \frac{\rho u_2^2}{g_c} \tag{215}$$

where L is the tube length and L' is the height from the tube entrance to the gas-liquid interface in the feed reservoir. P_a is the atmospheric pressure at the tube exit and P_{gas} is the gas pressure in the reservoir. Note that Equation 215 is derived for the case of Newtonian fluids. Loss factors for non-Newtonian tube flow are not well established; so for lack of a better approach, Equation 215 must do.

The greatest uncertainty in Equation 215 enters the correction term $(-1.12\,\rho u_2^2/g_c)$. However, this can be minimized by using very large L/D ratios. L/D ratios as high as 1000 have been used in some studies [76].

The effect of wall slippage can be quite pronounced. This is most often seen with slurries, where dispersed flocs move away from the wall region leaving a layer of dispersion medium adjacent to the fluid boundaries. This effect reduces the apparent viscosity near the wall. Slip effects can be accounted for through the following relation [77]:

$$\frac{Q}{\pi R^3 \tau_w} = \frac{C_s}{R} + \frac{1}{\tau_w^4} \int_0^{\tau_w} \tau_{rx}^2 f(\tau_{rx})\, d\tau_{rx} \tag{216}$$

where Q = volumetric flowrate (ft^3/s)
R = tube radius (ft)
τ_w = wall shear stress (lb_f/ft^2)
τ_{rx} = shear stress in x-direction on surfaces normal to r, where r is radius (lb_f/ft^2)

The term C_s is the slip coefficient, defined as

$$C_s = U_s/\tau_w \tag{217}$$

where U_s is the slip velocity (fps).

The following procedure may be used in evaluating the slip coefficient as a function of τ_w.

Step 1. For different tube diameters but constant L, prepare a plot of $Q/\pi R^3 \tau_w$ versus τ_w. If all the data fall on one curve, then $C_s = 0$ and no slippage has occurred. If separate curves are obtained, evaluate C_s by the following steps.

Step 2. Choose a value of τ_w and obtain values of $Q/\pi R^3 \tau_w$ for each tube diameter.

Step 3. Prepare a second plot of $Q/\pi R^3 \tau_w$ versus $1/R$ for the chosen τ_w value. The slope of the resultant curve is C_s.

Step 4. Repeat computations from Step 2 on for other values of τ_w.

Step 5. Prepare a plot of C_s versus τ_w so that a range of conditions may be corrected for.

Step 6. The measured volumetric flowrate from the capillary tube now may be corrected to the value it would have without slip using the following equation:

$$Q_{corrected} = Q_{measured} - C_s \tau_w \pi R^2 \tag{218}$$

Note that $\tau_w = D\Delta P/4L$ corresponds to $Q_{measured}$ in Equation 218.

Concentric Cylinder and Rotating Cylinder Techniques

A concentric cylinder rotary viscometer is designed to shear a fluid contained in the annulus between two concentric cylinders. During the experiment, one cylinder rotates while the other is fixed. Measurements made of the angular speed of the rotating cylinder and the torque applied to the stationary cylinder are used to construct the flow curve for the sheared fluid. Figure 182 illustrates the operation of the device.

The following equations may be used for computing the shear rate at the bob surface for three specific cases. The equations are based on a torque balance at the surface of the stationary bob while the cup rotates at steady angular velocity and are applicable to time-independent fluids only.

Kreiger and Maron [78] equation:

$$\left(\frac{du}{dr}\right)_b = \frac{4\pi N}{1 - Z^{-2}}\left\{1 + \frac{Z^2 - 1}{2Z^2}\left(1 + \frac{2}{3}\ln Z\right)\left(\frac{1}{n''} - 1\right)\right.$$

$$\left. + \frac{Z^2 - 1}{6Z^2}\ln Z\left[\left(\frac{1}{n''} - 1\right)^2 + \frac{d\left(\frac{1}{n''} - 1\right)}{d(\log T)}\right]\right\} \qquad (219)$$

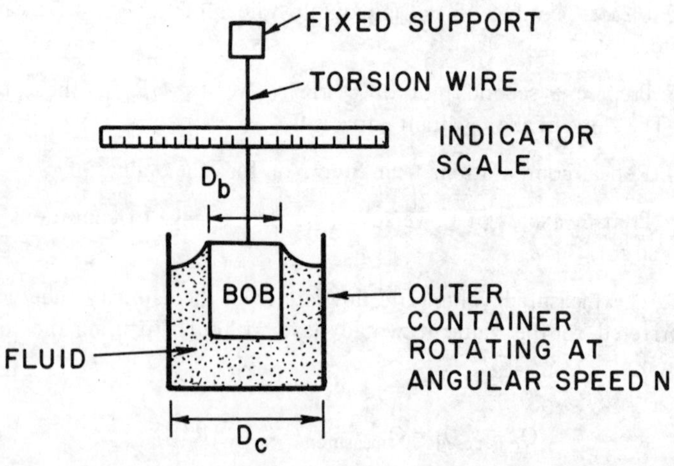

Figure 182. Illustrates the operational concept of the concentric rotary viscometer. The inner cylinder usually is called the bob and the outer is referred to as the cup. The bob remains stationary (it is supported by a torsion wire from a fixed support) as the cup rotates at a specified angular velocity. The torque induced in the torsion wire is measured on the scale indicator.

where T is the measured torque defined as

$$T = 2\pi R_b^2 \ell \tau_b \tag{220}$$

where R_b and ℓ are the bob radius and bob length, respectively, Z is the ratio of cup to bob radius, and n'' is the slope of logarithmic plot of torque, T, versus rotational speed, N. Equation 219 is applicable for $Z \leqslant 1.2$.

Calderbank and Moo-Young [79] equation:

$$\left(\frac{du}{dr}\right)_b = \frac{4\pi N}{1 - Z^{-2}} C_R \tag{221}$$

where

$$C_R = 1 + \frac{Z^2 - 1}{2Z^2}\left(\frac{1}{n''} - 1\right)\left[1 + \frac{2}{3}\ell n\, Z + \frac{1}{3}\gamma - \frac{1}{45}\gamma^3 + \frac{2}{945}\gamma^5 - \frac{1}{4725}\gamma^7 + \ldots\right] \tag{222}$$

and

$$\gamma = \left(\frac{1}{n''} - 1\right) \ell n\, Z \tag{223}$$

Equation 221 is applicable for Z values up to 1.75 and is based on a power law fluids relationship. Values of C_R were computed by Calderbank and Moo-Young [79]; some of these values are given in Table 40 for use with Equation 221.

The third useful relation is for the special case where $Z \rightarrow 1.0$, i.e., a very small annular gap:

$$\frac{du}{dr} = \frac{2\pi RN}{R_c - R_b} \tag{224}$$

where R is a mean diameter, $(R_c + R_b)/2$, and subscripts c, b refer to cup and bob, respectively. This type of close clearance viscometer should not be used with suspension where particle sizes are comparable to those of the annular space.

A modified version of the concentric cylinder viscometer is the rotating cylinder method in which the cup radius is, for practical purposes, extended

Table 40. C_R Values for Calderbank & Moo-Young [79]

	Cup to Bob Radius, Z			
η''	1.07	1.15	1.4	1.746
0.10	1.708	2.617	5.184	
0.20	1.287	1.622	2.593	3.615
0.40	1.102	1.213	1.511	1.826
0.60	1.045	1.091	1.212	1.335
0.80	1.017	1.034	1.077	1.119
1.00	1.000	1.000	1.000	1.000
1.50			0.903	0.853
2.00			0.857	0.785
3.00			0.812	0.720
4.00			0.791	0.689

effectively to infinity. The flow curve is constructed from measurements of the required torque to rotate a cylindrical rod at different known speeds. The radius of the cup is much larger than the rod so that wall effects on the shearing movement of the fluid are minimal. The shear rate at the surface of the rotating cylinder is

$$\left(-\frac{du}{dr}\right)_b = \frac{4\pi N}{n''} \tag{225}$$

Both techniques are subject to error from end effects. This is corrected for by replacing the bob length in Equation 220 by an effective bob length, ℓ_e, which is obtained through calibration of the instrument with a Newtonian material of known viscosity, μ_0:

$$\ell_e = \frac{T}{2\pi R_b^2 \mu_0 (du/dr)_b} \tag{226}$$

Cone-and-Plate Viscometers

Operation of the cone-and-plate viscometer is illustrated in Figure 183. As shown, the technique employs a flat, horizontal plate with an inverted cone whose apex barely touches the plate. The angle between the plate and cone surface is generally less than 1°. The fluid is smeared onto the surface of the

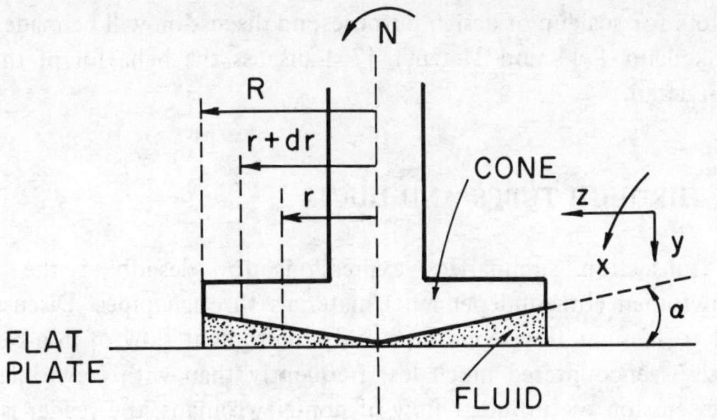

Figure 183. Illustrates operation of the cone-and-plate viscometer.

plate so that it fills the small gap between the cone-and-plate. Flow curves constructed from this technique are obtained from measurements of the torque needed to rotate the cone at different speeds. The magnitude of the shear rate at distance, r, from the cone centerline is

$$\text{Magnitude of Shear Rate} = \frac{2\pi rN}{r \tan \alpha} \simeq \frac{2\pi N}{\alpha} \tag{227}$$

The relationship between the measured torque, T, and shear stress is

$$T = \frac{2}{3}\pi R^3 \tau_{yx} \tag{228}$$

Values of τ_{yx} can be computed from Equation 228 for different measurements of T and plotted against shear rate to obtain the flow curve.

This type of viscometer can be used at very high shear rate measurements (typically several thousand s^{-1}). By contrast, rotating cylinder and concentric-cylinder rotary viscometers can provide only a limited shear range, typically less than 1–100 s^{-1}, depending on the fluid viscosity.

The foregoing discussions apply only to time-independent fluids. Time-dependent fluids include thixotropic and rheopectic fluids (see Section 1). Determination of hysteresis loops for these type fluids may be made with narrow-gap rotary viscometers. However, since limited use has been made of such plots for scaleup or design purposes, no discussion will be made of them here. Skelland [71] and Metzner [75] discuss the behavior of these type fluids in detail.

FLOW THROUGH TUBES AND DUCTS

This subsection summarizes expressions for describing the flow of non-Newtonian (time-independent) materials through pipes. Discussions are limited to laminar flow conditions only. Turbulent flow of non-Newtonian materials is encountered much less frequently than with Newtonian fluids. For a discussion of turbulent flow of non-Newtonians, the reader is referred to the literature [71,75,80-83]. A review of the theory of turbulence given in Section 1 is advisable before proceeding to these references.

In Section 1, the case of steady upward flow of a fluid in a vertical cylindrical tube was considered and a relationship for the volumetric flowrate developed. This relationship is restated below in a more general form:

$$\frac{Q}{\pi R^3} = \frac{8Q}{\pi D^3} = \frac{1}{4}\left(\frac{8U}{D}\right) = \frac{1}{\tau_w^3}\int_0^{\tau_w} \tau_{rw}^2\ f(\tau_{rx})\,d\tau_{rx} \tag{229}$$

where

$$f(\tau_{rx}) = -du/dr$$

Recall that the derivation does not account for slip at the wall. By defining the appropriate relationship for $f(\tau_{rx})$ from rheological characterizations (i.e., from the flow curve), Equation 229 may be used to determine the relationship between Q and frictional pressure drop. For time-dependent fluids, application of Equation 229 can be fairly straightforward.

We shall first consider those fluids having a yield stress (see Section 1). To review, fluids with a yield stress are generally characterized in terms of an internal structure in three dimensions, which is capable of resisting flow for values of shear stress less than the yield value, τ_y. For $\tau_{yx} > \tau_y$, the internal structure collapses, thus allowing shearing movement to take place. Several empirical models describing these materials were given in Section 1. The simplest model is the Bingham plastic model which contains two constants—"yield stress," τ_y, and "plastic viscosity," η. This model describes a linear flow curve. Bingham plastics are an idealized fluid; even so, many fluids can be approximated by this model over fairly wide shear rate range. The Bingham model states that

$$\tau_{rx} - \tau_y = \frac{\eta}{g_c}\left(-\frac{du}{dr}\right) \tag{230}$$

Buckingham [84] used this relation in Equation 229 to derive the following relation:

$$\frac{Q}{\pi R^3} = \frac{8Q}{\pi D^3} = \frac{1}{4}\left(\frac{8U}{D}\right) = \frac{g_c \tau_y}{4\eta}\left[1 - \frac{4}{3}\left(\frac{\tau_y}{\tau_w}\right) + \frac{1}{3}(\tau_y/\tau_w)^4\right] \tag{231}$$

Buckingham's equation cannot be solved directly for ΔP, the frictional pressure loss. Although there are exact procedures for direct application of Equation 231, approximate applications of Buckingham's equation are good for engineering purposes. Equation 231 can be approximated by

$$\frac{8U}{D} = \frac{g_c}{\eta}\left[\tau_w - \frac{4}{3}\tau_y\right] \tag{232}$$

It is obvious from Equation 232 that an experimental plot of 8U/D versus τ_w will provide estimates of η and τ_y from the slope and intercept on the τ_w axis, respectively.

Equation 232 applies only to laminar flow; hence, the Reynolds number should be checked before application. The Reynolds number should be based on an "apparent" viscosity, $N_{Re} = DU\rho/\mu_a$, where μ_a is defined as follows:

$$\mu_a = \frac{D\Delta P g_c/4L}{8U/D} = \frac{g_c \tau_w}{8U/D} \tag{233}$$

The apparent viscosity is defined as the viscosity that permits Poiseuille's equation to fit any set of laminar flow conditions for time-independent fluids, where Poiseuille's equation derived for upflow in a cylindrical tube is

$$\Delta P = \frac{8\mu LU}{g_c R^2} = \frac{32\mu LU}{g_c D^2} \tag{234a}$$

or

$$Q = \frac{\pi R^4 g_c \Delta P}{8\mu L} \tag{234b}$$

Note the term 8U/D used throughout our discussions is given various names by different investigators; among the more descriptive names are "average shear rate," "nominal shear rate at the pipe wall," "flow function," and "apparent shear rate at the pipe wall" [85-88].

Equation 233 can be rearranged in terms of τ_w:

$$\tau_w = \frac{D\Delta P}{4L} = \frac{D}{4}\,\frac{32\mu_a U}{g_c D^2} \tag{235}$$

and substituting in Buckingham's equation for τ_w (Equation 232),

$$\mu_a = \frac{\tau_y g_c D}{6U} + \eta \tag{236}$$

This expression provides an approximate apparent viscosity for a Bingham plastic.

There are other procedures for characterizing Bingham plastic flow in tubes. The more successful ones include the empirical approach of Ooyama and Ito [89], the Herschel-Bulkley models [90,91] and the Crowley-Kitzer model [92].

The velocity distributions derived from the principal models for fully developed flow in tubes are given below.

For Bingham plastic fluids,

$$u = \frac{g_c}{\eta}\left[\frac{\Delta P}{4L}(R^2 - r^2) - \tau_y(R - r)\right] \tag{237a}$$

when $r_p \leqslant r \leqslant R$, where r_p is the radius of the unsheared plug of fluid. For $0 \leqslant r \leqslant r_p$, u_p is obtained from Equation 237 with

$$r = 2\tau_y L/\Delta P \tag{237b}$$

From the Herschel-Bulkley model,

$$u = \frac{2g_c L}{\eta(m+1)\Delta P}\left[(\tau_w - \tau_y)^{m+1} - \left(\frac{r\Delta P}{2L} - \tau_y\right)^{m+1}\right] \tag{238}$$

The second class of non-Newtonians of general interest comprises those fluids without a yield stress. The majority of non-Newtonians fall into this class. These include pseudoplastics, more commonly known as power law fluids.

For power law fluids, the relationship between Q and ΔP for laminar flow in tubes is

$$\frac{Q}{\pi R^3} = \frac{n}{3n+1}\left(\frac{\tau_w g_c}{K}\right)^{1/n} \tag{239}$$

Equation 239 may be rearranged to the following forms:

$$\Delta P = 2K\left(\frac{3n+1}{n}\right)^n \frac{U^n L}{g_c R^{n+1}} \tag{240}$$

and

$$f = \frac{\tau_w g_c}{\rho U^2/2} = 16\Big/\left[\frac{D^n U^{2-n}\rho}{K}\, 8\left(\frac{n}{6n+2}\right)^n\right] \tag{241}$$

The flow behavior index, n, approaches zero for highly pseudoplastic fluids. For fluids exhibiting such behavior, Equation 240 shows that ΔP becomes insensitive to large changes in U for a tube of fixed diameter. Equation 241 defines the friction factor and Reynolds number for power law fluids.

The Ellis model [5,93] also has been somewhat successful in describing fluids without a yield stress. The Ellis model relates τ_{yx} to du/dy as follows:

$$\tau_{yx} = \frac{1}{A + B\,\tau_{yx}^{\gamma'-1}}\left(\frac{du}{dy}\right) \tag{242}$$

The empirical constants A and B are in units of $ft^2/s\text{-}lb_f$ and $ft^{2\gamma}/s\text{-}lb_f^{-\gamma'}$, respectively. The exponent γ' is dimensionless. The relationship between Q and ΔP based on this model is

$$\frac{Q}{\pi R^3} = \frac{A}{4}\tau_w + \frac{B}{\alpha+3}\tau_w^{\gamma'} \tag{243}$$

The velocity distributions derived from these models are as follows:
For power law fluids,

$$u = \left(\frac{\Delta P g_c}{2KL}\right)^{1/n} \frac{n}{n+1} \left(R^{(n+1)/n} - r^{(n+1)/n}\right) \tag{244}$$

or

$$u = U\left(\frac{3n+1}{n+1}\right)\left[1 - \left(\frac{r}{R}\right)^{(n+1)/n}\right] \tag{245}$$

For fluids based on the Ellis model,

$$u = \frac{A\Delta P}{4L}(R^2 - r^2) + \frac{B}{\alpha + 1}\left(\frac{\Delta P}{2L}\right)^{\gamma'}\left(R^{\gamma'+1} - r^{\gamma'+1}\right) \tag{246}$$

For illustration, velocity profiles were computed based on the power law expression (Equation 245) for different values of the flow behavior index. These are shown plotted in Figure 184. Note, as n approaches zero, the velocity profiles become flatter, approaching plug flow conditions.

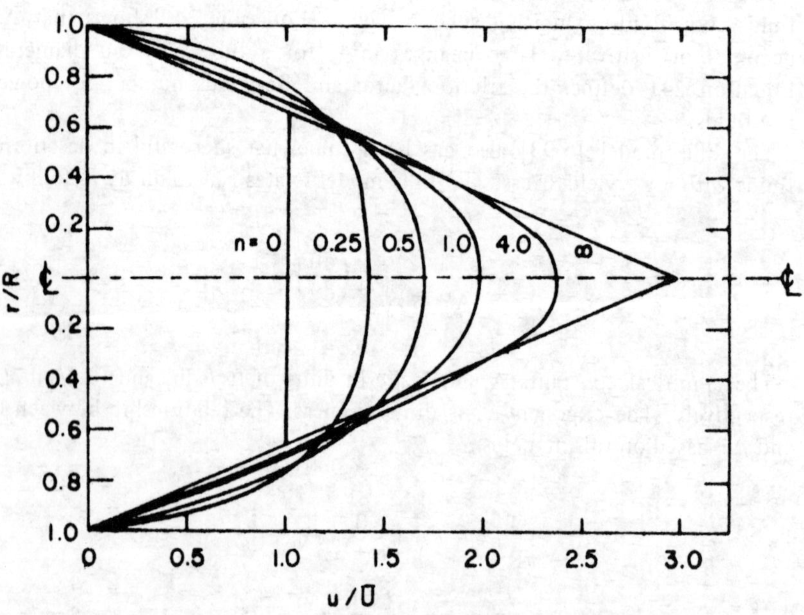

Figure 184. Computed velocity profiles for laminar flow of power law fluids in cylindrical tubes.

Flow Through Rectangular Pipes

Only a limited number of studies have been reported for the flow of non-Newtonians through rectangular ducts. These studies mainly have been with fluids without a yield stress [94-98].

Metzner [75] recommends the following relation for power law fluids:

$$Q = \frac{wb^2}{\Gamma}\left[\frac{wb\Delta Pg_c}{2LK(w+b)}\right]^{1/n} \tag{247}$$

w and b are the greater and smaller sides of the rectangular cross section. Parameter Γ is a function of the aspect ratio (aspect ratio = b/w). This relationship is reported by Metzner [75] in the form of a plot, as shown in Figure 185.

KINETIC ENERGY TERMS

To apply the total mechanical energy balance Equation 87, i.e., the general case of Bernoulli's equation, we must be able to evaluate the kinetic energy terms. The kinetic energy of a mass of any fluid flowing through a tube in unit time is

$$kE_t = \int_0^R \frac{U^2}{2g_c}\, 2\pi r\rho u dr = \frac{\pi\rho}{g_c}\int_0^R ru^3 dr \tag{248}$$

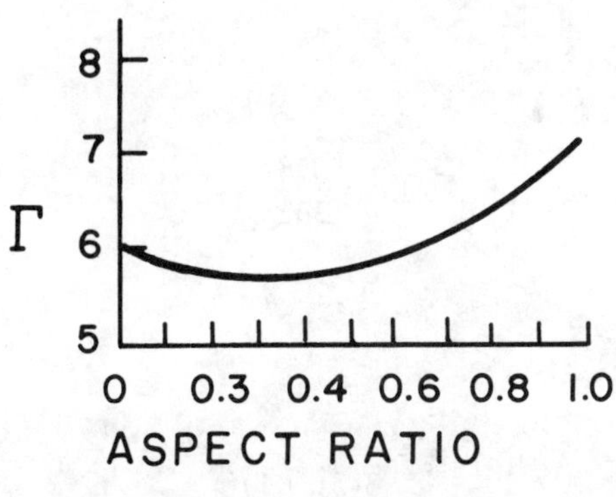

Figure 185. Shows relationship between Γ and aspect ratio for Equation 247 as reported by Metzner [75].

where kE_t it the kinetic energy of the fluid per unit time. The amount of fluid flowing in the cross section per unit time is $\pi R^2 U\rho$. Hence, the average kinetic energy per unit mass kE is

$$kE = kE_t/\pi R^2 U\rho = \frac{1}{g_c R^2 U}\int_0^R ru^3 dr \tag{249}$$

For Bingham plastic fluids it can be shown that the average kinetic energy per unit mass can be written as follows [88]:

$$kE = \frac{3}{560g_c}\left[\frac{1}{C^4 - 4C + 3}\right]\left(\frac{R\tau_y g_c}{\eta C}\right)^2 (35 - 152C + 224C^2 - 112C^3 + 70C^4$$
$$- 280C^5 + 392C^6 - 224C^7 + 47C^8) \tag{250}$$

where C is defined as $\tau_y/\tau_w = r_p/R$.

Within engineering accuracy, this expression can be approximated as follows [75]:

$$kE = \frac{U^2}{2g_c}(2 - C) \tag{251}$$

For power law fluids the average kinetic energy per unit mass is

$$kE = \frac{U^2}{g_c R^2}\left(\frac{3n + 1}{n + 1}\right)^3 \int_0^R \left[1 - \left(\frac{r}{R}\right)^{(n+1)/n}\right]^3 rdr \tag{252}$$

or, on integration,

$$kE = \frac{U^2}{2\alpha g_c} \tag{253}$$

where

$$\alpha = \frac{(2n + 1)(5n + 3)}{3(3n + 1)^2} \tag{254}$$

ENTRANCE/EXIT EFFECTS AND VARIOUS LOSSES

Only limited studies [88] have been made on the effects of flow through entrances, expansions, contractions and fittings for fluids having yield stress.

Entrance effects in tube flow result from a buildup of the boundary layer. At the tube entrance, the fluid velocity is uniform across the flow area and the boundary layer thickness is zero. The boundary layer increases along the tube in the direction of flow. The resistance to flow in the boundary layer must be accompanied by an acceleration of the axial core (this follows from continuity). Fully developed flow will occur only at points beyond where the boundary layer converges on the centerline. A good rule of thumb is that the entrance length is the distance from the inlet at which the centerline velocity is 99% of that for fully developed flow (this is discussed further in Chapters 1.5 and 2.9).

The pressure drop along the tube's entrance region is greater than that experienced along an equal length of pipe where the flow is fully developed. The higher ΔP is a result of the conversion of some pressure energy into kinetic energy as the fluid core undergoes acceleration, and of frictional losses caused by velocity gradients near the tube walls.

Theoretical studies of non-Newtonian entrance effects, primarily for power law fluids, have been reported by Bogue [99], Tomita [100,101], Collins and Schowalter [102] and Lemmon [103]. Experimental studies have been reported by Weltmann and Keller [104], Dodge and Metzner [105], LaNieve [106] and Cheng et al. [107].

Frictional losses associated with laminar flow through a sudden expansion in a tube can be obtained through application of the mechanical energy equation. Skelland [71] writes the following relation for power law fluids:

$$\Sigma F_{\text{expansion}} = \frac{3n+1}{2n+1} \frac{Q^2}{(\pi R_1^2)^2 g_c} \left[\frac{n+3}{2(5n+3)} \left(\frac{R_1}{R_2}\right)^4 - \left(\frac{R_1}{R_2}\right)^2 + \frac{3(3n+1)}{2(5n+3)} \right] \tag{255}$$

Laminar flow through sudden contractions and various pipe fittings has not been studied extensively. Of the limited studies [108,109] on flow through fittings such as elbows and through valves, Bingham plastic and power law fluids were found to experience pressure losses comparable to that of Newtonian fluids.

FLOWMETERING

Non-Newtonian fluids can pose some metering problems. As noted earlier, for highly pseudoplastic materials the pressure drop across a straight length of pipe is relatively insensitive to laminar flowrate. This excludes usage of flowmetering devices based on differential flow measurement principles such as venturis, orifice plates and flow nozzles. Flowmetering devices employed in various non-Newtonian services do include positive displacement meters (reciprocating piston, nutating disc, rotary vane), mechanical flowmeters (rotameters, turbine flowmeters), magnetic flowmeters and Doppler flowmeters. These are generalizations, however, as some of these devices have only limited application in non-Newtonian service. For example, rotameters and magnetic meters are used in the paper industry for metering pulp suspension; however, rotameters are limited to relatively clear fluids. Magnetic flowmeters may be used on opaque fluids, however, and, as an example, in the same industry are used in metering the flow of black liquors. The selection of the proper flowmeter should be based on the manufacturer's proven experience with the particular non-Newtonian material to be handled. The reader is referred to the book *Applied Fluid Flow Measurement* [9] for specific examples and performance data for different flowmetering devices in non-Newtonian service.

SCALEUP CONSIDERATIONS

The accuracy of predictions from application of generalized relationships largely depends on how closely the material can be characterized by a particular rheological model. Extrapolation to conditions of shear stress, shearing rate and temperature (see Section 1 for a discussion on temperature dependency) beyond those for which model constants were determined experimentally are dangerous. In fact, fluid behavior at extrapolated conditions may have new model constants or may even be characterized by another model.

It is also not advisable to infer properties of one non-Newtonian material from another that is seemingly similar. This has been observed from the author's own experience in the paper industry, where two black liquor samples from different digester cooks showed comparable viscosities at room temperature, but 100-fold differences in viscosity between the two fluids at the same elevated temperature and shear rate conditions. It is therefore not advisable to present so-called "typical" experimental values for model constants corresponding to a particular non-Newtonian material. Such data can be highly misleading.

In applying any of the relationships outlined in this chapter, be sure that the proposed conditions for model predictions correspond to the appropriate flow regime. The criterion for checking is the Reynolds number, $DU\rho/\mu_a$, where the apparent viscosity, μ_a, is defined in Equation 233. Values of N_{Re} below 2100 correspond to laminar flow.

For any time-independent fluid it can be shown that the relationship between $8Q/\pi D^3$ or $8U/D$ and τ_w (see Equation 229) is independent of the tube diameter and length, for constant temperature and fully developed laminar flow. This relationship can be established on a tube viscometer over the 8U/D range of interest for a specified operating temperature and used for estimating the pressure drop at any laminar flowrate for any pipe size. That is, experimental determination of $\Delta P/L$ in a prototype tube at a given U/D will provide prediction of the frictional pressure drop in a larger tube in which the fluid flows at the same U/D. This is true because of the uniqueness of the τ_w curve such that

$$\tau_{w\,\mathrm{prototype}} = \tau_{w\,\mathrm{actual}} \tag{256}$$

As such, the following relation follows:

$$\frac{(\Delta P/L)_{\mathrm{prototype}}}{(\Delta P/L)_{\mathrm{actual}}} = \frac{D_{\mathrm{actual}}}{D_{\mathrm{prototype}}} \tag{257}$$

In obtaining scaleup predictions for larger pipe diameters, perform tube viscometer experiments in several prototype tube sizes. Multiple tests will detect the presence of thixotropic or rheopectic characteristics in the fluid if they exist. If either of these characteristics exists, then separate relationships will be obtained for each tube diameter and length, as shown in Figure 186.

The shear stress curve also may be constructed from data obtained from a rotational viscometer, as noted earlier. The rotational viscometer will provide data for the flow curve (τ_w versus $-du/dr$), which defines $f(\tau_{rx})$ in Equation 229. Once $f(\tau_{rx})$ is defined, the integral in Equation 229 can be evaluated analytically to provide a plot of 8U/D versus τ_w.

The reader may wish to examine some of the solved study problems in Chapter 2.12 under the subsection entitled "Problems on Non-Newtonian Flows," p. 594. These techniques are outlined with example problems.

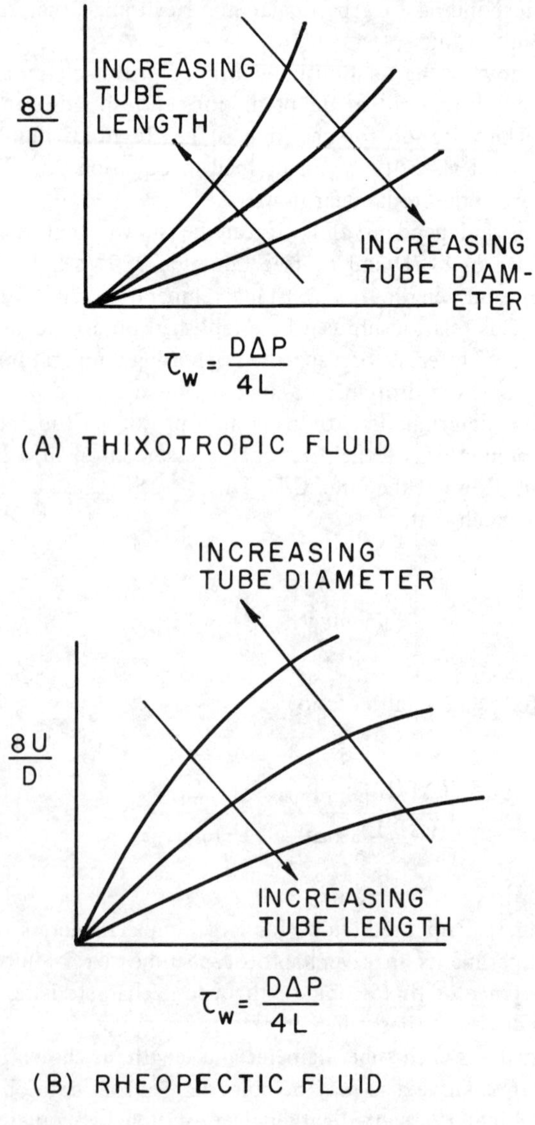

Figure 186. Shows the effects of tube dimensions on thixotropic and rheopectic fluids.

CHAPTER 2.11

OPEN CHANNEL FLOWS

Open channel flow is the flow of fluids in a channel whose geometry is such that one liquid surface is free of solid boundaries. Typically, such flows occur in rivers, canals and in channels or pipes that are not flowing full. Open channel flows are encountered in many industrial applications, primarily in the form of process effluent streams. Examples in which such flows occur include breweries, slaughterhouses, paper pulp mills, chemical plants, oil refineries and a multitude of processing plants. This chapter discusses the characteristics of open channel flows and outlines calculations for estimating free surface profiles and critical flow parameters. Also discussed in this chapter are flow measurement techniques for open channel flows.

VELOCITY DISTRIBUTIONS

Consider a straight channel having a constant and symmetrical cross section for a sufficient length upstream. The velocity profile as a function of depth would appear similar to that illustrated in Figure 187, where the point of maximum velocity exists at some distance down from the free surface. Profile (A) represents the vertical velocity distribution on the centerline of a rectangular channel in which the depth is one-half the breadth. Curve (B) shows the vertical distribution of the mean velocity, i.e., each point on the curve is the average velocity along the horizontal across the section at that level.

Secondary circulation is known to exist in channel flow as shown in Figure 188. Actually, the phenomenon is well known for full flow in non-circular pipes as well. The solid lines in Figure 188 are called *isovels*, which are lines drawn through points of equal velocity. The arrows indicate a double spiral secondary circulation. The number of "cells" of circulation varies with the proportions of the cross section. The relative strength of these circulation

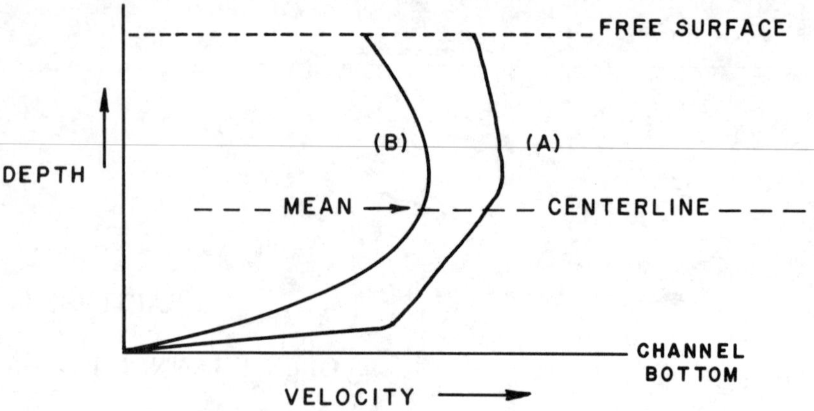

Figure 187. Typical velocity profile for flow in a channel of rectangular cross section.

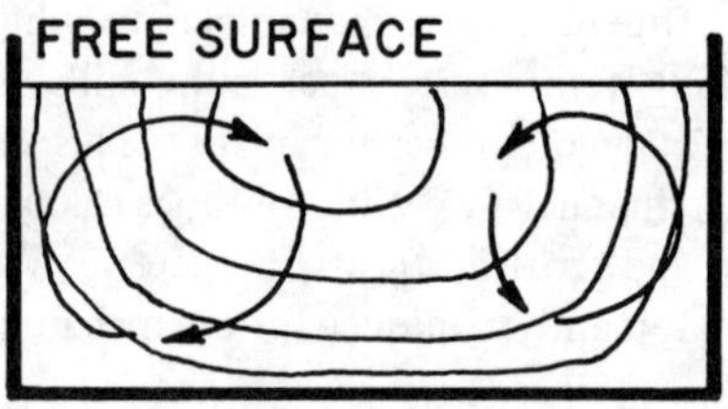

Figure 188. Shows double spiral secondary circulation in open channel flow.

cells is sensitive to the presence of bends or disturbances capable of producing transitions. Generally, flow around a bend results in a single spiral. The faster-moving fluid near the free surface moves toward the outside of the bend, attempting to balance its radial acceleration with that of slower-moving layers near the bed. Spiral flow in bends is called "secondary circulation of the first kind." When it occurs in a straight channel, it is known as "secondary circulation of the second kind."

PRINCIPAL FLOW RELATIONSHIPS

From boundary layer theory it can be shown that the mean velocity can be located at a distance of 37% of the total fluid level above the channel bottom. Velocity measurements obtained at this depth over a number of verticals can be averaged to obtain a mean velocity over the cross section. Another simple approach to obtaining the mean velocity in a vertical is to average measurements obtained at 0.2h and 0.8h, where h is the height of the fluid surface above the channel bottom (referred to as the "bed").

From an engineering standpoint, flow capacity is of prime importance in open channels. Once a value for the mean fluid velocity has been determined for a given channel cross section, the discharge through that section can be established. Early work to establish engineering formulas was entirely empirical. The first relation developed for open channel flow was the Chezy formula. The basis for the Chezy formula is the following expression:

$$\text{Flow Resistance} = C' pLU^2 \tag{258}$$

where C' = friction coefficient
p = length of the wetted perimeter of the channel
L = length of the channel

If the channel is inclined at an angle, θ, from the horizontal, then the slope, s, of the fluid level is $\sin\theta = h/L$, where h denotes the full level over a finite fluid element length, L. Figure 189 defines the system under consideration.

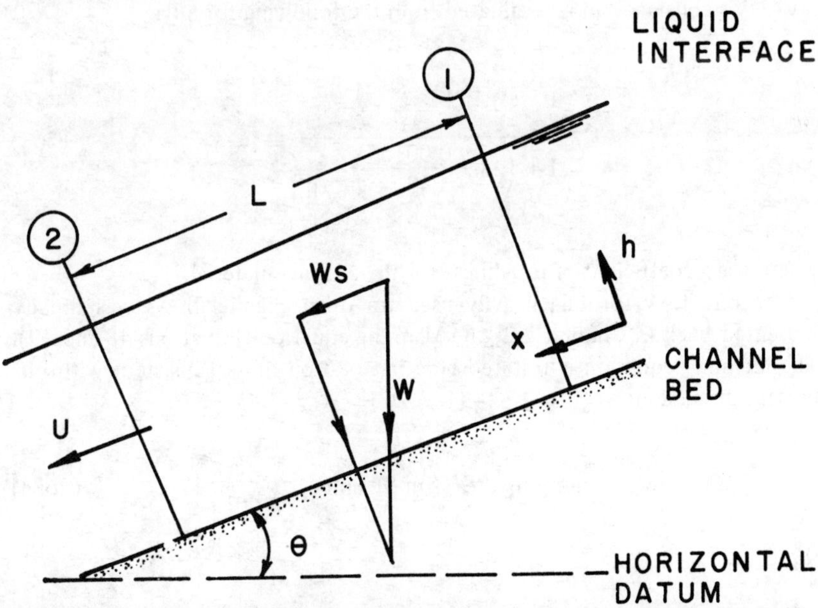

Figure 189. Defines terms for a sloping channel used in the derivation of the Chezy formula.

Defining W as the weight of a finite volume of fluid designated between points 1 and 2, then

$$Ws = ALWS \tag{259}$$

where A is the cross-sectional area of the channel flow and W is the specific weight of the fluid.

WS is actually a force causing flow, hence, equating Equations 258 and 259, and, rearranging,

$$U^2 = \frac{W}{K}\frac{A}{p}S \tag{260}$$

Equation 260 is the Chezy formula, which is more commonly written as

$$u = C\sqrt{h_D S} \tag{261}$$

where h_D is the hydraulic mean depth = A/p, and C is the Chezy constant = $\sqrt{W/K}$.

Chezy's constant can be evaluated from the following formula:

$$C = \frac{23 + \frac{0.00155}{S} + \frac{1}{\epsilon}}{1 + \left(23 + \frac{0.00155}{S}\right)\frac{\epsilon}{\sqrt{h_D}}} \tag{262}$$

where ϵ is a coefficient of roughness for the channel material.

A second flow formula heavily used for open channel flows was already presented back in Chapter 2.2–the Manning equation (Equations 48 and 49). The Manning equation is restated here in notation more consistent with our present discussion:

$$u = \frac{1}{\epsilon} h_D^{2/3} S^{1/2} \text{ in SI units} \tag{263a}$$

or

$$u = \frac{1.49}{\epsilon} h_D^{2/3} S^{1/2} \text{ in British Units} \tag{263b}$$

Of the two relationships, the Manning formula is to be preferred for accuracy.

Unlike full pipe flow, the roughness element, ϵ, for open channel flows is sensitive to the hydraulic state of the channel. Usually, assigning a reliable value for ϵ will depend on experience and judgment. Table 41 gives a typical range of values for ϵ for different straight channel conditions. For nonstraight channels, ϵ values should be increased by 30% [110].

The simple flow expressions presented in this subsection form the starting basis for describing open channel flow and channel discharge. However, the complexity of analysis in describing the characteristics of an open channel flow situation depends on the nature of the flow. There are three general cases that are frequently encountered in engineering; these are uniform flow in channels, nonuniform flows and unsteady flows in open channels.

UNIFORM FLOWS IN CHANNELS

For uniform, fully developed flow in open channels, the Chezy and Manning formulas provide an estimate of the mean fluid velocity from which the channel's volumetric discharge rate can be computed from a knowledge of the flow area, A. Using the Manning equation, volumetric discharge rate is

$$Q = \frac{1}{\epsilon} A h_D^{2/3} S^{1/2} \tag{264}$$

Table 41. Range of Typical Values for Roughness Coefficients for Open Channel Flows [110]

Channel Description	Range of ϵ (in.)
Steel Channels	0.011–0.017
Planned Timber, Joints Flush	0.010–0.014
Sawn Timber, Joints Uneven	0.011–0.015
Concrete	0.011–0.017
Brickwork	0.012–0.018
Excavated Channels	
(earth)	0.016–0.030
(gravel)	0.022–0.030
(rock cut, smooth)	0.025–0.040
(rock cut, jagged)	0.035–0.060
Natural Channels	
(clean, regular)	0.025–0.040
(rocky, brushwood, debris)	0.050–0.150
Flood plains	
(short grass pasture)	0.025–0.035
(brushwood)	0.035–0.070

Civil engineers prefer to call the group $Ah_D^{2/3}$, the *section factor*, as it infers information on the geometry of the cross section. The group $1/\epsilon\ Ah_D^{2/3}$ is referred to as the *conveyance*, K′, and reflects not only the channel geometry but the relative roughness of the channel. These definitions become useful when long or advanced computations are undertaken.

Depth

For flow in a rectangular cross-section channel, the depth of liquid corresponding to uniform flow for a specified discharge and bed slope will be constant at all points. The fluid depth for this special case is referred to as the normal depth, h_0. For nonrectangular channels of uniform flow, normal depth is defined as the maximum depth in the cross section. The normal depth meets the criterion of

$$\frac{dh}{dx} = 0 \tag{265}$$

For the special case of a wide channel, $b \gg h_0$, where b is the channel width, the normal depth can be estimated directly from the Chezy formula:

$$Q = AC\sqrt{h_0 S} = bh_0 C\sqrt{h_0 S} \tag{266}$$

whence it follows that since the hydraulic mean depth approaches the depth for a wide channel,

$$h_0 = \sqrt[3]{\frac{Q^2}{b^2 S C^2}} \tag{267}$$

Defining q as the discharge per unit width,

$$q = Q/b \tag{268}$$

then

$$h_0 = \sqrt[3]{q^2/sC^2} \tag{269}$$

The Manning formula gives the following relation for wide channels:

$$h_0 = \left(\frac{\epsilon q}{S^{1/2}}\right)^{3/5} \tag{270}$$

For the more general case where h_0 is comparable to channel with b, the definition of the hydraulic radius, referred to as the hydraulic mean depth by civil engineers, becomes useful:

$$h_D = \frac{bh_0}{b + 2h_0} \qquad (271)$$

And using the Manning relationship,

$$Q = \frac{1}{\epsilon}(bh_0)(h_D)^{3/2}\sqrt{S} \qquad (272)$$

Note that if uniform flow does not exist in the channel there may be a number of possible fluid levels for a specified discharge. In such cases, the solution to Equation 272 becomes one of trial and error.

For nonrectangular, cross-sectional channels, the following procedure may be used for determining the normal depth in a channel for uniform flow:

Step 1. Restate the Manning formula as follows:

$$\frac{Q}{\sqrt{S}} = \frac{1}{\epsilon}\, Ah_D^{2/3} = K' \qquad (273)$$

Compute values of K' for a suitable range of h up to the maximum depth in the cross section.

Step 2. Prepare a plot of conveyance, K', versus depth, as shown in Figure 190.

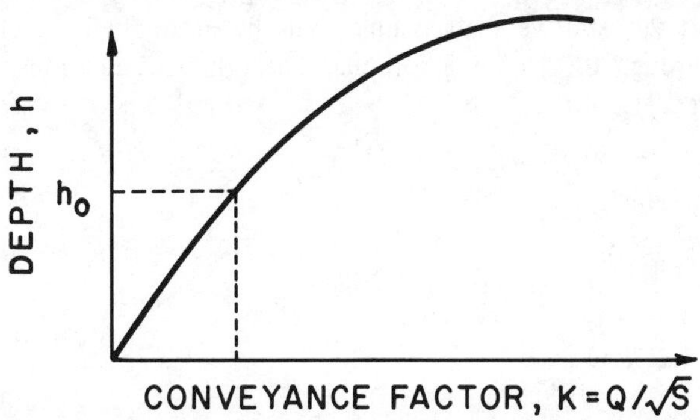

Figure 190. Plot of conveyance versus flow depth to determine normal channel depth from Manning formula.

Step 3. Using the plot, read off the normal depth for the K′ value corresponding to the specified Q and S values.

Channel Geometry

Channel geometry has a direct impact on the cost of the fluid transport system. In Chapter 2.10 it was pointed out that the engineer's primary objective is to design and operate a process at minimum cost to maximize profits or production. The game is still played the same for open channel flows, where the cost of construction is proportional to the channel's cross-sectional area of flow. Hence, the most economical design will be based on a channel geometry that requires the minimum cross-sectional area for a specified discharge rate. Examining the Manning flow formula, this becomes an exercise in maximizing the term A^5/p^2, or since A is often fixed, we attempt to minimize p. Obviously, a semicircular channel would provide the most economical or best hydraulic section; however, this often presents problems from a fabrication standpoint.

The two most common geometries employed are rectangular and trapezoidal-shaped channels. Table 42 gives formulas for the cross-sectional area of flow and wetted perimeter, as well as the best hydraulic depth for these two geometries. The best hydraulic depth is determined by minimizing p with respect to depth, i.e., the best hydraulic depth exists for $dp/dh = 0$. The details of these derivations are left to the reader.

Partially Flowing Pipes

Our discussion of uniform flow would not be complete without considering partially flowing pipes. This situation is encountered most frequently in sewage pipes and storm drainage lines. The system under consideration is illustrated in Figure 191. Note that the angle subtended by the wetted perimeter is 2θ, and we shall assume that the mean fluid velocity can be obtained through the Manning formula. The following relationships follow from Figure 191:

Wetted perimeter:

$$p = 2R\theta \tag{274}$$

Cross-sectional flow area:

$$A = R^2\theta - R^2 \frac{\sin 2\theta}{2} \tag{275}$$

Table 42. Geometric Definitions for Rectangular and Trapezoidal Channels

Channel Description	Cross-Sectional Area of Flow	Wetted Perimeter	Best Hydraulic Depth
Rectangular Channel 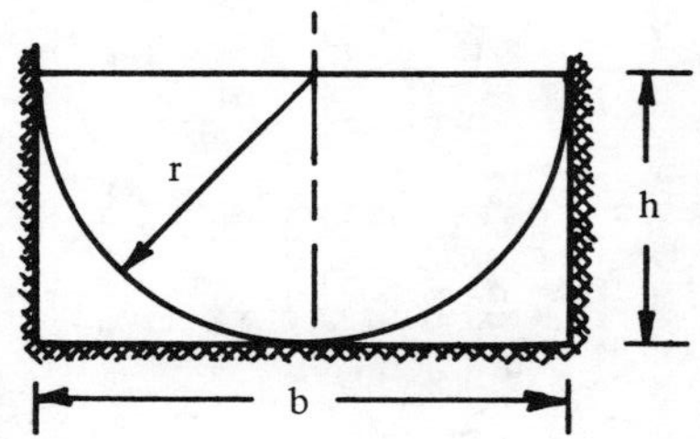	$A = bh$	$P = 2h + b$ $= 2h + \frac{A}{h}$	$H_D = \frac{bh}{2h + b} = \frac{2h^2}{4h}$ $= h/2$
Trapezoidal Channel 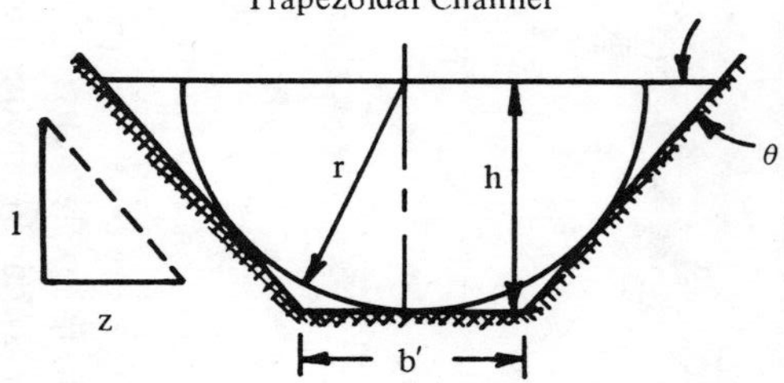	$A = h(zh + b')$ $= h\left(zh + p - 2h\sqrt{(z^2 + 1)}\right)$	$P = 2h\sqrt{(z^2 + 1)} + b'$ Where z is the side slope; if z is a variable and 'A' and 'h' are fixed, then $P = \frac{A}{h} - zh + 2h\sqrt{(z^2 + 1)}$	$H_D = h/2$, with $\theta = 60°$ for the best hydraulic section

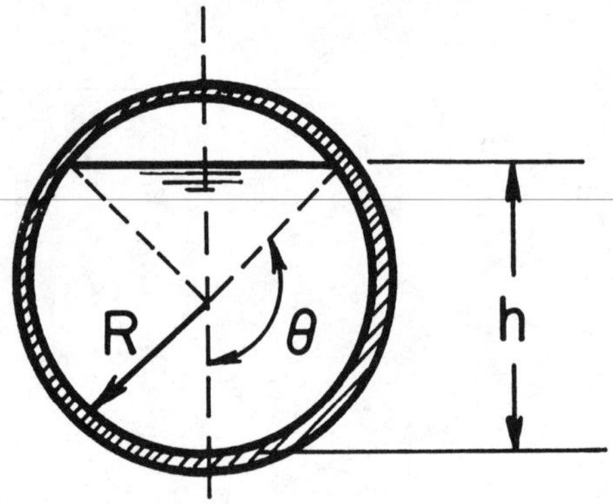

Figure 191. Defines system of partial flow in a circular pipe.

Hydraulic mean depth:

$$h_D = A/p = \frac{R}{2} - \frac{R \sin 2\theta}{4\theta} \tag{276}$$

Differentiating the Manning formula (Equation 263), an expression for the maximum discharge rate may be derived as outlined in the following steps:

$$\frac{dQ}{d\theta} = 0$$

whence

$$\frac{d(A^5/p^2)}{d\theta} = 0 \ ,$$

$$\frac{5A^4}{p^2} \cdot \frac{dA}{d\theta} - 2\frac{A^5}{p^3} \cdot \frac{dp}{d\theta} = 0$$

which simplifies to

$$5p\frac{dA}{d\theta} - 2A\frac{dp}{d\theta} = 0 \tag{277}$$

Substituting expressions for A and p (Equations 274 and 275) into Equation 277 and combining terms,

$$3\theta - 5\theta \cos 2\theta + \sin 2\theta = 0 \qquad (288)$$

Equation 288 becomes satisfied when $\theta = 151°12'$. Hence, the maximum fluid discharge will occur when the water depth is 0.938D (from Equation 276), where D is the pipe diameter.

Combining Equations 264, the Manning expression for volumetric flowrate, and Equation 275 and combining terms, the relationship between Q and θ is obtained:

$$Q = \frac{1}{\epsilon} \frac{D^{8/3}}{10.08} \left(\theta - \frac{\sin 2\theta}{2}\right)\left(1 - \frac{\sin 2\theta}{2\theta}\right)^{2/3} S^{1/2} \qquad (289)$$

For full pipe flow, h = D and $\theta = \pi$, from which the volumetric discharge rate for full pipe flow becomes

$$Q_f = \frac{\pi D^{8/3} S^{1/2}}{10.08\epsilon} = \frac{D^{8/3} S^{1/2}}{3.21\epsilon} \qquad (290)$$

For the maximum discharge rate, where h = 0.938D,

$$Q_{max} = \frac{\pi D^{8/3} S^{1/2}}{9.38\epsilon} = \frac{D^{8/3} S^{1/2}}{2.98\epsilon} \qquad (291)$$

And for the pipe running half full,

$$Q_{1/2} = \frac{\pi D^{8/3} S^{1/2}}{20.16\epsilon} = \frac{D^{8/3} S^{1/2}}{6.41\epsilon} \qquad (292)$$

NONUNIFORM FLOWS: GRADUALLY VARIED FLOWS

Nonuniform flow in channels is characterized by a changing fluid level in the direction of flow. This change can occur either gradually or suddenly. This somewhat arbitrary distinction of the rate of level change as a function of distance along the flow axis enables us to define specific situations that are encountered frequently.

Channel flows undergoing gradual changes normally involve small changes in depths. Consequently, the concern here is with relatively long channels. Since this type flow is encountered most often in natural channels such as rivers, its interest to chemical engineers is limited. Therefore, we shall only give it a superficial treatment.

From our system definition illustrated in Figure 192(A), the Bernoulli equation is written to obtain the total energy (head) of the flowing fluid:

$$\Delta H = \frac{\Delta U^2}{2g} + h + Z \tag{293}$$

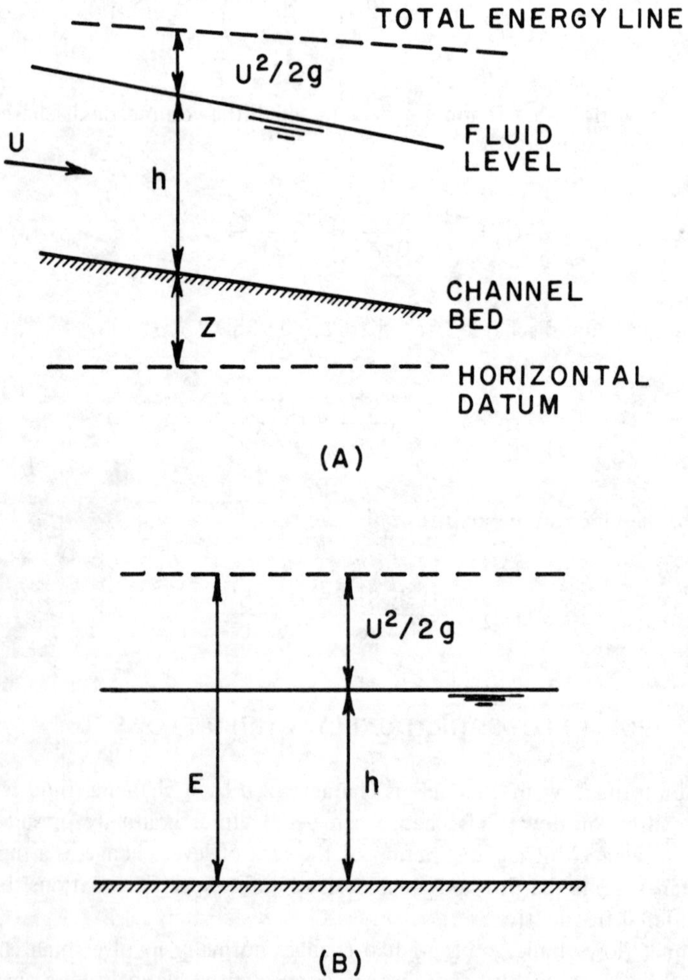

Figure 192. Defines system for nonuniform flow.

Note from Figure 192(A) that if a length of channel is considered in profile, a line may be drawn above the fluid surface to denote the total energy of flow. This is called the *total energy line*. That is, each point on this line is $U^2/2g$ above the fluid surface. For an ideal fluid there would be no energy losses and the total energy line would be horizontal. For a real fluid of uniform flow, the line is parallel with the fluid surface and channel bed. In the case of nonuniform flow considered here, the total energy line, the line describing the fluid's free surface profile, and the channel bed all may be of different slopes.

Specific energy, also called the *specific head*, refers to the fluid's energy for a cross section. With Z = 0 and noting that U = Q/A = q/b for a rectangular channel, Equation 293 can be restated as follows:

$$E = \frac{q^2}{2gh^2} + h \tag{294}$$

where E is the specific energy shown by Equation 294 to be the sum of the kinetic energy of the fluid and its depth (refer to Figure 192(B)). Specific energy is defined with respect to the horizontal datum. Hence, if the channel bottom is sloped, then the specific energy of a specified discharge may vary along the flow axis. Variations in specific energy may not necessarily be accompanied by changes in the total energy of flow. The two are related as follows:

$$H = E + Z \tag{295}$$

For flows in which the velocity distribution is highly nonuniform, the kinetic energy correction term is needed to correct the specific energy:

$$E = \alpha \frac{U^2}{2g} + h \tag{296}$$

The relationship between specific energy and depth is shown in Figure 193, which is called the specific energy curve. This is a plot of E versus h for a constant discharge per unit channel width q. At any depth, h, the specific energy is the sum of kinetic energy, curve (A), and potential energy, curve B. Note that for a particular value of the specific energy there are two possible depths, h_1 and h_2. The point of minimum specific energy, E_{min}, is called the *critical depth*, h_c. At E_{min}, the dimensionless group, the Froude number ($N_{Fr} = U/\sqrt{gh}$) has a value of unity. The velocity at this point is referred to

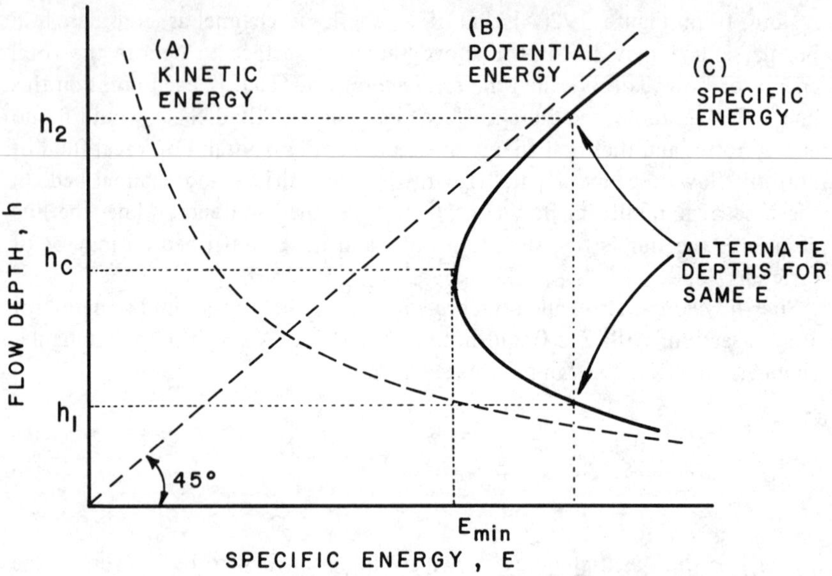

Figure 193. Specific energy curve at constant q for open channel flow.

as the critical velocity. At depths greater than h_c, the mean velocity is referred to as being *subcritical* (also called streaming or tranquil flow) since $U < U_c$. Similarly, at depths less than h_c, $U > U_c$ and the flow is referred to as *supercritical*.

Equation 296 written in its more general form for a channel of any geometry is

$$E = \frac{Q^2}{2gA^2} + h \tag{297}$$

Differentiating this expression with respect to depth and noting that at $dE/dh = 0$, $h = h_c$, the following is obtained:

$$\frac{Q^2}{gA_c^3}\frac{dA}{dh} = 1 \tag{298}$$

where A_c is the critical flow area. Note that $dA/dh \simeq \Delta A/\Delta h$ and, if Δh is taken to be a thin sliver of fluid in the channel flow near the surface, we can approximate the flow area of the finite fluid element as a rectangle such that $\Delta A/\Delta h \simeq b$. Equation 298, the equation for critical flow, becomes

$$\frac{Q^2 b_c}{g A_c^3} = 1 \qquad (299)$$

where subscript c refers to the critical condition. Since $Q = U_c A_c$, then Equation 299 can be rearranged to give

$$u_c = \sqrt{\frac{g A_c}{b_c}} \qquad (300a)$$

or

$$u_c = \sqrt{g h_{mc}} \qquad (300b)$$

where h_{mc} is the mean critical depth of an irregularly shaped channel.

For a rectangular channel showing constant depth, Equation 297 can be developed in terms of the flow per unit width to give the following relationships:

$$h_c = \sqrt[3]{q^2/g} \qquad (301)$$

$$u_c = q/h_c = \sqrt[3]{gq} = \sqrt{g h_c} \qquad (302)$$

These two relations can be combined through h_c and, when substituted back into Equation 297, give

$$E_c = \frac{U_c^2}{2g} + \underbrace{\frac{2U_c^2}{2g}}_{h_c} \qquad (303a)$$

or

$$h_c = \frac{2}{3} E_c \qquad (303b)$$

The last expression states that the critical flow depth is two-thirds the critical specific energy. The above discussion has been based on the assumption that the discharge in the channel is constant. However, the reverse also may be true; that is, the specific energy may be fixed and q may be a function of depth. In this case, it follows that at the critical depth the discharge is at a maximum. Foregoing the mathematics, it can be shown through a derivation similar to the above that

$$q_{max} = \frac{2}{3} \frac{\sqrt{2g}}{\sqrt{3}} E_c^{3/2} , \qquad (304)$$

for the critical depth given by Equation 303b.

Surface Profile/Classification

Proper classification of the surface profile is essential to analyzing any open channel flow problem. For water flows in open channels the emphasis is placed in establishing the proper free surface slope for a particular flow situation. The surface profile slope may be classified in reference to the critical slope, S_c. If the slope is less than S_c, the flow at the normal depth is referred to as being *mild*. For supercritical flow, $S > S_c$, the slope is said to be *steep*.

The critical slope is essentially the slope at which the critical and normal depths coincide. Since the flow at the critical depth is uniform, then the bed and total energy line must have the same slope. From the Manning formula, an expression for the critical slope is

$$S_c = \frac{gh_c\epsilon^2}{h_{mc}^{4/3}} \qquad (305)$$

Twelve free-surface profiles have been classified to describe gradually varied flow situations [111]. These are summarized in Table 43. Examples of each flow situation are included in the table, showing the critical depth line and the normal depth line. Table 43 can serve as a guide to classifying profiles for different open channel flow situations encountered.

Computational Methods

Computational methods enable detailed calculations of the free surface profile of the flowing fluid. In outlining computational methods we shall follow the analysis proposed by Sellin [110] and refer the reader to this source for details of the derivation of the general equation for gradually varied flow. Starting with the total energy head expression for gradually varied nonuniform flow (Equation 293), Sellin applied the following assumptions:

1. The slope of the channel bed, S, is considered small so that depth, h, may be considered a normal distance, i.e., $\cos\theta = 1$ (refer to Figure 189).
2. The velocity distribution across the channel is uniform, i.e., the results for uniform flow are applied locally to gradually varied flow.
3. The total energy head, H, and the bed elevation, Z, are measured positive above a horizontal datum.
4. The relationship between change in depth and change in cross-sectional area can be approximated by $dA/dh = b$, where b is the channel width.

The general equation of gradually varied flow is

$$\frac{dh}{dx} = \frac{S - j}{1 - \dfrac{Q^2 b}{gA^3}} \tag{306}$$

where j is the slope of the total energy line.

For a rectangular, cross-sectional channel, Equation 306 becomes

$$\frac{dh}{dx} = \frac{S - j}{1 - q^2/gh^3} \tag{307}$$

where $A = bd$, and $q = Q/b$.

From assumption (2), we can introduce the Manning formula to obtain an expression for j:

$$j = \frac{U^2 \epsilon^2}{h_{Dm}^{4/3}} \tag{308}$$

where h_{Dm} is the hydraulic mean depth. Substituting into Equation 307, we obtain a workable expression for a rectangular channel:

$$\frac{dh}{dx} = \frac{S - \dfrac{U^2}{C^2 h_{Dm}}}{1 - U^2/gh} \tag{309}$$

Table 43. Free Surface Profiles as

— — — — — Denotes
------------- Denotes
——▶ Denotes

Slope Description	Profile Classification
Horizontal Slope	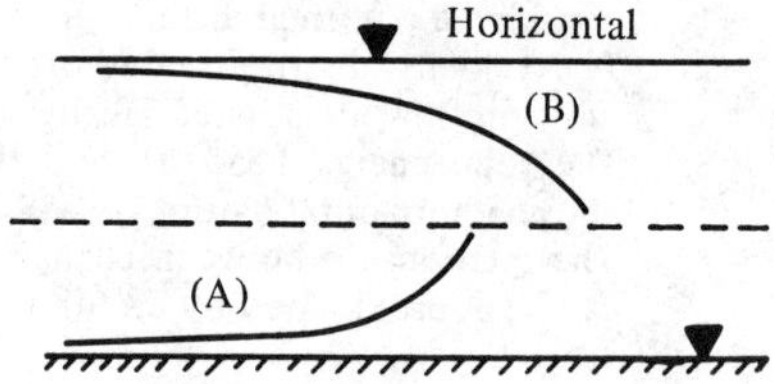
Mild Slope ($S < S_C$)	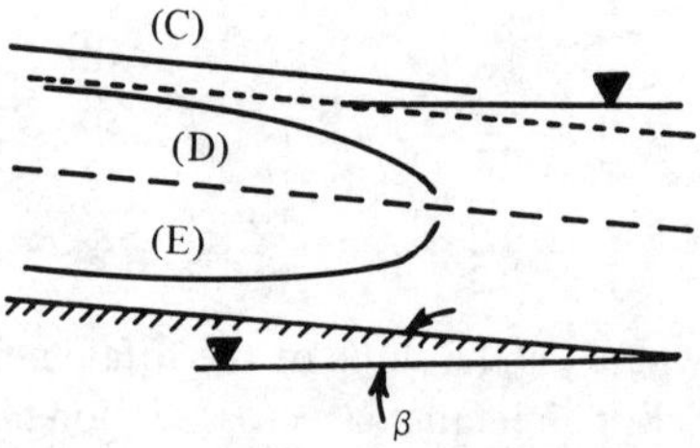
Critical Slope ($S = S_C$)	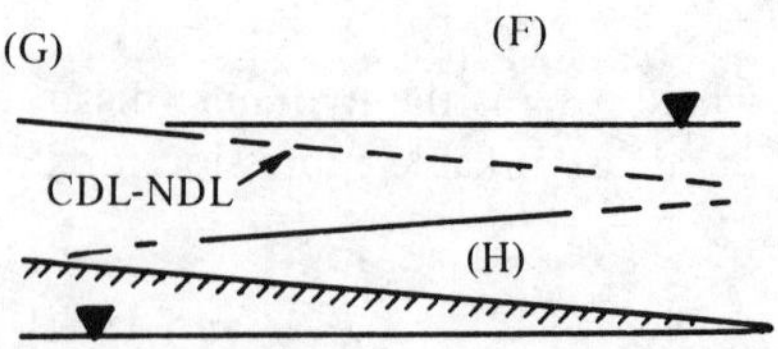

Reported by Bakhmeteff [111]

critical depth line (CDL).
normal depth (NDL).
direction of flow.

Examples	Comments

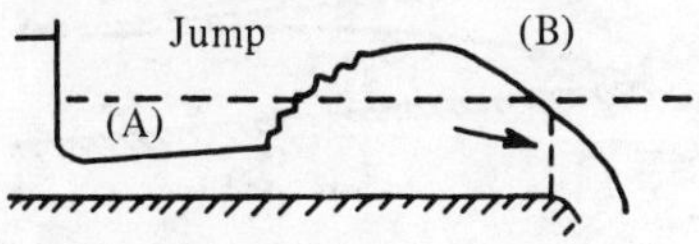

Flow under a sluice gate

Here, $h_0 \to \infty$, so uniform flow cannot exist for $h > h_0, h_c$. In example, profiles (A) and (B) are joined by a hydraulic jump. Profile (B) constitutes a *drawdown* curve; the profile passes through the critical depth again at the end of the hydraulic jump.

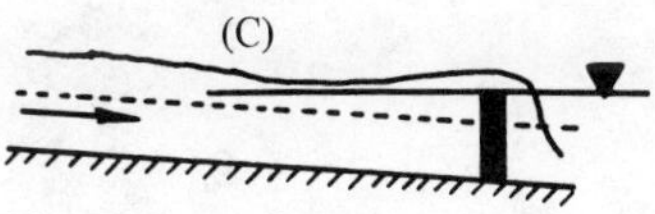

Flow over a weir

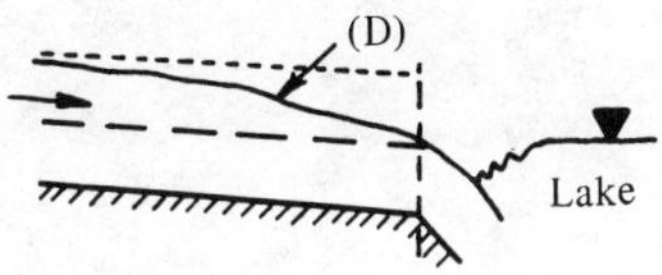

Flow into lake or reservoir junction

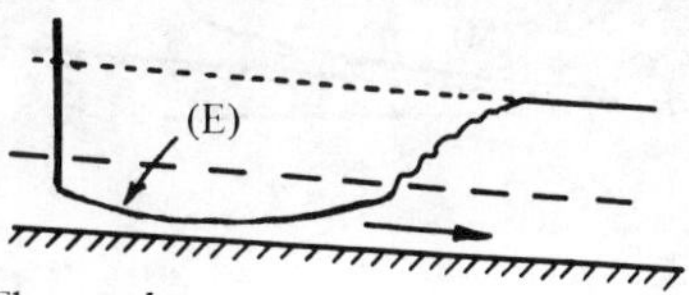

Flow under a sluice gate

Profile (C) is referred to as the *backwater curve,* commonly observed in channels of changing section or flow over a weir. Curve (A) is asymptotic, both to the normal depth line upstream and the horizontal downstream. The (D) profile will occur at the junction with a reservoir if the fluid level is below the level of the normal depth in the channel exit. Profile (E) occurs below steep spillways and sluices. Profiles (D) and (E) are generally short compared to (C).

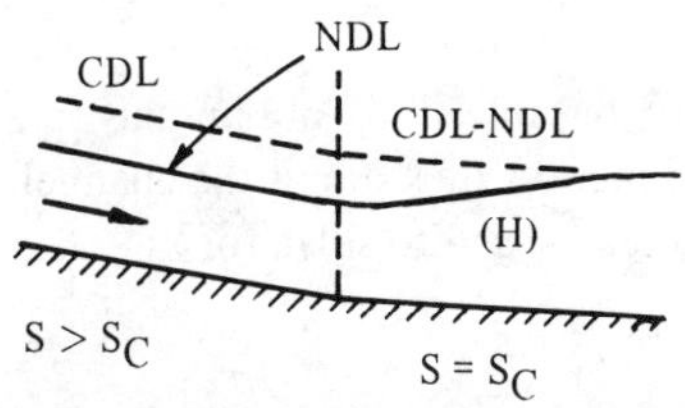

Flow through channel of changing bed slope.

Profiles (F) and (H) are close to being straight horizontal lines. Theoretically, profile (G) corresponds to uniform flow at the normal depth. At the critical slope, the fluid surface oscillates about h_0 for uniform flow. This gives rise to a series of stationary surface waves, CDL-critical depth line.

Table 43,

Slope Description	Profile Classification
Steep Slope ($S > S_C$)	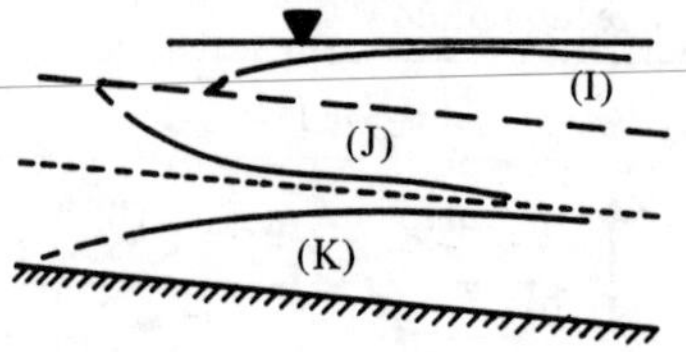
Adverse Slope ($S < 0$)	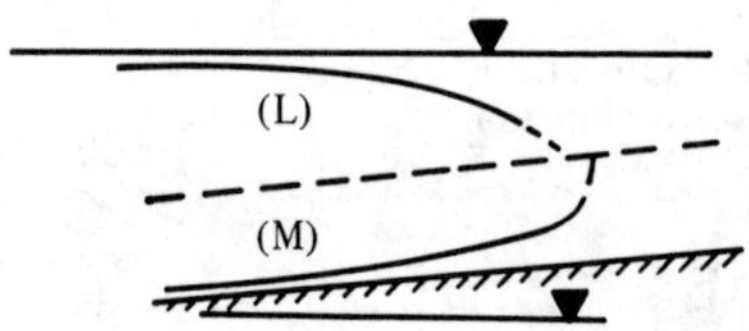

The gradually varied flow expression may be solved to obtain the free surface profile, that is, the fluid depth as a function of x along the channel. The general expression, Equation 306, can be rearranged to solve for x:

$$x = \int dx = \int \left(\frac{1 - B}{S - d} \right) dh \tag{310}$$

continued

Examples	Comments

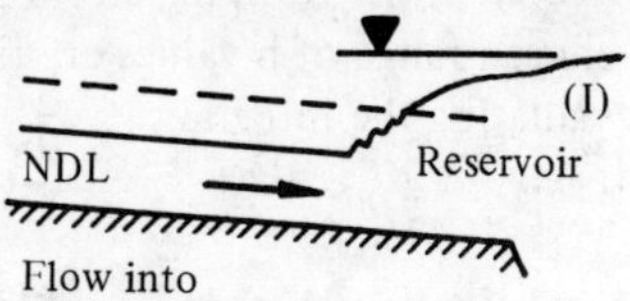

Flow into reservoir

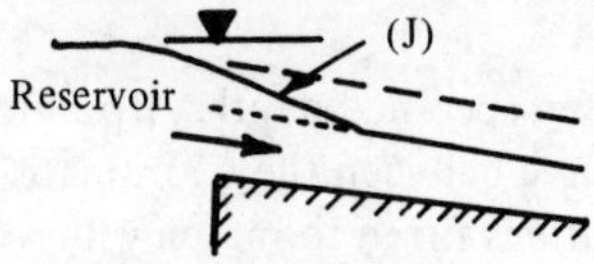

Flow from reservoir

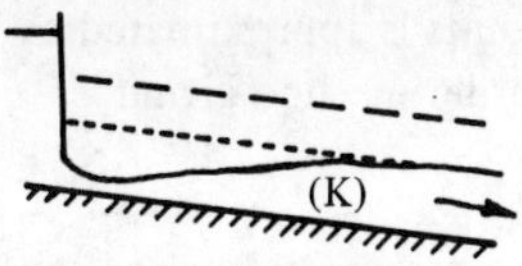

Flow under sluice gate

Such profiles are observed in spillways and chutes. Profiles are short and asymptotic downstream. Profile (I) is normally found downstream from a jump. Profile (J) is a continuation of a hydraulic drop formed when the channel steepens or when flow exists a reservoir. Profile (K) is formed below a sluice gate or below the junction with a steeper channel.

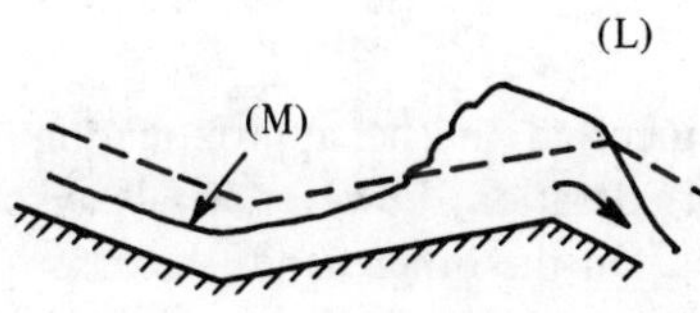

$S > S_C$ $\quad$ $S < 0$

Flow over an increasing channel

Uniform flow cannot exist in the direction shown. Profiles (L) and (M) normally occur together and are joined by a jump on the apron below a spillway. Profile (L) may also be found in front of profile (J) (see above) where a steep channel runs out of a reservoir. These profiles are very short.

where

$$B = \frac{Q^2 b}{g A^3} \quad \text{and} \quad d = \frac{\epsilon^2 Q^2}{A^2 h_{Dm}^{4/3}}$$

Note that B and d are functions of h, since A, b, h_{Dm} are functions of h. Hence, Equation 310 is actually $x = \int f(h)\,dh$.

If f(h) is a simple function, although usually it is not, the integral may be directly integrated.

Simpson's rule may be used to evaluate quadratic functions numerically. In applying Simpson's rule, Equation 310 is treated in the form dx/dh = f(h), whereby values of dx/dh corresponding to a chosen range of h values are first computed. Simpson's rule is then applied to evaluating the integral

$$\Delta x = \int_{h_1}^{h} \left(\frac{dx}{dh}\right) dh \tag{311}$$

whereby Δx is a distance along the channel between specified depths, h_1 and h_2, i.e., Equation 311 gives the area under the curve between the coordinate h_1 and h_2 for the curve dx/dh versus h. A quadratic is fitted to match values at h_0, h_1, h_2, at h_2, h_3, h_4, and so forth. The number of intervals must be even, and the area under each successive three points is approximated by the area under the quadratic. Hence, this is a second-order method giving

$$\text{Area under curve} = \frac{h_n - h_0}{3n} \left[f(h_0) + 4f(h_1) + 2f(h_2) + 4f(h_3) + 2f(h_4) \right.$$

$$\left. + \ldots + 4f(h_{n-1}) + f(h_n)\right] \tag{312}$$

where f(h) = dx/dh. Note that the error incurred in the approximation technique depends on the order of the approximation; however, it always decreases as the number of intervals increases. On the other hand, the error accumulation in the summation process in Simpson's rule increases with increasing n. Thus, the relative error must have a minimum for some value of n.

Various numerical techniques for treating this problem are outlined in the literature [112,113]. Sellin [110] outlines one called the *direct step* method, which is straightforward but limited to prismatic channels only. The steps to this method as outlined by Sellin are as follows:

Step 1. Evaluate h_0 and h_c values.

Step 2. Classify the type profile (use Table 43).

Step 3. Divide the depth, h, into a convenient number of intervals of equal size.

Step 4. Tabulate hydraulic properties of the channel for each interval.

Step 5. Compute mean values for h_D and j for each interval.

Step 6. Compute Δx values between adjacent intervals.

Step 7. Sum Δx values to obtain the distance of each interval and to evaluate h's over the total length of the profile.

Although the method may be used for hand computations, it is rather tedious. Such a method is better adapted to a short program on the computer.

NONUNIFORM FLOWS: RAPIDLY VARIED FLOWS

The distinction between gradually varied and rapidly varied flows is made based on the criterion of streamline curvature. In the former situation, only small changes in depth are observed over the channel length; hence, energy losses due to channel boundary friction are the only significant means of energy dissipation. In the case of rapidly varied flows, energy losses become complicated by the appearance of hydraulic jumps.

A hydraulic jump is a stationary surge or shock wave in which the speed of advance of the wave front (celerity) is equalized by the velocity of flow in the opposite direction. The position of the hydraulic jump will depend on a combination of the upstream and downstream flow conditions. The form or shape, as well as the performance of the jump, will depend on the upstream and downstream depths and the velocity of the approaching flow. The Froude number, $N_{Fr} = U_1/\sqrt{gh_1}$, is used to characterize the jump. Note that subscript 1 denotes conditions immediately before the jump. There are three types of hydraulic jumps: undular, transitional and direct (Figure 194).

The undular jump is described as having a flow that first expands slowly and then oscillates. Hence, this type jump has the form of a smooth initial wave followed by a series of waves of decreasing amplitude. Energy losses associated with this type jump are small.

A transitional jump produces an increasing rate of energy dissipation resulting in the first few wave crests breaking, and isolated "reverse rollers" form on the upstream facing the slope of these waves below the crest. Once waves begin to break, any waves formerly downstream of them become lost in surface agitation.

Direct hydraulic jumps occur at higher Froude numbers and are often characterized by high turbulence and, consequently, high energy losses. Such jumps consist of a single monoclinical wave covered by an intensely turbulent reverse roller, which normally shows a froth (called "white water") due to air entrainment. Figure 194 lists the range of Froude numbers in which the different hydraulic jumps occur.

A hydraulic jump can be induced to the flow for a variety of reasons. Basically, it provides a means of dissipating excess energy, which can prove

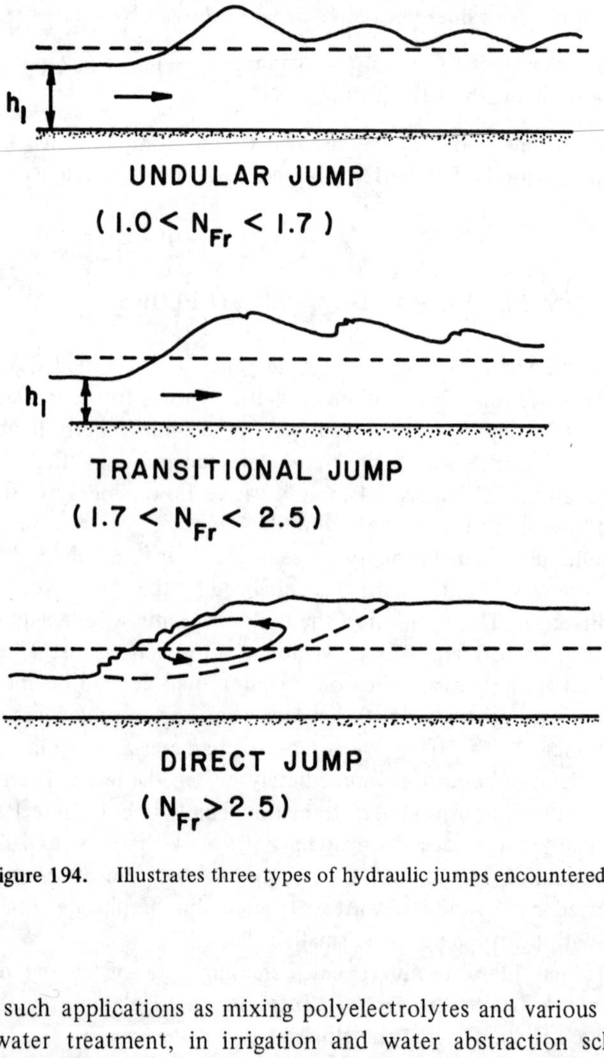

Figure 194. Illustrates three types of hydraulic jumps encountered.

useful in such applications as mixing polyelectrolytes and various chemicals in wastewater treatment, in irrigation and water abstraction schemes by raising downstream water levels, and in handling flood discharge carried from reservoirs by spillways.

An analysis of hydraulic jumps is given by Sellin [110] and Bakhmeteff [111]. Their description of a hydraulic jump is in terms of the *specific force* function, $\bar{f}(h)$:

$$\bar{f}(h) = \frac{Q^2}{gA} + AZ \tag{313}$$

The specific force function is shown plotted against depth, along with the specific energy curve in Figure 195. Both relationships exhibit a minimum at the critical depth. Note that for any $\bar{f}(h)$ value two depths are possible: one depth, h_1, corresponds to a depth less than the critical depth; the other, h_2, corresponds to a depth greater than the critical.

A stable hydraulic jump may exist between two depths referred to as *conjugate depths*, where the corresponding energies are E_1 and E_2 in Figure 195. The difference between these two energy states, ΔE, corresponds to the specific energy loss associated with the hydraulic jump. Note that $E_1 > E_2$, hence ΔE is negative, meaning that energy is dissipated in a hydraulic jump, not added.

The following limiting conditions can be observed from Figure 195: as the height of the jump decreases, i.e., $h_2/h_1 \rightarrow 1$, then $\Delta E \rightarrow 0$.

For a wide horizontal channel, Sellin [110] gives the following relationship for the height of the jump:

$$\frac{h_2}{h_1} = \frac{1}{2}[\sqrt{1 + 8N_{Fr}^2} - 1] \tag{314}$$

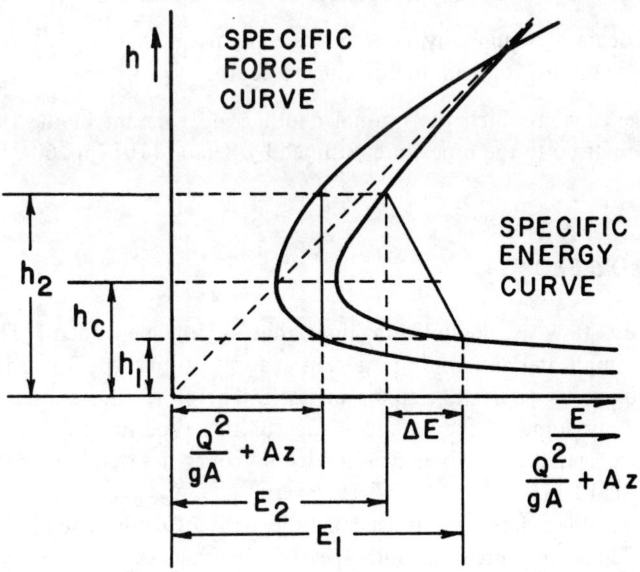

Figure 195. Shows plots of specific energy and specific force for a hydraulic jump.

Other important physical characteristics of a hydraulic jump that may be described from momentum considerations and Figure 195 are the energy loss, the length of the jump and the location of the jump along the channel axis.

For a wide channel, the energy loss may be computed from the following expression:

$$\Delta E = \frac{(h_2 - h_1)^3}{4h_1 h_2} \tag{315}$$

The length of the jump typically lies between the limits $S < \ell_2/h_2 < 6.5$, where ℓ is the length of the hydraulic jump.

If a stable jump occurs it will form at a point along the channel axis where suitable conjugate depths exist. The jump's location will depend on whether the normal depth is present in the channel flow. If the normal depth forms one of the limits of the jump, the following calculation procedure may be used to estimate the location of the jump:

Step 1. Compute the conjugate depth from Equation 314 for steep slopes or from a parallel form of that equation for mild slopes.

Step 2. Compute the free surface profile of the varied flow forming the other limit of the jump (refer back to Table 43 for profile descriptions).

Step 3. Locate the jump by selecting the point on that profile where the depth is the same as the required conjugate depth.

For the case in which the normal depth is not present in the flow, the reader is referred to the procedure outlined by Sellin [110] (pp. 60-61).

UNSTEADY FLOW

Unsteady flow is nonuniform flow that is time-dependent. The time variable complicates any kind of rigorous analysis of this type flow and would indeed require a separate chapter in itself. Therefore, this subsection will describe only some of the characteristics of this type flow. For a more in-depth coverage, the reader may refer to any one of several sources in the literature [113-116].

Unsteady flows are characterized on the basis of disturbances at the free surface. There are three general types of disturbances: (1) solitary waves, (2) monoclinical waves, and (3) surges. These are illustrated in Figure 196.

Solitary waves (Figure 196(A)) are a form of gradually varied flow and are characterized by a single smooth profiled wave. This type wave travels along the channel with a constant velocity of propagation (called "celerity"), C.

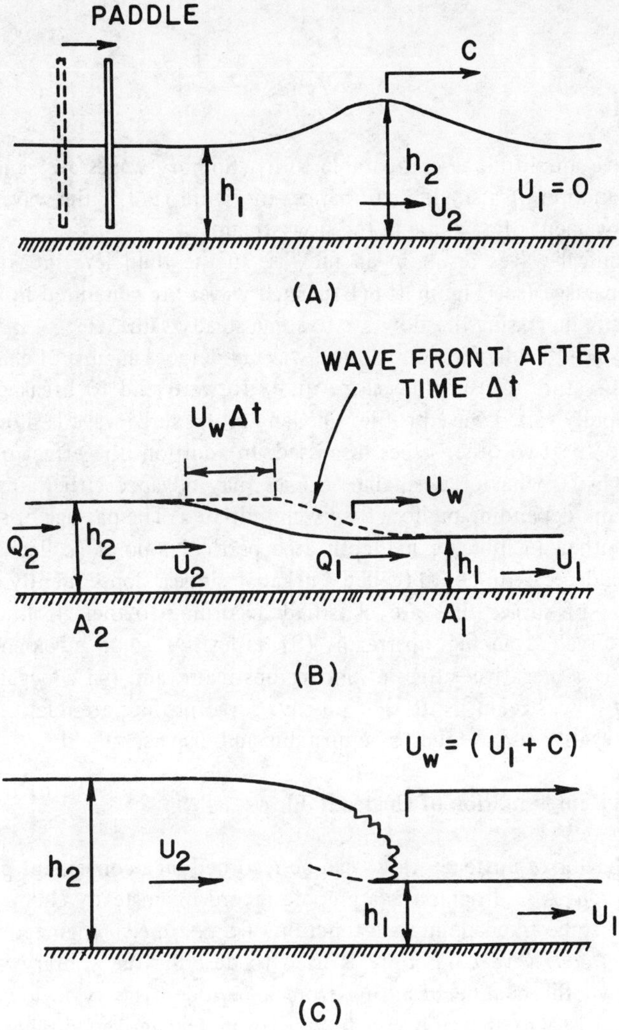

Figure 196. Three types of interfacial phenomena observed in unsteady open channel flows: (A) paddle generation of a solitary wave for water flow; (B) monoclinical rising wave; and (C) channel surge wave.

The wave amplitude gradually diminishes as the flow energy is dissipated by friction. The celerity can be estimated from the following:

$$C = \sqrt{\frac{g}{2}(h_1 + h_2)} \tag{316a}$$

or

$$C = \sqrt{gh} \tag{316b}$$

if the wave amplitude is comparable to h_1. Solitary waves are generated by the introduction of a single disturbance. In Figure 196(A) the wave is caused by the movement of a paddle in the flow stream.

Monoclinical waves result in an increase in the fluid level at some point after its passage (see Figure 196(B)). Such waves are generated in a channel by gradually increasing the flowrate to some steady value, Q_2.

A surge results when a monoclinical wave undergoes an abrupt change in its surface curvature, causing the slope at its forward end to break down and form a rapidly varied wave profile. The length of a surge wave is short in comparison to the two other types discussed. In addition, the effect of channel friction on its behavior is smaller. Surges may advance either upstream or downstream, depending on how they were initiated. The passage of surges can produce either an increase in depth at a particular point (called a positive surge) or a decrease in depth (called a negative surge). Consequently, there are four types of surges that are classified according to their frontage: (1) a positive surge advancing upstream; (2) a positive surge advancing downstream; (3) a negative surge advancing upstream; and (4) a negative surge advancing downstream. Note that positive surge profiles are relatively stable, whereas negative surges tend to be unstable and disperse quickly.

Graphical Representation of Unsteady Flows

The behavior of unsteady flows can be recorded on a convenient plot called a *hydrograph*. A hydrograph is a graphic record of unsteady channel or river flows past some fixed point as a function of time. In civil engineering jargon, the term *flood wave* commonly is used to describe the solitary wave that passes down the channel resulting from a period of heavy inflow. A flood hydrograph is a hydrograph constructed from the derived discharge past a gauging station during the passage of a flood wave. Figure 197 illustrates a typical unit hydrograph. Such plots are used typically for monitoring river flows and are correlated with rainfall data. The reader may refer to the literature [117-119] for a discussion of the graphical methods and examples of recording river flows.

FLOW THROUGH A SLUICE GATE

A sluice gate is a structure used in controlling the flow in canals or rivers. It consists of an opening in a water retaining wall or dam that stretches across

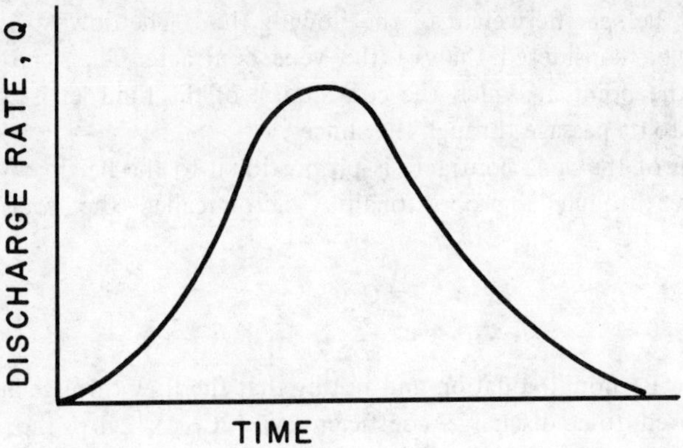

Figure 197. Typical unit hydrograph for characterizing unsteady flows in rivers and channels.

the channel. Water flows through the opening because of the head on the upstream side where the depth is always greater than the height of the opening. At the downstream side, the fluid jet discharging from the sluice opening usually is exposed to the atmosphere and may be partially or totally submerged by a hydraulic jump.

The horizontal force exerted on the flow by the sluice gate is illustrated by the pressure distribution curve in Figure 198. The force is the area under the distribution diagram depicting conditions on the upstream side of the sluice. If there is no energy loss, then the specific energy before and after the hydraulic jump on the discharge side of the sluice is

$$h_1 + \frac{q^2}{2gh_1^2} = h_2 + \frac{q^2}{2gh_2^2} = E_{1,2} \tag{317}$$

or

$$\frac{2q^2}{g} = \frac{4h_1^2h_2^2}{(h_1 + h_2)} \tag{318}$$

From momentum considerations, the specific force exerted on the fluid can be evaluated in terms of two depths:

$$F' = \frac{W}{2}\frac{(h_1 - h_2)^3}{(h_1 + h_2)} \tag{319}$$

where W is the specific weight of the flowing fluid. The downstream depth, h_2, should be considered that of the vena contracta. The vena contracta represents the point at which the contraction of the fluid jet is completed, subsequent to its passage through the sluice.

The depth of the vena contracta is proportional to the height of the sluice opening, y, through a proportionality factor called the *coefficient of contraction*:

$$h_2 = C_c y \tag{320}$$

Through the Bernoulli equation and noting that the flow through an opening can be related to a discharge coefficient, $q = C_D y \sqrt{2gh_1}$, the following expression is derived:

$$C_D = \frac{g}{y\sqrt{2gh_1}} = \frac{C_c}{\sqrt{1 + C_c y/h_1}} \tag{321}$$

For the type sluice illustrated in Figure 198, a typical value of C_c is 0.61.

Our discussions have assumed that the sluice gate is discharging freely. If the normal depth of the downstream channel is larger than the conjugate depth at the vena contracta, the hydraulic jump will not reach an equilibrium state and will tend to move upstream until it is stopped by the sluice. In this state the sluice becomes "drowned," and the fluid velocity through the sluice now becomes a function of the difference in levels between the upstream

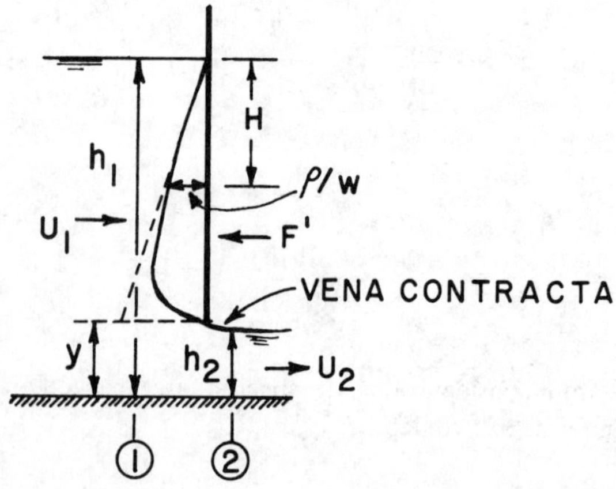

Figure 198. Illustrates pressure distribution over a sluice gate.

and downstream. Henry [120] has correlated a family of curves for C_D as a function of h_1/y, with downstream depth of submergence, h_3/y, as a parameter.

FLOW OVER WEIRS

Weirs are the simplest and least expensive devices that can be used in measuring flows in open channels. A weir is an obstruction built across an open channel over which the liquid flows. There is an opening or notch on the weir plate over which the fluid can flow. The edge or surface over which the fluid passes is referred to as the crest of the weir. In most applications this edge is cut with a sharp upstream corner such that the liquid tends to spring clear of the crest; hence, the name sharp-crested weirs. So-called broad-crested weirs will not be discussed because of their limited use. The layer of fluid flowing over the weir opening is referred to as the nappe. To ensure accurate measurements, the crest must be positioned high enough so that the nappe discharges freely to the atmosphere. If submergence of the nappe occurs, a reduction in the accuracy of the flow measurement will result and special corrections must be made. Discharge rates are established by obtaining measurements of the vertical distance from the weir crest to the fluid-free surface in the reservoir upstream from the crest. This depth of fluid is called the head of the weir. Measurements should be made upstream from the weir at a distance of at least four times the head. This minimizes the effect of surface drawdown that takes place at the crest. As a rule of thumb, the minimum head should be approximately 0.2 feet to prevent the nappe from clinging to the crest. Figure 199 illustrates the principle of flow measurement over a sharp-crested weir.

The basic classes of weirs are rectangular, trapezoidal, and triangular or V-notch designs.

With rectangular weirs, the bottom edge of the opening is the crest and the side edges are referred to as the sides, or weir ends. There are two types of rectangular weirs. One type consists of a box-like opening where the weir ends are a distance at least twice the maximum head on the weir from the side of the channel. This type weir is referred to as a *contracted weir* because a contraction of the fluid results, with the nappe forming a jet that is narrower than the weir opening. Hence, flow through this type weir is with end contractions (Figure 200).

The second type of rectangular weir suppresses these contractions by extending the weir across the entire channel width. The sides of the flume serve as the ends of the weir. The nappe does not undergo contraction from the width of the channel. This type rectangular weir is referred to as a *suppressed weir*. The flow through it is said to be without end contraction.

Vents normally are placed under the nappe on both sides of the weir box in this design to obtain adequate aeration of the nappe.

The trapezoidal weir, also known as the Cipolletti weir, is most often employed in irrigation applications. Its operation is illustrated in Figure 201. It resembles a contracted rectangular weir except that the sides incline outwardly at a slope of 1 horizontal to 4 vertical. The effect of this slope is to cause the fluid discharge to occur as though it were without end contraction. This design has the advantage that no correction for end contraction is

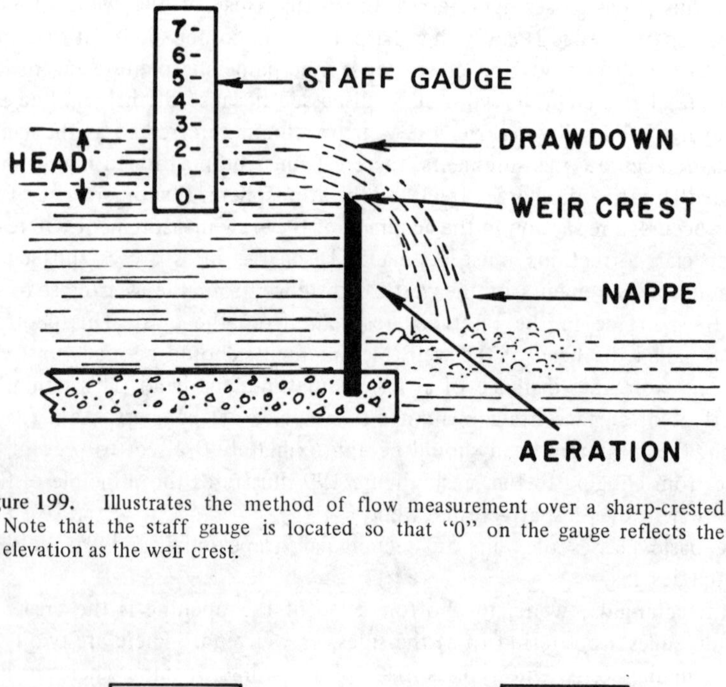

Figure 199. Illustrates the method of flow measurement over a sharp-crested weir. Note that the staff gauge is located so that "0" on the gauge reflects the same elevation as the weir crest.

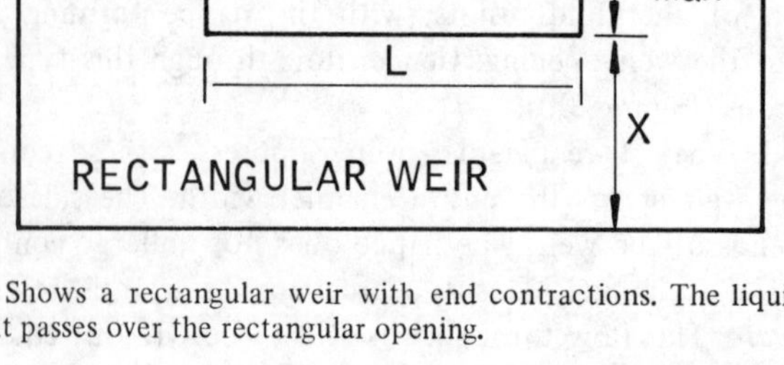

Figure 200. Shows a rectangular weir with end contractions. The liquid flow is contracted as it passes over the rectangular opening.

needed. Its disadvantage is that measurement accuracy is inherently less than that achievable with a rectangular suppressed or V-notch weir.

Triangular or V-notch weirs provide accurate flow measurement generally for smaller flows than do rectangular weirs. These designs consist of a weir plate containing a V-shaped notch with its apex pointing downward (refer to Figure 202). The apex angle can vary and typical angles are 22½, 45, 60 and 90°. Since this type weir has no crest length, considerably smaller flows than for a rectangular weir are represented by a given head. This configuration weir should always be used when frequent low flows are included in the overall range to be measured.

In applications involving small watershed and agricultural hydrology studies, 120° V-notch weirs are often used as they are capable of accommodating a wide range of flows. Some installations employ modified V-notch weirs with a central 90° notch for very precise measurement of low discharges. A shallower sloped crest of 120° or 150° is used for flood discharges.

Figure 201. Cipolletti weir used for applications in stream gauging, irrigation, domestic water supplies, sewage and in general flow measurement (courtesy of Leupold and Stevens, Inc., Beaverton, OR).

Discharge Formulas

Idealized relationships for discharge rates over weirs can be developed from a simplified analysis. For the rectangular weir shown in Figure 200, we may treat the nappe as a freely falling body, in which case the fluid velocity flowing over the weir is

$$u_h = \sqrt{2gh} \tag{322}$$

where h is an incremental height of liquid (head) over the notch. This assumes that fluid approach velocities are negligible.

The ideal volumetric flowrate can be expressed as

$$dQ_i = u_h dA \tag{323}$$

where dA = Ldh.

These equations are combined to give

$$dQ_i = 2gh(Ldh) \tag{324}$$

and, integrating between the limits of 0 and H,

$$Q_i = \frac{2}{3} L(2g)^{1/2}H^{3/2} \tag{325}$$

Equation 325 states that the ideal volumetric flowrate over a rectangular weir is proportional to the 3/2 power of the liquid head.

In practice, empirical or semiempirical relations are used to estimate actual flow conditions, one of the reasons being that the approach velocity does

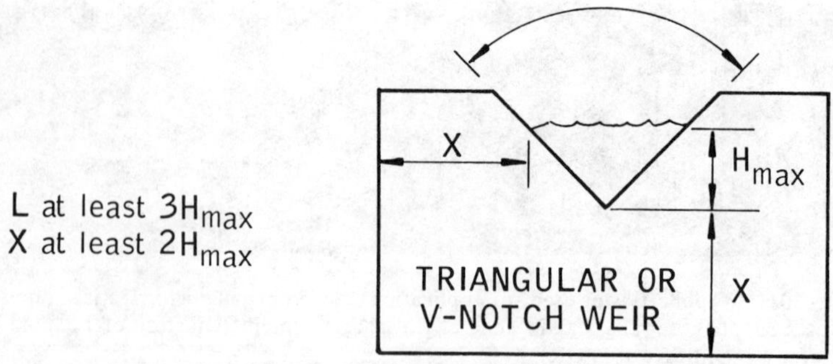

Figure 202. Illustrates a V-notch weir.

have a significant effect on flowrate and should not be neglected. The Francis formula is one of the most widely used empirical expressions for flow over rectangular weirs:

$$Q_a = 3.33\left(L - \frac{nH}{10}\right)\left[\left(H - \frac{u_a^2}{2g}\right)^{3/2} - \left(\frac{u_a^2}{2g}\right)^{3/2}\right] \tag{326}$$

where u_a is the fluid's approach velocity and n is the number of lateral contractions of the weir. Lateral contractions arise when the weir plate is not constructed over the entire width of the approach channel. For the case illustrated in Figure 200, n = 0 and, hence, Equation 326 reduces to

$$Q_a = 3.33\,LH^{3/2} \tag{327}$$

A similar expression for the ideal discharge rate over a V-notch may be developed:

$$Q_i = \frac{4L}{15H}(2g)^{1/2}H^{5/2} \tag{328}$$

See Figure 202 for a definition of terms.

Table 44 gives empirical formulas for computing volumetric discharge rates for water flow for different weir types. Tabulated weir tables can be found in the literature [121,122].

FLOW THROUGH FLUMES

Flumes are to open channels what the venturi tube is to a pipe. Flumes are specially designed open channel flow sections that provide a restriction in area that produces an increase in fluid velocity. They may be modified for use in circular pipes such as sewers. The most widely used flume is the so-called Parshall flume. A plan view of the Parshall flume is shown in Figure 203. Note that the size of the flume is established by the throat width, W. The converging upstream section of the flume accelerates the flow, thus reducing the deposition of solids, which could affect the accuracy of the flow measurement. For free flow, the constricted throat and downwardly sloping floor cause the flow to pass through the critical depth (usually near the crest of the flume). The critical depth control allows determination of free flow discharge by a single depth measurement upstream from the point of critical depth. The measurement point is located upstream from the throat about two-thirds the distance of the converging section.

Parshall flumes can experience two types of flow: free flow and submerged

Table 44. Volumetric Discharge Formulas for Different Weir Configurations

For rectangular weirs having end contractions,

$$Q(\text{cfs}) = 3.33(L - 0.2H)H^{3/2} \tag{329c}$$

$$Q(\text{mgd}) = 0.6463\,Q(\text{cfs}) \tag{329b}$$

where L = length of weir crest (ft)
H = head (ft)

For Cipolletti (trapezoidal) weirs,

$$Q(\text{cfs}) = 3.367\ LH^{3/2} \tag{330}$$

For rectangular (suppressed) weirs without end contractions,

$$Q(\text{cfs}) = 3.33\ LH^{3/2} \tag{331}$$

For 30° V-notch weirs,

$$Q(\text{cfs}) = 0.670H^{5/2} \tag{332}$$

For 45° V-notch weirs,

$$Q(\text{cfs}) = 1.035H^{5/2} \tag{333}$$

For 60° V-notch weirs,

$$Q(\text{cfs}) = 1.443H^{5/2} \tag{334}$$

For 90° V-notch weirs,

$$Q(\text{cfs}) = 2.50H^{5/2} \tag{335}$$

For 120° V-notch weirs,

$$Q(\text{cfs}) = 4.33H^{5/2} \tag{336}$$

flow. Free flow takes place when there is insufficient tail water depth to reduce the discharge rate. The latter occurs when the downstream levels are sufficiently above the flume crest to cause a backwater effect, which reduces the discharge rate. Submerged flows require that depth measurements be made both upstream and downstream. Submerged flows should be avoided if possible. Recall from our earlier discussions that the formation of a hydraulic jump downstream can serve as an indication of free flow conditions.

Flumes generally require relatively smooth, near-laminar flow for proper operation. As such, they should not be installed near a sharp change in slope or near obstructions in the channel. The user must determine whether the flow is laminar at least ten pipe diameters upstream of the proposed location in sewer applications.

Table 45 gives empirical formulas for computing volumetric discharge rates for water flow for different weir types. Table 46 gives some tabulated values of typical dimensions and capacities for water flow through Parshall flumes.

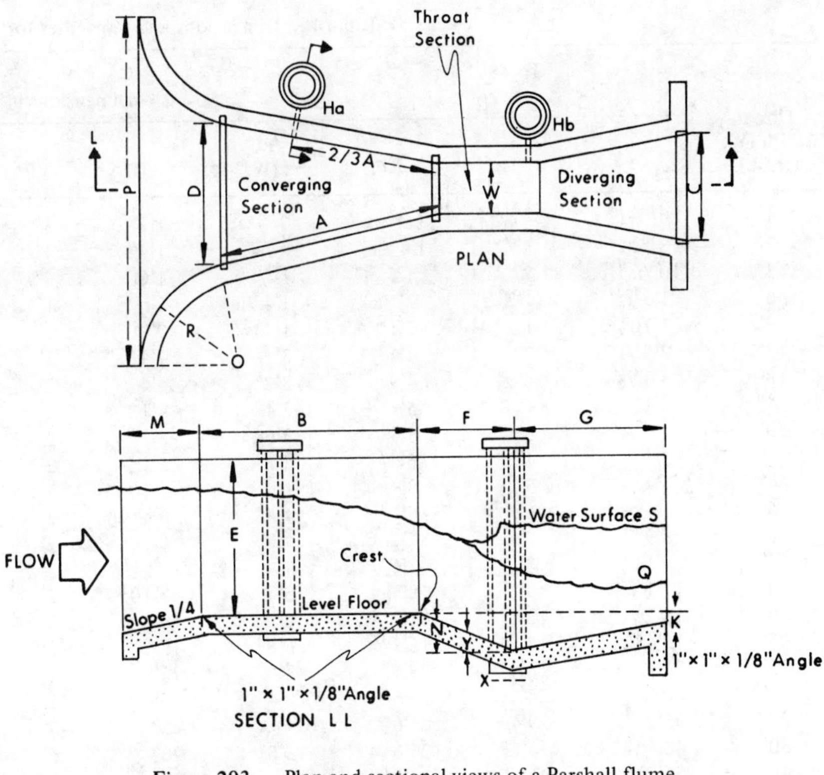

Figure 203. Plan and sectional views of a Parshall flume.

Table 45. Volumetric Discharge Formulas for Different Sizes of Parshall Flumes

For throat widths ranging from 1 to 18 inches,

$$1\text{ in., } Q(\text{cfs}) = 0.338H^{1.55} \quad (337a)$$

$$2\text{ in., } Q(\text{cfs}) = 0.676H^{1.55} \quad (337b)$$

$$3\text{ in., } Q(\text{cfs}) = 0.922H^{1.547} \quad (337c)$$

$$6\text{ in., } Q(\text{cfs}) = 2.06H^{1.58} \quad (337d)$$

$$9\text{ in., } Q(\text{cfs}) = 3.07H^{1.53} \quad (337e)$$

For throat widths ranging from 1 to 18 feet,

$$Q(\text{cfs}) = 4WH^{1.522W^{0.26}} \quad (338)$$

For throat widths greater than 10 feet,

$$Q(\text{cfs}) = (3.6875W + 2.5)H^{1.6} \quad (339)$$

Table 46. Dimensions and Capacities for

Throat Width (W) (ft/in.)	Dimensions in					
	A	B	C	2/3 C or 2/3 (W/244)	D	E
1"	6 19/32"	3 21/32"	1'2 9/32"	9 17/32"	1'2"	3"
2"	8 13/32"	5 5/16"	1'4 5/16"	10 7/8"	1'4"	4 1/2"
3"	10 3/16"	7"	1'6 3/8"	1'1/4"	1'6"	6"
6"	1'3 1/2"	1'3 1/2"	2'7/16"	1'4 5/16"	2'	1'
9"	1'10 5/8"	1'3"	2'10 5/8"	1'11 1/8"	2'10"	1'
12"	2'9 1/4"	2'	4'6"	3'	4'4 7/8"	2'
18"	3'4 3/8"	2'6"	4'9"	3'2"	4'4 7/8"	2'
24"	3'11 1/2"	3'	5'	3'4"	4'10 7/8"	2'
30"	4'6 3/4"	3'	5'4 1/4"	3'6 3/4"	5'3"	2'
3'	5'1 7/8"	4'	5'6"	3'8"	5'4 3/4"	2'
4'	6'4 1/4"	5'	6'	4'	5'10 5/8"	2'
5'	7'6 5/8"	6'	6'6"	4'4"	6'4 1/2"	2'
6'	8'9"	7'	7'	4'8"	6'10 3/8"	2'
7'	9'11 3/8"	8'	7'6"	5'	7'4 1/4"	2'
8'	11'1 3/4"	9'	8'	5'4"	7'10 1/8"	2'
10'	15'7 1/4"	12'	14'3 1/4"	6'	14'	3'
12'	18'4 3/4"	14'8"	16'3 3/4"	6'8"	16'	3'
15'	25'	18'4"	25'6"	7'8"	25'	4'
20'	30'	24'	25'6"	9'4"	25'	6'
25'	35'	29'4"	25'6"	11'	25'	6'
30'	40'4 3/4"	34'8"	26'6 1/4"	12'8"	26'	6'
40'	50'9 1/2"	45'4"	27'6 1/2"	16'	27'	6'
50'	60'9 1/2"	56'8"	27'6 1/2"	19'4"	27'	6'

FLOW MEASUREMENT IN SEWERS

Flow measurements of open sewage flow are best made through a manhole. Usually a manhole flow measuring installation is temporary and, as such, instrumentation should be portable. Flow data may be obtained using a calibrated weir or flume. The relationship between discharge rate and fluid level is termed *rating*, and rating curves and tables for standard weirs and flumes are available from manufacturers.

Any one of the weirs described previously may be installed for temporary flow studies at a manhole. Figure 204 illustrates such an installation. The disadvantage of this approach is that the area immediately upstream of the weir may be filled with sludge and grit, thus adversely affecting the accuracy of flow measurements. Frequent cleaning of the weir box structure is recommended.

Flumes often perform better in monitoring raw sludge flows as they tend to be self-cleaning devices. In addition, they operate with a smaller head

Parshall Flumes (use with Figure 203)

Feet and Inches						Free Flow Capacities			
						Minimum		Maximum	
F	G	H	K	X	Y	cfs	mgd	cfs	mgd
8″	6″	1⅛″	¾″	5/16″	½″	0.01	0.006	0.2	0.13
10″	8″	1 11/16″	⅞″	⅝″	1″	0.02	0.012	0.4	0.26
1′	1′3″	2¼″	1″	1″	1½″	0.03	0.02	0.6	0.39
2″	1′6″	4½″	3″	2″	3″	0.05	0.03	2.9	1.9
1′6″	2′	4½″	3″	2″	3″	0.1	0.06	5.1	3.3
3′	3′	9″	3″	2″	3″	0.4	0.26	16	10
3′	3′	9″	3″	2″	3″	0.5	0.32	24	15
3′	3′	9″	3″	2″	3″	0.7	0.46	33	21
3′	3′	9″	3″	2″	3″	0.8	0.52	41	26
3′	3′	9″	3″	2″	3″	1.0	0.65	50	32
3′	3′	9″	3″	2″	3″	1.3	0.84	68	44
3′	3′	9″	3″	2″	3″	2.2	1.4	86	56
3′	3′	9″	3″	2″	3″	2.6	1.7	104	67
3′	3′	9″	3″	2″	3″	4.1	2.6	121	78
3′	3′	9″	3″	2″	3″	4.6	2.0	140	90
6′	4′	1′1½″	6″	1′	9″	6.0	3.9	200	129
8′	5′	1′1½″	6″	1′	9″	8.0	5.2	360	226
10′	6′	1′6″	9″	1′	9″	8.0	5.2	600	388
12′	7′	2′3″	1′	1′	9″	10.0	6.5	1000	646
13′	7′	2′3″	1′	1′	9″	15.0	9.7	1200	775
14′	7′	2′3″	1′	1′	9″	15.0	9.7	1500	969
16′	7′	2′3″	1′	1′	9″	20.0	13.0	2000	1293
20′	7′	2′3″	1′	1′	9″	25.0	16.0	3000	1939

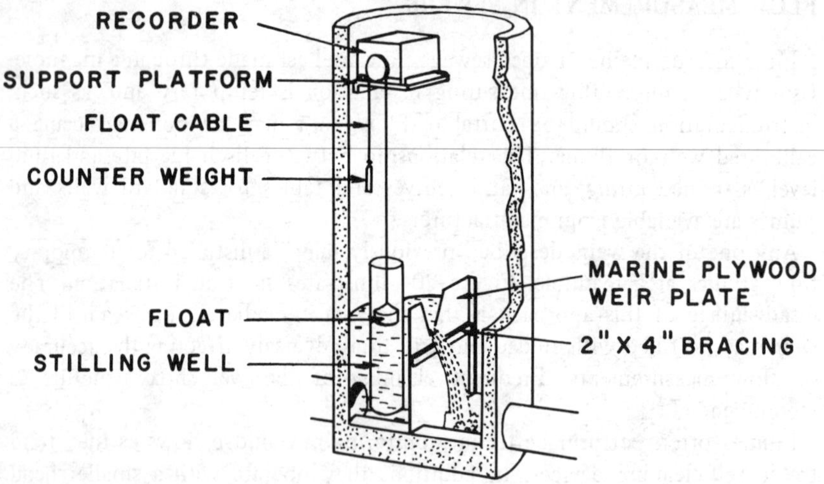

Figure 204. Manhole installation of a weir for measuring sewer surcharges (courtesy of Leupold and Stevens, Inc., Beaverton, OR).

loss than do weirs. Parshall flumes are often installed in new manhole measurement installations. For installation in an existing sewer line, a venturi-type flume is most often placed directly in the U-shaped channel at the manhole bottom or in the outlet sewer pipe.

After the weir or flume has been installed and the rating determined from manufacturer's data, the discharge-level relationship for the flow can be obtained. Naturally, the simplest method of determining discharge is simply to measure the level with a scale or depth gauge. By observing a staff gauge at periodic intervals, a reasonably good measurement can be obtained of the average discharge. For most installations, however, it is advisable to install some type of automatic gauging device and recording instrument. For a discussion on level measuring instrumentation the reader is referred to the literature [123].

CHAPTER 2.12

PROBLEMS AND SELECTED SOLUTIONS

This chapter is a compilation of problems the author has posed in undergraduate and graduate chemical engineering course work. As such, it was prepared mainly with engineering students in mind; however, practising engineers may wish to peruse these questions, particularly some of the solved problems. There are six sets of problems covering:

- Applications of basic flow definitions
- Problems related to pipe flow
- Problems on piping system analysis
- Problems on pumping service and system evaluation
- Problems on non-Newtonian flows
- Problems on open channel flows

Because the engineering and scientific communities have been slow to adopt a universal system of units, problems presented here are in English, cgs and SI units. We believe that in this transition phase of converting to universal standards engineers must be familiar with all the unit systems. Consequently, having solved problems in this section with different units, the engineer/student will attach a physical significance to the units applied to the solution.

A APPLICATION OF FLOW DEFINITIONS

A.1 Material Balance Problem

For the flow system in Figure 1A, develop an expression for computing the *mass* flowrate through line 2.

1. Compute the volumetric flowrate in line 2 using the expression developed.
2. Compute the average fluid velocity in each line.
3. For water flow, compute the Reynolds number in each line.

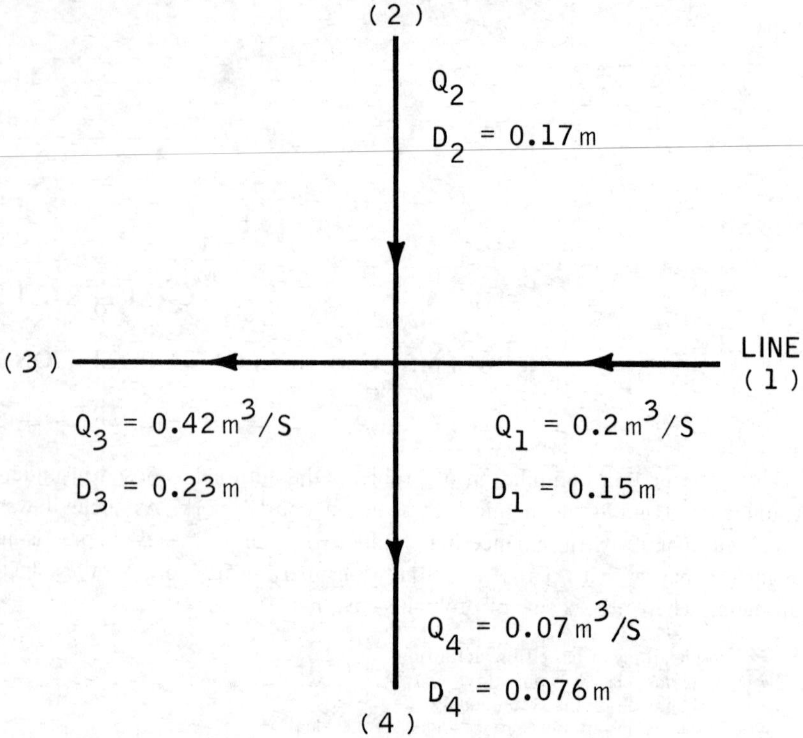

Figure 1A. System for material balance problem.

A.2 Mass Balance Application

A process line has three metering streams at various stations along the line where different concentrations of sulfuric acid are being added to the main flow. Each stream has a different volumetric rate, i.e., Q_1, Q_2, Q_3. Downstream, a sampling stream removes a portion of the flow at a continuous rate, Q_4. Develop an expression to determine the mass flowrate of sulfuric acid downstream from the sampling point.

A.3 Reynolds Number Determination

A 7.6-cm i.d. tube discharges water into a 19.1-cm tube. The average velocity in the large-diameter tube is 15 m/s. Determine

1. the volumetric flowrate
2. the mass flowrate

3. the average fluid velocity in the 7.6-cm tube
4. the Reynolds number in the smaller tube

Answer: 1. 526 m^3/s, 2. 0.041 kg/s, 3. 37.5 m/s, 4. 26,200.

A.4 Reynolds Number/Friction Factor Determination

A 2-in. standard wall tube (smoothed wall) is used for transporting water at a rate of 6.7 ft^3/min. Determine the Reynolds number and friction factor.

Answer: $N_{Re} = 8.96 \times 10^4$; f = 0.018

A.5 Reynolds Number/Friction Factor Determination

A 5-year-old cast iron pipe (3.5-in. nominal diameter) is used for recycling spent cooling tower water (92°F) at a rate of 15 ft^3/min. Determine the Reynolds number and friction factor.

A.6 Application of Equivalent Diameter

A smooth pipe is two-thirds full with water flowing at a rate of 80 m^3/min. The pipe diameter and various dimensions are shown in Figure 2A. Determine the Reynolds number and friction factor.

Answer: To determine a Reynolds number, an equivalent diameter must be used. The equivalent diameter is defined as four times the hydraulic radius, where

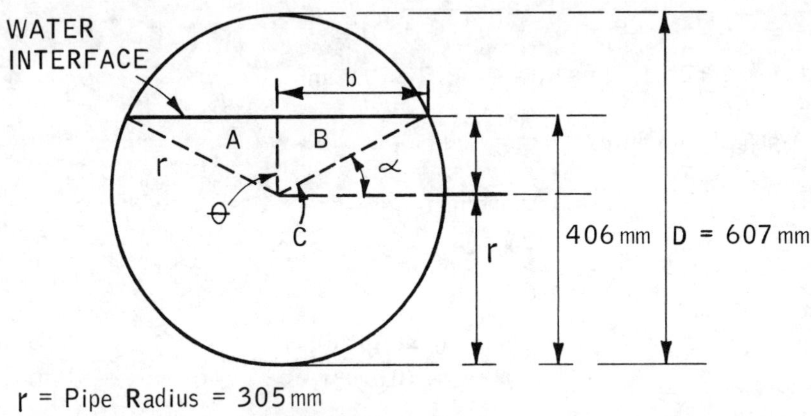

Figure 2A. Partial pipe flow.

$$R_H = \frac{\text{Cross-Sectional Area of Flow}}{\text{Wetted Perimeter}}$$

From Figure 2A, the following dimensions are computed:

$$\text{Depth of Water} = \frac{2}{3}(609 \text{ mm}) = 406 \text{ mm}$$

$$\cos\theta = \frac{406 - r}{r} = \frac{102}{305} = 0.33$$

$$\theta = 70.46^\circ$$

$$\alpha = 90^\circ - 70.46 = 19.5^\circ$$

$$\text{Area "C"} = \frac{\pi D^2}{4} \frac{180 + (2 \times 19.5)}{360}$$

$$= \frac{\pi(609)^2}{4}(0.61) \times 10^{-4} = 1773 \text{ cm}^3$$

$$b = \sqrt{r^2 - (102)^2} = \sqrt{(305)^2 - (102)^2} = 287 \text{ mm}^2$$

Cross-sectional area of flow = "A" + "B" + "C"

$$= 145.8 + 145.8 + 1773$$

$$= 2065 \text{ cm}^2$$

Wetted perimeter = $\pi D(0.61)$

$$= \pi(609)(0.61) = 1164 \text{ mm} = 116.4 \text{ cm}$$

$$R_H = \frac{2065}{116.4} = 17.7 \text{ cm}$$

Equivalent diameter = $4R_H = 4 \times 17.7 = 70.8$ cm

$$\text{Average fluid velocity } u = \frac{Q}{\text{Area}}$$

$$= \frac{80 \text{ m}^3/\text{min}}{2065 \text{ cm}^2}$$

$$u = 6.46 \text{ m/s}$$

$$N_{Re} = \frac{D_H u \rho}{\mu} = \frac{(70.8)(6.46)(1)}{(0.0108)} \times 100$$

$$= 4.2 \times 10^6$$

From Moody plot, f = 0.0093.

A.7 Application of Equivalent Diameter to Irregularly Shaped Conduits

Water is flowing at a rate of 900 gpm through the following systems:

1. an elliptical conduit whose major axis is 3 ft and minor axis is 2 ft;
2. an annulus whose inner diameter is 1 ft and outer diameter 2.5 ft; and,
3. a rectangular channel of dimensions 5 ft × 3 ft.

Compute the Reynolds number and friction factors for each of the flow cases, assuming full conduit flow.

A.8 Equivalent Diameter Problem

Repeat the Reynolds number and friction factor calculations for parts 1 and 3 of Problem A.7 if only two-thirds of the conduit is filled with flowing water.

A.9 Unit Conversions

Convert the following to SI units from base units definitions:

1. 30 psi
2. 1 lb_m/ft-s
3. 1 poundal-seconds/ft^2
4. 70 lb_m/ft^3
5. 60,000 Btu/hr

A.10 Application of Continuity Equation

A 0.12-m i.d. pipe is connected to a 0.25-m i.d. pipe. The average fluid velocity in the small pipe is 18 m/s.

1. Determine the average velocity in the larger pipe.
2. Determine the mass flowrate.
3. Determine the volumetric flowrate.

B PROBLEMS ON PIPE FLOW

B.1 Friction Loss and Shear Stress Problem

Water is flowing at a rate of 0.02 m^3/s through a 13-cm i.d. tube that is 70 m long.

1. Estimate the head loss per unit length of tube.
2. Estimate the frictional pressure drop across the length of tube.
3. Compute the fluid's shear stress at the tube wall.

Answers: 1. 341 N/m^2 per m.
2. 0.024 Pa.
3. 10.9 N/m^2.

B.2 Pressure Drop in Pipe Flow

Bunker C fuel oil at 30°C is flowing through a straight section of 76-mm i.d. steel pipe, 38 m long at a rate of 12 ℓ/s. Calculate the pressure drop in (a) Pascals, and (b) psi.

B.3 Flow Through an Elbow

Fuel oil (specific gravity = 0.82, kinematic viscosity = 0.035 cm^2/s), is flowing through a 50-mm i.d. steel pipe. Calculate the total pressure drop over 70 m of straight pipe and a 90° elbow. What fraction of the total pressure drop is caused by the elbow?

B.4 Flow Through a Coil

Cooling water is flowing through a cylindrical pipe coil, as shown in Figure 1B, at a rate of 370 gph. The coil consists of 2-in. i.d. stainless-steel tubing. Determine the pressure drop from point A to point B. Note there are 6.5 coils, i.e., twenty-six 90° bends in the coil.

B.5 Steam Flow Through a Coil

Repeat Problem B.4 for saturated steam at 170 psig flowing at a rate of 2800 lb_m/hr.

B.6 Estimating Relative Roughness from Flow Data

A 3-in. horizontal pipeline is carrying a 45 wt% sulfuric acid solution at 20°C. The flowrate is 60 gpm, and the pressure differential between two points 200 ft apart is 2.9 psi.

1. Estimate the relative roughness of the pipe.
2. Prepare a plot of the velocity profile in the pipe. On the same figure, plot the velocity profile for the same flow in a smooth pipe.
3. For the data prepared in part (2) obtain average velocities by numerically integrating over each profile.

B.7 Shear Stress Calculations

A Newtonian fluid flowing through a 5-in. (127 mm) i.d. pipe has a velocity profile defined by the following expression: $u = 0.6(1 - (r/R)^2)$ fps, where r

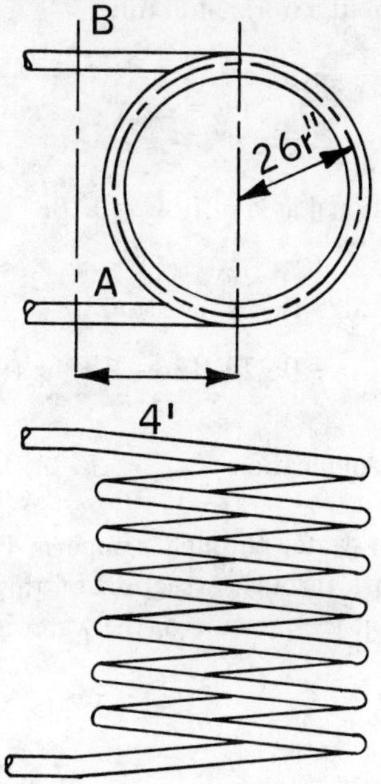

Figure 1B. Water flow through a cooling coil.

is the radial distance from the pipe centerline and R is the pipe radius. The viscosity of the fluid is 3.2×10^{-4} lb_m-s/ft^2.

1. Prepare a plot of the velocity profile and comment on its shape.
2. Prepare a plot of the shear stress profile.

B.8 Determination of Fluid Pulling Force

Determine the force of the fluid pulling an 800-ft length of the 5-in. i.d. pipe in Problem B.7. Relate the force to the frictional pressure drop in the pipe.

Answer: From Problem B.7, the shear stress at the wall of the tube is 9.22×10^{-4} lb_f/ft^2.

Force at Wall:

$$F_W = A_s\tau = (2\pi R \times L)\tau$$

$$F_W = 2\pi(0.21)(800)(9.22 \times 10^{-4}) = 0.97 \; lb_f$$

From a force balance about a horizontal tube,

$$F_W = \Delta PA$$

where A is the cross-sectional area of flow = 0.136 ft^2. Hence,

$$\Delta P = F_W/A$$
$$= 0.97/0.136 = 7.1\ \text{lb}_f/\text{ft}^2$$

B.9 Conservation Law Application

The mass flowrate of water through a pipe is 103 kg/s. Determine how much fluid energy flows through a section of this pipe in a period of 20 minutes if at this section the pressure of the water is 125 kPa and its velocity is 2.5 m/s.

B.10 Pipe Flow Problem

No. 6 fuel oil is flowing at a rate of 200 gpm through a 4-in.-diameter Schedule 40 carbon-steel pipe.

1. Determine the Reynolds number and classify the flow as laminar or turbulent.
2. Determine the head loss per 1000 ft of piping.
3. Determine the fluid shear stress at the wall.

B.11 Comparison of Friction Factor Calculations

A 5-in. diameter cast iron pipe is 1800 ft long and carries water at a rate of 1.8 cfs. Determine the friction factor and the head loss in the pipe.

1. first by the appropriate friction factor equation,
2. then by using the Moody diagram.

Compare your results and comment.

B.12 Flow from a Reservoir

A 6-in. PVC pipe, 1800 ft long, is used for conveying water between two reservoirs. The surface elevations between the two reservoirs differ by 190 ft. Determine the volumetric flowrate.
(Hint: This requires a trial and error solution for Q.)

B.13 Application of Empirical Formulas

Resolve Problem B.11 for the head loss using the Hazen-Williams and Manning formulas.

B.14 Calculations of Flow Parameters

Gasoline (specific gravity $\simeq$ 0.7, viscosity $\simeq$ 0.55 cp) is flowing through a new 7-in. level pipeline 2600 ft long. The average velocity in the pipeline is 15 fps:

1. Calculate the friction factor and Reynolds number.
2. Calculate the velocity gradient at the wall.
3. Calculate the wall shear stress.
4. Calculate the thickness of the laminar layer.
5. Calculate U/U_{max}.
6. Prepare a plot of the velocity profile.

B.15 Location of Average Velocity

Water is flowing through a straight section of 7-in. i.d. pipe that is 1 mile long. The static pressure at one end of the line is 135 psig and 55 psig at the other end.

1. If the Reynolds number is 200,000, determine at what point (across the diameter) the velocity is equal to the average stream velocity.
2. What is the friction loss if the flow is 75,000 gph?
3. What is the shear stress at the wall if the flow is 75,000 gph?

C PROBLEMS ON PIPING SYSTEM ANALYSIS

C.1 Pressure Drop Across Orifice Meter

Water is flowing at a rate of 1.23 m^3/hr through a 10-cm i.d. tube. If an orifice plate is installed in the line, what is the pressure differential across the corner taps? The orifice plate opening is 82% of the cross-sectional area of the pipe and its discharge coefficient is 0.975.

Answer: 0.32 N/m^2.

C.2 Flow Coefficient Determination for Orifice Meter

A square-edged orifice of 2.5-in. diameter is installed in a 5-in. Schedule 40 pipe. A mercury manometer is connected between the flange taps. Determine the theoretical calibration constant for the meter when used on 14°C water. Note that the discharge coefficient for the flow range is constant.

Answer: From a handbook on physical properties the following information is obtained:

$$\text{Density of Water } \rho_w = 62.37 \text{ lb}_m/\text{ft}^3$$

$$\text{Specific Gravity of Mercury} = S_{Hg} = 13.57$$

$$\text{The density of Hg under water} = 62.37(13.57 - 1.00)$$

$$= 784 \text{ lb}_m/\text{ft}^3$$

The differential pressure across the flange taps:

$$\Delta P = \frac{\Delta h_m \rho}{12 \times 144} = \frac{\Delta h_m (784)}{12 \times 144} = 0.454$$

The volumetric flowrate is

$$Q = 236\, d_0^2 C \sqrt{\frac{\Delta P}{\rho}}$$

where d_0 is the orifice diameter in inches and Q is in gpm.

Combining these two expressions and substituting values with C = 0.62

$$Q = 236 \times (2.5)^2 \times 0.62 \sqrt{\frac{0.454\, \Delta h_m}{62.3}}$$

$$Q = 78.1 \sqrt{\Delta h_m}, \text{ calibration constant}$$

C.3 Flowrate Determination Through Orifice

For Problem C.2 determine the volumetric flowrate for water flow if the mercury manometer deflection is 3.0 in.

C.4 Flow Determination Through Orifice

Bunker C fuel oil at 85°F is flow through a 4-in. Schedule 40 pipe. A 2.25-in. i.d. square-edged orifice is used to meter the flow. If the pressure differential between the flange taps is 0.6 psi, what is the flowrate in ft^3/min?

C.5 Flow Through a Venturi Meter

A venturi meter is being used to measure the flowrate of a fluid through a 7.6-cm i.d. line. The maximum flowrate that can be expected is 0.53 m^3/hr. The specific gravity and viscosity of the fluid are 1.2 and 0.97 mPa-s, respectively.

1. Determine the Reynolds number upstream of the venturi.
2. Determine the dimensions of the venturi if the pressure drop across the meter at the maximum flow is 62.1 kPa.
3. Determine the throat Reynolds number.

Answers:

1. 151,200.
2. throat diameter = 32.2 mm = d
   b = 32.2 mm
   c = d/2 = 16.1 mm
   r_2 = 3.5d = 112.6 mm
   r_1 = 1.375D = 104.9 mm, refer to Figure 17(B) for definitions
3. Throat Reynolds number = 358,100.

C.6 Application of Pitot Tube

Water is flowing at a rate of 1500 gpm in a 4-ft-diameter smooth pipe. A pitot tube with an opening of 0.01 mm^2 is inserted into the flowing stream. The *impact pressure* is measured with the center of the pitot tube located at 1.8, 1.3, 1.0 and 0.51 mm from the wall. Calculate the velocity corresponding to the measured impact pressure and prepare a plot of the velocity profile.

C.7 Application of Nutating Disc Meter

A nutating disc meter is used for measuring water flow in a 1-in. nominal size line. The maximum pressure drop across the meter should not exceed 23 psi. Determine the minimum flowrate that can be expected.

Answer: 29 gpm.

C.8 Pressure Loss Through a Valve

Fuel oil at 85°F is flowing through a wide open 5-in., 150-lb steel gate valve at a rate of 300 gpm. The face-to-face dimension of the valve is 13 in. Determine the pressure drop across the valve.

Answer: The viscosity and specific gravity of the oil are 2500 cp and 0.92, respectively.

For 5-in. Schedule 40 pipe, inside diameter D = 5.047 in.
= 0.421 ft.

$$N_{Re} = \frac{Du\rho}{\mu} = \frac{4\,Q\rho}{\pi\,D\mu}$$

where $Q = 300 \text{ gpm} \times 2.228 \times 10^{-3} = 0.668 \text{ ft}^3/\text{s}$
$\mu = 2500 \text{ cp} \times 6.72 \times 10^{-4} = 1.68 \text{ lb}_m/\text{ft-s}$
$\rho = 0.92 \times 62.3 \text{ lb}_m/\text{ft}^3 = 57.3 \text{ lb}_m/\text{ft}^3$

$$N_{Re} = \frac{4(0.668)(57.3)}{\pi\,(0.421)(1.68)} = 68.9$$

Since $N_{Re} < 2000$, the flow is laminar. The following equation is recommended for estimating the pressure drop across the valve:

$$\Delta P = 2.28 \times 10^{-5}\left(\frac{L}{D}\right)\frac{\mu Q}{d^3}$$

where D is in ft, d in in., μ in cp, Q in gpm, ΔP in psi and L/D is equivalent length in pipe diameters (see Table 13, p. 233). The above equation is derived from the Darcy formula and includes the resistance coefficient for this type valve. The student should go through the derivation.

Since conditions are laminar, a correction factor should be applied to the equivalent length since L/D values are based on turbulent conditions:

$$\left(\frac{L}{D}\right)_{\ell} = \left(\frac{L}{D}\right)_{t}\frac{N_{Re}}{1000}$$

From Table 13, p. 233, $(L/D)_t = 13$ for a fully open gate valve. Hence,

$$L = \left(\frac{L}{D}\right)_{t}\frac{N_{Re}}{1000}(D)$$

$$L = 13\left(\frac{68.9}{1000}\right)(0.421) = 0.38 \text{ ft}$$

Note that the actual L = 13/12 = 1.08. When the actual length to valve flow path exceeds the equivalent length, manufacturers recommend that the larger number be used. Hence,

$$\Delta P = 2.28 \times 10^{-5} \left(\frac{1.08}{0.421}\right)\left(\frac{2500 \times 300}{(5.047)^3}\right)$$

$$\Delta P = 0.34 \text{ psi}$$

C.9 Pressure Losses Through Piping and Valve

Fuel oil No. 6 is flowing at a rate of 400 barrels per hour through 700 ft of horizontal 6-in. Schedule 40 pipe. A 6-in. globe valve is used for flow control. Calculate the pressure loss through the piping and valve when the valve is fully open. (Note there are 52 gallons in a barrel.)

C.10 Flowrate and Pressure Drop Determination

Water is flowing in a 51-mm i.d. smooth pipeline. A straight stretch of this line is 91 m in length and contains a single gate valve. During normal operation the valve is left in the half-open position and the total head loss is known to be about 5.5 m.

1. Determine the average fluid velocity and the volumetric flowrate through the pipeline.
2. Determine what fraction of the total head loss is contributed by the valve.

Answer: 1. Part (1) involves a trial and error solution to the equation

$$H_T = 5.5 = f\frac{L}{D}\frac{u^2}{2g} + K_L\frac{u^2}{2g}$$

The Moody plot (Figure 1, p. 167) is used as follows: choose a value for the Reynolds number and obtain f from the plot. Compute u from N_{Re}. With values for u and f from a guess of N_{Re}, the RHS of the head loss equation is solved. If the value matches the LHS, the guess is correct; if not, a new value of N_{Re} is chosen and the computations repeated until convergence within a specified accuracy is obtained.

$$Q = 3.39 \ \ell/\text{s}$$

2. 0.146 or 15%.

C.11 Equivalent Length Determination

A 900-ft section of 8-in. Schedule 40 cast iron pipe has a fully open globe valve. The flowrate averages about 950 gpm. Determine the length of the equivalent pipe.

C.12 Flow from a Reservoir

The flow from a reservoir is controlled by a gate valve. The reservoir maintains a constant head of 22 ft. The piping layout is shown in Figure 1C.

1. Determine the maximum flowrate in gpm.
2. Comment on the type of flow control obtained with this arrangement.

(Hint: Part (1) requires a trial and error solution similar to Problem C.10.)

C.13 Application of Bernoulli's Theorem

Water is flowing through the piping system shown in Figure 2C at a rate of 650 gpm.

1. Determine the average fluid velocity in the two size lines.
2. Determine the pressure drop between points (1) and (2).

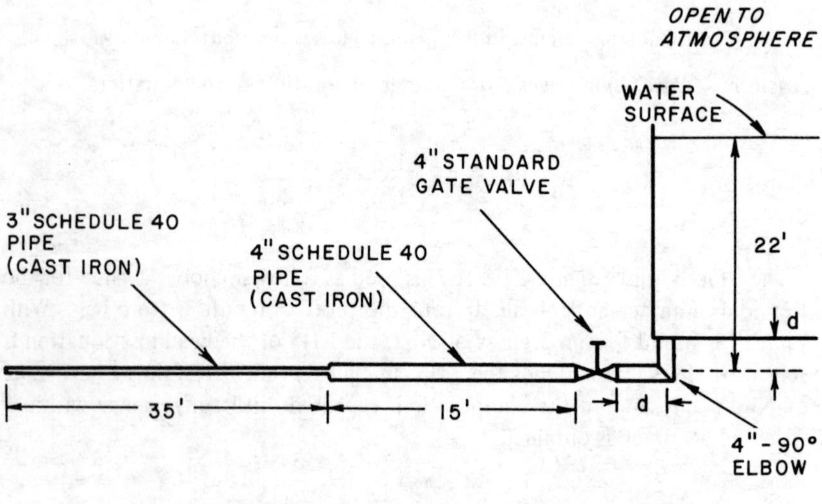

Figure 1C. Piping arrangement for flow from a reservoir.

C.14 Losses from Expansions and Contractions

Crude oil (specific gravity 60°F = 0.81) is flowing through the piping shown in Figure 3C. Pipe diameters shown are nominal and all sections are Schedule 40. The flowrate through pipe A is 45 gpm. Compute the following:

1. the average linear velocity of the fluid in each section of pipe,
2. the mass flowrate in each pipe;
3. the mass velocity in each pipe; and
4. the pressure drop over the entire piping section.

Answers:

1. $U_A = 7.09$ fps, $U_B = 1.95$ fps, $U_C = 4.30$ fps.
2. 304 lb_m/min.

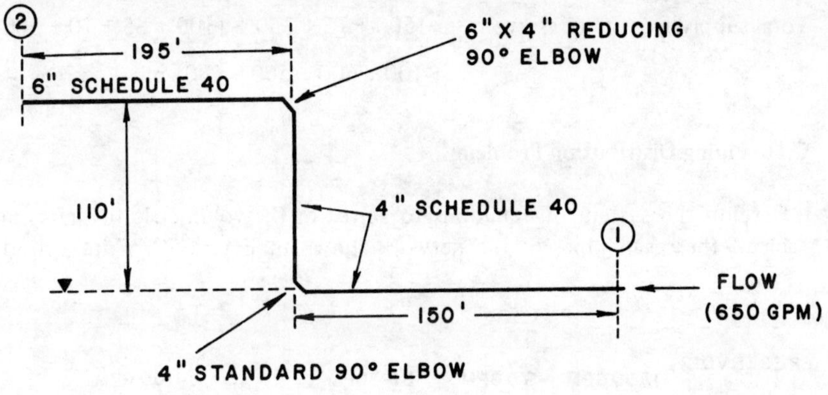

Figure 2C. Piping arrangement for Problem C.13.

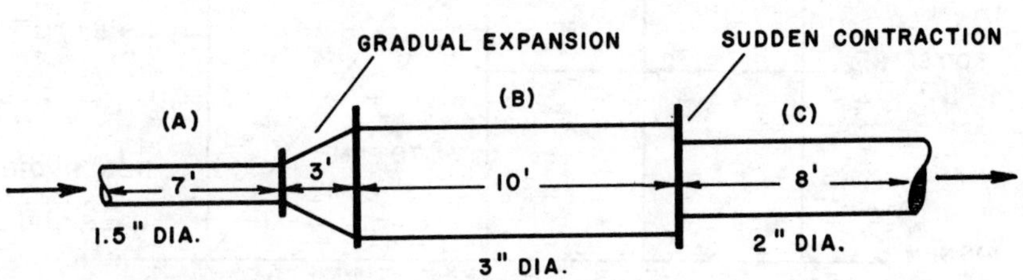

Figure 3C. Section of pipe with expansion and contraction.

3. Pipe A 358 lb_m/ft^2-s.
   Pipe B 98.6 lb_m/ft^2-s.
   Pipe C 217.3 lb_m/ft^2-s.
4. For the student to solve.

C.15 Water Distribution System Analysis

A water distribution system is supplied by two reservoirs. Figure 4C shows the arrangement of demands and consumptions by means of arrows. Determine the total volumetric flowrate that must be supplied by the two reservoirs to meet the demands.

Answer: Since the flow is steady and incompressible, then the total net inflow must be zero (i.e., $\Sigma Q_i = 0$).

Hence, from a simple material balance,

$$\text{Total supply from two reservoirs} = -1500 + 75 + 350 + 1100 + 85 + 70 + 300 + 100 + 600 - 500 = 680 \text{ gpm}$$

C.16 Piping Distribution Problem

Develop the system of equations to solve for the volumetric flowrates in each of the seven pipes of the network shown in Figure 5C. Write a short

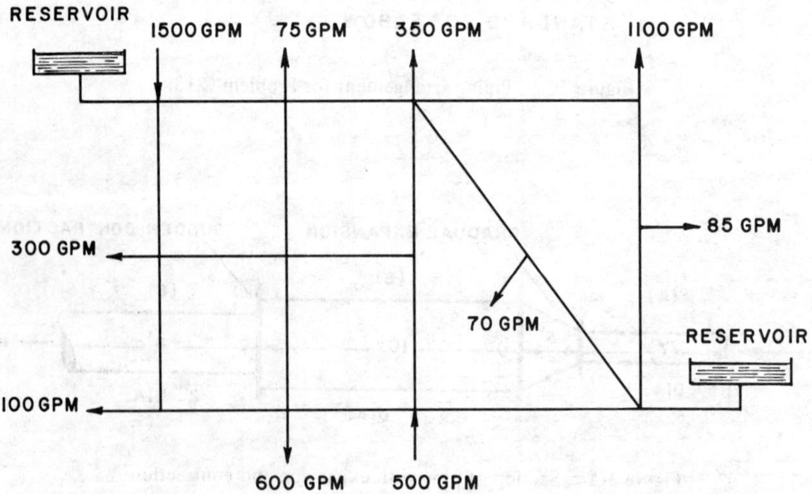

Figure 4C. Water distribution system supplied by two reservoirs.

computer program and solve for these flowrates in gpm and cfs. All piping is cast iron Schedule 40 with $\epsilon = 0.010$ in.

C.17 Analysis of Piping Network

A pipe network with six junctions is shown in Figure 6C. All piping is cast iron Schedule 40 pipe with $\epsilon = 0.0102$ in. The piping between junctions (1) and (2) has an 8-in. globe valve fully open, as does the piping between junctions (2) and (5). The piping between junctions (3) and (6) has a double disc gate valve that is one-half open. The piping between junctions (2) and (3) has an orifice meter attached to a flow indicator. Applying linear theory method,

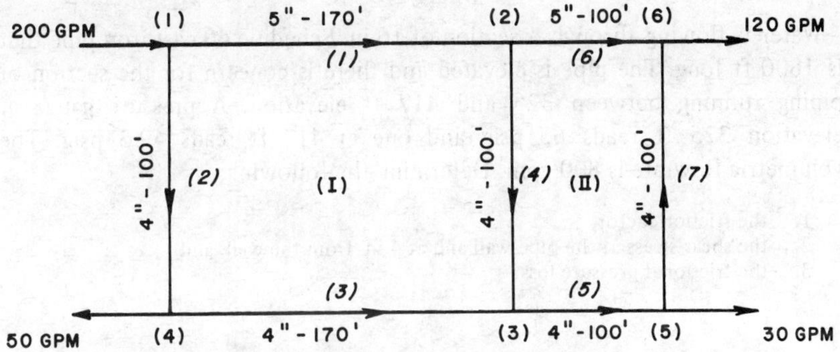

Figure 5C. Piping distribution problem.

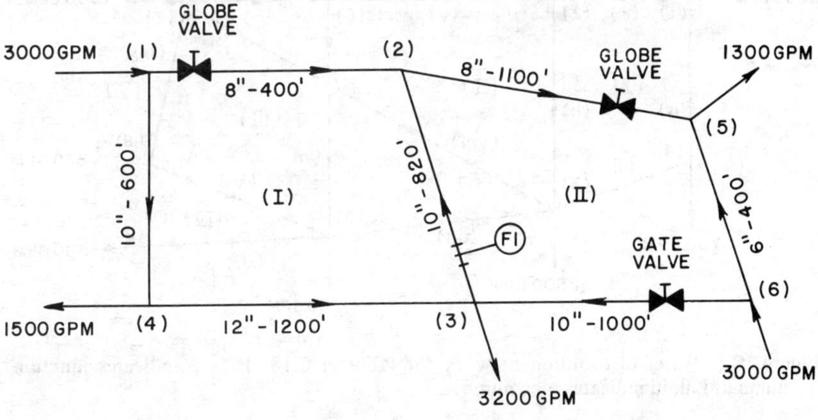

Figure 6C. Piping system network for Problem C.17.

develop a series of equations to compute the volumetric flowrate and head loss in each section of piping. Prepare a computer program and solve. In implementing the method, the first values used for loss coefficients and exponential terms may be obtained from the Hazen-Williams equation.

C.18 Analysis of Piping Network

A water distribution network is shown in Figure 7C. Line sizes and lengths are given in Table 47. Prepare a computer program to compute the flowrate and head loss in each pipe. Use the linear theory method to solve your system of equations, which considers the flowrates as unknowns.

C.19 Head Loss Problem

Water is flowing through a section of 16-in. Schedule 40 cast iron pipe that is 1800 ft long. The pipe is elevated and there is concern for the section of piping running between 325 and 417 ft elevation. A pressure gauge at elevation 325 ft reads 65 psig and one at 417 ft reads 49.3 psig. The volumetric flowrate is 800 gpm. Determine the following:

1. the friction factor,
2. the shear stress at the pipe wall and at 4 in. from the wall, and
3. the frictional pressure loss.

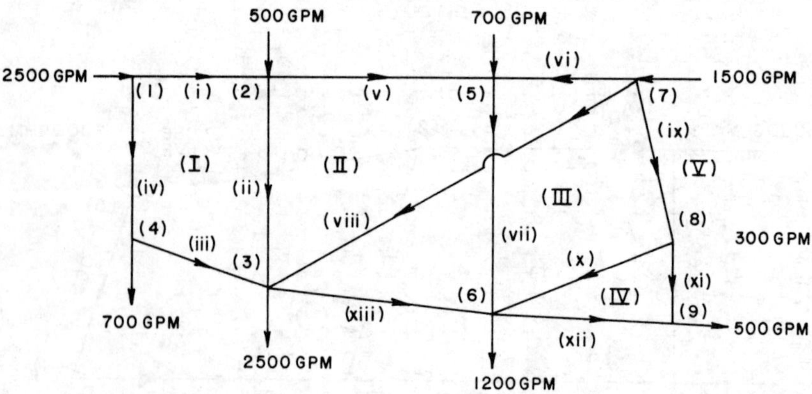

Figure 7C. Water distribution network for Problem C.18. 1, 2, 3 indicates junction number. i, ii, iii indicates pipe number.

Table 47. Line Sizes and Lengths

Pipe Number	Nominal[a] Diameter (in.)	Length (ft)	Roughness ϵ (in.)
i	24	1000	0.014
ii	18	1800	0.014
iii	24	1100	0.014
iv	18	1200	0.014
v	24	2500	0.014
vi	24	1200	0.014
vii	18	2000	0.014
viii	16	3500	0.014
ix	18	1400	0.014
x	10	1750	0.01
xi	18	600	0.0095
xii	24	1600	0.013
xiii	24	1800	0.013

[a]All piping is cast iron Schedule 40.

C.20 Increasing Flow by Replacing Old Pipe

Field tests on a 12-year-old horizontal water main showed that the wall roughness increased to 0.007 ft. The pipe diameter is 14 in. and presently carries 55 gpm.

1. Determine the increase in flow at the same power input resulting from replacing the line with a new cast iron pipe of the same diameter.
2. Compute the frictional pressure losses for the new and old piping at the present flowrate and compare.

C.21 Selecting Optimum Piping Material

City engineers have been commissioned to determine the most economical pipe material to install as an addition to the city's water main. The plan calls for approximately 15 miles of 10-in.-diameter piping that would deliver water at an average rate of 1500 gpm. Cast iron and wrought iron piping are being considered. The cost of installing 10-in. cast iron piping will run \$25/ft and has an estimated service life of 10 years.

The installed cost for wrought iron piping is \$43/ft, but its service life is estimated to be 17 years. With both types of pipe the roughness is estimated at 1.6 times the roughness of new pipe and may be considered constant over the service life of the material. Also to be considered is the cost of power for overcoming friction, which is presently 6 ¢/hp-hr and is expected to double over the next 10 years. Determine which pipe would be most economical to install.

D PROBLEMS ON PUMPING SERVICE AND SYSTEM EVALUATION

D.1 Determining Total Head Requirements

A pump must supply water at a rate of 100 gpm. The total friction in the piping, including entrance losses, etc., is 15 ft at that capacity. The pump takes suction from a closed vessel, which is under a vacuum of 29 inches of mercury. The height of water in the tank is 3 ft above the floor. Water flows to the pump through a 3-in. pipe and is discharged through a 2-in. pipe. The water is finally discharged at an elevation of 10 ft above the floor.

Answer: The water level in the feed tank is used as the arbitrary reference point for static head measurements. Application of the Bernoulli equation shows that the pump must overcome a total friction loss of 55 ft. Hence, the total head required for the pump is 55 ft. Details are left to the student.

D.2 Estimating Energy Costs for Pumping

A wastewater treatment plant uses three vertical turbine pumps, each capable of 3600 USGPM against 265 ft of head. Three pumps operate in parallel for 40% of the year and two pumps in parallel for the balance of the year. Average power costs for the station are about 6 ¢/kWh. Determine the annual power costs for this pumping operation if the pumps typically operate at 82% efficiency.

Answer: For 82% efficiency,

$$\text{hp} = \frac{\left(3600 \dfrac{\text{gal}}{\text{min}}\right)\left(8.35 \dfrac{\text{gal}}{\text{lb}_\text{m}}\right)(265 \text{ ft})}{(0.82)\left(33{,}000 \dfrac{\text{ft-lb}_\text{m}}{\text{min}} \text{ per hp}\right)}$$

$$= 294.4 \text{ hp}$$

The efficiency of a 300-hp induction motor will be roughly 93%. Accounting for another 3% losses for the transformer and switchgear, the overall electrical efficiency will be about 90%.

Hence, the number of kilowatts per pump equals

$$\frac{(294.4\ \text{hp})\left(0.746\ \dfrac{\text{kW}}{\text{hp}}\right)}{0.9} = 244\ \text{kW}$$

There are 8760 hours in a year. Hence, for three pumps in operation for 40% of the year,

$$3 \times 8760 \times 0.40 \times 244 = 2.57 \times 10^6\ \text{kWh}$$

and for two pumps in operation for 60% of the year,

$$2 \times 8760 \times 0.60 \times 244 = 2.57 \times 10^6\ \text{kWh}$$

$$\begin{aligned}\text{The total power consumption} &= 2.57 \times 10^6 + 2.57 \times 10^6\\ &= 5.14 \times 10^6\ \text{kWh/yr}\end{aligned}$$

$$\begin{aligned}\text{The annual power costs} &= 0.06\ \frac{\$}{\text{kWh}} \times 5.14 \times 10^6\ \text{kWh}\\ &= \$308{,}400\end{aligned}$$

D.3 Effect of Increased Efficiency on Pumping Costs

Repeat Problem D.2 for pump efficiencies of 83% and 85%. Are the savings significant?

D.4 Determining NPSH

A vertical turbine pump is located at 5000 ft above sea level. The pump has a capacity for 3500 gpm and operates at a maximum temperature of 110°F. The friction head and entrance losses in the suction piping amount to 0.13 ft. The water level is never less than 10 ft above the first-stage impeller. From the information provided in Tables 48 and 49, determine the NPSH.

Answer:

$$\text{Available NPSH} = H_a + H_s - H_f - H_{vp}$$

where H_a = absolute pressure on the surface of the liquid in the suction well. At 5000 ft elevation, interpolating from the first table,

Table 48. Atmospheric Pressure at Various Altitudes [124]

Altitude (ft)	Barometric Reading (in. Hg)	(mm Hg)	Atmospheric Pressure (psia)	($ft H_2O$)
−1000	31.0	788	15.2	35.2
Sea Level	29.9	760	14.7	33.9
+1000	28.9	734	14.2	32.8
2000	27.8	706	13.7	31.5
4000	25.8	655	12.7	29.2
6000	24.0	610	11.8	27.2
8000	22.2	564	10.9	25.2

Table 49. Various Properties of Water

Temperature (°F)	(°C)	Absolute Vapor Pressure (psia)	($ft H_2O$)	Specific Weight (lb/ft^3)	Specific Gravity	Absolute Viscosity (cp)
32	0.00	0.088	0.20	62.42	1.0016	1.79
40	4.44	0.122	0.28	62.43	1.0018	1.54
60	15.56	0.256	0.59	62.37	1.0008	1.12
80	26.67	0.507	1.2	62.22	0.9984	0.86
100	37.78	0.949	2.2	62.00	0.9949	0.77
120	48.89	1.693	3.9	61.71	0.9903	0.56
160	71.11	4.741	11.2	61.01	0.9790	0.40
210	98.89	14.123	33.9	59.88	0.9609	0.29

$$H_a = 28.2 \text{ ft}$$

where H_s = static elevation of the liquid above the centerline of the pump (i.e., to the entrance of the first stage impeller) = 10 ft
H_f = friction head and entrance losses in the suction piping = 0.13 ft
H_{vp} = absolute vapor pressure of fluid at the pumping temperature

From the second table, for 110°F,

$$H_{vp} = 3.0 \text{ ft}$$

Hence,

$$\text{Available NPSH} = 28.2 + 10 - 0.13 - 3.0 = 35.1 \text{ ft}$$

D.5 Determining Brake Horsepower

Crude oil at 25° API at 14°C is being pumped through a 14-in. Schedule 40 steel pipe at a rate of 2600 bbl/hr. The pipeline is 400 miles long and the discharge is 3500 ft above the pump inlet. The oil has a viscosity of 95 SSU. If the pump operates at 72% efficiency, determine the total discharge head at the pump and the brake horsepower.

D.6 Determining Brake Horsepower

Water is being pumped at a rate of 250 gpm to an elevation of 320 ft. The piping system is shown in Figure 1D. All piping is 3-in. Schedule 40 and the roughness factor is 0.092 in. Determine the pump's brake horsepower. The pump operates at 73% efficiency.

Answer: Starting with the Bernoulli equation and noting that $u_1 = u_2$, the equation can be rewritten to determine the pump head:

$$144\,\frac{\Delta P}{\rho} = \Delta Z + h_L$$

Inside diameter of pipe = 3.068 in. = 0.256 ft

Area $= \frac{1}{4}\,\pi\,(0.256)^2 = 0.0515\ \text{ft}^2$

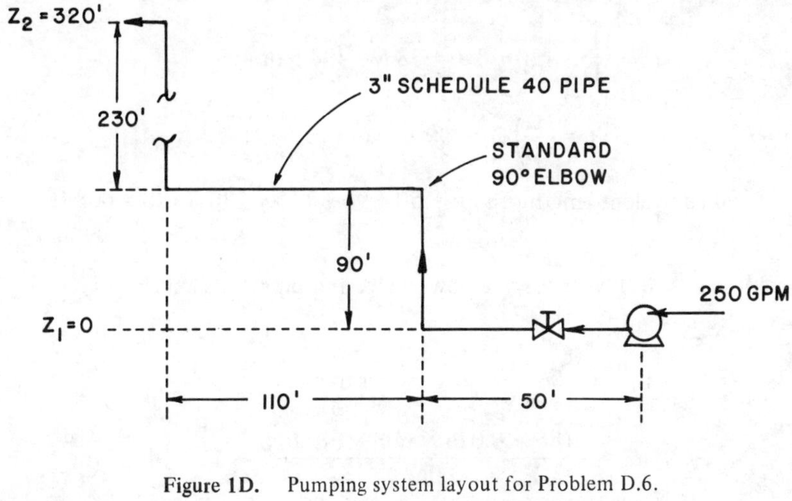

Figure 1D. Pumping system layout for Problem D.6.

$u = 250 \text{ gpm}/(7.48 \times 60 \times 0.0515) = 10.82 \text{ fps}$

$\rho = 62.3 \text{ lb/ft}^3$

$\mu = 0.95 \text{ cp} = 6.72 \times 10^{-4} \text{ lb}_m/\text{ft-s}$

$$N_{Re} = \frac{Du\rho}{\mu} = \frac{0.256 \times 10.82 \times 62.3}{6.38 \times 10^{-4}} = 2.705 \times 10^5$$

From the Moody plot, f = 0.026.

From Figure 1D, there are four 90° standard elbows:

$$L/D = 30 \times 4 = 120$$

For a gate valve (fully open),

$$L/D = 340$$

Exit losses may be approximated as K = f(L/D); K $\simeq$ 1.0. Hence,

$$\frac{L}{D} = \frac{K}{f} = \frac{1}{0.026} = 38.5$$

For elbows, valve and exit loss:

$$\frac{L}{D} = 120 + 340 + 38.5 = 498.5 \text{ or } 499$$

$$L = \left(\frac{L}{D}\right) D = 499 \times 256 \text{ ft} = 128 \text{ ft}$$

Total equivalent length of pipe = 50 + 90 + 110 + 230 + 128 = 608 ft.

Head loss due to flow through elbows, valve and pipe, plus exit losses:

$$h_L = 0.186 \frac{fLu^2}{d}, \text{ where d is in inches.}$$

$$h_L = \frac{0.186 \times 0.026 \times 608 \times (10.8)^2}{3.068} = 112.2 \text{ ft}$$

Total Pump Head $H_t = 320 - 0 + 112.2 = 432.2$ ft.

$$\text{Brake hp} = \frac{QH_t\rho}{247{,}000\eta}, \text{ where } \eta \text{ is pump efficiency}$$

$$= \frac{250 \times 432 \times 62.3}{247{,}000 \times 0.73} = 37.3 \text{ hp}$$

D.7 Storage Tank Pumping Problem

A solution of specific gravity 1.5 is pumped from a storage tank through a 4-in. Schedule 40 steel pipe. The velocity in the suction line is 6 fps and the distance between the pump and storage tank is 20 ft. The pump discharges through a 3-in. Schedule 40 cast iron pipe to an overhead tank. The end of the discharge pipe is 70 ft above the level of the solution in the storage tank.

1. Determine the pressure that the pump must develop.
2. Determine the pump's brake horsepower.

D.8 Pumping from a Reservoir

Carbon tetrachloride at 65°F is being pumped at a rate of 55 gpm from a holding tank. The tank is open to the atmosphere. The gauge pressure at the end of the discharge line is 65 psig. The discharge is 30 ft and the pump suction is 6 ft above the level in the holding tank. The discharge line is 2-in. Schedule 40 stainless steel pipe. A globe valve (fully open) is located immediately after the pump. The pump efficiency is 63%. There are four 90° standard elbows in the line. Determine the following:

1. the developed head of the pump,
2. the total power input, and
3. the net positive suction head (NPSH).

D.9 Effect of Impeller Speed on Flow

A centrifugal pump delivers 300 gpm of water when its impeller speed is 1800 rpm and the pressure drop across the pump is 27 psi. Determine the pump's capacity and the developed head in the pump if the speed is reduced to 1250 rpm. Assume that the pump operation is ideal.

D.10 Discussion Problem

Define the purpose of a mechanical seal. What are the additional advantages of a double seal? What are the disadvantages?

E PROBLEMS ON NON-NEWTONIAN FLOWS

E.1 Flow Curve Determination for a Non-Newtonian Fluid

The rheological properties of an aqueous suspension of paper pulp were studied using the capillary tube viscometer technique. Flow measurements were made at 24°C using three different tube diameters. The density of the fluid was 1.3 g/cm^3. The data obtained are given in Table 50. Construct a flow curve from these data.

E.2 Parameters Determination for a Non-Newtonian

From the flow curve determined for Problem E.1 (the plot of τ_w vs 8U/D), obtain values for the properties parameters K′ and n′ in Equation 213. (Hint: the flow curve plot for the data in Problem E.1 may be nonlinear, indicating that the material is not a power law fluid over the range of shear stress tested. Use of tangent-drawing procedures may be necessary to obtain K′ and n′: tangents should be drawn to the curve at various 8U/D to obtain corresponding values of n′ from the tangent slope, and K′ from the tangent intercept at 8U/D equal to unity.)

E.3 Calculating Capillary Tube Diameter from Flow Measurements

Measurements of the rate of flow of a viscous fluid through a capillary tube can be used to determine the diameter of the tube. Given the following data, determine the diameter of the capillary.

length of capillary tube = 620 mm
kinematic viscosity $\simeq 15 \times 10^{-5}$ m^2/s
specific gravity of fluid = 0.93

Table 50. Capillary Tube Viscometer Technique Data

Tube i.d. (cm)	Tube Length (cm)	Tube Length to Diameter Ratio	Mass Flowrate (g/s)	Frictional Pressure Drop (kPa)
0.15	14	93	0.2	3200
0.15	14	93	0.02	1200
0.30	28	93	0.46	1950
0.30	28	93	0.10	860
0.40	28	70	1.2	1410

mass rate of flow through the tube = 5.4×10^{-3} kg/s
pressure drop across the capillary tube = 6.5×10^{5} N/m^2

Assume the capillary is oriented horizontally.

E.4 Non-Newtonian Film Flow

Derive an expression for the thickness of a film of non-Newtonian material flowing down a vertical plate in terms of its mass rate per unit width of wall:

1. Assume the material to be a Bingham plastic.
2. Assume the material to be a power law fluid.

E.5 Derivation of Ellis Tube Flow Equation

For fluids with a yield stress flowing through cylindrical tubes, the general expression for the local value of linear velocity u at radius r in the unsheared portion between the solid plug and tube wall is

$$u = \int_0^u du = \int_R^r \left(\frac{du}{dr}\right) dr = \int_r^R \left(-\frac{du}{dr}\right) dr \ , \ \text{for } r_P \leqslant r \leqslant R \ .$$

Assuming the material follows an Ellis model fluid, derive an expression for the velocity distribution for laminar flow. Ignore any slip between the fluid and the wall.

Answer:

$$u = \frac{A\Delta P}{4L}(R^2 - r^2) + \frac{B}{\alpha + 1}\left(\frac{\Delta P}{2L}\right)^{\alpha} (R^{\alpha+1} - r^{\alpha+1})$$

where A is in $ft^2\ s^{-1}\ lb_f^{-1}$, B in $ft^{2\alpha}\ s^{-1}\ lb_f^{-\alpha}$, and α is dimensionless.

E.6 Laminar Flow Between Flat Plates

Consider the steady flow of fluid between stationary, flat parallel plates separated by some distance, δ, as shown in Figure 1E. Develop an expression for the shear stress distribution and comment.

Answer: From a force balance,

$$(\tau_{hx})_{h=\delta/2} = \tau_w = \frac{\delta \Delta P}{2L}$$

$$\tau_{hx} = \frac{h\Delta P}{L}$$

$$\tau_{hx} = \tau_w \frac{h}{\delta/2}$$

The expression indicates a linear stress distribution from zero at the midplane to τ_w at the plate surface, regardless of the flow regime.

E.7 Velocity Distribution for Flow Between Flat Plates

For fluids having a yield stress, the general expression for the local value of the linear velocity u at distance h from the midplane in the sheared portion between the solid plug and plate surface is

$$u = \int_0^u du = \int_{\delta/2}^h \left(\frac{du}{dh}\right) dh = \int_h^{\delta/2} \left(-\frac{du}{dh}\right) dh\ , \quad \text{for } h_P \leqslant h \leqslant \delta/2$$

where δ and h are defined in Figure 1E.

Develop an expression for the velocity distribution for laminar flow between plates assuming a power law fluid.

Answer:

$$u = \left(\frac{\Delta P g_c}{KL}\right)^{1/n} \frac{n}{n+1} \left[\left(\frac{\delta}{2}\right)^{(n+1)/n} - h^{(n+1)/n}\right]$$

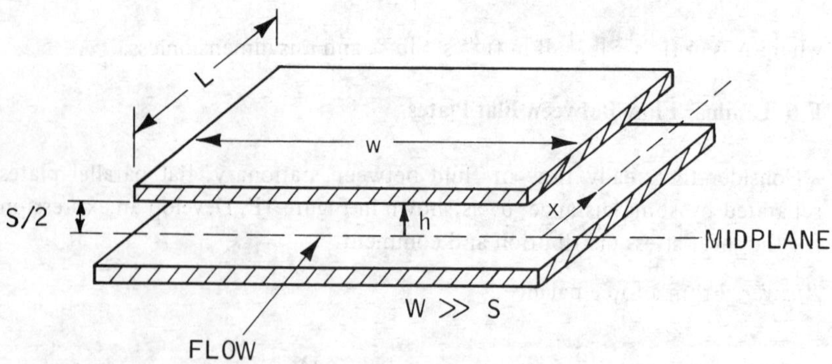

Figure 1E. Steady flow between plates.

E.8 Derivation of Couette Flow Expression

Consider the parallel plate system in Figure 1E. If the bottom plate is stationary and the top plate moves at some steady velocity, u_x, in the x-direction, develop an expression for the velocity profile over h for the fluid between the two plates. Assume that the fluid between the plates is a power law fluid.

Answer: A steady-state force balance on the fluid element dwdhdx gives

$$\frac{d\tau_{yx}}{dh} = \frac{dP}{dx}$$

where h = 0 at the bottom plate. For a power law fluid,

$$\frac{K}{g_c}\frac{d}{dy}\left(\frac{du}{dy}\right)^n = -\frac{dP}{dx}$$

and, on integration,

$$u = \frac{\left[\frac{g_c}{K}\frac{dP}{dx}h + C\right]^{(n+1)/n} - C^{(n+1)/n}}{\left(\frac{n+1}{n}\right)\frac{g_c}{K}\frac{dP}{dx}}$$

where C is a constant of integration.

E.9 Flow Through a Rectangular Conduit

A power law fluid is flowing steadily through a rectangular cross-sectional conduit that is 6 m long. The dimensions of the cross section are 7.6 cm × 4.5 cm. Determine the fluids volumetric flowrate if the maximum frictional pressure drop per unit length of pipe is 2200 N/m. The geometric term Γ is 5.9 (value for Γ from Metzer [75], Section 7, p. 28).

E.10 Pumping a Power Law Fluid

A power law fluid is being pumped at a mass rate of 600 kg/hr from a large storage vessel to an elevated receiving tank. A sketch of the pipe network and line dimensions are given in Figure 2E. The specific gravity of the fluid is 1.2

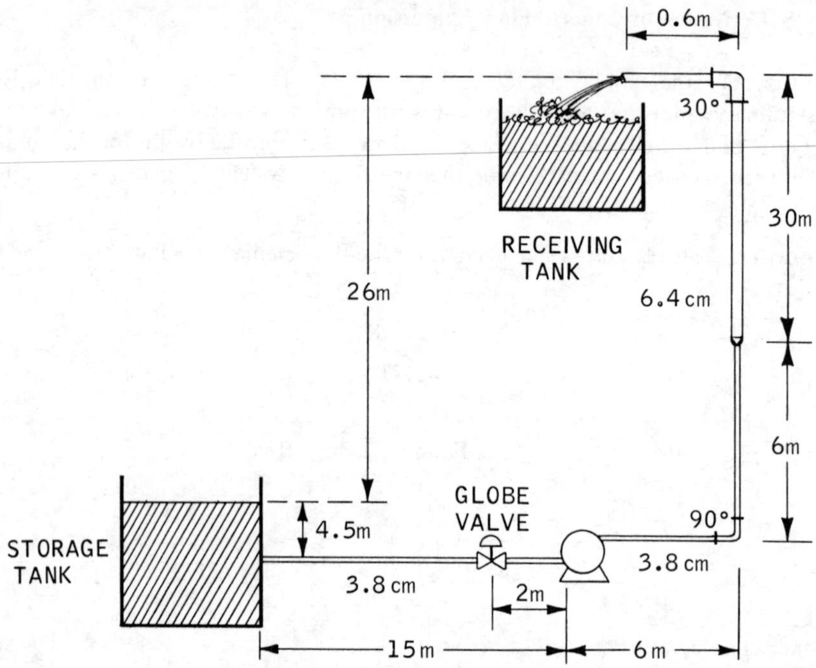

Figure 2E. Schematic of power law fluid pumping problem.

and the rheological parameters of the fluid at the process temperature are as follows:

$$\text{Fluid consistency index } K = 150 \text{ N/m-s}^{n-2}$$

$$\text{Flow behavior index } n = 0.52$$

1. Compute the horsepower required by the pump if the overall pump efficiency is 60%.
2. Compute the average fluid velocity in the two sizes of pipe.
3. Compute the centerline velocity and the velocity at a distance of 0.5R from the axis for both tube sizes.
4. Check the applicability of the rheological parameters used in obtaining solutions to the above.

E.11 Flow of Bingham Plastic Fluids

A Bingham plastic is flowing through a 20-cm-diameter pipe at a rate of 60 kg/s. The specific gravity of the fluid is 1.3 and the rheological parameters are as follows:

$$\text{Yield Stress } \tau_y = 24 \text{ N/m}^2$$

$$\text{Plastic Viscosity } \eta = 34 \text{ Pa-s}$$

1. Is the flow laminar or turbulent?
2. Determine the frictional pressure drop per unit length of pipe.
3. Determine the flowrate if only 40% of the pressure gradient in part 2 were found.
4. Check the applicability of the rheological constants used in computing the above.

(Hint: Standard friction factor correlations may be used with the modified Reynolds number.)

E.12 Film Flow of a Bingham Plastic Fluid

A particular brand of paint is found to be approximated by a Bingham plastic. Develop expressions for (1) the film thickness, and (2) the fluid velocity profile, in terms of the mass flowrate and physical properties when the paint is applied to a vertical wall. Assume that the painter applies steady strokes with his brush such that the flow may be approximated as laminar film flow down a vertical flat surface.

E.13 Film Flow of a Power Law Fluid

Repeat problem E.12 for a power law fluid paint. Also, derive expressions for painting a surface inclined at some angle, θ, from the horizontal.

E.14 Pumping a Bingham Plastic

Repeat Problem E.10 for a Bingham plastic material having the following rheological parameters:

$$\tau_y = 4.2 \text{ mPa} \quad , \quad \eta = 1.1 \text{ Pa-s}$$

E.15 Partial Pipe Flow of a Non-Newtonian Fluid

A power law fluid is freely flowing down an inclined pipeline such that the pipe is only partially full. Starting with the overall energy balance, develop an expression that will enable the free surface profile (i.e., depth) to be computed. Assume a mean velocity across the fluid depth for the derivation. The system is defined in Figure 3E.

Answer: (Hint: treat the system as an open channel flow.)

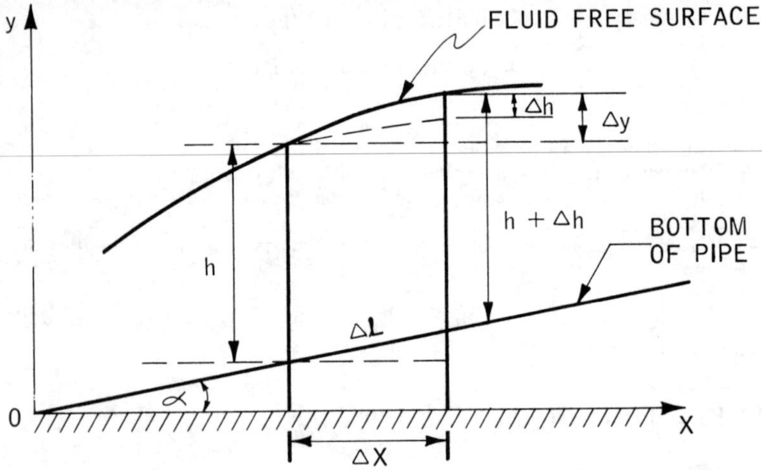

Figure 3E. System orientation for partial pipe flow problem (Problem E.15).

$$\frac{dh}{dl} = -\frac{\frac{2K}{\rho gC}\left(\frac{3n+1}{n}\right)^n u_m^n + \sin\alpha}{1 - u_m^2/\beta gh}$$

where h is the fluid depth,

$$C = [D_H/2]^{n+1} \; ,$$

where

$$D_H = \frac{4bh}{b + 2h} \quad \text{(hydraulic diameter)}$$

and

$$\beta = \frac{(2n+1)(5n+3)}{3(3n+1)^2} \quad \text{(kinetic energy correction factor for laminar flow)}$$

The expression can be integrated using a Runga-Kutta method to compute depth, h, as a function of channel length, ℓ.

E.16 Partial Pipe Flow

Repeat Problem E.15 for a Bingham plastic fluid.

E.17 Kinetic Energy Expression

Derive an expression for the average kinetic energy per unit mass in the laminar flow of a power law fluid between flat parallel plates.

E.18 Velocity Profile Derivation

Develop an expression for the velocity profile of a non-Newtonian fluid flowing through a cylindrical tube based on the Herschel-Bulkley model.

E.19 Nonisothermal Laminar Flow Derivation

The Arrhenius relationship describes the relationship between Newtonian viscosity and temperature as follows:

$$\mu = ae^{E/R_G T}$$

where
E = the activation energy of flow per mole
R_G = the gas constant per mole
T = the absolute temperature
a = is a constant characteristic of the fluid

1. Develop an expression for describing temperature effects on shear stress for a Newtonian fluid.
2. Repeat part (1) for a power law fluid.

Answer:

$$1.\quad \tau_{rx} = \frac{a}{g_c}\left(-\frac{du}{dr}\right)\exp\left(\frac{E}{R_G T}\right)$$

$$2.\quad \tau_{rx} = \frac{k_\sigma}{g_c}\left[\left(-\frac{du}{dr}\right)\exp\left(\frac{E}{R_G T}\right)\right]^n = \frac{k_\sigma}{g_c}S_r^n$$

where k_σ, E and n are constants independent of temperature, and the term S_r is referred to as the "reduced shear rate."

E.20 Laminar Flow through Sudden Expansions

Figure 4E illustrates flow through a sudden expansion in a tube. Derive an expression for the frictional pressure drop caused by the expansion for a power law fluid.

Answer: Section 2 of the expansion shown in Figure 4E is located immediately beyond the expansion. Hence, we may assume that the pressure drop between sections 1 and 2 can be neglected (i.e., $p_2 = p_1$). As shown in the figure, the velocity profile entering the expansion at section 2 will be same as the fully developed profile at section 1. The static pressure drop across the expansion may be obtained by applying the linear momentum equation for steady flow:

$$\Sigma g_c F_x = g_c(P_2 - P_3)\pi R_2^2 = -\int_0^{R_1} \rho u_2^2 2\pi r dr + \int_0^{R_2} \rho u_3^2 2\pi r dr$$

The velocity distribution for a power law fluid flowing through a tube was given in Chapter 2.10, Equation 245. Combining Equation 245 with the above expression (along with some rearranging), gives the following:

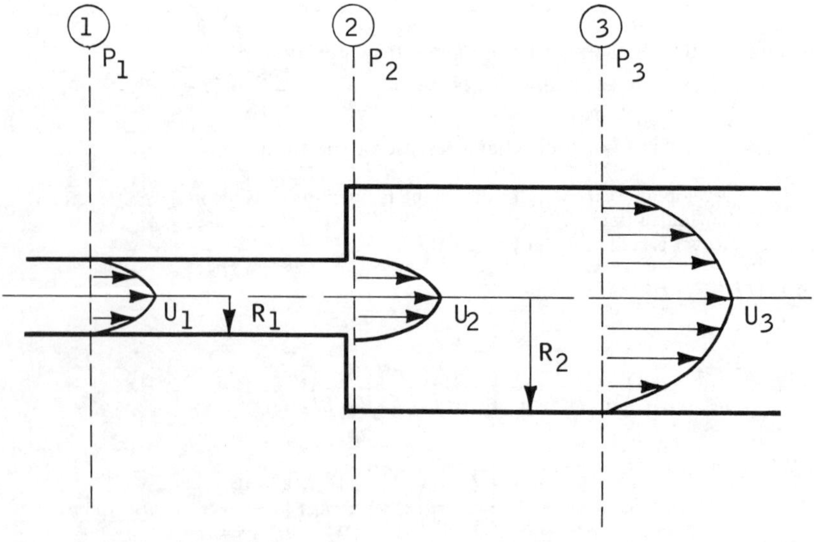

Figure 4E. Shows laminar flow through a sudden expansion in a tube.

$$\frac{\Delta P_{1\text{-}3}}{\rho} = \frac{\Delta P_{2\text{-}3}}{\rho} = \frac{1}{g_c}\left(\frac{3n+1}{2n+1}\right)\frac{Q^2}{(\pi R_1^2)^2}\left[\left(\frac{R_1}{R_2}\right)^4 - \left(\frac{R_1}{R_2}\right)^2\right]$$

F PROBLEMS ON OPEN CHANNEL FLOWS

F.1 Partially Flowing Pipe

Water is flowing in a horizontal sewer line and found to fill only two-thirds of the pipe. A cross-sectional view of the system is shown in Figure 2A. From this information, do the following:

1. Determine the volumetric flowrate in gpm.
2. Determine the pressure drop per foot of pipe.
3. Convert your answers to SI units.

The pipe material is ceramic and is only three years old.

F.2 Partial Pipe Flow in an Inclined Line

For Problem F.1 determine the volumetric flowrate if the pipeline is inclined 9° from the horizontal and the length of pipe is 1000 ft.

F.3 Flow in a Rectangular Duct

The level in a reservoir is controlled by a concrete rectangular overflow aqueduct. The aqueduct has dimensions of 30 ft high × 15 ft wide. The floor of the aqueduct is heavily deposited with debris and plant growth. The maximum height that the reservoir reached in a record rainfall is noted in Figure 1F. Determine the discharge rate in gpm and in cfs.

F.4 Computing Backwater Profile

Water is flowing through a trapezoidal cross-sectional channel as shown in Figure 2F. At some point, a depth of 9 ft is produced in the channel when the discharge is 700 gpm. The channel's bottom width $b' = 25$ ft. The channel has equal side slopes defined by the angle in Figure 2F. The channel surface is relatively smooth and a Manning exponential coefficient of 0.012 may be used. The channel's slope is 0.003. Determine the resulting water surface profile and its total length if the limit occurs when the depth equals $(h_0 + 0.1)$ ft.

Notes on Answer:

This involves a lengthy solution by hand calculations. An outline of the solution is given below and the student may perform the actual calculations by writing a short computer program:

Step 1. Determine h_0 and h_c.

Step 2. Classify the type profile.

Step 3. Divide the change of depth between h_1 (9 ft) and (h_0 + 0.1) into convenient increments.

Step 4. Tabulate hydraulic properties for each section.

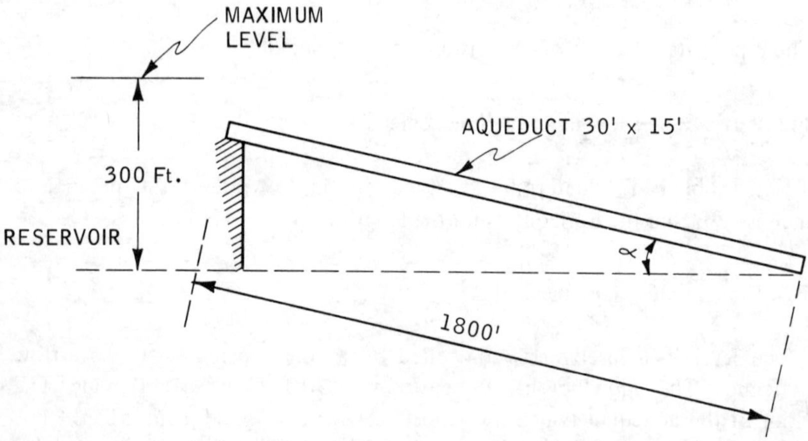

Figure 1F. Overflow from a reservoir.

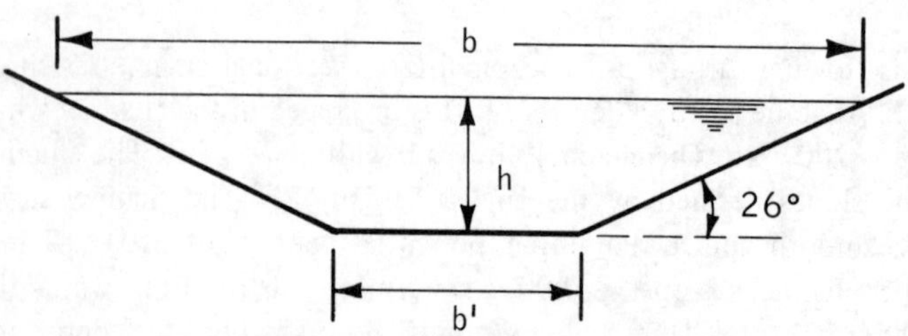

Figure 2F. Trapezoidal channel cross section for Problem F.4.

Step 5. Compute mean hydraulic depth, h_{D_m}, and the slope of the total energy line between adjacent sections.

Step 6. Compute Δx values between adjacent sections (where x is the distance along the channel bed).

Step 7. Sum the Δx values to obtain the distance of each section from h_1 and the total length of the profile.

Determination of normal and critical depths will require a trial and error solution. Before performing the calculation procedure, it is convenient to express the hydraulic properties of the cross section in terms of the single variable, depth (h).

Surface Width b $= 25 + 2h \tan(90\text{-}26^\circ)$

$= 25 + 4.1h$

Cross-Sectional Area $= 25h + 2\left(\frac{1}{2}(2.05h)h\right)$

$= h(25 + 2.05h)$

Wetted Perimeter $= 25 + 2(h/\sin 26^\circ)$

$= 25 + 4.56h$

Hydraulic Mean Depth $h_{D_m} = \dfrac{\text{Cross-Sectional Area}}{\text{Wetted Perimeter}}$

$= \dfrac{h(25 + 2.05h)}{(25 + 4.56h)}$

F.5 Reservoir Water Consumption

A reservoir has a maximum capacity of 53 million gallons (mmgal) and its average depth is approximately 30 ft. Monthly data on reservoir capacity and average rainfall/snow accumulations are given in Table 51. Prepare a hydrograph of the monthly consumption and determine the daily average and average annual consumptions.

F.6 Channel Flow of a Non-Newtonian Fluid

Repeat Problem E.15 assuming the fluid to be described by a Bingham plastic model.

Table 51. Monthly Data on Reservoir Capacity and Average Rainfall/Snow Accumulations

Month	Reservoir Capacity (MMgal)	Average Rainfall Accumulation (in.)
January	43	0
February	43	11
March	45	12.5
April	45.5	3.1
May	39	0
June	41	3
July	39	0.5
August	35	0
September	29	0.2
October	37	3.2
November	46	3.7
December	48	7.1

F.7 Discussion Problem

Describe the physical characteristics of a hydraulic jump. What are the different types of hydraulic jumps?

F.8 Effects of Errors in Values of the Hydraulic Depth and Slope

Determine the extent of error that can be introduced to the velocity calculation from the Chezy formula if x% error is introduced in the value of h_D (the mean hydraulic depth) or s (the channel slope).

Answer: The Chezy formula is

$$u = c\sqrt{h_D S}$$

whence

$$\frac{du}{dh} = \frac{c}{2}\sqrt{\frac{S}{h_D}}$$

or

$$du = c/2\sqrt{S/h_D}\cdot dh$$

Substituting for c,

$$\frac{du}{u} = \frac{1}{2} \cdot \frac{dh_D}{h_D}$$

From this expression it can be seen that an error of x% in the value of h_D will produce an error of ½x% in the computed velocity. A similar analysis results in the same error for the bed slope, S.

F.9 Water Flow Over a Weir

A rectangular weir is used to measure water flow from a reservoir. The width of the weir is 11 ft and the height of the water surface over the weir is 4 ft. The approach area is 86 ft^2.

1. Compute the mass flowrate assuming the velocity of approach to be negligible.
2. Determine the approach velocity head. Was the assumption used in part (1) valid?

F.10 Submerged Efflux Under a Sluice Gate

Figure 3F illustrates a submerged efflux sluice gate. Determine the flowrate when h = 4 in., h_1 = 3.5 ft, h_3 = 22 in., and the width of the sluice L = 3 ft. Convert your answer to SI.

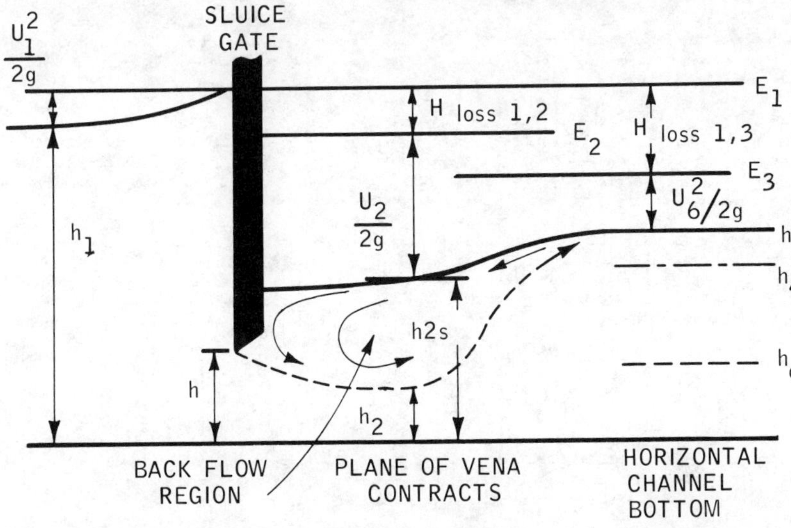

Figure 3F. Submerged efflux under a sluice gate. Refer to Problem F.10.

NOMENCLATURE FOR SECTION 2

Chapter	Symbol	Definition	English Units	SI Units
2.1	a, b	Characteristic length defined in Table 2	ft	m
	c	Speed of light	ft/s	m/s
	C_p	Specific heat at constant pressure	Btu/(lb)(°F)	J/(kg-°K)
	d	Diameter defined in Table 2	ft	m
	D	Diameter or characteristic length	ft	m
	D_H	Equivalent diameter	ft	m
	D_ν	Diffusivity	ft^2/hr	m^2/s
	E_M	Mechanical energy	ft-lb_f/lb_m	N-m/kg
	f	Friction factor	–	–
	F	Frictional head loss	ft-lb_f/lb_m	Pa-m^3/kg
	g	Acceleration of gravity	32.2 ft/s^2	9.81 m/s^2
	g_c	Conversion factor	32.174 ft-lb_m/(lb_f-s^2)	1
	G	Mass velocity	lb/(hr-ft^2)	kg/(s-m^2)
	h	Heat transfer coefficient	Btu/(hr)(ft^2)(°F)	W/(m^2-°K)
	I_H	Heat transfer factor	–	–
	j_M	Mass transfer factor	–	–
	k	Parameter defined in Table 2	–	–
	k	Thermal conductivity	Btu-ft/(hr)(ft^2)(°F)	(J-m)/(s-m^2-°C)
	k_c	Mass transfer coefficient	lb-moles/(hr)(ft^2)(Δconc)	g-moles/(s)(cm^2)(Δconc)
	L	Pipe length	ft	m
	m	Mass	lb_m	kg
	N_{Br}	Brinkman number	–	–

Nomenclature for Section 2, continued

Chapter	Symbol	Definition	English Units	SI Units
	N_{Eu}	Euler number	–	–
	N_{Fo}	Fourier number	–	–
	N_{Fr}	Froude number	–	–
	N_{Gr}	Grashof number	–	–
	N_{Gz}	Graetz number	–	–
	N_{Kn}	Knudsen number	–	–
	N_{Lc}	Lewis number	–	–
	N_{Ma}	Mach number	–	–
	N_{Nu}	Nusselt number	–	–
	N_{Pe}	Péclet number	–	–
	N_{Pr}	Prandtl number	–	–
	N_{Re}	Reynolds number, see Equation 2	–	–
	N_{Sc}	Schmidt number	–	–
	N_{Sh}	Sherwood number	–	–
	N_{St}	Stanton number	–	–
	N_{We}	Weber number	–	–
	P	Pressure	lb_f/ft^2	Pa
	R_H	Hydraulic radius, see Equation 4	ft	m
	t	Time	s, hr	s, hr
	T	Temperature	°F, °R	°C, °K
	u	Fluid velocity in x-direction	ft/s	m/s
	U	Average fluid velocity	ft/s	m/s
	v	Fluid velocity in y-direction	ft/s	m/s
	w	Fluid velocity in z-direction	ft/s	m/s
	x	Abscissa of coordinate system	ft	m
	y	Ordinate of coordinate system	ft	m
	z	Elevation	ft	m
	α	Kinetic energy correction term, see Equation 1	–	–
	ϵ	Roughness	ft	m
	λ_m	Molecular mean free path	ft, in.	m
	μ	Viscosity	$lb_m/(ft\text{-}s)$	cp
	ν	Kinematic viscosity, μ/ρ	ft^2/s	m^2/s

Nomenclature for Section 2, continued

Chapter	Symbol	Definition	English Units	SI Units
	ρ	Density	lb/ft^3	kg/m^3
	σ	Surface tension	lb_f/ft	N/m^2
	τ	Shear stress	lb_f/ft^2	Pa
	Ω	Potential energy	ft/lb_m	m/kg
2.2 (new symbols introduced)	C	Coefficient of roughness, see Equation 46	–	–
	d, d_H	Diameter and hydraulic diameter	in.	cm
	D_J	Jacobian matrix of flow equations	–	–
	F_ℓ	Head loss term	$ft\text{-}lb_f/lb_m$	$Pa\text{-}m^3/kg$
	F(x)	Head loss function	–	–
	h_f, h_L	Frictional head loss	ft	m
	H_n	System of equations for computing head losses	–	–
	K	Resistance coefficient, see Figures 3-5	–	–
	L_{eq}	Equivalent length of fittings	ft	m
	m	Exponential term in Newton-Raphson method	–	–
	n	Exponential term in Hardy-Cross method, see Equation 53	–	–
	$(\Delta P)_e$	Pressure drop due to elevation change	$lb_f/in.^2$	Pa
	$(\Delta P)_f$	Frictional pressure drop	$lb_f/in.^2$	Pa
	$(\Delta P)_k$	Pressure drop due to kinetic energy change	$lb_f/in.^2$	Pa
	$(\Delta P)_t$	Total pressure drop	$lb_f/in.^2$	Pa
	Q	Volumetric flowrate	gpm or ft^3/s	m^3/s
	S	Slope of energy line, h_L/L, see Equation 46	–	–
	W	Mass flowrate	klb_m/hr	kg/s
	X	Variable	–	–
	β	Flow correction factor, see Table 8	–	–
2.3 (new symbols introduced)	A	Flow area	ft^2	m^2
	A_B	Area of front face of float	ft^2	m^2
	C, C_v	Flow or discharge coefficient	–	–
	C′	Discharge coefficient for rotameters, see Equation 84	–	–

Nomenclature for Section 2, continued

Chapter	Symbol	Definition	English Units	SI Units
	C_D	Drag coefficient	–	–
	ΔH	Total head of flowing fluid	ft	m
	I	Current	amps(A)	kA
	k	Discharge coefficient for orifice meter	–	–
	M	Velocity of approach factor for venturis, see Equation 74	–	–
	n	Defined in Equation 86	–	–
	N	Number of point traverses, see Equation 86	–	–
	S_g	Specific gravity	–	–
	T_a	Absolute temperature	°R	°K
	ΔT	Temperature difference	°F	°C
	u_m	Mean fluid velocity	ft/s	m/s
	u_0	Point velocity	ft/s	m/s
	$\bar{u}$	Mean local velocity of fluid	ft/s	m/s
	Y	Expansion factor, see Equations 76, 79, 80	–	–
	β	Ratio of throat diameter to inlet diameter	–	–
	γ	Ratio of specific heat at constant pressure to specific heat at constant volume, C_p/C_v	–	–
	η	Superheat correction factor for steam flow, see Equation 69	–	–
	ν_2	Specific volume	ft^3/lb_m	m^3/kg
	$\bar{\nu}$	Flow quantity for gases, see Equation 67	ft^3/hr	m^3/s
	ξ	Parameter related to rotameter tube taper	–	–
	ρ_B	Density of bob or float, see Equation 82	lb/ft^3	kg/m^3
	ρ_f	Fluid density	lb/ft^3	kg/m^3
	ϕ	Volume of float, see Equation 82	ft^3	m^3
	Ω_w	Resistance	ohms(Ω)	Ω

Nomenclature for Section 2, continued

Chapter	Symbol	Definition	English Units	SI Units
2.4 (new symbols introduced)	E_P	Pump energy	ft-lb_f/lb_m	N-m/kg
	H_f	Friction head loss	ft	m
2.5 (new symbols introduced)	BHP	Break horsepower, see Equation 104	hp	W
	C	Velocity of sound in liquid	ft/s	m/s
	D	Pump impeller diameter	in.	cm
	H_{wh}	Maximum instantaneous head increase due to water hammer, see Equation 100	ft	m
	N	Rotational speed of pump	rpm	s^{-1}
	N_s	Pump specific speed	fps	m/s
	NPSH	Net positive suction head	psia	Pa
	NPSHA	Net positive suction head available	psia	Pa
	S	Static head	ft	m
	USGPM	U.S. gallons per minute	–	–
2.6 (new symbols introduced)	$b_k(\varphi)$	Piston acceleration or delay	–	–
	C	Parameter dependent on compressibility of liquid	–	–
	d_v, D_v	Dimensions defined in Figure 97	in.	mm
	F_K	Cross section of piston	ft^2	m^2
	F_R	Cross section of piping	ft^2	m^2
	h	Stroke length	in.	cm
	h_0	Limiting stroke length	in.	cm
	h_{100}	Full stroke length (100% efficiency)	in.	cm
	h_{Pk}	Geodetic extension of pump's working space	in.	cm
	n	Stroke frequency	cycles per sec	Hz
	P_F	Pressure due to resilience	$lb_f/in.^2$	bar
	ΔP_E	Entry pressure losses	$lb_f/in.^2$	bar
	ΔP_{E2M}	Mass pressure during flow phase	$lb_f/in.^2$	bar
	ΔP_{E2r}	Frictional pressure loss during flow phase	$lb_f/in.^2$	bar
	Q_{mth}	Mean theoretical flowrate	ft^3/s	m^3/s

Nomenclature for Section 2, continued

Chapter	Symbol	Definition	English Units	SI Units
	$V_k(\varphi)$	Piston speed	ft/s	m/s
	ΔV	Change in fluid volume due to compressibility	ft^3	m^3
	η_E	Elasticity level (volumetric efficiency-%)	–	–
	η_F	Volumetric efficiency, %	–	–
	η_G	Quality level (volumetric efficiency-%)	–	–
	λ	Rod ratio	–	–
	ξ	Drag coefficient	–	–
	φ	Crank angle	radians(rad)	rad
2.8 (new symbols introduced)	D_t	Optimum blade tip diameter	ft	m
	h_s	Suction lift	ft	m
	$\Delta\widetilde{H}_f$	Change in enthalpy	Btu/lb_m	J
	k_1, k_2	Constants in Equation 140	–	–
	K_n	Nondimensional specific speed, see Equation 134	–	–
	P_a	Atmospheric pressure	lb_f/ft^2	Pa
	P_c	Critical pressure	lb_f/ft^2	Pa
	P_v	Vapor pressure	lb_f/ft^2	Pa
	S_N	Suction specific speed	rev/min	s^{-1}
	S'_N	Suction specific speed defined by Equation 132	rev/min	s^{-1}
	$\overline{V}$	Volume	ft^3	m^3
	β	Fluid angle, defined in Figure 176	deg	rad
	λ	Latent heat	Btu/lb_m	J/kg
	ν	Specific volume	ft^3/lb_m	m^3/kg
	σ	Cavitation coefficient, see Figure 119	–	–
	σ_b	Blade cavitation coefficient, see Equation 130	–	–
	σ_{crit}	Critical cavitation coefficient	–	–
	σ_{Th}	Thoma cavitation coefficient	–	–
	ϕ	Parameter defined by Equation 140; $= \sqrt{\frac{1}{2} k_2/k_1}$	–	–

Nomenclature for Section 2, continued

Chapter	Symbol	Definition	English Units	SI Units
	ψ	Ratio of vapor to liquid volume, see Equation 122	–	–
	ω	Angular velocity of pump impeller	rad/s	rad/s
2.9 (new symbols introduced)	A_p	Projected area of body on plane normal to flow direction	ft^2	m^2
	A_w	Projected area of immersed body	ft^2	m^2
	$c_{1,2,3\cdots}$	Constants in Equation 194	–	–
	C	Variable for Prandtl friction factor equation; see Table 38	–	–
	C_2	Constant or coefficient, units variable	–	–
	f, f_D	Drag coefficient, analogous to friction factor	–	–
	f'	Local drag coefficient for turbulent boundary layer, see Equation 177	–	–
	F	Force	lb_f	N
	P_θ	Static pressure in boundary layer	lb_f/ft^2	Pa
	r	Radius	ft	m
	u	Local velocity	ft/s	m/s
	u_{max}	Velocity at edge of boundary layer	ft/s	m/s
	U	Time-averaged velocity of undisturbed flowing field	ft/s	m/s
	δ	Thickness of fluid film; boundary layer thickness	in.	cm
	Δ	Function defined by Equation 196	deg^{-1}	deg^{-1}
	ϵ_{ad}	Admissible height of roughness projections	in.	cm
	η	Function defined by Equation 153	–	–
	θ	Angle	deg	rad
	τ_w	Wall shear stress	lb_f/ft^2	Pa
	ϕ	Function defined by Equation 154	–	–

Nomenclature for Section 2, continued

Chapter	Symbol	Definition	English Units	SI Units
	$\phi', \phi'', \ldots$	First, second and higher derivatives of the function ϕ; $\phi(0)$, $\phi(1)$ are values of ϕ when independent variable has values of 0 and 1, respectively	–	–
	ψ	Stream function	–	–
	Ω	Force potential	ft^2/s	m^2/s
2.10 (new symbols introduced)	A	Empirical constant in Equation 242	$ft^2/s\text{-}lb_f$	$m^2/s\text{-}N$
	b	Smaller side of rectangular cross section	ft	m
	B	Empirical constant in Equation 242	$ft^{2\gamma}/s\text{-}lb_f^{-\gamma}$	$m^{2\gamma}/s\text{-}N$
	C	Parameter in Equation 250, τ_y/τ_w	–	–
	C_R	Function defined by Equation 222	–	–
	C_S	Slip coefficient, see Equation 217	–	–
	ΣF	Sum of frictional forces	lb_f	N
	K′	Variable defined in Equation 213	$lb_f\text{-}s^{n'}\ ft^{-2}$	$N\text{-}s^{n'}\ m^{-2}$
	kE	Kinetic energy per unit mass	$ft\text{-}lb_f/lb_m$	N-m/kg
	kE_t	Kinetic energy per unit time	$ft\text{-}lb_f/s$	N-m/s
	ℓ_e	Effective bob length	ft	m
	L, L′	Lengths defined in Equation 215	ft	m
	n	Flow behavior index	–	–
	n′	Variable defined in Equation 208	–	–
	n″	Slope of logarithmic plot of torque vs rotational speed	–	–
	N	Rotational speed	rps	s^{-1}
	P_a	Atmospheric pressure	lb_f/ft^2	Pa
	P_{gas}	Gas pressure in reservoir	lb_f/ft^2	Pa
	r_p	Radius of unsheared plug of fluid flowing in tube	ft	m
	R	Tube radius	ft	m
	R_b	Bob radius	ft	m

Nomenclature for Section 2, continued

Chapter	Symbol	Definition	English Units	SI Units
	R_c	Cup radius	ft	m
	U	Average fluid velocity	ft/s	m/s
	u_p	Velocity of unsheared plug of fluid flowing in tube	ft/s	m/s
	w	Greater side of rectangular cross section	ft	m
	Z	Ratio of viscometer cup to bob radius, R_c/R_b	–	–
	α	Kinetic energy correction factor defined in Equation 254	–	–
	γ	Function defined by Equation 223	–	–
	γ'	Exponential term in Equation 242	–	–
	Γ	Function defined in Equation 247	–	–
	η	Plastic viscosity	$lb_m/ft\text{-}s$	Pa-s
	μ_a	Apparent viscosity	$lb_m/ft\text{-}s$	Pa-s
	τ_y	Yield shear stress	lb_f/ft^2	Pa
	τ_{rx}, τ_{yx}	Shear stress in x-direction on surfaces normal to r and y, respectively	lb_f/ft^2	Pa
	T	Torque	$ft\text{-}lb_f$	N-m
2.11	A_c	Critical flow area	ft^2	m^2
(new symbols introduced)	b	Channel width	ft	m
	c	Wave velocity of propagation, celerity	ft/s	m/s
	C	Chezy constant, see Equation 262	–	–
	C′	Friction coefficient defined in Chezy formula, Equation 258	–	–
	C_c	Coefficient of contraction	–	–
	E	Specific energy; fluid's energy for a cross section	ft	m
	ΔE	Energy loss	ft	m
	h	Depth	ft	m
	h_c	Critical depth	ft	m
	h_D	Mean hydraulic depth, A/P	ft	m
	h_0	Normal depth, corresponding to uniform flow	ft	m

Nomenclature for Section 2, continued

Chapter	Symbol	Definition	English Units	SI Units
	h_{mc}	Mean critical flow depth	ft	m
	H	Total or average head	ft	m
	j	Slope of total energy line, see Equation 308	–	–
	K	Resistance coefficient	–	–
	K′	Conveyance, $1/\epsilon\ Ah_D^{2/3}$	$ft^{5/3}$	$m^{5/3}$
	L	Channel length or weir length	ft	m
	p	Length of wetted perimeter	ft	m
	q	Discharge per unit width	gpm/ft	$m^3/s/m$
	S	Slope of fluid level	–	–
	S_c	Critical slope	–	–
	w	Width of flume throat	ft	m
	W	Specific weight of fluid	lb_m/ft^3	kg/m^3
	X	Distance along channel axis of flow	ft	m
	θ	Angle	deg	rad
Subscripts	a	Absolute or apparent		
	b	Bob or blade		
	c	Cup		
	crit	Critical condition		
	e	Elevation		
	eq	Equivalent		
	f	Fluid or frictional		
	J	Jacobian		
	k	Kinetic energy		
	ℓ, L	Liquid		
	m	Mean value		
	max	Maximum		
	n	Number of iteration or point value		
	o	Point condition		
	p	Projected		
	t	Total		
	v	Vapor		
	w	Wetted, wall, wire		
	wh	Water hammer		
	1,2	Upstream, downstream		

REFERENCES FOR SECTION 2

1. Colebrook, C. F., and D. White. *J. Inst. Civil Eng. (London)* 11:133 (1938-1939).
2. Moody, LF. "Friction Factors for Pipe Flow," *Trans. ASME* 66 (1944).
3. Hutchinson, F. W. "Friction Losses in Round Aluminum Ducts," *Trans. ASHVE* 59:127-138.
4. Knudsen, J. G., and D. L. Katz. *Fluid Dynamics and Heat Transfer* (New York: McGraw-Hill Book Co., 1958).
5. Bird, R. B., W. E. Stewart and E. N. Lightfoot. *Transport Phenomena* (New York: John Wiley & Sons, Inc., 1960).
6. *Chem. Eng.* 75(13):198-199 (1968).
7. Crane Co. "Flow of Fluids Through Valves, Fittings and Pipe," Tech. Paper No. 410 (1970).
8. Simpson, L. L. "Process Piping: Functional Design," *Chem. Eng.* 76(8):167-181 (1969).
9. Cheremisinoff, N. P. *Applied Fluid Flow Measurement* (New York: Marcel Dekker, Inc., 1979).
10. *Fluid Meters: Their Theory and Application*, 5th ed., (New York: American Society of Mechanical Engineers, 1959).
11. *Flowmeter Computation Handbook* (New York: American Society of Mechanical Engineers, 1961).
12. Perry, R. H., and C. H. Chilton, Eds. *Chemical Engineers' Handbook*, 5th ed. (New York: McGraw-Hill Book Co., 1973).
13. Krassik, I. J., et al. *Pump Handbook* (New York: McGraw-Hill Book Co., 1976).
14. Vetter, G., and H. Fritsch. "Layout and Rating of Piping for Oscillating Displacement Pumps, Pipes, Construction of Piping, Conveyance of Piping" (1972), pp. 300-313.
15. Vetter, G., H. Fritsch and R. Lange. "Metering Within Millimeter Range Against High Pressure," paper presented at the Pump Convention, Karlsruhe, Germany, 1978.
16. Dettinger, W. W. "Performance Losses in Suction Piping of Piston Pumps," *Process Technol.* Brochure 4:1-6 (1968).
17. Vetter, G., H. Fritsch and A. Muller. "Influences on Metering Accuracy of Oscillating Displacement Pumps," *Process Eng.* 1:1-11 (1974).
18. Vetter, G., and O. Bohm. "Electronic Testing of Oscillating Displacement Pumps," *Pompes-Pumps-Pumpen* 1-6 (1968).
19. Vetter, G. "Accuracy of Piston Type Metering Pumps," *CIT* No. 4:267-272 (1963).

20. Vetter, G., and H. Fritsch. "Researching Cavitation in Oscillating Displacement Pumps," *CIT* 271-278 (1969).
21. Vetter, G. "Fundamentals and Concepts of Volumetric Metering," *VDI Reports* No. 169 (1971).
22. Dettinger, W. "Economic Efficiency of Displacement and Centrifugal Pumps," *Chem. Facilities + Processes* 3-8 (1976).
23. Vetter, G. "Volumetric Liquid Metering with Oscillating Metering Pumps," *Process Eng.* 1:1-16 (1975).
24. Korner, J. P., and M. Kopl. "Equipment for High-Pressure Processes in Chemical Engineering," *Tech. Information* 6 (1976).
25. LEWA Information D8-830.
26. Morrison, Crossland and Parry. "Fatigue Strength of Cylinders with CROSS BORES," *J. Mech. Eng. Sci.* 13 (1959).
27. Vetter, G., H. Fritsch and A. Muller. "Leakproof Oscillating Metering and Discharge Pumps for Process Operation," paper presented at the Pump Convention, Karlsruhe, Germany, 1978.
28. LEWA Information: Micro-Metering Pump, FLMk D1-220.
29. Vetter, G. DBP (German Federal Patent) No. 1.800.018.
30. Vetter, G., R. Hessenberger and E. Holzapfel. "Design and Application of Metering Pumps in the Nuclear Technology," *Nucl. Technol.* 577-589 (1969).
31. "Properties and Installation of Metering Pumps," LEWA Information No. D10-011d (1976).
32. Harvey, E. N., W. D. McElroy and A. H. Whiteley. "On Cavity Formation in Water," *J. Appl. Phys.* 18:162-172 (1947).
33. Plesset, M. S. "The Tensile Strength of Liquids," Cavitation State of Knowledge, *Trans. ASME* 15-25 (1969).
34. Jarman, P. "Measurements of Sonoluminescence from Pure Liquids and Some Aqueous Solutions," *Proc. Phys. Soc.* 73:628 (1959).
35. Wheeler, W. H. "Indentation of Metals by Cavitation," *Trans. ASME*, Series D, 82(N1):184-194 (1960).
36. Harrison, M. "Experimental Study of Single Bubble Cavitation Noise," *J. Acoust. Soc. Am.* 24:P776 (1952).
37. Pearsall, I. S. *Cavitation*, M&B Monographs-Mechanical Engineering, Mills & Boon Ltd., London (1972).
38. Holl, J. W., and G. F. Wislicenus. "Scale Effects on Cavitation," *J. Bas. Eng. Trans. ASME*, Series D, 83:385-398 (1961).
39. Kermeen, R. W., J. T. McGraw and B. R. Parkin. "Mechanism of Cavitation Inception and Related Scale Effects Problem," *Trans. ASME* 77:533-541 (1955).
40. Calehuff, G. L., and G. F. Wislicenus. "ORL Investigation of Scale Effects on Hydrofoil Cavitation," Pennsylvania State University ORL Report TM 19-4212-03 (February 1956).
41. Stepanoff, A. J. *Pumps and Blowers* (New York: John Wiley & Sons, Inc., 1965).
42. Ward, T., and M. Sutton. "A Review of the Literature on Cavitation in Centrifugal Pumps with Various Liquids," TN892.BHRA, Cranfield (April 1967).
43. Pearsall, I. S. "Acoustic Detection of Cavitation," Symp. Vibrations in Hydraulic Pumps and Turbines, *Inst. Mech. Eng. Proc.* 181(Pt 3A) (1966).

44. Hobbs, J. M., A. Laird and W. C. Brunton. "Laboratory Evaluation of the Vibratory Cavitation Erosion Test," NEL Report 271, National Engineering Laboratory (1967).
45. Thoma, D. "Die Kavitation bei Wasser-Turbinen," *VDI Verlag, Berlin* 65-74 (1926).
46. *Standards of the Hydraulic Institute*, 11th ed., Hydraulic Institute, New York (1965).
47. Pearsall, I. S., and G. Scobie. "Pumps for Low Suction Pressures," paper No. 2, presented at the Symposium on Pumping Problems, Inst. Chem. Eng. and Inst. Mech. Eng., Swansea, March 25 (1970).
48. Blasius, H. *Z. Math. Phys.* 56:1 (1908).
49. Howarth, L. *Proc. Roy. Soc. (London)* 164A:547 (1938).
50. von Karman, T. *Z. Angew. Math. Mech.* 1:233 (1921).
51. Prandtl, L. *NACA TN 452* (1928).
52. Pohlhausen, K. *Z. Angew. Math. Mech.* 1:252 (1921).
53. Janour, Z. *NACA TN 1316* (1951).
54. von Karman, T. *NACA TN 1092* (1946).
55. Hansen, M. *NACA TN 585* (1930).
56. Burgers, J. M. *Proc. Int. Congr. Appl. Mech.*, 1st Cong., Delft, The Netherlands (1924), p. 113.
57. van der Hegge-Zijnen, B. G. *Measurements of the Velocity Distribution in the Boundary Layer Along a Plane Surface*, Thesis, University of Delft, The Netherlands (1924).
58. Falkner, V. M. *Aircraft Eng.* 15:65 (1943).
59. Schlichting, H. *Boundary Layer Theory* (New York: McGraw-Hill Book Co., 1955).
60. Prandtl, L. *Ergeb. Aerodyn. Versuchanstalt Göttingen* 3:1 (1927).
61. Shultz-Grunow, F. *NACA TN 986* (1941).
62. Schlichting, H. *NACA TN 1218* (1949).
63. van der Hegge-Zijnen, B. G. *Verhandel. Koninkl. Akad. Wetenschap. Amsterdam Afdeel. Natuurk.* 31:499 (1928).
64. Lamb, H. *Phil. Mag.* 21:112 (1911).
65. Wieselsberger, C. *Ergeb. Aerodyn. Versuchanstalt Göttingen* 2:22 (1923).
66. Wieselsberger, C. *Ergeb. Aerodyn. Versuchanstalt Göttingen* 1:120 (1923).
67. Delany, K., and N. E. Sorenson. *NACA TN 3038* (1953).
68. Millikan, C. B. *Trans. ASME* 54:APM-3 (1932).
69. Tomotika, S. *Brit. Aeronaut. Research Comm., R&M* 1678 (1935).
70. Oseen, C. W. *Arkiv. Mat. Astron. Fysik* 6:75 (1910); 9:1 (1913).
71. Skelland, A. H. P. *Non-Newtonian Flow and Heat Transfer* (New York: John Wiley & Sons, Inc., 1967).
72. Bowen, R. L. *Chem. Eng.* 68:119-122 (August 21, 1961).
73. Severs, E. T., and J. M. Austin. *Ind. Eng. Chem.* 46:2369-2375 (1954).
74. Van Wazer, J. R., J. W. Lyons, K. Y. Kim and R. E. Colwell. *Viscosity and Flow Measurement–A Laboratory Handbook of Rheology* (New York: Wiley-Interscience, 1963).
75. Metzner, A. B. *Handbook of Fluid Dynamics*, V. L. Streeter, Ed. (New York: McGraw-Hill Book Co., 1961), Section 7.
76. Thomas, D. G. *Ind. Eng. Chem.* 55(11):22 (1963).
77. Mooney, M. *J. Rheol.* 2:210 (1931).

78. Kreiger, I. M., and S. H. Maron. *J. Appl. Phys.* 25:72 (1954).
79. Calderbank, P. H., and M. B. Moo-Young. *Trans. Inst. Chem. Eng. (London)* 37:26-33 (1959).
80. Alves, G. E., D. F. Boucher and R. L. Pigford. *Chem. Eng. Prog.* 48:385-393 (1952).
81. Dodge, D. W., and A. B. Metzner. *AIChE. J.* 5:189-204 (1959).
82. Dodge, D. W., and A. B. Metzner. *AIChE. J.* 8:143 (1962).
83. Hanks, R. W., and E. B. Christiansen. *AIChE. J.* 8:467-471 (1962).
84. Buckingham, E. *Proc. Am. Soc. Test. Mat.* 21:1154-1161 (1921).
85. Severs, E. T., and J. M. Austin. *Ind. Eng. Chem.* 46:2369 (1954).
86. Symonds, F. L., A. J. Rosenthal and E. H. Shaw. *Ind. Eng. Chem.* 47:2463 (1955).
87. Bowen, R. L. *Chem. Eng.* 243 (June 12, 1961).
88. McMillan, E. L. *Chem. Eng. Prog.* 44:537 (1948).
89. Ooyama, Y., and S. Ito. *Chem. Eng.* 14:96-100 (1950).
90. Herschel, W. H., and R. Bulkley. *Proc. Am. Soc. Test. Mat.* xxvi:621 (1926).
91. Herschel, W. H., and R. Bulkley. *Kolloid-Z.* xxxix:291 (1926).
92. Crowley, P. R., and A. S. Kitzes. *Ind. Eng. Chem.* 49:888-892 (1957).
93. Reiner, M. *Deformation, Strain, and Flow* (New York: Wiley-Interscience, 1960), p. 246.
94. Wheeler, J. A., and E. H. Wissler. *AIChE. J.* 11:207-216 (1965).
95. Schechter, R. S. *AIChE. J.* 7:445 (1961).
96. Robinson, W. D. B.Ch.Eng. Thesis, University of Delaware, Newark, DE (1958).
97. Beyer, C. E., and F. E. Towsley. *J. Colloid. Sci.* 7:236 (1952).
98. DeHaven, E. S. *Ind. Eng. Chem.* 51:813-816 (1959).
99. Bogue, D. C. *Ind. Eng. Chem.* 51:874-878 (1959).
100. Tomita, Y. *Soc. Chem. Eng. (Japan)* 23:525-529 (1959).
101. Tomita, Y. *Soc. Mech. Eng. (Japan)* 4:13 (February 1961).
102. Collins, M., and W. R. Schowalter. *AIChE J.* 9:804-809 (1963).
103. Lemmon, H. E. Ph.D. Thesis, University of Utah, Salt Lake City, UT (1963).
104. Weltmann, R. N., and T. A. Keller. *NACA TN 3889* (1957).
105. Dodge, D. W., and A. B. Metzner. *AIChE J.* 5:189-204 (1959).
106. LaNieve, H. L., III. M.S. Thesis, University of Tennessee, Knoxville, TN (1963).
107. Cheng, D. C. H., D. J. Ray and F. H. H. Valentin. *Trans. Inst. Chem. Eng. (London)* 43:T176-T186 (1965).
108. Dombrowski, N., P. Eisenklam and R. P. Fraser. *J. Inst. Fuel* 30:198, 399 (1957).
109. Head, V. P. *TAPPI* 35:260 (1952).
110. Sellin, R. H. J. *Flow in Channels* (London: Macmillan St. Martins Press, 1969).
111. Bakhmeteff, B. A. *Hydraulics of Open Channels*, Engineering Societies Monographs (New York: McGraw-Hill Book Co., 1932).
112. Woodward, S. M., and C. J. Posey. *Hydraulics of Steady Flow in Open Channels* (New York: John Wiley & Sons, Inc., 1941), pp. 94-109.
113. King, H. W., and E. F. Brater. *Handbook of Hydraulics,* 5th ed. (New York: McGraw-Hill Book Co., 1963), pp. 8-48–8-66.
114. Chow, V. T. *Open Channel Hydraulics* (New York: McGraw-Hill Book Co., 1959).

115. Henderson, F. M. *Open Channel Flows* (London: Collier, 1966).
116. Morris, H. M. *Applied Hydraulics in Engineering* (New York: Ronald Press, 1963).
117. Sherman, L. K. "The Unit Hydrograph Method," in *Hydrology*, O. E. Meinzer, Ed. (New York: Dover Publications, Inc., 1942).
118. Kuiper, E. *Water Resources Development* (London: Butterworths, 1965), pp. 94-101.
119. Gilcrest, B. R. "Flood Routing," in *Engineering Hydraulics*, H. Rouse, Ed. (New York: John Wiley & Sons, Inc., 1949).
120. Henry, H. R. "Characteristics of Sluice Gate Discharge," M.Sc. Thesis, State University of Iowa (1949), published as a discussion of paper by Albertson, M. L., et al. "Diffusion of Submerged Jets," *Proc. ASCE* 1541-1548 (December 1949).
121. *Stevens Water Resources Data Book* (Beaverton, OR: Leypold & Stevens, Inc., 1978).
122. Cheremisinoff, N. P., and P. N. Cheremisinoff. *Unit Conversions and Formulas Manual* (Ann Arbor, MI: Ann Arbor Science Publishers, Inc., 1980).
123. Cheremisinoff, N. P. *Process Level Instrumentation and Control* (New York: Marcel Dekker, Inc., 1981).
124. *Hydraulic Handbook*, 5th ed., Fairbanks Morse Pump & Electric Div., Kansas.

APPENDIX A

GENERAL CONVERSION FACTORS AND PHYSICAL PROPERTIES DATA

Data and conversion factors given in this Appendix were compiled by:

Durametallic Corp.
Kalamazoo, MI

CONTENTS

Table A1. Decimal Equivalents

Fraction	Decimal	Fraction	Decimal
1/64	0.015625	33/64	0.515625
1/32	0.03125	17/32	0.53125
3/64	0.046875	35/64	0.546875
1/16	0.0625	9/16	0.5625
5/64	0.078125	37/64	0.578125
3/32	0.09375	19/32	0.59375
7/64	0.109375	39/64	0.609375
1/8	0.125	5/8	0.625
9/64	0.140625	41/64	0.640625
5/32	0.15625	21/32	0.65625
11/64	0.171875	43/64	0.671875
3/16	0.1875	11/16	0.6875
13/64	0.203125	45/64	0.703125
7/32	0.21875	23/32	0.71875
15/64	0.234375	47/64	0.734375
1/4	0.250	3/4	0.750
17/64	0.265625	49/64	0.765625
9/32	0.28125	25/32	0.78125
19/64	0.296875	51/64	0.796875
5/16	0.3125	13/16	0.8125
21/64	0.328125	53/64	0.828125
11/32	0.34375	27/32	0.84375
23/64	0.359375	55/64	0.859375
3/8	0.375	7/8	0.875
25/64	0.390625	57/64	0.890625
13/32	0.40625	29/32	0.90625
27/64	0.421875	59/64	0.921875
7/16	0.4375	15/16	0.9375
29/64	0.453125	61/64	0.953125
15/32	0.46875	31/32	0.96875
31/64	0.484375	63/64	0.984375
1/2	0.500	1	1.000

Table A2. Areas of Circumferences of Circles

$$\text{Circumference} = 3.14159265359 \times D$$
$$\text{Area} = 0.7853981634 \times D^2$$

	0″		1″		2″		3″	
Diameter	Circumference	Area	Circumference	Area	Circumference	Area	Circumference	Area
0			3.1416	0.7854	6.2832	3.1416	9.425	7.0686
1/64	0.04909	0.000192	3.1907	0.8101	6.3323	3.1909	9.474	7.1424
1/32	0.09817	0.000767	3.2398	0.8352	6.3814	3.2405	9.523	7.2166
3/64	0.1473	0.001726	3.2889	0.8607	6.4304	3.2906	9.572	7.2912
1/16	0.1963	0.003068	3.3379	0.8866	6.4795	3.3410	9.621	7.3662
5/64	0.2454	0.004794	3.3870	0.9129	6.5286	3.3918	9.670	7.4415
3/32	0.2945	0.006903	3.4361	0.9396	6.5777	3.4430	9.719	7.5173
7/64	0.3436	0.009396	3.4852	0.9666	6.6268	3.4946	9.768	7.5934
1/8	0.3927	0.01227	3.5343	0.9940	6.6759	3.5466	9.817	7.6699
9/64	0.4419	0.01553	3.5835	1.0218	6.7250	3.5989	9.867	7.7468
5/32	0.4909	0.01917	3.6325	1.0500	6.7741	3.6516	9.916	7.8241
11/64	0.5400	0.02320	3.6816	1.0786	6.8231	3.7048	9.965	7.9017
3/16	0.5890	0.02761	3.7306	1.1075	6.8722	3.7583	10.014	7.9798
13/64	0.6381	0.03240	3.7797	1.1369	6.9213	3.8121	10.063	8.0582
7/32	0.6872	0.03758	3.8288	1.1666	6.9704	3.8664	10.112	8.1370
15/64	0.7363	0.04314	3.8779	1.1967	7.0195	3.9211	10.161	8.2162
1/4	0.7854	0.04909	3.9270	1.2272	7.0686	3.9761	10.210	8.2958
17/64	0.8345	0.05541	3.9761	1.2577	7.1177	4.0315	10.259	8.3757
9/32	0.8836	0.06213	4.0252	1.2893	7.1668	4.0873	10.308	8.4561
19/64	0.9327	0.06922	4.0743	1.3209	7.2158	4.1435	10.357	8.5368

Table A2, continued

Diameter	0″		1″		2″		3″	
	Circumference	Area	Circumference	Area	Circumference	Area	Circumference	Area
5/16	0.9817	0.07670	4.1233	1.3530	7.2649	4.2000	10.407	8.6179
21/64	1.0308	0.08456	4.1724	1.3854	7.3140	4.2570	10.456	8.6994
11/32	1.0799	0.09281	4.2215	1.4182	7.3631	4.3143	10.505	8.7813
23/64	1.1290	0.1014	4.2706	1.4513	7.4122	4.3720	10.554	8.8636
3/8	1.1781	0.1104	4.3197	1.4849	7.4613	4.4301	10.603	8.9462
25/64	1.2272	0.1198	4.3688	1.5188	7.5104	4.4886	10.652	9.0292
13/32	1.2763	0.1296	4.4179	1.5532	7.5595	4.5475	10.701	9.1126
27/64	1.3254	0.1398	4.4670	1.5879	7.6085	4.6067	10.750	9.1964
7/16	1.3744	0.1503	4.5160	1.6230	7.6576	4.6664	10.799	9.2806
29/64	1.4235	0.1613	4.5651	1.6584	7.7067	4.7264	10.848	9.3652
15/32	1.4726	0.1726	4.6142	1.6943	7.7558	4.7868	10.897	9.4501
31/64	1.5217	0.1843	4.6633	1.7305	7.8049	4.8476	10.946	9.5354
1/2	1.5708	0.1963	4.7124	1.7671	7.8540	4.9088	10.996	9.6212
33/64	1.6199	0.2088	4.7615	1.8042	7.9031	4.9703	11.045	9.7072
17/32	1.6690	0.2217	4.8106	1.8415	7.9522	5.0322	11.094	9.7937
35/64	1.7181	0.2349	4.8597	1.8793	8.0012	5.0946	11.143	9.8806
9/16	1.7671	0.2485	4.9087	1.9175	8.0503	5.1573	11.192	9.9678
37/64	1.8162	0.2625	4.9578	1.9560	8.0994	5.2203	11.241	10.0554
19/32	1.8653	0.2769	5.0069	1.9949	8.1485	5.2838	11.290	10.1435
39/64	1.9144	0.2916	5.0560	2.0342	8.1976	5.3477	11.339	10.2318

5/8	1.9635	0.3068	5.1051	2.0739	8.2467	5.4119	11.388	10.3206
41/64	2.0126	0.3223	5.1542	2.1140	8.2958	5.4765	11.437	10.4098
21/32	2.0617	0.3382	5.2033	2.1545	8.3449	5.5415	11.486	10.4994
43/64	2.1108	0.3545	5.2524	2.1953	8.3939	5.6069	11.536	10.5893
11/16	2.1598	0.3712	5.3014	2.2365	8.4430	5.6727	11.585	10.6796
45/64	2.2089	0.3883	5.3505	2.2782	8.4921	5.7388	11.634	10.7703
23/32	2.2580	0.4057	5.3996	2.3201	8.5412	5.8054	11.683	10.8614
47/64	2.3071	0.4236	5.4487	2.3623	8.5903	5.8723	11.732	10.9528
3/4	2.3562	0.4418	5.4978	2.4053	8.6394	5.9396	11.781	11.0447
49/64	2.4053	0.4604	5.5469	2.4484	8.6885	6.0073	11.830	11.1369
25/32	2.4544	0.4794	5.5960	2.4929	8.7376	6.0753	11.879	11.2295
51/64	2.5035	0.4987	5.6451	2.5359	8.7866	6.1438	11.928	11.3236
13/16	2.5525	0.5185	5.6941	2.5802	8.8357	6.2126	11.977	11.4159
53/64	2.6016	0.5386	5.7432	2.6248	8.8848	6.2819	12.026	11.5096
27/32	2.6507	0.5591	5.7923	2.6699	8.9339	6.3515	12.075	11.6038
55/64	2.6998	0.5800	5.8414	2.7153	8.9830	6.4215	12.125	11.6983
7/8	2.7489	0.6013	5.8905	2.7612	9.0321	6.4918	12.174	11.7933
57/64	2.7980	0.6230	5.9396	2.8074	9.0812	6.5626	12.223	11.8885
29/32	2.8471	0.6450	5.9887	2.8540	9.1303	6.6337	12.272	11.9842
59/64	2.8962	0.6675	6.0377	2.9010	9.1793	6.7052	12.321	12.0803
15/16	2.9452	0.6903	6.0868	2.9483	9.2284	6.7771	12.370	12.1768
61/64	2.9943	0.7135	6.1359	2.9961	9.2775	6.8494	12.419	12.2736
31/32	3.0434	0.7371	6.1850	3.0442	9.3266	6.9221	12.468	12.3708
63/64	3.0925	0.7610	6.2341	3.0927	9.3757	6.9952	12.517	12.4684

Table A2, continued

	4″		5″		6″		7″	
Diameter	Circumference	Area	Circumference	Area	Circumference	Area	Circumference	Area
0	12.566	12.566	15.708	19.635	18.850	28.274	21.991	38.485
1/64	12.615	12.665	15.757	19.753	18.889	28.422	22.040	38.656
1/32	12.665	12.763	15.806	19.881	18.948	28.570	22.089	38.829
3/64	12.714	12.863	15.855	20.005	18.997	28.718	22.138	39.002
1/16	12.763	12.962	15.904	20.129	19.046	28.866	22.187	39.175
5/64	12.812	13.062	15.953	20.253	19.095	29.015	22.237	39.348
3/32	12.861	13.162	16.002	20.378	19.144	29.165	22.286	39.522
7/64	12.910	13.263	16.052	20.503	19.193	29.315	22.335	39.696
1/8	12.959	13.364	16.101	20.629	19.242	29.465	22.384	39.871
9/64	13.008	13.465	16.150	20.755	19.291	29.615	22.433	40.046
5/32	13.057	13.567	16.199	20.881	19.340	29.766	22.482	40.222
11/64	13.106	13.669	16.248	21.008	19.390	29.917	22.531	40.398
3/16	13.155	13.772	16.297	21.135	19.439	30.069	22.580	40.574
13/64	13.205	13.875	16.346	21.263	19.488	30.221	22.629	40.750
7/32	13.254	13.978	16.395	21.391	19.537	30.374	22.678	40.927
15/64	13.303	14.082	16.444	21.519	19.586	30.526	22.727	41.105
1/4	13.352	14.186	16.493	21.648	19.635	30.680	22.777	41.282
17/64	13.401	14.291	16.542	21.777	19.684	30.833	22.826	41.461
9/32	13.450	14.396	16.592	21.906	19.733	30.987	22.875	41.639
19/64	13.499	14.501	16.641	22.036	19.782	31.141	22.924	41.818

5/16	13.548	14.607	16.690	22.166	19.831	31.296	22.973	41.997
21/64	13.597	14.713	16.739	22.297	19.880	31.451	23.022	42.177
11/32	13.646	14.819	16.788	22.428	19.929	31.607	23.071	42.357
23/64	13.695	14.926	16.837	22.559	19.979	31.763	23.120	42.537
3/8	13.744	15.033	16.886	22.691	20.028	31.919	23.169	42.718
25/64	13.794	15.141	16.935	22.823	20.077	32.076	23.218	42.899
13/32	13.843	15.249	16.984	22.955	20.126	32.233	23.267	43.081
27/64	13.892	15.357	17.033	23.088	20.175	32.390	23.317	43.263
7/16	13.941	15.466	17.082	23.221	20.224	32.548	23.366	43.445
29/64	13.990	15.575	17.131	23.355	20.273	32.706	23.415	43.628
15/32	14.039	15.684	17.181	23.489	20.322	32.865	23.464	43.811
31/64	14.088	15.794	17.230	23.623	20.371	33.024	23.513	43.995
1/2	14.137	15.904	17.279	23.758	20.420	33.183	23.562	44.179
33/64	14.186	16.015	17.328	23.893	20.469	33.343	23.611	44.363
17/32	14.235	16.126	17.377	24.029	20.519	33.503	23.660	44.548
35/64	14.284	16.237	17.426	24.165	20.568	33.663	23.709	44.733
9/16	14.334	16.349	17.475	24.301	20.617	33.824	23.758	44.918
37/64	14.383	16.461	17.524	24.438	20.666	33.985	23.807	45.104
19/32	14.432	16.574	17.573	24.575	20.715	34.147	23.856	45.290
39/64	14.481	16.687	17.622	24.713	20.764	34.309	23.906	45.477
5/8	14.530	16.800	17.671	24.850	20.813	34.472	23.955	45.664
41/64	14.579	16.914	17.721	24.989	20.862	34.634	24.004	45.851
21/32	14.628	17.028	17.770	25.127	20.911	34.798	24.053	46.039
43/64	14.677	17.142	17.819	25.266	20.960	34.961	24.102	46.227
11/16	14.726	17.257	17.868	25.406	21.009	35.125	24.151	46.415

Table A2, continued

Diameter	4″ Circumference	4″ Area	5″ Circumference	5″ Area	6″ Circumference	6″ Area	7″ Circumference	7″ Area
45/64	14.775	17.372	17.917	25.546	21.058	35.289	24.200	46.604
23/32	14.824	17.488	17.966	25.686	21.108	35.454	24.249	46.793
47/64	14.873	17.604	18.015	25.826	21.157	35.619	24.298	46.983
3/4	14.923	17.721	18.064	25.967	21.206	35.785	24.347	47.173
49/64	14.972	17.837	18.113	26.108	21.255	35.951	24.396	47.363
25/32	15.021	17.954	18.162	26.250	21.304	36.117	24.446	47.554
51/64	15.070	18.072	18.211	26.392	21.353	36.283	24.495	47.745
13/16	15.119	18.190	18.261	26.535	21.402	36.450	24.544	47.937
53/64	15.168	18.308	18.310	26.678	21.451	36.618	24.593	48.129
27/32	15.217	18.427	18.359	26.821	21.500	36.787	24.642	48.321
55/64	15.266	18.546	18.408	26.964	21.549	36.954	24.691	48.514
7/8	15.315	18.665	18.457	27.109	21.598	37.122	24.740	48.707
57/64	15.364	18.785	18.506	27.252	21.648	37.291	24.789	48.900
29/32	15.413	18.906	18.555	27.398	21.697	37.461	24.838	49.094
59/64	15.463	19.026	18.604	27.543	21.746	37.630	24.887	49.288
15/16	15.512	19.147	18.653	27.688	21.795	37.800	24.936	49.483
61/64	15.561	19.268	18.702	27.834	21.844	37.971	24.985	49.678
31/32	15.610	19.390	18.751	27.981	21.893	38.142	25.035	49.874
63/64	15.659	19.512	18.800	28.127	21.942	38.313	25.084	50.069

Table A3. Conversion Factors

To convert from ...	To ...	Multiply by ...
Atmospheres	Inches of mercury, at 0°C	29.92
Atmospheres	Pounds/square inch	14.70
Btu	Foot-pounds	778.3
Btu	Horsepower-hours	3.931×10^{-4}
Btu/hour	Foot-pounds/second	0.2162
Centimeters	Inches	0.3937
Cubic centimeters	Cubic feet	3.531×10^{-5}
Cubic centimeters	Cubic inches	0.06102
Cubic centimeters	Gallons (U.S. liquid)	2.642×10^{-4}
Cubic centimeters	Quarts (U.S. liquid)	1.057×10^{-3}
Cubic feet	Cubic centimeters	28320.0
Cubic feet	Cubic inches	1728.0
Cubic feet	Gallons (U.S. liquid)	7.48052
Cubic feet/minute	Cubic centimeters/second	472.0
Cubic feet/second	Gallons/minute	448.831
Cubic inches	Cubic centimeters	16.39
Degrees/second	Revolutions/minute	0.1667
Dynes	Kilograms	1.020×10^{-6}
Dynes	Pounds	2.248×10^{-6}
Feet of water	Inches of mercury	0.8826
Feet of water	Pounds/square inch	0.4335
Feet/minute	Centimeters/second	0.5080
Foot-pounds	Horsepower-hours	5.050×10^{-7}
Foot-pounds/minute	Foot-pounds/second	0.01667
Foot-pounds/second	Horsepower	1.818×10^{-3}
Gallons	Cubic centimeters	3785.0
Gallons	Cubic feet	0.1337
Gallons	Cubic inches	231.0
Gallons (U.S.)	Gallons (Imperial)	0.83267
Gallons/minute	Cubic feet/hour	8.0208
Horsepower	Foot-pounds/second	550.0
Inches	Centimeters	2.540
Inches of mercury	Kilograms/square centimeter	0.03453
Inches of mercury	Pounds/square inch	0.4912
Liters	Cubic inches	61.02
Liters	Gallons (U.S. liquid)	0.2642
Liters/minute	Gallons/second	4.403×10^{-3}
Millimeters	Inches	0.03937
Pounds/square inch	Atmospheres	0.06804
Pounds/square inch	Feet of water	2.307
Pounds/square inch	Inches of mercury	2.036
Pounds/square inch	Kilograms/square meter	703.1
Radians	Degrees	57.30
Radians	Minutes	3438.0
Radians	Seconds	2.063×10^{5}
Radians/second	Revolutions/minute	9.549
Radians/second	Revolutions/second	0.1592
Revolutions	Radians	6.283
Square centimeters	Square inches	0.1550
Square inches	Square centimeters	6.452
Square inches	Square millimeters	645.2
Square millimeters	Square inches	1.550×10^{-3}
Temperature (°F) −32	Temperature, °C	5/9

Table A4.

Centistokes	Seconds Redwood No. 1	Seconds Saybolt Universal

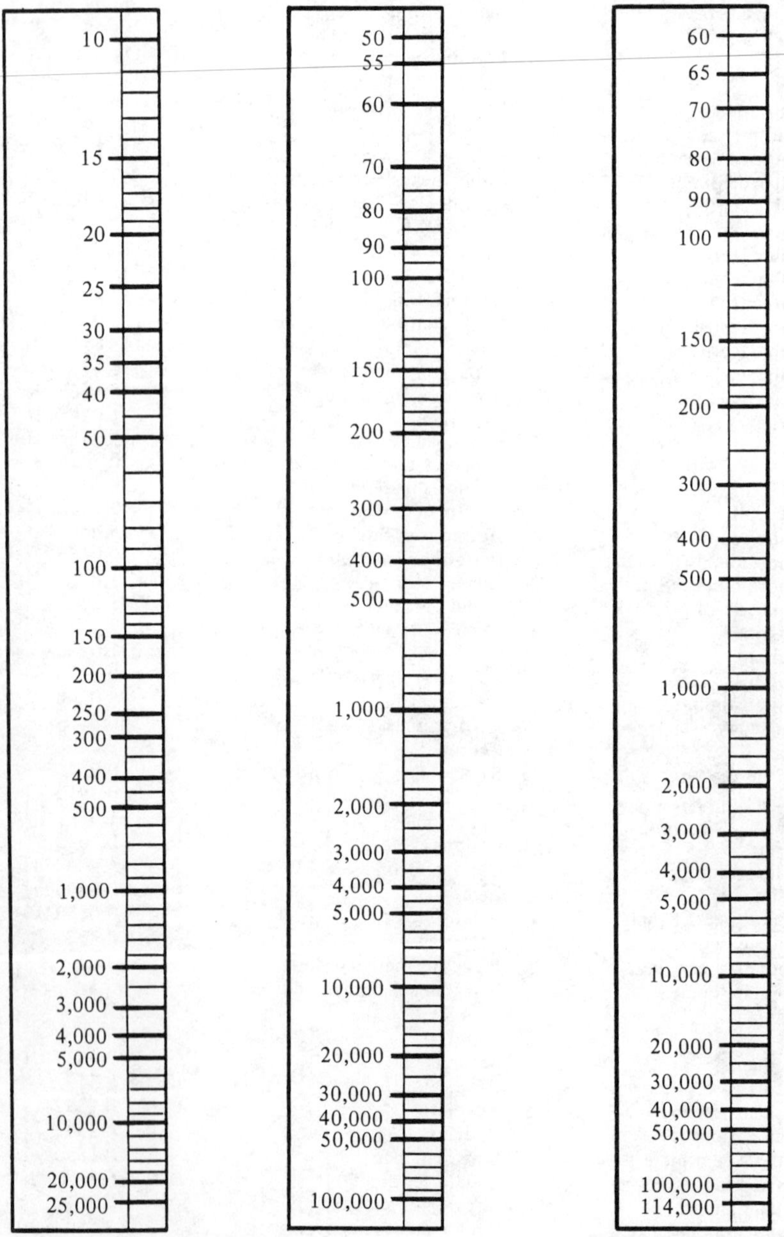

[a]Centipoise = centistokes × specific gravity.

Viscosity Conversion[a]

Seconds Saybolt Furol	Degrees Engler	Seconds Ford Cup No. 4

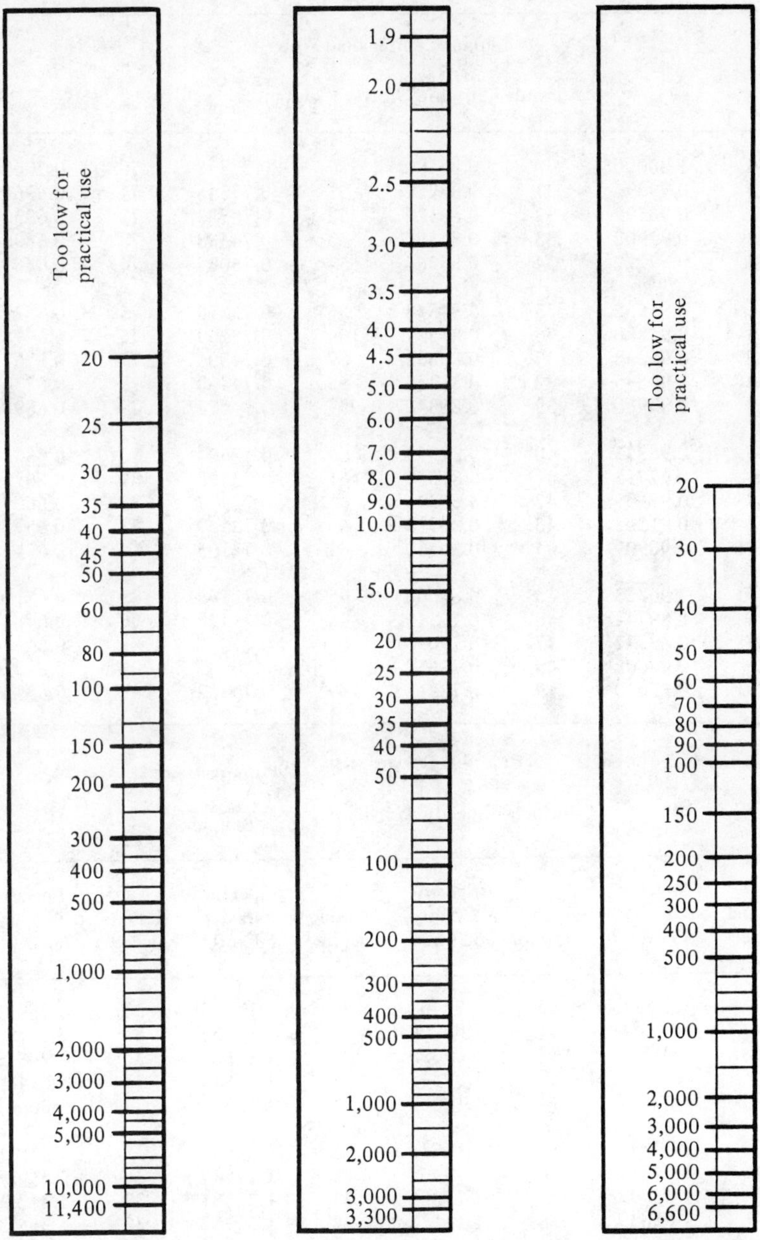

Table A5. Specific Gravity Conversion

Relationship Between Baume Degrees and Specific Gravity

Liquids Lighter than Water

$$\text{Formula–Specific Gravity} = \frac{140}{130 - {}^\circ\text{Baume}}$$

10	1.00000	30	0.87500	50	0.77778	70	0.70000
11	0.99291	31	0.86957	51	0.77348	71	0.69652
12	0.98592	32	0.86420	52	0.76923	72	0.69307
13	0.97902	33	0.85890	53	0.76503	73	0.68966
14	0.97222	34	0.85366	54	0.76087	74	0.68627
15	0.96552	35	0.84848	55	0.75676	75	0.68293
16	0.95890	36	0.84337	56	0.75269	76	0.67961
17	0.95238	37	0.83832	57	0.74866	77	0.67633
18	0.94595	38	0.83333	58	0.74468	78	0.67308
19	0.93960	39	0.82840	59	0.74074	79	0.66986
20	0.93333	40	0.82353	60	0.73684	80	0.66667
21	0.92715	41	0.81871	61	0.73298	81	0.66351
22	0.92105	42	0.81395	62	0.72917	82	0.66038
23	0.91503	43	0.80925	63	0.72539	83	0.65728
24	0.90909	44	0.80460	64	0.72165	84	0.65421
25	0.90323	45	0.80000	65	0.71795	85	0.65117
26	0.89744	46	0.79545	66	0.71428	86	0.64815
27	0.89172	47	0.79096	67	0.71066	87	0.64516
28	0.88608	48	0.78652	68	0.70707	88	0.64220
29	0.88050	49	0.78212	69	0.70352	89	0.63927

Liquids Heavier than Water

$$\text{Formula–Specific Gravity} = \frac{145}{145 - {}^\circ\text{Baume}}$$

Baume Degrees	Specific Gravity 60°–60°F	Baume Degrees	Specific Gravity 60°–60°F	Baume Degrees	Specific Gravity 60°–60°F	Baume Degrees	Specific Gravity 60°–60°F
0	1.00000	10	1.07407	20	1.16000	30	1.26087
1	1.00694	11	1.08209	21	1.16935	31	1.27193
2	1.01399	12	1.09023	22	1.17886	32	1.28319
3	1.02113	13	1.09848	23	1.18852	33	1.29464
4	1.02837	14	1.10687	24	1.19835	34	1.30631
5	1.03571	15	1.11538	25	1.20833	35	1.31818
6	1.04317	16	1.12403	26	1.21849	36	1.33028
7	1.05072	17	1.13281	27	1.22881	37	1.34259
8	1.05839	18	1.14173	28	1.23932	38	1.35514
9	1.06618	19	1.15079	29	1.25000	39	1.36792

Table A5, continued

Relationship between Baume Degrees and Specific Gravity

Liquids Heavier than Water

$$\text{Formula–Specific Gravity} = \frac{145}{145 - {}^\circ\text{Baume}}$$

Baume Degrees	Specific Gravity 60°–60°F	Baume Degrees	Specific Gravity 60°–60°F	Baume Degrees	Specific Gravity 60°–60°F	Baume Degrees	Specific Gravity 60°–60°F
40	1.38095	50	1.52632	60	1.70588	70	1.93333
41	1.39423	51	1.54255	61	1.72619	71	1.95946
42	1.40777	52	1.55914	62	1.74699	72	1.98630
43	1.42157	53	1.57609	63	1.76829	73	2.01389
44	1.43564	54	1.59341	64	1.79012	74	2.04225
45	1.45000	55	1.61111	65	1.81250	75	2.07143
46	1.46465	56	1.62921	66	1.83544	76	2.10145
47	1.47959	57	1.64773	67	1.85897	77	2.13235
48	1.49485	58	1.66667	68	1.88312	78	2.16418
49	1.51042	59	1.68605	69	1.90789	79	2.19697

Relation of API Hydrometer Scale to Specific Gravity

Degrees API	Specific Gravity	Degrees API	Specific Gravity	Degrees API	Specific Gravity	Degrees API	Specific Gravity
10	1.0000	30	0.8762	50	0.7796	70	0.7022
11	0.9930	31	0.8708	51	0.7753	71	0.6987
12	0.9861	32	0.8654	52	0.7711	72	0.6953
13	0.9792	33	0.8602	53	0.7669	73	0.6919
14	0.9725	34	0.8550	54	0.7628	74	0.6886
15	0.9659	35	0.8498	55	0.7587	75	0.6852
16	0.9593	36	0.8448	56	0.7547	76	0.6819
17	0.9529	37	0.8398	57	0.7507	77	0.6787
18	0.9465	38	0.8348	58	0.7467	78	0.6754
19	0.9402	39	0.8299	59	0.7428	79	0.6722
20	0.9340	40	0.8251	60	0.7389	80	0.6690
21	0.9279	41	0.8203	61	0.7351	81	0.6659
22	0.9218	42	0.8156	62	0.7313	82	0.6628
23	0.9159	43	0.8109	63	0.7275	83	0.6597
24	0.9100	44	0.8063	64	0.7238	84	0.6566
25	0.9042	45	0.8017	65	0.7201	85	0.6536
26	0.8984	46	0.7972	66	0.7165	86	0.6506
27	0.8927	47	0.7927	67	0.7128	87	0.6476
28	0.8871	48	0.7883	68	0.7093	88	0.6446
29	0.8816	49	0.7839	69	0.7057	89	0.6417

Table A6.

Brinell	Brinell	Rockwell C	Rockwell A	Rockwell B	Shore Scleroscope
10-mm Ball 3000-kg Diameter of Impression		Brale 150-kg Load	Brale 60-kg Load	1/16″ Ball 100-kg Load	
–	–	80	92.0	–	–
–	–	79	91.5	–	–
–	–	78	91.0	–	–
–	–	77	90.5	–	–
–	–	76	90.0	–	–
–	–	75	89.5	–	–
–	–	74	89.0	–	–
–	–	73	88.5	–	–
–	–	72	88.0	–	–
–	–	71	87.0	–	–
2.20	780	70	86.5	–	106
–	760	69	86.0	–	103
2.25	745	68	85.5	–	100
–	725	67	85.0	–	97
2.30	712	66	84.5	–	95
–	697	65	84.0	–	93
2.35	682	64	83.5	–	91
–	668	63	83.0	–	89
2.40	653	62	82.5	–	87
–	640	61	81.5	–	86
2.45	627	60	81.0	–	84
–	615	59	80.5	–	82
2.50	601	58	80.0	–	81
–	590	57	79.5	–	80
2.55	578	56	79.0	–	78
–	567	55	78.5	–	77
2.60	555	54	78.0	120	75
–	545	53	77.5	120	74
2.65	534	52	77.0	119	72
2.70	514	51	76.5	119	70
2.75	495	50	76.0	118	67
2.80	477	49	75.5	117	65
–	470	48	74.5	117	64
2.85	461	47	74.0	116	63
2.90	444	46	73.5	115	61
2.95	429	45	73.0	115	59
3.00	415	44	72.5	114	57
–	408	43	72.0	114	56
3.05	401	42	71.5	113	55
3.10	388	41	71.0	112	54
3.15	375	40	70.5	111	52
3.20	363	39	70.0	111	51
3.25	352	38	69.5	110	49
–	346	37	69.0	110	48
3.30	341	36	68.5	109	47

Hardness Conversion

Brinell	Brinell	Rockwell C	Rockwell A	Rockwell B	Shore Scleroscope
10-mm Ball 3000-kg Diameter of Impression		Brale 150-kg Load	Brale 60-kg Load	1/16″ Ball 100-kg Load	
3.35	331	35	68.0	109	46
3.40	321	34	67.5	108	45
3.45	311	33	67.0	108	44
3.50	302	32	66.5	107	43
3.55	293	31	66.0	106	42
3.60	285	30	65.5	105	40
3.65	277	29	65.0	104	39
3.70	269	28	64.5	104	38
3.75	262	27	64.0	103	38
–	258	26	63.5	102	37
3.80	255	25	63.0	102	37
3.85	248	24	62.5	101	36
3.90	241	23	62.0	101	35
–	238	22	61.5	100	34
3.95	235	21	61.0	99	34
4.00	229	20	60.0	98	33
4.05	223	–	59.5	97	32
4.10	217	–	59.0	96	31
4.15	212	–	58.0	95	31
4.20	207	–	57.5	94	30
4.25	202	–	57.0	93	30
4.30	197	–	56.5	92	29
4.35	192	–	56.0	91	28
4.40	187	–	55.5	90	28
4.45	183	–	55.0	89	27
4.50	179	–	54.0	88	27
4.55	174	–	53.5	87	26
4.60	170	–	53.0	86	26
4.65	166	–	52.5	85	25
4.70	163	–	52.0	84	25
4.75	159	–	51.0	83	24
4.80	156	–	50.5	82	24
4.85	153	–	50.0	81	23
4.90	149	–	49.5	80	23
4.95	146	–	49.0	79	22
5.00	143	–	48.5	78	22
5.05	140	–	48.0	77	21
5.10	137	–	47.0	76	21
5.15	134	–	46.5	75	21
5.15	134	–	46.0	74	21
5.20	131	–	45.5	73	20
5.20	131	–	45.0	72	20
5.25	128	–	44.5	71	20
5.30	126	–	44.0	70	20

Table A7. Temperature Conversion Fahrenheit–Celsius[a]

(Centigrade To Fahrenheit–Fahrenheit To Centigrade)

-459.4° to 0°			1° to 60°			61° to 290°			300° to 890°			900° to 3000°		
C	Centigrade or Fahrenheit	F	C	Centigrade or Fahrenheit	F	C	Centigrade or Fahrenheit	F	C	Centigrade or Fahrenheit	F	C	Centigrade or Fahrenheit	F
-273	-459.4		-17.2	1	33.8	16.1	61	141.8	149	300	572	482	900	1652
-268	-450		-16.7	2	35.6	16.7	62	143.6	154	310	590	488	910	1670
-262	-440		-16.1	3	37.4	17.2	63	145.4	160	320	608	493	920	1688
-257	-430		-15.6	4	39.2	17.8	64	147.2	166	330	626	499	930	1706
-251	-420		-15.0	5	41.0	18.3	65	149.0	171	340	644	504	940	1724
-246	-410		-14.4	6	42.8	18.9	66	150.8	177	350	662	510	950	1742
-240	-400		-13.9	7	44.6	19.4	67	152.6	182	360	680	516	960	1760
-234	-390		-13.3	8	46.4	20.0	68	154.4	188	370	698	521	970	1778
-229	-380		-12.8	9	48.2	20.6	69	156.2	193	380	716	527	980	1796
-223	-370		-12.2	10	50.0	21.1	70	158.0	199	390	734	532	990	1814
-218	-360		-11.7	11	51.8	21.7	71	159.8	204	400	752	538	1000	1832
-212	-350		-11.1	12	53.6	22.2	72	161.6	210	410	770	549	1020	1868
-207	-340		-10.6	13	55.4	22.8	73	163.4	215	420	788	560	1040	1904
-201	-330		-10.0	14	57.2	23.3	74	165.2	221	430	806	571	1060	1940
-196	-320		-9.4	15	59.0	23.9	75	167.0	227	440	824	582	1080	1976
-190	-310		-8.9	16	60.8	24.4	76	168.8	232	450	842	593	1100	2012
-184	-300		-8.3	17	62.6	25.0	77	170.6	238	460	860	604	1120	2048
-179	-290		-7.8	18	64.4	25.6	78	172.4	243	470	878	616	1140	2084
-173	-280		-7.2	19	66.2	26.1	79	174.2	249	480	896	627	1160	2120
-169	-273	-459.4	-6.7	20	68.0	26.7	80	176.0	254	490	914	638	1180	2156

−168	−270	−454	−6.1	21	69.8	27.2	81	177.8	260	500	932	649	1200	2192
−162	−260	−436	−5.6	22	71.6	27.8	82	179.6	266	510	950	660	1220	2228
−157	−250	−418	−5.0	23	73.4	28.3	83	181.4	271	520	968	671	1240	2264
−151	−240	−400	−4.4	24	75.2	28.9	84	183.2	277	530	986	682	1260	2300
−146	−230	−382	−3.9	25	77.0	29.4	85	185.0	282	540	1004	693	1280	2336
−140	−220	−364	−3.3	26	78.8	30.0	86	186.8	288	550	1022	704	1300	2372
−134	−210	−346	−2.8	27	80.6	30.6	87	188.6	293	560	1040	732	1350	2462
−129	−200	−328	−2.2	28	82.4	31.1	88	190.4	299	570	1058	760	1400	2552
−123	−190	−310	−1.7	29	84.2	31.7	89	192.2	304	580	1076	788	1450	2642
−118	−180	−292	−1.1	30	86.0	32.2	90	194.0	310	590	1094	816	1500	2732
−112	−170	−274	−0.6	31	87.8	32.8	91	195.8	316	600	1112	843	1550	2822
−107	−160	−256	0.0	32	89.6	33.3	92	197.6	321	610	1130	871	1600	2912
−101	−150	−238	0.6	33	91.4	33.9	93	199.4	327	620	1148	899	1650	3002
−96	−140	−220	1.1	34	93.2	34.4	94	201.2	332	630	1166	927	1700	3092
−90	−130	−202	1.7	35	95.0	35.0	95	203.0	338	640	1184	954	1750	3182
−84	−120	−184	2.2	36	96.8	35.6	96	204.8	343	650	1202	982	1800	3272
−79	−110	−166	2.8	37	98.6	36.1	97	206.6	349	660	1220	1010	1850	3362
−73	−100	−148	3.3	38	100.4	36.7	98	208.4	354	670	1238	1038	1900	3452
−68	−90	−130	3.9	39	102.2	37.2	99	210.2	360	680	1256	1066	1950	3542
−62	−80	−112	4.4	40	104.0	37.8	100	212.0	366	690	1274	1093	2000	3632
−57	−70	−94	5.0	41	105.8	43	110	230	371	700	1292	1121	2050	3722
−51	−60	−76	5.6	42	107.6	49	120	248	377	710	1310	1149	2100	3812
−46	−50	−58	6.1	43	109.4	54	130	266	382	720	1328	1177	2150	3902
−40	−40	−40	6.7	44	111.2	60	140	284	388	730	1346	1204	2200	3992
−34	−30	−22	7.2	45	113.0	66	150	302	393	740	1364	1232	2250	4082

Table A7, continued

−459.4° to 0°			1° to 60°			61° to 290°			300° to 890°			900° to 3000°		
C	Centigrade or Fahrenheit	F	C	Centigrade or Fahrenheit	F	C	Centigrade or Fahrenheit	F	C	Centigrade or Fahrenheit	F	C	Centigrade or Fahrenheit	F
−29	−20	−4	7.8	46	114.8	71	160	320	399	750	1382	1260	2300	4172
−23	−10	14	8.3	47	116.6	77	170	338	404	760	1400	1288	2350	4262
−17.8	0	32	8.9	48	118.4	82	180	356	410	770	1418	1316	2400	4352
			9.4	49	120.2	88	190	374	416	780	1436	1343	2450	4442
			10.0	50	122.0	93	200	392	421	790	1454	1371	2500	4532
			10.6	51	123.8	99	210	410	427	800	1472	1399	2550	4622
			11.1	52	125.6	100	212	413.6	432	810	1490	1427	2600	4712
			11.7	53	127.4	104	220	428	438	820	1508	1454	2650	4802
			12.2	54	129.2	110	230	446	443	830	1526	1482	2700	4892
			12.8	55	131.0	116	240	464	449	840	1544	1510	2750	4982
			13.3	56	132.8	121	250	482	454	850	1562	1538	2800	5072
			13.9	57	134.6	127	260	500	460	860	1580	1566	2850	5162
			14.4	58	136.4	132	270	518	466	870	1598	1593	2900	5252
			15.0	59	138.2	138	280	536	471	880	1616	1621	2950	5342
			15.6	60	140.0	143	290	554	477	890	1634	1649	3000	5432

[a] Locate temperature in middle column. If in degrees Centigrade, read Fahrenheight equivalent in right-hand column; if in degrees Fahrenheit, read Centigrade equivalent in left-hand column.

$F^\circ = \frac{9}{5} C^\circ + 32^\circ$ $\qquad$ $C^\circ = \frac{5}{9} (F^\circ - 32^\circ)$

Table A8. Length Conversion Millimeters to Inches

Millimeters	0	1	2	3	4	5	6	7	8	9
0	0.00000	0.03937	0.07874	0.11811	0.15748	0.19685	0.23622	0.27559	0.31496	0.35433
10	0.39370	0.43307	0.47244	0.51181	0.55118	0.59055	0.62992	0.66929	0.70866	0.74803
20	0.78740	0.82677	0.86614	0.90551	0.94488	0.98425	1.02362	1.06299	1.10236	1.14173
30	1.18110	1.22047	1.25984	1.29921	1.33858	1.37795	1.41732	1.45669	1.49606	1.53543
40	1.57480	1.61417	1.65354	1.69291	1.73228	1.77165	1.81102	1.85039	1.88976	1.92913
50	1.96850	2.00787	2.04724	2.08661	2.12598	2.16535	2.20472	2.24409	2.28346	2.32283
60	2.36220	2.40157	2.44094	2.48031	2.51968	2.55905	2.59842	2.63779	2.67716	2.71653
70	2.75590	2.79527	2.83464	2.87401	2.91338	2.95275	2.99212	3.03149	3.07086	3.11023
80	3.14960	3.18897	3.22834	3.26771	3.30708	3.34645	3.38582	3.42519	3.46456	3.50393
90	3.54330	3.58267	3.62204	3.66141	3.70078	3.74015	3.77952	3.81889	3.85826	3.89763
100	3.93700	3.97637	4.01574	4.05511	4.09448	4.13385	4.17322	4.21259	4.25196	4.29133
110	4.33070	4.37007	4.40944	4.44881	4.48818	4.52755	4.56692	4.60629	4.64566	4.68503
120	4.72440	4.76377	4.80314	4.84251	4.88188	4.92125	4.96062	4.99999	5.03936	5.07873
130	5.11810	5.15747	5.19684	5.23621	5.27558	5.31495	5.35432	5.39369	5.43306	5.47243
140	5.51180	5.55117	5.59054	5.62991	5.66928	5.70865	5.74802	5.78739	5.82676	5.86613

Table A8, continued

Millimeters	0	1	2	3	4	5	6	7	8	9
150	5.90550	5.94487	5.98424	6.02361	6.06298	6.10235	6.14172	6.18109	6.22046	6.25983
160	6.29920	6.33857	6.37794	6.41731	6.45668	6.49605	6.53542	6.57479	6.61416	6.65353
170	6.69290	6.73227	6.77164	6.81101	6.85038	6.88975	6.92912	6.96849	7.00786	7.04723
180	7.08660	7.12597	7.16534	7.20471	7.24408	7.28345	7.32282	7.36219	7.40156	7.44093
190	7.48030	7.51967	7.55904	7.59841	7.63778	7.67715	7.71652	7.75589	7.79526	7.83463
200	7.87400	7.91337	7.95274	7.99211	8.03148	8.07085	8.11022	8.14959	8.18896	8.22833
210	8.26770	8.30707	8.34644	8.38581	8.42518	8.46455	8.50392	8.54329	8.58266	8.62203
220	8.66140	8.70077	8.74014	8.77951	8.81888	8.85825	8.89762	8.93699	8.97636	9.01573
230	9.05510	9.09447	9.13384	9.17321	9.21258	9.25195	9.29132	9.33069	9.37006	9.40943
240	9.44880	9.48817	9.52754	9.56691	9.60628	9.64565	9.68502	9.72439	9.76376	9.80313
250	9.84250	9.88187	9.92124	9.96061	9.99998	10.03935	10.07872	10.11809	10.15746	10.19683
260	10.23620	10.27557	10.31494	10.35431	10.39368	10.43305	10.47242	10.51179	10.55116	10.59053
270	10.62990	10.66927	10.70864	10.74801	10.78738	10.82675	10.86612	10.90549	10.94486	10.98423
280	11.02360	11.06297	11.10234	11.14171	11.18108	11.22045	11.25982	11.29919	11.33856	11.37793
290	11.41730	11.45667	11.49604	11.53541	11.57478	11.61415	11.65352	11.69289	11.73226	11.77163
300	11.81100	11.85037	11.88974	11.92911	11.96848	12.00785	12.04722	12.08659	12.12596	12.16533

Table A9. Pressure Conversion Table Kilogram per Square Centimeter to Pounds per Square Inch

(1 kg/cm^2 = 14.233 P.S.I.)

kg/cm^2	kg 0	1	2	3	4	5	6	7	8	9	10	11	12	13	14
0.0		14.22	28.45	42.67	56.89	71.11	85.34	99.56	113.8	128.0	142.2	156.5	170.7	184.9	199.1
0.1	1.4223	15.65	29.87	44.09	58.31	72.54	86.76	100.98	115.2	129.4	143.7	157.9	172.1	186.3	200.5
0.2	2.8446	17.07	31.29	45.51	59.74	73.96	88.18	102.41	116.6	130.9	145.1	159.3	173.5	187.7	202.0
0.3	4.2669	18.49	32.71	46.94	61.16	75.38	89.61	103.83	118.1	132.3	146.5	160.7	174.9	189.2	203.4
0.4	5.6892	19.91	34.14	48.36	62.58	76.80	91.03	105.25	119.5	133.7	147.9	162.1	176.4	190.6	204.8
0.5	7.1115	21.34	35.56	49.78	64.00	78.23	92.45	106.67	120.9	135.1	149.3	163.6	177.8	192.0	206.2
0.6	8.5338	22.76	36.98	51.20	65.43	79.65	93.87	108.10	122.3	136.5	150.8	165.0	179.2	193.4	207.7
0.7	9.9561	24.18	38.40	52.63	66.49	81.07	95.29	109.52	123.7	138.0	152.2	166.4	180.6	194.9	209.1
0.8	11.3784	25.60	39.82	54.05	68.27	82.49	96.72	110.94	125.2	139.4	153.6	167.8	182.0	196.3	210.5
0.9	12.8007	27.02	41.25	55.47	69.69	83.92	98.14	112.36	126.6	140.8	155.0	169.3	183.5	197.7	211.9
	15	**16**	**17**	**18**	**19**	**20**	**21**	**22**	**23**	**24**	**25**	**26**	**27**	**28**	**29**
0.0	213.4	227.6	241.8	256.0	270.2	284.5	298.7	312.9	327.1	341.4	355.6	369.8	384.0	398.2	412.5
0.1	214.8	229.0	243.2	257.4	271.7	285.9	300.1	314.3	328.6	342.8	357.0	371.2	385.4	399.7	413.9
0.2	216.2	230.4	244.6	258.9	273.1	287.3	301.5	315.8	330.0	344.2	358.4	372.6	386.9	401.1	415.3
0.3	217.6	231.8	246.1	260.3	274.5	288.7	303.0	317.2	331.4	345.6	359.8	374.1	388.3	402.5	416.7
0.4	219.0	233.3	247.5	261.7	275.9	290.2	304.4	318.6	332.8	347.0	361.3	375.5	389.7	403.9	418.2
0.5	220.5	234.7	248.9	263.1	277.4	291.6	305.8	320.0	334.2	248.5	362.7	376.9	391.1	405.4	419.6
0.6	221.9	236.1	250.3	264.6	278.8	293.0	307.2	321.4	335.7	349.9	364.1	378.3	392.6	407.8	421.0
0.7	223.3	237.5	251.7	266.0	280.2	294.4	308.6	322.9	337.1	351.3	365.5	379.8	394.0	408.2	422.4
0.8	224.7	239.0	253.2	267.4	281.6	295.8	310.1	324.3	338.5	352.7	367.0	381.2	395.4	409.6	423.9
0.9	226.2	240.4	254.6	268.8	283.0	297.3	311.5	325.7	339.9	354.2	368.4	382.6	396.8	411.0	425.3

Table A9, continued

kg/cm^2	30	31	32	33	34	35	36	37	38	39	40	41	42	43	44
0.0	426.7	440.9	455.1	469.4	483.6	497.8	512.0	526.3	540.5	554.7	568.9	583.1	597.4	611.6	625.8
0.1	428.1	442.3	456.6	470.8	485.0	499.2	513.4	527.7	541.9	556.1	570.3	584.6	598.8	613.0	627.2
0.2	429.5	443.8	458.0	472.2	486.4	500.7	514.9	529.1	543.3	557.5	571.8	586.0	600.2	614.4	628.7
0.3	431.0	445.2	459.4	473.6	487.9	502.1	516.3	530.5	544.7	559.0	573.2	587.4	601.6	615.9	630.1
0.4	432.4	446.6	460.8	475.1	489.3	503.5	517.7	531.9	546.2	560.4	574.6	588.8	603.1	617.3	631.5
0.5	433.8	448.0	462.3	476.5	490.7	504.9	519.1	533.4	547.6	561.8	576.0	590.3	604.5	618.7	632.9
0.6	435.2	449.5	463.7	477.9	492.2	506.3	520.6	534.8	549.0	563.2	577.4	591.7	605.9	620.1	634.4
0.7	436.7	450.9	465.1	479.3	493.5	507.8	522.0	536.2	550.4	564.7	578.9	593.1	607.3	621.6	635.8
0.8	438.1	452.3	466.5	480.7	495.0	509.2	523.4	537.6	551.9	566.1	580.3	594.5	608.7	623.0	637.2
0.9	439.5	453.7	467.9	482.2	496.4	510.6	524.8	539.1	553.3	567.5	581.7	595.9	610.2	624.4	638.6
	45	46	47	48	49	50	51	52	53	54	55	56	57	58	59
0.0	640.0	654.3	668.5	682.7	696.9	711.1	725.4	739.6	753.8	768.0	782.3	796.5	810.7	824.9	839.2
0.1	641.5	655.7	669.9	684.1	698.4	712.6	726.8	741.0	755.2	769.5	783.7	797.9	812.1	826.4	840.6
0.2	642.9	657.1	671.3	685.6	699.8	714.0	728.2	742.4	756.7	770.9	785.1	799.3	813.6	827.8	842.0
0.3	644.3	658.5	672.8	687.0	701.2	715.4	729.6	743.9	758.1	772.3	786.5	800.8	815.0	829.2	843.4
0.4	645.7	660.0	674.2	688.4	702.6	716.8	731.1	745.3	759.5	773.7	788.0	802.2	816.4	830.6	844.9
0.5	647.2	661.4	675.6	689.8	704.0	718.3	732.5	746.7	760.9	775.2	789.4	803.6	817.8	832.1	846.3
0.6	648.6	662.8	677.0	691.2	705.5	719.7	733.9	748.1	762.4	776.6	790.8	805.0	819.2	833.5	847.7
0.7	650.0	664.2	678.4	692.7	706.9	721.1	735.3	749.6	763.8	778.0	792.2	806.4	820.7	834.9	849.1
0.8	651.4	665.6	679.9	694.1	708.3	722.5	736.8	751.0	765.2	779.4	793.6	807.9	822.1	836.3	850.5
0.9	652.8	667.0	681.3	695.5	709.7	724.0	738.2	752.4	766.6	780.8	795.1	809.3	823.5	837.7	852.0

	60	61	62	63	64	65	66	67	68	69	70	71	72	73	74
0.0	853.4	868	882	896	910	924	939	953	967	981	996	1010	1024	1038	1053
0.1	854.8	869	883	897	912	926	940	954	969	983	997	1011	1025	1040	1054
0.2	856.2	870	885	899	913	927	942	956	970	984	998	1013	1027	1041	1055
0.3	857.7	872	886	900	915	929	943	957	971	986	1000	1014	1028	1043	1057
0.4	859.1	873	888	902	916	930	944	959	973	987	1001	1016	1030	1044	1058
0.5	860.5	875	889	903	917	932	946	960	974	988	1003	1017	1031	1045	1060
0.6	861.9	876	890	905	919	933	947	961	976	990	1004	1018	1033	1047	1061
0.7	863.3	878	892	906	920	934	949	963	977	991	1006	1020	1034	1048	1062
0.8	864.8	879	893	907	922	936	950	964	979	993	1007	1021	1035	1050	1064
0.9	866.2	880	895	909	923	937	952	966	980	994	1008	1022	1037	1051	1065

	75	76	77	78	79	80	81	82	83	84	85	86	87	88	89
0.0	1067	1081	1095	1109	1124	1138	1152	1166	1181	1195	1209	1223	1237	1252	1266
0.2	1070	1084	1098	1112	1126	1141	1155	1169	1183	1198	1212	1226	1240	1254	1269
0.4	1072	1087	1101	1115	1129	1144	1158	1172	1186	1200	1215	1229	1243	1257	1272
0.6	1075	1089	1104	1118	1132	1146	1161	1175	1189	1203	1217	1232	1246	1260	1274
0.8	1078	1092	1107	1121	1135	1149	1163	1178	1192	1206	1220	1235	1249	1263	1277

	90	91	92	93	94	95	96	97	98	99
0.0	1280	1294	1309	1323	1337	1351	1365	1380	1394	1408
0.2	1283	1297	1311	1326	1340	1354	1368	1382	1397	1411
0.4	1286	1300	1314	1328	1343	1357	1371	1385	1400	1414
0.6	1289	1303	1317	1331	1345	1360	1374	1388	1402	1417
0.8	1291	1306	1320	1334	1348	1363	1377	1391	1405	1419

Table A10. Volume Conversion

Cubic Inches to Cubic Centimeters

Cubic Inches	0	1	2	3	4	5	6	7	8	9
0	–	16.3871	32.7741	49.1612	65.548	81.935	98.322	114.709	131.097	147.484
10	163.871	180.258	196.645	213.032	229.419	245.806	262.193	278.580	294.967	311.354
20	327.741	344.128	360.515	376.902	393.290	409.677	426.064	442.451	458.838	475.225
30	491.612	508.00	524.39	540.77	557.16	573.55	589.93	606.32	622.71	639.10
40	655.48	671.87	688.36	704.64	721.03	737.42	753.80	770.19	786.58	802.97
50	819.35	835.74	852.13	868.51	884.90	901.29	917.68	934.06	950.45	966.84
60	983.22	999.61	1016.00	1032.39	1048.77	1065.16	1081.55	1097.93	1114.32	1130.71
70	1147.09	1163.48	1179.87	1196.26	1212.64	1229.03	1245.42	1261.80	1278.19	1294.58
80	1310.97	1327.35	1343.74	1360.13	1376.51	1392.90	1409.29	1425.67	1442.06	1458.45
90	1474.84	1491.22	1507.61	1524.00	1540.38	1556.77	1573.16	1589.55	1605.93	1622.32

Cubic Centimeters to Cubic Inches

Cubic Centimeter	0	1	2	3	4	5	6	7	8	9
0	–	0.061024	0.122047	0.183071	0.244095	0.305119	0.366142	0.427166	0.488190	0.54921
10	0.61024	0.67126	0.73228	0.79331	0.85433	0.91536	0.97638	1.03740	1.09843	1.15945
20	1.22047	1.28150	1.34252	1.40355	1.46457	1.52559	1.58662	1.64764	1.70866	1.76969
30	1.83071	1.89174	1.95276	2.01378	2.07481	2.13583	2.19685	2.25788	2.31890	2.37993
40	2.44095	2.50197	2.56300	2.62402	2.68504	2.74607	2.80709	2.86812	2.92914	2.99016
50	3.05119	3.11221	3.17323	3.23426	3.29528	3.35631	3.41733	3.47835	3.53938	3.60040
60	3.66142	3.72245	3.78347	3.84450	3.90552	3.96654	4.02757	4.08859	4.14961	4.21064
70	4.27166	4.33269	4.39371	4.45473	4.51576	4.57678	4.63780	4.69883	4.75985	4.82088
80	4.88190	4.94292	5.0039	5.0650	5.1260	5.1870	5.2480	5.3091	5.3701	5.4311
90	5.4921	5.5532	5.6142	5.6752	5.7362	5.7973	5.8583	5.9193	5.9803	6.0414

Gallons to Liters

Gallons	0	1	2	3	4	5	6	7	8	9
0	–	4.5460	9.0919	13.6379	18.1839	22.7298	27.2758	31.8217	36.3677	40.9137
10	45.4596	50.006	54.552	59.098	63.643	68.189	72.735	77.281	81.827	86.373
20	90.919	95.465	100.011	104.557	109.103	113.649	118.195	122.741	127.287	131.833
30	136.379	140.925	145.471	150.017	154.563	159.109	163.655	168.201	172.747	177.293
40	181.839	186.384	190.930	195.476	200.022	204.568	209.114	213.660	218.206	222.752
50	227.298	231.844	236.390	240.936	245.482	250.028	254.574	259.120	263.666	268.212
60	272.758	277.304	281.850	286.396	290.942	295.488	300.493	304.580	309.125	313.671
70	318.217	322.763	327.309	331.855	336.401	340.947	345.493	350.039	354.585	369.131
80	363.877	368.223	372.769	377.315	381.861	386.407	390.953	395.499	400.045	404.591
90	409.137	413.683	418.229	422.775	427.321	431.866	436.412	440.958	445.504	450.050

Liters to Gallons

Liters	0	1	2	3	4	5	6	7	8	9
0	–	0.219975	0.439951	0.65993	0.87990	1.09988	1.31985	1.53983	1.75980	1.97978
10	2.19975	2.41973	2.63970	2.85968	3.07965	3.29963	3.51960	3.73958	3.95956	4.17953
20	4.39951	4.61948	4.83946	5.0594	5.2794	5.4994	5.7194	5.9393	6.1593	6.3793
30	6.5993	6.8192	7.0392	7.2592	7.4792	7.6991	7.9191	8.1391	8.3591	8.5790
40	8.7990	9.0190	9.2390	9.4589	9.6789	9.8989	10.1189	10.3388	10.5588	10.7788
50	10.9988	11.2187	11.4387	11.6587	11.8787	12.0986	12.3186	12.5386	12.7586	12.9785
60	13.1985	13.4185	13.6385	13.8584	14.0784	14.2984	14.5184	14.7383	14.9583	15.1783
70	15.3983	15.6182	15.8382	16.0582	16.2782	16.4981	16.7181	16.9381	17.1581	17.3780
80	17.5980	17.8180	18.0380	18.2579	18.4779	18.6979	18.9179	19.1379	19.3578	19.5778
90	19.7978	20.0178	20.2377	20.4577	20.6777	20.8977	21.1176	21.3376	21.5576	21.7776

Table A11. Unit Volume Conversion

U.S. Gallons per Minute to Liters per Minute

U.S. GPM	0	1	2	3	4	5	6	7	8	9
0	00	3.78	7.57	11.35	15.14	18.92	22.71	26.49	30.28	34.06
10	37.85	41.63	45.42	49.20	52.99	56.77	60.56	64.34	68.13	71.91
20	75.70	79.48	83.27	87.05	90.84	94.62	98.40	102.19	105.97	109.76
30	113.54	117.33	121.11	124.90	128.68	132.47	136.25	140.04	143.82	147.61
40	151.39	155.18	158.96	162.75	166.53	170.32	174.10	177.89	181.67	185.46
50	189.24	193.02	196.81	200.59	204.38	208.16	211.95	215.73	219.52	223.30
60	227.09	230.87	234.66	238.44	242.23	246.01	249.80	253.58	257.37	261.15
70	264.94	268.72	272.51	276.29	280.08	283.86	287.64	291.43	295.21	299.00
80	302.78	306.57	310.35	314.14	317.92	321.71	325.49	329.28	333.06	336.85
90	340.63	344.42	348.20	351.99	355.77	359.56	363.34	367.13	370.91	374.70
100	378.48	382.26	386.05	389.83	393.62	397.40	401.19	404.97	408.76	412.54
110	416.33	420.11	423.90	427.68	431.47	435.25	439.04	442.82	446.61	450.39
120	454.18	457.96	461.75	465.53	469.32	473.10	476.88	480.67	484.45	488.24
130	492.02	495.81	499.59	503.38	507.16	510.95	514.73	518.52	522.30	526.09
140	529.87	533.66	537.44	541.23	545.01	548.80	552.58	556.37	560.15	563.94
150	567.72	571.50	575.29	579.07	582.86	586.64	590.43	594.21	598.00	601.78

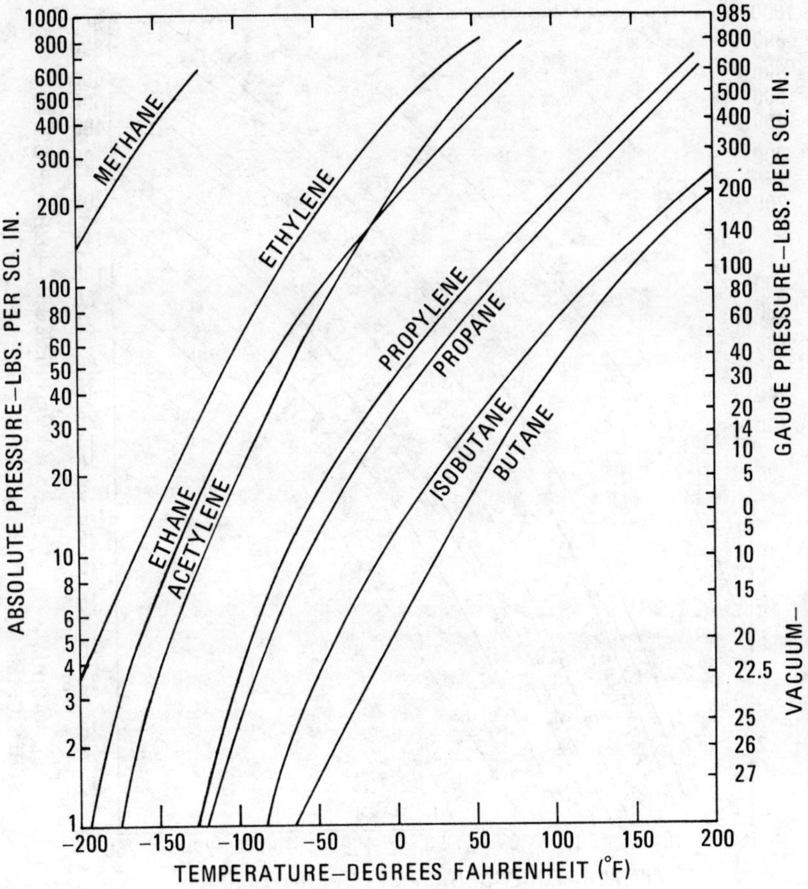

Figure A1. Vapor pressures of light hydrocarbons.

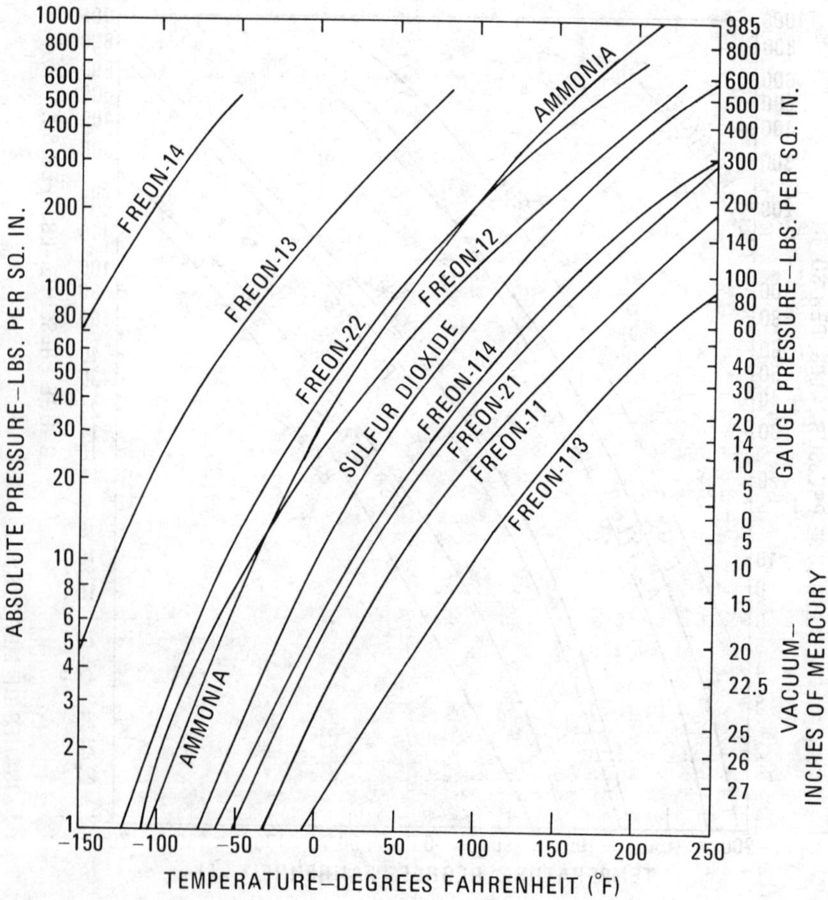

Figure A2. Vapor pressures of refrigerants.

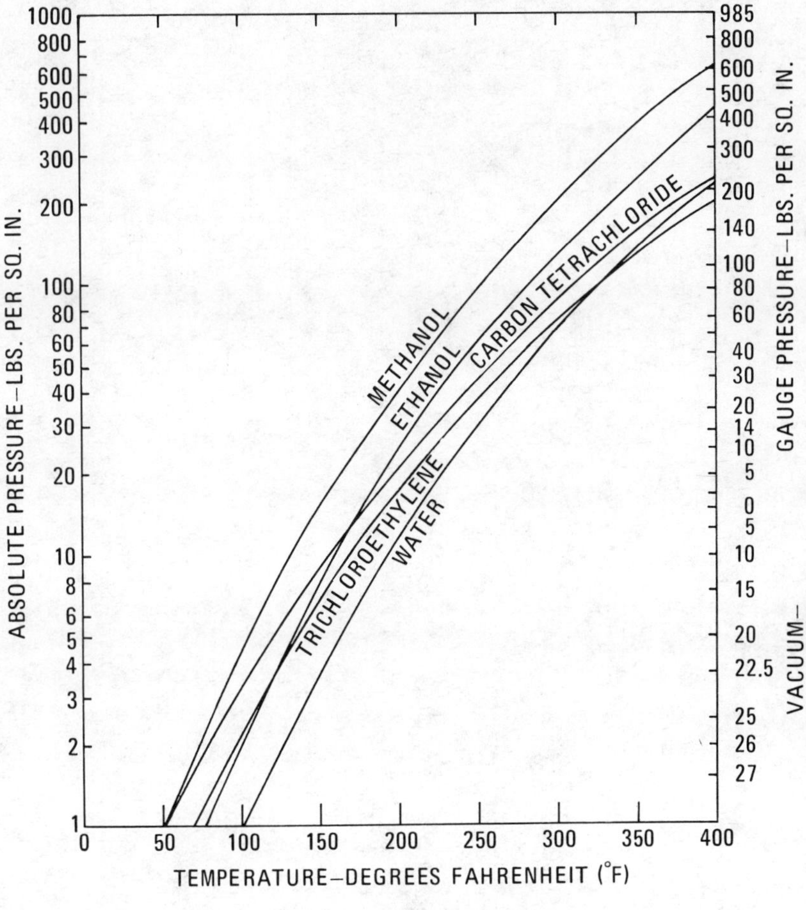

Figure A3. Vapor pressures of common liquids.

APPENDIX B

TABLES OF RECOMMENDED SI UNITS AND CONVERSION FACTORS

The tables contained in this Appendix were reprinted from API Publication 2564–*Manual of Petroleum Measurement Standards*: Chapter 15, "Guidelines for the Use of the International System of Units (SI) in the Petroleum and Allied Industries," 2nd edition (December 1980), pp. 7-31, courtesy of the American Petroleum Institute, 2101 L Street, Northwest, Washington, DC.

CONTENTS

NOTES ON USE OF TABLES

Metric units recommended for general use are given under the heading "API preferred metric unit." Other units that also may be needed are shown in the "other allowable" column. Preferred units do not preclude the use of other multiples or submultiples, as the choice of such unit-multiple is governed by the magnitude of the numerical value.

Notation used conforms to SI practice, i.e., groups of three digits to the left or right of the decimal marker are separated by spaces. No commas or other triad spacers are used. E-notation is used for convenience because it is a standard method of display in many calculators and because of the inability of computers to print out or transmit superscripts. An asterisk (*) denotes that all of the succeeding digits would be zeroes. If a conversion factor ends in zero but does not have an asterisk, then any subsequent digits would not necessarily be zeroes.

For example, then:

$$3.056\ 0 \text{ E+00} = 3.056\ 0 \times 10^{0} = 3.056\ 0$$

$$3.056\text{*E-01} = 3.056\ 000 \times 10^{-1} = 0.305\ 600\ 0$$

$$9.290\ 304 \text{ E+02} = 9.290\ 304 \times 10^{2} = 929.030\ 4$$

Table B1. Nomenclature for Conversion Tables

Symbol	Name	Quantity	Definition: In Terms of Other Units	Definition: In Terms of Base Units	Type of Unit
A	Ampere	Electric current	–	–	Base
a	Annum (year)	Time	365 d	$3.153\,600 \times 10^7$s	Allowable
bar	Bar	Pressure	10^5 Pa	10^5 kg/(m·s^2)	Allowable
Bq	Becquerel	Activity (of a radionuclide)	–	1 s^{-1}	Derived
C	Coulomb	Quantity of electricity, electric charge	–	1 A·s	Derived
°C	Degree Celsius	Celsius temperature	–	–	Derived
cd	Candela	Luminous intensity	–	–	Base
cP	Centipoise	Dynamic viscosity	1 mPa·s	10^{-3} kg/(m·s)	Allowable
cSt	Centistokes	Kinematic viscosity	1 mm^2/s	10^{-6} m^2/s	Allowable
d	Day	Time	24 hr	8.640×10^4s	Allowable
F	Farad	Capacitance	1 C/V	1 s^4 A^2/(m^2·kg)	Derived
g	Gram	Mass	–	10^{-3} kg	Allowable (submultiple of base unit)
Gy	Gray	Absorbed dose, specific energy imparted, kerma, absorbed dose index	1 J/kg	1 m^2/S^2	Derived
h	Hour	Time	60 min	3.6×10^3 s	Allowable
H	Henry	Inductance	1 Wb/A	1 m^2·kg/(s^2·A^2)	Derived
ha	Hectare	Area	–	10^4 m^2	Allowable
Hz	Hertz	Frequency (of periodic phenomenon)	–	1 s^{-1}	Derived
J	Joule	Work, energy, quantity of heat	1 N·m	1 m^2·kg/s^2	Derived
K	Kelvin	Thermodynamic temperature	–	–	Base
kg	Kilogram	Mass	–	–	Base
kn	Knot	Velocity	1.852 km/h	5.144 444 m^2/s	Allowable
L	Liter	Volume	1 dm^3	10^{-3} m^3	Allowable

Table B1, continued

Symbol	Name	Quantity	Definition: In Terms of Other Units	Definition: In Terms of Base Units	Type of Unit
lm	Lumen	Luminous flux	–	1 cd·sr	Derived
lx	Lux	Illumination	1 lm/m^2	1 cd·sr/m^2	Derived
m	Meter	Length	–	–	Base
min	Minute	Time	–	60 s	Allowable
mol	Mole	Amount of substance	–	–	Base
N	Newton	Force	–	1 m·kg/s^2	Derived
naut. mi	Nautical mile	Distance	1.852 km	1.852 × 10^3 m	Allowable
Pa	Pascal	Pressure	1 N/m^2	1 kg/(m·s^2)	Derived
r	Revolution	Angular displacement	360°	2π rad	Allowable
rad	Radian	Plane angle	–	–	Supplementary
S	Siemens	Conductance	1 A/V	1 s^3·A^2/(m^2·kg)	Derived
s	Second	Time	–	–	Base
sr	Steradian	Solid angle	–	–	Supplementary
Sv	Sievert	Dose equivalent	1 J/kg	1 m^2/s^2	Derived
t	Metric ton	Mass	1 Mg	10^{3}kg	Allowable
T	Tesla	Magnetic flux density	1 Wb/m^2	1 kg/(s^2·A)	Derived
V	Volt	Electric potential, potential difference, electromotive force	1 W/A	1 m^2·kg/(s^3·A)	Derived
W	Watt	Power, radiant flux	1 J/s	1 m^2·kg/s^3	Derived
Wb	Weber	Magnetic flux	1 V/s	1 m^2·kg/(s^2·A)	Derived
Ω	Ohm	Electric resistance	1 V/A	1 m^2·kg/(s^3·A^2)	Derived
°	Degree	Plane angle	–	π/180 rad	Allowable
′	Minute	Plane angle	(1/60)°	2.908 882 × 10^{-4} rad	Allowable
″	Second	Plane angle	(1/60)′	4.848 137 × 10^{-6} rad	Allowable

Table B2. Tables of Recommended SI Units and Conversion Factors

Quantity	SI Unit	Customary Unit	Metric Unit		Conversion Factor (multiply quantity expressed in customary units by factor to get metric equivalent)	
			API Preferred	Other Allowable		
Space, Time						
Length	m	naut. mi	km		1.852*	E+00
				naut. mi	1	
		mi	km		1.609 344*	E+00
		mi (U.S. statute)	km		1.609 347	E+00
		Chain	m		2.011 684	E+01
		Rod	m		5.029 210	E+00
		Fathom	m		1.828 804	E+00
		m	m		1	
		yard	m		9.144*	E–01
		ft	m		3.048*	E–01
		ft (U.S. survey)	m		3.048 006	E–01
		link	m		2.011 684	E–01
		in.	mm		2.54*	E+01
				cm	2.54*	E+00
		cm	mm		1.0*	E+01
				cm	1	
		mm	mm		1	
		mil	μm		2.54*	E+01
		micron (μ)	μm		1	

Table B2, continued

Quantity	SI Unit	Customary Unit	Metric Unit: API Preferred	Metric Unit: Other Allowable	Conversion Factor (multiply quantity expressed in customary units by factor to get metric equivalent)
Space, Time (continued)					
Surface Texture	m	μin.	μm		2.54* E–02
		nm	μm		1.0* E–03
Length/Length	m/m	ft/mi	m/km		1.893 939 E–01
Length/Volume	m/m^3	ft/U.S. gal	m/m^3		8.051 964 E+01
		ft/ft^3	m/m^3		1.076 391 E+01
		ft/bbl	m/m^3		1.917 134 E+00
Length/Temperature	m/K	see Temperature, Pressure, Vacuum			
Area	m^2	mi^2	km^2		2.589 988 E+00
		mi^2 (U.S. statute)	km^2		2.589 998 E+00
		ha	m^2		1.0* E+04
		acre	ha		4.046 873 E–01
				m^2	4.046 873 E+03
		sq chain	m^2		4.046 873 E+02
		sq rod	m^2		2.529 295 E+01
Area		yd^2	m^2		8.361 274 E–01
		ft^2	m^2		9.290 304* E–02
		ft^2 (U.S. survey)	m^2		9.290 341 E–02

		in.^{2}	mm^2		6.451 6*	E+02
				cm^2	6.451 6*	E+00
		cm^2	mm^2		1.0*	E+02
				cm^2	1	
		mm^2	mm^2		1	
Area/Volume	m^2/m^3	ft^2/in.^{3}	m^2/cm^3		5.669 291	E–03
Area/Mass	m^2/kg	cm^2/g	m^2/kg		1.0*	E–01
Volume, Capacity	m^3	m^3	km^3		4.168 182	E+00
		acre·ft	m^3		1.233 489	E+03
				ha·m	1.233 489	E–01
		m^3	m^3		1	
		yd^3	m^3		7.645 549	E–01
		bbl (42 U.S. gal)	m^3		1.589 873	E–01
		ft^3	m^3		2.831 685	E–02
			dm^3	L	2.831 685	E+01
		Can. gal	m^3		4.546 09*	E–03
			dm^3	L	4.546 09*	E+00
		U.K. gal	m^3		4.546 092	E–03
			dm^3	L	4.546 092	E+00
		U.S. gal	m^3		3.785 412	E–03
			dm^3	L	3.785 412	E+00
		L	dm^3	L	1	
		U.K. qt	dm^3	L	1.136 523	E+00
		U.S. qt	dm^3	L	9.463 529	E–01
		U.K. pt	dm^3	L	5.682 615	E–01
		U.S. pt	dm^3	L	4.731 765	E–01

Table B2, continued

Quantity	SI Unit	Customary Unit	Metric Unit: API Preferred	Metric Unit: Other Allowable	Conversion Factor (multiply quantity expressed in customary units by factor to get metric equivalent)
Space, Time (continued)					
Volume, Capacity	m^3	U.K. fl oz	cm^3		2.841 308 E+01
		U.S. fl oz	cm^3		2.957 353 E+01
		in.^{3}	cm^3		1.638 706 E+01
		mL	cm^3		1
Volume/Length (linear displacement)	m^3/m	bbl/in.	m^3/m		6.259 342 E+00
		bbl/ft	m^3/m		5.216 119 E–01
		ft^3/ft	m^3/m		9.290 304* E–02
		U.S. gal/ft	dm^3/m	L/m	1.241 933 E+01
Volume/Mass	m^3/kg	see Density, Specific Volume, Concentration, Dosage			
Plane Angle	rad	rad	rad		1
		deg (°)	rad		1.745 329 E–02
				°	1
		min (′)	rad		2.908 882 E–04
				′	1
		sec (″)	rad		4.848 137 E–06
				″	1

Solid Angle	sr	sr	sr		1	
Time	s	million years (MY)	Ma		1	
		yr	a		1	
		wk	d		7.0*	E+00
		d	d		1	
		h	h		1	
				min	6.0*	E+01
		min	s		6.0*	E+01
				h	1.666 667	E–02
				min	1	
		s	s		1	
		millimicrosecond	ns		1	
Mass, Amount of Substance						
Mass	kg	U.K. ton (long ton)	Mg	t	1.016 047	E+00
		U.S. ton (short ton)	Mg	t	9.071 847	E–01
		U.K. cwt	kg		5.080 235	E+01
		U.S. cwt	kg		4.535 924	E+01
		kg	kg		1	
		lb	kg		4.535 924	E–01
		oz (troy)	g		3.110 348	E+01
		oz (avdp)	g		2.834 952	E+01
		g	g		1	
		grain	mg		6.479 891	E+01
		mg	mg		1	
		μg	μg		1	

Table B2, continued

Quantity	SI Unit	Customary Unit	Metric Unit: API Preferred	Metric Unit: Other Allowable	Conversion Factor (multiply quantity expressed in customary units by factor to get metric equivalent)	
Mass, Amount of Substance (continued)						
Mass/Length	kg/m	see Mechanics				
Mass/Area	kg/m^2	see Mechanics				
Mass/Volume	kg/m^3	see Density, Specific Volume, Concentration, Dosage				
Mass/Mass	kg/kg	see Density, Specific Volume, Concentration, Dosage				
Amount of Substance	mol	ft^3 (60°F, 1 atm)	kmol		1.195 29	E–03
		ft^3 (60°F, 14.73 $lbf/in.^2$)	kmol		1.198 06	E–03
		m^3 (0°C, 1 atm)	kmol		4.461 53	E–02
		m^3 (15°C, 1 atm)	kmol		4.229 28	E–02
		m^3 (20°C, 1 atm)	kmol		4.157 15	E–02
		m^3 (25°C, 1 atm)	kmol		4.087 43	E–02
Heating Value, Entropy, Heat Capacity						
Heating Value (mass basis)	J/kg	Btu/lb	MJ/kg		2.326 000	E–03
			kJ/kg	J/g	2.326 000	E+00
				kW·h/kg	6.461 112	E–04
		cal/g	kJ/kg	J/g	4.184*	E+00
		cal/lb	J/kg		9.224 141	E+00

Heating Value (mole basis)	J/mol	kcal/g mol	kJ/kmol		4.184*	E+03
		Btu/lb mol	MJ/kmol		2.326 000	E–03
			kJ/kmol		2.326 000	E+00
Heating Value (volume basis-solids and liquids)	J/m^3	therm/U.S. gal	MJ/m^3	kJ/dm^3	2.787 163	E+04
			kJ/m^3		2.787 163	E+07
				kW·h/dm^3	7.742 119	E+00
		therm/U.K. gal	MJ/m^3	kJ/dm^3	2.320 798	E+04
			kJ/m^3		2.320 798	E+07
				kW·h/dm^3	6.446 660	E+00
		therm/Can. gal	MJ/m^3	kJ/dm^3	2.320 799	E+04
			kJ/m^3		2.320 799	E+07
				kW·h/dm^3	6.446 663	E+00
		Btu/U.S. gal	MJ/m^3	kJ/dm^3	2.787 163	E–01
			kJ/m^3		2.787 163	E+02
				kW·h/m^3	7.742 119	E–02
		Btu/U.K. gal	MJ/m^3	kJ/dm^3	2.320 800	E–01
			kJ/m^3		2.320 800	E+02
				kW·h/m^3	6.446 660	E–02
		Btu/Can. gal	MJ/m^3	kJ/dm^3	2.320 799	E–01
			kJ/m^3		2.320 799	E+02
				kW·h/m^3	6.446 663	E–02
		Btu/ft^3	MJ/m^3	kJ/dm^3	3.725 895	E–02
			kJ/m^3		3.725 895	E+01
				kW·h/m^3	1.034 971	E–02
		kcal/m^3	MJ/m^3	kJ/dm^3	4.184*	E–03
			kJ/m^3		4.184*	E+00
		cal/mL	MJ/m^3		4.184*	E+00
		ft·lbf/U.S. gal	kJ/m^3		3.581 692	E–01

Table B2, continued

Quantity	SI Unit	Customary Unit	Metric Unit: API Preferred	Metric Unit: Other Allowable	Conversion Factor (multiply quantity expressed in customary units by factor to get metric equivalent)	
Heating Value, Entropy, Heat Capacity (continued)						
Heating Value (volume basis–gases)	J/m³	cal/mL	kJ/m³	J/dm³	4.184*	E+03
		kcal/m³	kJ/m³	J/dm³	4.184*	E+00
		Btu/ft³	kJ/m³	J/dm³	3.725 895	E+01
Specific Entropy	J/(kg·K)	Btu/(lb·°R)	kJ/(kg·K)	J/(g·K)	4.186 8*	E+00
		cal/(g·K)	kJ/(kg·K)	J/(g·K)	4.184*	E+00
		kcal/(kg·°C)	kJ/(kg·K)	J/(g·K)	4.184*	E+00
Specific Heat Capacity (mass basis)	J/(kg·K)	kW·h/(kg·°C)	kJ/(kg·K)	J/(g·°C)	3.6*	E+03
		Btu/(lb·°F)	kJ/(kg·K)	J/(g·°C)	4.186 8*	E+00
		kcal/(kg·°C)	kJ/(kg·K)	J/(g·°C)	4.184*	E+00
Molar Heat Capacity	J/(mol·K)	Btu/(lb mol·°F)	kJ/(kmol·K)	J/(g·°C)	4.186 8*	E+00
		cal/(g mol·°C)	kJ/(kmol·K)	J/(g·°C)	4.184*	E+00
Temperature, Pressure, Vacuum						
Temperature (absolute)	K	°R	K		5/9	
		K	K		1	
Temperature (traditional)	K	°F	°C		(°F – 32)/1.8	
		°C	°C		1	

Temperature (difference)	K	°F	K	°C	5/9	
		°C	K	°C	1	
Temperature/Length (geothermal gradient)	K/m	°F/100 ft	mK/m		1.822 689	E+01
Length/Temperature (geothermal step)	m/K	ft/°F	m/K		5.486 4*	E–01
Pressure	Pa	atm (14.696 lbf/in.^{2} or 760 mm Hg at 0°C)	MPa		1.013 250*	E–01
			kPa		1.013 250*	E+02
				bar	1.013 250*	E+00
		bar	MPa		1.0*	E–01
			kPa		1.0*	E+02
				bar	1	
		at (kgf/cm^2) (technical atmosphere)	Mpa		9.806 650*	E–02
			kPa		9.806 650*	E+01
				bar	9.806 650*	E–01
		lbf/in.^{2} (psi)	MPa		6.894 757	E–03
			kPa		6.894 757	E+00
				bar	6.894 757	E–02
		in Hg at 60°F	kPa		3.376 85	E+00
		in Hg at 32°F	kPa		3.386 38	E+00
		in H_2O at 39.2°F	kPa		2.490 82	E–01
		in H_2O at 60°F	kPa		2.488 4	E–01
		mm Hg at 0°C (torr)	kPa		1.333 22	E–01
		cm H_2O at 4°C	kPa		9.806 38	E–02
		lbf/ft^2(psf)	kPa		4.788 026	E–02
		μm Hg at 0°C	Pa		1.333 22	E–01
		μbar	Pa		1.0*	E–01
		dyn/cm^2	Pa		1.0*	E–01

Table B2, continued

Quantity	SI Unit	Customary Unit	Metric Unit: API Preferred	Metric Unit: Other Allowable	Conversion Factor (multiply quantity expressed in customary units by factor to get metric equivalent)
Temperature, Pressure, Vacuum (continued)					
Vacuum, Draft	Pa	in Hg at 60°F	kPa		3.376 85 E+00
		in H_2O at 39.2°F	kPa		2.490 82 E–01
		in H_2O at 60°F	kPa		2.488 4 E–01
		mm Hg at 0°C (torr)	kPa		1.333 22 E–01
		cm H_2O at 4°C	kPa		9.806 38 E–02
Liquid Head	m	ft	m		3.048* E–01
		in.	mm		2.54* E+01
Pressure Drop/Length	Pa/m	psi/ft	kPa/m		2.262 059 E+01
		psi/100 ft	kPa/m		2.262 059 E–01
		psi/mi	kPa/km		4.284 203 E+00
Density, Specific Volume, Concentration, Dosage					
Density (gases)	kg/m^3	lb/ft^3	kg/m^3		1.601 846 E+01
Density (liquids)	kg/m^3	lb/U.S. gal	kg/m^3		1.198 264 E+02
				kg/dm^3	1.198 264 E–01
		lb/U.K. gal	kg/m^3		9.977 633 E+01
				kg/dm^3	9.977 633 E–02
		lb/ft^3	kg/m^3		1.601 846 E+01
				kg/dm^3	1.601 846 E–02

Quantity	SI unit	Customary unit	Metric unit	Alternate unit	Factor	Exponent
		g/cm^3	kg/m^3		1.0*	E+03
				kg/dm^3	1	
		kg/L	kg/m^3		1.0*	E+03
		°API	kg/m^3		Use tables	
Density (solids)	kg/m^3	lb/ft^3	kg/m^3		1.601 846	E+01
				kg/dm^3	1.601 846	E–02
Specific Volume (gases)	m^3/kg	ft^3/lb	m^3/kg		6.242 796	E–02
				dm^3/kg	6.242 796	E+01
Specific Volume (liquids)	m^3/kg	ft^3/lb	m^3/kg		6.242 796	E–02
				dm^3/kg	6.242 796	E+01
		U.K. gal/lb	m^3/kg		1.022 242	E–02
				dm^3/kg	1.022 242	E+01
		U.S. gal/lb	m^3/kg		8.345 404	E–03
				dm^3/kg	8.345 404	E+00
Molar Volume	m^3/mol	L/g mol	m^3/kmol		1	
		ft^3/lb mol	m^3/kmol		6.242 796	E–02
Specific Volume (clay yield)	m^3/kg	bbl/U.S. ton	m^3/Mg	m^3/t	1.752 535	E–01
		bbl/U.K. ton	m^3/Mg	m^3/t	1.564 763	E–01
Yield (shale distillation)	m^3/kg	bbl/U.S. ton	dm^3/Mg	dm^3/t	1.752 535	E+02
		bbl/U.K. ton	dm^3/Mg	dm^3/t	1.564 763	E+02
		U.S. gal/U.S. ton	dm^3/Mg	dm^3/t	4.172 702	E+00
		U.S. gal/U.K. ton	dm^3/Mg	dm^3/t	3.725 627	E+00
Concentration (mass/mass)	kg/kg	wt %	kg/kg		1.0*	E–02
			g/kg		1.0*	E+01
		wt ppm	mg/kg		1	

Table B2, continued

Quantity	SI Unit	Customary Unit	Metric Unit: API Preferred	Metric Unit: Other Allowable	Conversion Factor (multiply quantity expressed in customary units by factor to get metric equivalent)
Density, Specific Volume, Concentration, Dosage (continued)					
Concentration (mass/volume)	kg/m^3	lb/bbl	kg/m^3	g/dm^3	2.853 010 E+00
		g/U.S. gal	kg/m^3		2.641 720 E–01
		g/U.K. gal	kg/m^3		2.199 692 E–01
		lb/1000 U.S. gal	g/m^3	mg/dm^3	1.198 264 E+02
		lb/1000 U.K. gal	g/m^3	mg/dm^3	9.977 633 E+01
		gr/U.S. gal	g/m^3	mg/dm^3	1.711 806 E+01
		lb/1000 bbl	g/m^3	mg/dm^3	2.853 010 E+00
		mg/U.S. gal	g/m^3	mg/dm^3	2.641 720 E–01
		gr/100 ft^3	mg/m^3		2.288 352 E+01
		gr/ft^3	mg/m^3		2.288 352 E+03
Concentration (volume/volume)	m^3/m^3	bbl/bbl	m^3/m^3		1
		ft^3/ft^3	m^3/m^3		1
		bbl/(acre·ft)	dm^3/m^3	L/m^3	1.288 923 E–01
		U.K. gal/ft^3	dm^3/m^3	L/m^3	1.605 437 E+02
		U.S. gal/ft^3	dm^3/m^3	L/m^3	1.336 806 E+02
		mL/U.S. gal	dm^3/m^3	L/m^3	2.641 720 E–01
		mL/U.K. gal	dm^3/m^3	L/m^3	2.199 692 E–01
		Vol %	m^3/m^3		1.0* E–02
		Vol ppm	cm^3/m^3		1
			dm^3/m^3	L/m^3	1.0* E–03

Quantity	SI unit	Customary unit	SI metric unit	Other allowable	Conversion factor	
		U.K. gal/1000 bbl	cm^3/m^3		2.859 406	E+01
		U.S. gal/1000 bbl	cm^3/m^3		2.380 952	E+01
Concentration (mole/volume)	mol/m^3	lb mol/U.S. gal	$kmol/m^3$		1.198 264	E+02
		lb mol/U.K. gal	$kmol/m^3$		9.977 633	E+01
		lb mol/ft^3	$kmol/m^3$		1.601 846	E+01
		std ft^3(60°F, 1 atm)/bbl	$kmol/m^3$		7.518 18	E–03
Concentration (volume/mole)	m^3/mol	U.S. gal/1000 std ft^3 (60°F/60°F)	$dm^3/kmol$	L/kmol	3.166 93	E+00
		bbl/million std ft^3 (60°F/60°F)	$dm^3/kmol$	L/kmol	1.330 11	E–01
Facility Throughput, Capacity						
Throughput (mass basis)	kg/s	million lb/yr	Mg/a	t/a	4.535 924	E+02
		U.K. ton/yr	Mg/a	t/a	1.016 047	E+00
		U.S. ton/yr	Mg/a	t/a	9.071 847	E–01
		U.K. ton/d	Mg/d	t/d	1.016 047	E+00
				t/h, Mg/h	4.233 529	E–02
		U.S. ton/d	Mg/d	t/d	9.071 847	E–01
				t/h, Mg/h	3.779 936	E–02
		U.K. ton/h	Mg/h	t/h	1.016 047	E+00
		U.S. ton/h	Mg/h	t/h	9.071 847	E–01
		lb/h	kg/h		4.535 924	E–01
Throughput (volume basis)	m^3/s	bbl/d	m^3/a		5.803 036	E+01
			m^3/d		1.589 873	E–01
			m^3/h		6.624 471	E–03
		ft^3/d	m^3/h		1.179 869	E–03
			m^3/d		2.831 685	E–02

Table B2, continued

Quantity	SI Unit	Customary Unit	Metric Unit: API Preferred	Metric Unit: Other Allowable	Conversion Factor (multiply quantity expressed in customary units by factor to get metric equivalent)
Facility Throughput, Capacity (continued)					
		bbl/h	m^3/h		1.589 873 E–01
		ft^3/h	m^3/h		2.831 685 E–02
		U.K. gal/h	m^3/h		4.546 092 E–03
				L/h	4.546 092 E+00
		U.S. gal/h	m^3/h		3.785 412 E–03
				L/h	3.785 412 E+00
		U.K. gal/min	m^3/h		2.727 655 E–01
				L/min	4.546 092 E+00
		U.S. gal/min	m^3/h		2.271 247 E–01
				L/min	3.785 412 E+00
Throughput (mole basis)	mol/s	lb mol/h	kmol/h		4.535 924 E–01
				kmol/s	1.259 979 E–04
Pipeline Capacity	m^3/m	bbl/mi	m^3/km		9.879 013 E–02
Flowrate					
Flowrate (mass basis)	kg/s	U.K. ton/min	kg/s		1.693 412 E+01
		U.S. ton/min	kg/s		1.511 975 E+01
		U.K. ton/h	kg/s		2.822 353 E–01
		U.S. ton/h	kg/s		2.519 958 E–01

		million lb/d	kg/s		5.249 912 E+00
		U.K. ton/d	kg/s		1.175 980 E–02
		U.S. ton/d	kg/s		1.049 982 E–02
		million lb/yr	kg/s		1.438 332 E–02
		U.K. ton/yr	kg/s		3.221 864 E–05
		U.S. ton/yr	kg/s		2.876 664 E–05
		lb/s	kg/s		4.535 924 E–01
		lb/min	kg/s		7.559 873 E–03
		lb/h	kg/s		1.259 979 E–04
Flowrate (volume basis)	m^3/s	bbl/d	dm^3/s		1.840 131 E–03
		ft^3/d	dm^3/s		3.277 413 E–04
		bbl/h	dm^3/s		4.416 314 E–02
		ft^3/h	dm^3/s		7.865 791 E–03
		U.K. gal/h	dm^3/s		1.262 803 E–03
		U.S. gal/h	dm^3/s		1.051 503 E–03
		U.K. gal/min	dm^3/s		7.576 820 E–02
		U.S. gal/min	dm^3/s		6.309 020 E–02
		ft^3/min	dm^3/s		4.719 474 E–01
		ft^3/s	dm^3/s		2.831 685 E+01
Flowrate (mole basis)	mol/s	lb mol/s	kmol/s		4.535 924 E–01
		lb mol/h	kmol/s		1.259 979 E–04
		million scf/sd	kmol/s		1.383 449 E–02
Flowrate/Length (mass basis)	kg/(s·m)	lb/(s·ft)	kg/(s·m)		1.488 164 E+00
		lb/(h·ft)	kg/(s·m)		4.133 789 E–04
Flowrate/Length (volume basis)	m^2/s	U.K. gal/(min·ft)	m^2/s	$m^3/(s \cdot m)$	2.485 833 E–04
		U.S. gal/(min·ft)	m^2/s	$m^3/(s \cdot m)$	2.069 888 E–04

Table B2, continued

Quantity	SI Unit	Customary Unit	Metric Unit: API Preferred	Metric Unit: Other Allowable	Conversion Factor (multiply quantity expressed in customary units by factor to get metric equivalent)
Flowrate (continued)					
		U.K. gal/(h·in.)	m^2/s	$m^3/(s \cdot m)$	4.971 667 E–05
		U.S. gal/(h·in.)	m^2/s	$m^3/(s \cdot m)$	4.139 776 E–05
		U.K. gal/(h·ft)	m^2/s	$m^3/(s \cdot m)$	4.143 055 E–06
		U.S. gal/(h·ft)	m^2/s	$m^3/(s \cdot m)$	3.449 814 E–06
Flowrate/Area (mass basis)	$kg/(s \cdot m^2)$	$lb/(s \cdot ft^2)$	$kg/(s \cdot m^2)$		4.882 428 E+00
		$lb/(h \cdot ft^2)$	$kg/(s \cdot m^2)$		1.356 230 E–03
Flowrate/Area (volume basis)	m/s	$ft^3/(s \cdot ft^2)$	m/s	$m^3/(s \cdot m^2)$	3.048* E–01
		$ft^3/(min \cdot ft^2)$	m/s	$m^3/(s \cdot m^2)$	5.08* E–03
		U.K. gal/(h·in.^{2})	m/s	$m^3/(s \cdot m^2)$	1.957 349 E–03
		U.S. gal/(h·in.^{2})	m/s	$m^3/(s \cdot m^2)$	1.629 833 E–03
		U.K. gal/(min·ft^2)	m/s	$m^3/(s \cdot m^2)$	8.155 621 E–04
		U.S. gal/(min·ft^2)	m/s	$m^3/(s \cdot m^2)$	6.790 972 E–04
		U.K. gal/(h·ft^2)	m/s	$m^3/(s \cdot m^2)$	1.359 270 E–05
		U.S. gal/(h·ft^2)	m/s	$m^3/(s \cdot m^2)$	1.131 829 E–05
Flowrate/Pressure Drop (productivity index)	$m^3/(s \cdot Pa)$	bbl/(d·psi)	$m^3/(d \cdot kPa)$		2.305 916 E–02

Energy, Work, Quantity of Heat, Power						
Energy, Work, Quantity of Heat	J	quad	EJ		1.055 056	E+00
			TW·h		2.930 711	E+02
		therm	MJ		1.055 056	E+02
			kJ		1.055 056	E+05
				kW·h	2.930 711	E+01
		U.S. tonf·mi	MJ		1.431 744	E+01
		hp·h	MJ		2.684 520	E+00
			kJ		2.684 520	E+03
				kW·h	7.456 999	E–01
		ch·h or CV·h	MJ		2.647 796	E+00
			kJ		2.647 796	E+03
				kW·h	7.354 99	E–01
		kW·h	MJ		3.6*	E+00
			kJ		3.6*	E+03
		Chu	kJ		1.899 101	E+00
				kW·h	5.275 280	E–04
		Btu	kJ		1.055 056	E+00
				kW·h	2.930 711	E–04
		kcal	kJ		4.184*	E+00
		cal	kJ		4.184*	E–03
		ft·lbf	kJ		1.355 818	E–03
		J	kJ		1.0*	E–03
		$lb \cdot ft^2/s^2$(ft·pdl)	kJ		4.214 011	E–05
		erg	J		1.0*	E–07

Table B2, continued

Quantity	SI Unit	Customary Unit	Metric Unit: API Preferred	Metric Unit: Other Allowable	Conversion Factor (multiply quantity expressed in customary units by factor to get metric equivalent)
Energy, Work, Quantity of Heat Power (continued)					
Impact Energy	J	kgf·m	J		9.806 650* E+00
		ft·lbf	J		1.355 818 E+00
Work/Length	J/m	U.S. tonf·mi/ft	MJ/m		4.697 322 E+01
Surface Energy	J/m^2	erg/cm^2	mJ/m^2		1.0* E+00
Power	W	quad/yr	EJ/a		1.055 056 E+00
				GW	3.345 561 E+01
		erg/a	TW		3.170 979 E–27
			GW		3.170 979 E–24
		million Btu/h	MW		2.930 711 E–01
		ton of refrigeration	kW		3.516 853 E+00
		Btu/s	kW		1.055 056 E+00
		kW	kW		1
		hydraulic horse-power-hhp	kW		7.460 43 E–01
		hp (electric)	kW		7.46* E–01
		hp (550 ft·lbf/s)	kW		7.456 999 E–01
		ch or CV	kW		7.354 99 E–01
		Btu/min	kW		1.758 427 E–02
		ft·lbf/s	kW		1.355 818 E–03

		kcal/h	W		1.162 222 E+00
		Btu/h	W		2.930 711 E−01
		ft·lbf/min	W		2.259 697 E−02
Power/Area	W/m^3	$Btu/(s \cdot ft^2)$	kW/m^2		1.135 653 E+01
		$cal/(h \cdot cm^2)$	kW/m^2		1.162 222 E−02
		$Btu/(h \cdot ft^2)$	kW/m^3		3.154 591 E−03
Heat Flow Unit-hfu (geothermics)		$\mu cal/(s \cdot cm^2)$	mW/m^2		4.184* E+01
Heat Release Rate, Mixing Power	W/m^3	hp/ft^3	kW/m^3		2.633 414 E+01
		$cal/(h \cdot cm^3)$	kW/m^3		1.162 222 E+00
		$Btu/(s \cdot ft^3)$	kW/m^3		3.725 895 E+01
		$Btu/(h \cdot ft^3)$	kW/m^3		1.034 971 E−02
Heat Generation Unit-hgu (radioactive rocks)		$cal/(s \cdot cm^3)$	$\mu W/m^3$		4.184* E+12
Cooling Duty (machinery)	W/W	Btu/(bhp·h)	W/kW		3.930 148 E−01
Mass Fuel Consumption	kg/J	lb/(hp·h)	mg/J	kg/MJ	1.689 659 E−01
				kg/(kW·h)	6.082 774 E−01
Volume Fuel Consumption	m^3/J	$m^3/(kW \cdot h)$	dm^3/MJ	mm^3/J	2.777 778 E+02
				$dm^3/(kW \cdot h)$	1.0* E+03
		U.S. gal/(hp·h)	dm^3/MJ	mm^3/J	1.410 089 E+00
				$dm^3/(kW \cdot h)$	5.076 321 E+00
		U.K. pt/(hp·h)	dm^3/MJ	mm^3/J	2.116 809 E−01
				$dm^3/(kW \cdot h)$	7.620 512 E−01

Table B2, continued

Quantity	SI Unit	Customary Unit	Metric Unit: API Preferred	Metric Unit: Other Allowable	Conversion Factor (multiply quantity expressed in customary units by factor to get metric equivalent)
Energy, Work, Quantity of Heat Power (continued)					
Fuel Consumption (automotive)	m^3/m	U.K. gal/mi	$dm^3/100$ km	L/100 km	2.824 811 E+02
		U.S. gal/mi	$dm^3/100$ km	L/100 km	2.352 146 E+02
		mi/U.S. gal	km/dm^3	km/L	4.251 437 E–01
		mi/U.K. gal	km/dm^3	km/L	3.540 060 E–01
Mechanics					
Velocity (linear), Speed	m/s	knot	km/h		1.852 E+00
				knot	1
		mi/h	km/h		1.609 344* E+00
		m/s	m/s		1
		ft/s	m/s		3.048* E–01
				cm/s	3.048* E+01
				m/ms	3.048* E–04
		ft/min	m/s		5.08* E–03
				cm/s	5.08* E–01
		ft/h	mm/s		8.466 667 E–02
				cm/s	8.466 667 E–03
		ft/d	mm/s		3.527 778 E–03
				m/d	3.048* E–01

		in./s	mm/s		2.54*	E+01
				cm/s	2.54*	E+00
		in./min	mm/s		4.233 333	E–01
				cm/s	4.233 333	E–02
Velocity (angular)	rad/s	r/min	rad/s		1.047 198	E–01
		r/min		r/min	1	
		r/s	rad/s		6.283 185	E+00
		r/s		r/s	1	
		deg/min	rad/s		2.908 882	E–04
		deg/s	rad/s		1.745 329	E–02
Reciprocal Velocity	μs/m	μs/ft	μs/m		3.280 840	E+00
Acceleration (linear)	m/s^2	ft/s^2	m/s^2		3.048*	E–01
		gal (cm/s^2)	m/s^2		1.0*	E–02
Acceleration (angular)	rad/s^2	rad/s^2	rad/s^2		1	
		rpm/s	rad/s^2		1.047 198	E–01
Corrosion Rate	mm/a	in./yr (ipy)	mm/a		2.54*	E+01
Momentum	kg·m/s	lb·ft/s	kg·m/s		1.382 550	E–01
Force	N	U.K. tonf	kN		9.964 016	E+00
		U.S. tonf	kN		8.896 443	E+00
		kgf (kp)	N		9.806 650*	E+00
		lbf	N		4.448 222	E+00
		N	N		1	
		pdl	mN		1.382 550	E+02
		dyn	mN		1.0*	E–02

Table B2, continued

Quantity	SI Unit	Customary Unit	Metric Unit: API Preferred	Metric Unit: Other Allowable	Conversion Factor (multiply quantity expressed in customary units by factor to get metric equivalent)
Mechanics (continued)					
Bending Moment, Torque	N·m	U.S. tonf·m	kN·m		2.711 636 E+00
		kgf·m	N·m		9.806 650* E+00
		lbf·ft	N·m		1.355 818 E+00
		lbf·in.	N·m		1.129 848 E–01
		pdl·ft	N·m		4.214 011 E–02
Bending Movement, Length	N·m/m	lbf·ft/in.	N·m/m		5.337 866 E+01
		kgf·m/m	N·m/m		9.806 650* E+00
		lbf·in./in.	N·m/m		4.448 222 E+00
Moment of Inertia	$kg·m^2$	$lb·ft^2$	$kg·m^2$		4.214 011 E–02
		$in.^4$	cm^4		4.162 314 E+01
Stress	Pa	$U.S. tonf/in.^2$	MPa	N/mm^2	1.378 951 E+00
		kgf/mm^2	MPa	N/mm^2	9.806 650* E+00
		$U.S. tonf/ft^2$	MPa	N/mm^2	9.576 052 E–02
		$lbf/in.^2$ (psi)	MPa	N/mm^2	6.894 757 E–03
		lbf/ft^2 (psf)	kPa		4.788 026 E–02
		dyn/cm^2	Pa		1.0* E–01
Yield Point, Gel Strength (drilling fluid)		$lbf/100 ft^2$	Pa		4.788 026 E+01

Quantity	SI unit	Customary unit	Metric unit	Other metric unit	Conversion factor	Exponent
Mass/Length	kg/m	lb/ft	kg/m		1.488 164	E+00
Mass/Area Structural Loading, Bearing Capacity (mass basis)	kg/m^2	U.S. ton/ft^2	Mg/m^2	t/m^2	9.764 855	E+00
		lb/ft^2	kg/m^2		4.882 428	E+00
Modulus of Elasticity	Pa	lbf/in.^{2} (psi)	MPa	N/mm^2	6.894 757	E–03
Section Modulus	m^3	in.^{3}	cm^3		1.638 706	E+01
Coefficient of Thermal Expansion	m/(m·K)	in./(in.°F)	mm/(mm·K)	mm/(mm·°C)	5.555 556	E–01
Transport Properties						
Diffusivity	m^2/s	ft^2/s	mm^2/s		9.290 304*	E+04
		cm^2/s	mm^2/s		1.0*	E+02
		ft^2/h	mm^2/s		2.580 64*	E+01
Thermal Resistance	K·m^2/W	°C·m^2·h/kcal	K·m^2/kW		8.604 208	E+02
		°F·ft^2·h/Btu	K·m^2/kW		1.761 102	E+02
Heat Flux	W/m^2	Btu/(h·ft^2)	kW/m^2		3.154 591	E–03
Thermal Conductivity	W/(m·K)	cal/(s·cm^2·°C/cm)	W/(m·K)	W/(m^2·°C/m)	4.184*	E+02
		Btu/(h·ft^2·°F/ft)	W/(m·K)	W/(m^2·°C/m)	1.730 735	E+00
		kcal/(h·m^2·°C/m)	W/(m·K)	W/(m^2·°C/m)	1.162 222	E+00
		Btu/(h·ft^2·°F/in.)	W/(m·K)	W/(m^2·°C/m)	1.422 279	E–01
		cal/(h·cm^2·°C/cm)	W/(m·K)	W/(m^2·°C/m)	1.162 222	E–01
Heat Transfer Coefficient	W/(m^2·K)	cal/(s·cm^2·°C)	kW/(m^2·K)		4.184*	E+01
		Btu/(s·ft^2·°F)	kW/(m^2·K)		2.044 175	E+01
		cal/(h·cm^2·°C)	kW/(m^2·K)		1.162 222	E–02
		Btu/(h·ft^2·°F)	kW/(m^2·K)		5.678 263	E–03
		Btu/(h·ft^2·°R)	kW/(m^2·K)		5.678 263	E–03
		kcal/(h·m^2·°C)	kW/(m^2·K)		1.162 222	E–03

Table B2, continued

Quantity	SI Unit	Customary Unit	Metric Unit		Conversion Factor (multiply quantity expressed in customary units by factor to get metric equivalent)
			API Preferred	Other Allowable	
Transport Properties (continued)					
Volumetric Heat Transfer Coefficient	W/(m^3·K)	Btu/(s·ft^3·°F)	kW/(m^3·K)		6.706 611 E+01
		Btu/(h·ft^3·°F)	kW/(m^3·K)		1.862 947 E–02
Surface Tension	N/m	dyn/cm	mN/m		1.0* E+00
Viscosity (dynamic)	Pa·s	lbf·s/in.^{2}	mPa·s	cP	6.894 757 E+06
		lbf·s/ft^2	mPa·s	cP	4.788 026 E+04
		kgf·s/m^2	mPa·s	cP	9.806 650* E+03
		dyn·s/cm^2	Pa·s		1.0* E–01
			mPa·s	cP	1.0* E+02
		P	Pa·s		1.0* E–01
			mPa·s	cP	1.0* E+02
		cP	mPa·s	cP	1.0* E+00
Viscosity (kinematic)	m^2/s	ft^2/s	mm^2/s	cSt	9.290 304* E+04
		in.^{2}/s	mm^2/s	cSt	6.451 6* E+02
		ft^2/h	mm^2/s	cSt	2.580 64* E+01
		m^2/h	mm^2/s	cSt	2.777 778 E+02
		cm^2/s	mm^2/s	cSt	1.0* E+02
		St	mm^2/s	cSt	1.0* E+02
		cSt	mm^2/s	cSt	1.0* E+00

Permeability	m^2	D	μm^2	D	1.0*	E+00
		mD	μm^2	D	1.0*	E–03
Electricity, Magnetism						
Admittance	S	S	S		1	
Capacitance	F	μF	μF		1	
Capacity, Storage Battery	C	A·h	kC		3.6*	E+00
Charge Density	C/m^3	C/mm^3	C/mm^3		1	
Conductance	S	S	S		1	
		℧	S		1	
Conductivity	S/m	S/m	S/m			
		m℧/m	mS/m		1	
Current Density	A/m^2	A/mm^2	A/mm^2		1	
		$A/in.^2$	A/mm^2		1.550 003	E–03
Displacement	C/m^2	C/cm^2	C/cm^2		1	
Electric Charge	C	C	C		1	
Electric Current	A	A	A		1	
Electric Dipole Moment	C·m	C·m	C·m		1	
Electric Field Strength	V/m	V/m	V/m		1	
Electric Flux	C	C	C		1	
Electric Polarization	C/m^2	C/cm^2	C/cm^2		1	

Table B2, continued

Quantity	SI Unit	Customary Unit	Metric Unit		Conversion Factor (multiply quantity expressed in customary units by factor to get metric equivalent)
			API Preferred	Other Allowable	
Electricity, Magnetism (continued)					
Electric Potential	V	V	V		1
		mV	mV		1
Electromagnetic Moment	$A \cdot m^2$	$A \cdot m^2$	$A \cdot m^2$		1
Electromotive Force	V	V	V		1
Flux of Displacement	C	C	C		1
Frequency	Hz	cycles/s	Hz		1
Impedance	Ω	Ω	Ω		1
Linear Current Density	A/m	A/mm	A/mm		1
Magnetic Dipole Moment	Wb·m	Wb·m	Wb·m		1
Magnetic Field Strength	A/m	A/mm	A/mm		1
		oersted	A/m		7.957 747 E+01
Magnetic Flux	Wb	mWb	mWb		1
		maxwell	nWb		1.0* E+01
Magnetic Flux Density	T	mT	mT		1
		gauss	T		1.0* E–04
		gamma	nT		1

Magnetic Induction	T	mT	mT	1
Magnetic Moment	$A \cdot m^2$	$A \cdot m^2$	$A \cdot m^2$	1
Magnetic Polarization	T	mT	mT	1
Magnetic Potential Difference	A	A	A	1
Magnetic Vector Potential	Wb/m	Wb/mm	Wb/mm	1
Magnetization	A/m	A/mm	A/mm	1
Modulus of Admittance	S	S	S	1
Modulus of Impedance	Ω	Ω	Ω	1
Mutual Inductance	H	H	H	1
Permeability	H/m	μH/m	μH/m	1
Permeance	H	H	H	1
Permittivity	F/m	μF/m	μF/m	1
Potential Difference	V	V	V	1
Quantity of Electricity	C	C	C	1
Reactance	Ω	Ω	Ω	1
Reluctance	H^{-1}	H^{-1}	H^{-1}	1
Resistance	Ω	Ω	Ω	1
Resistivity	Ω·m	Ω·cm Ω·m	Ω·cm Ω·m	1 1

Table B2, continued

Quantity	SI Unit	Customary Unit	Metric Unit: API Preferred	Metric Unit: Other Allowable	Conversion Factor (multiply quantity expressed in customary units by factor to get metric equivalent)	
Electricity, Magnetism (continued)						
Self Inductance	H	mH	mH		1	
Surface Density of Charge	C/m^2	mC/m^2	mC/m^2		1	
Susceptance	S	S	S		1	
Volume Density of Charge	C/m^3	C/mm^3	C/mm^3		1	
Acoustics, Light, Radiation						
Acoustical Energy	J	J	J		1	
Acoustical Intensity	W/m^2	W/cm^2	W/m^2		1.0*	E+04
Acoustical Power	W	W	W		1	
Activity (of a radionuclide)	Bq	curie	Bq		3.7*	E+10
Illuminance, Illumination	lx	footcandle (fc)	lx		1.076 391	E+01
		lm/ft^2	lx		1.076 391	E+01
Irradiance	W/m^2	W/m^2	W/m^2		1	
Light Exposure	lx·s	fc·s	lx·s		1.076 391	E+01

Luminance	cd/m^2	cd/m^2	cd/m^2	1	
		footlambert (fL)	cd/m^2	3.426 259	E+00
Luminous Efficacy	lm/W	lm/W	lm/W	1	
Luminous Exitance	lm/m^2	lm/m^2	lm/m^2	1	
		lm/ft^2	lm/m^2	1.076 391	E+01
Luminous Flux	lm	lm	lm	1	
Luminous Intensity	cd	cd	cd	1	
Quantity of Light	lm·s	talbot	lm·s	1	
Radiance	W/(m^2·sr)	W/(m^2·sr)	W/(m^2·sr)	1	
Radiant Energy	J	J	J	1	
Radiant Flux	W	W	W	1	
Radiant Intensity	W/sr	W/sr	W/sr	1	
Radiant Power	W	W	W	1	
Sound Pressure	Pa	Pa	Pa	1	
Wavelength	m	Å	nm	1.0*	E–01

APPENDIX C

EQUATIONS OF MOTION AND CONTINUITY IN RECTANGULAR, CYLINDRICAL AND SPHERICAL COORDINATES

Table C1. Equation of Continuity

Coordinate System	Axes	Equation
Rectangular	x-, y-, z-	$\frac{\partial \rho}{\partial t} + \frac{\partial}{\partial x}(\rho u) + \frac{\partial}{\partial y}(\rho v) + \frac{\partial}{\partial z}(\rho w) = 0$
Cylindrical	r-, θ-, y-	$\frac{\partial \rho}{\partial t} + \frac{1}{r}\frac{\partial}{\partial r}(\rho r u_r) + \frac{1}{r}\frac{\partial}{\partial \theta}(\rho u_\theta) + \frac{\partial}{\partial y}(\rho w) = 0$.
Spherical	r-, θ-, ϕ-	$\frac{\partial \rho}{\partial t} + \frac{1}{r^2}\frac{\partial}{\partial r}(\rho r^2 u_r) + \frac{1}{r \sin\theta}\frac{\partial}{\partial \theta}(\rho u_\theta \sin\theta) + \frac{1}{r \sin\theta}\frac{\partial}{\partial \phi}(\rho u_\phi) = 0$.

Table C2. General Equation of Motion

Coordinate System	Component	Equation
Rectangular (x, y, z)	x-	$\rho\left(\frac{\partial u}{\partial t} + u\frac{\partial u}{\partial x} + v\frac{\partial u}{\partial y} + w\frac{\partial u}{\partial z}\right) = -\frac{\partial P}{\partial x} - \left(\frac{\partial \tau_{xx}}{\partial x} + \frac{\partial \tau_{yx}}{\partial y} + \frac{\partial \tau_{zx}}{\partial z}\right) + \rho g_x$
	y-	$\rho\left(\frac{\partial v}{\partial t} + u\frac{\partial v}{\partial x} + v\frac{\partial v}{\partial y} + w\frac{\partial v}{\partial z}\right) = -\frac{\partial P}{\partial y} - \left(\frac{\partial \tau_{xy}}{\partial x} + \frac{\partial \tau_{yy}}{\partial y} + \frac{\partial \tau_{zy}}{\partial z}\right) + \rho g_y$
	z-	$\rho\left(\frac{\partial w}{\partial t} + u\frac{\partial w}{\partial x} + v\frac{\partial w}{\partial y} + w\frac{\partial w}{\partial z}\right) = -\frac{\partial P}{\partial z} - \left(\frac{\partial \tau_{xz}}{\partial x} + \frac{\partial \tau_{yz}}{\partial y} + \frac{\partial \tau_{zz}}{\partial z}\right) + \rho g_z$
Cylindrical (r, θ, y)	r-	$\rho\left(\frac{\partial u_r}{\partial t} + u_r\frac{\partial u_r}{\partial r} + \frac{u_\theta}{r}\frac{\partial u_r}{\partial \theta} - \frac{u_\theta^2}{r} + u_y\frac{\partial u_r}{\partial y}\right) = -\frac{\partial P}{\partial r} - \left(\frac{1}{r}\frac{\partial}{\partial r}(r\tau_{rr}) + \frac{1}{r}\frac{\partial \tau_{r\theta}}{\partial \theta} - \frac{\tau_{\theta\theta}}{r} + \frac{\partial \tau_{ry}}{\partial y}\right) + \rho g_r$

θ-
$$\rho\left(\frac{\partial u_\theta}{\partial t} + u_r\frac{\partial u_\theta}{\partial r} + \frac{u_\theta}{r}\frac{\partial u_\theta}{\partial \theta} + \frac{u_r u_\theta}{r} + u_y\frac{\partial u_\theta}{\partial y}\right) = -\frac{1}{r}\frac{\partial P}{\partial \theta} - \left(\frac{1}{r^2}\frac{\partial}{\partial r}(r^2\tau_{r\theta}) + \frac{1}{r}\frac{\partial \tau_{\theta\theta}}{\partial \theta} + \frac{\partial \tau_{\theta y}}{\partial y}\right) + \rho g_\theta$$

y-
$$\rho\left(\frac{\partial u_y}{\partial t} + u_r\frac{\partial u_y}{\partial r} + \frac{u_\theta}{r}\frac{\partial u_y}{\partial \theta} + u_y\frac{\partial u_y}{\partial y}\right) = -\frac{\partial P}{\partial y} - \left(\frac{1}{r}\frac{\partial}{\partial r}(r\tau_{ry}) + \frac{1}{r}\frac{\partial \tau_{\theta y}}{\partial \theta} + \frac{\partial \tau_{yy}}{\partial y}\right) + \rho g_y$$

Spherical (r, θ, ϕ)

r-
$$\rho\left(\frac{\partial u_r}{\partial t} + u_r\frac{\partial u_r}{\partial r} + \frac{u_\theta}{r}\frac{\partial u_r}{\partial \theta} + \frac{u_\phi}{r\sin\theta}\frac{\partial u_r}{\partial \phi} - \frac{u_\theta^2 + u_\phi^2}{r}\right) = -\frac{\partial P}{\partial r} - \left(\frac{1}{r^2}\frac{\partial}{\partial r}(r^2\tau_{rr}) + \frac{1}{r\sin\theta}\frac{\partial}{\partial \theta}(\tau_{r\theta}\sin\theta)\right.$$
$$\left. + \frac{1}{r\sin\theta}\frac{\partial \tau_{r\phi}}{\partial \phi} - \frac{\tau_{\theta\theta} + \tau_{\phi\phi}}{r}\right) + \rho g_r$$

θ-
$$\rho\left(\frac{\partial u_\theta}{\partial t} + u_r\frac{\partial u_\theta}{\partial r} + \frac{u_\theta}{r}\frac{\partial u_\theta}{\partial \theta} + \frac{u_\phi}{r\sin\theta}\frac{\partial u_\theta}{\partial \phi} + \frac{u_r u_\theta}{r} - \frac{u_\phi^2\cot\theta}{r}\right)$$
$$= -\frac{1}{r}\frac{\partial P}{\partial \theta} - \left(\frac{1}{r^2}\frac{\partial}{\partial r}(r^2\tau_{r\theta}) + \frac{1}{r\sin\theta}\frac{\partial}{\partial \theta}(\tau_{\theta\theta}\sin\theta) + \frac{1}{r\sin\theta}\frac{\partial \tau_{\theta\phi}}{\partial \phi} + \frac{\tau_{r\theta}}{r} - \frac{\cot\theta}{r}\tau_{\phi\phi}\right) + \rho g_\theta$$

φ-
$$\rho\left(\frac{\partial u_\phi}{\partial t} + u_r\frac{\partial u_\phi}{\partial r} + \frac{u_\theta}{r}\frac{\partial u_\phi}{\partial \theta} + \frac{u_\phi}{r\sin\theta}\frac{\partial u_\phi}{\partial \phi} + \frac{u_\phi u_r}{r} + \frac{u_\theta u_\phi}{r}\cot\theta\right)$$
$$= -\frac{1}{r\sin\theta}\frac{\partial P}{\partial \phi} - \left(\frac{1}{r^2}\frac{\partial}{\partial r}(r^2\tau_{r\phi}) + \frac{1}{r}\frac{\partial \tau_{\theta\phi}}{\partial \phi} + \frac{1}{r\sin\theta}\frac{\partial \tau_{\phi\phi}}{\partial \phi} + \frac{\tau_{r\phi}}{r} + \frac{2\cot\theta}{r}\tau_{\theta\phi}\right) + \rho g_\phi$$

INDEX